Liquid Cell Electron Microscopy

The first book on the topic, with each chapter written by pioneers in the field, this essential resource details the fundamental concepts, applications, and future developments of liquid cell electron microscopy.

This book describes the techniques that have been developed to image liquids in both transmission and scanning electron microscopes, including general strategies for examining liquids, closed and open cell electron microscopy, experimental design, resolution, and electron beam effects. A wealth of practical guidance is provided, and applications are described in areas such as electrochemistry, corrosion and batteries, nanocrystal growth, biomineralization, biomaterials and biological processes, beam-induced processing, and fluid physics. The book also looks ahead to the future development of the technique, discussing technical advances that will enable higher resolution, analytical microscopy, and even holography of liquid samples.

This is essential reading for researchers and practitioners alike.

Frances M. Ross is based at the IBM Thomas J. Watson Research Center, where she has built a program around a microscope with deposition and focused ion beam capabilities and developed closed liquid cell microscopy to image electrochemical processes. Previously she worked at the National Center for Electron Microscopy, Lawrence Berkeley National Laboratory, and she has also been a Visiting Scientist at Lund University and an Adjunct Professor at Arizona State University. She received the UK Institute of Physics Boys Medal, the MRS Outstanding Young Investigator Award, and the MSA Burton Medal, holds an Honorary Doctorate from Lund University, and is a Fellow of APS, AAAS, MRS, MSA, and AVS.

Advances in Microscopy and Microanalysis

Microscopic visualization techniques range from atomic imaging to visualization of living cells at near nanometer spatial resolution, and advances in the field are fueled by developments in computation, image detection devices, labeling, and sample preparation strategies. Microscopy has proven to be one of the most attractive and progressive research tools available to the scientific community, and remains at the forefront of research in many disciplines, from nanotechnology to live cell molecular imaging.

This series reflects the diverse role of microscopy, defining it as any method of imaging objects of micrometer scale or less, and includes both introductory texts and highly technical and focused monographs for researchers and practitioners in materials and the life sciences.

Series Editors
Patricia Calarco, *University of California, San Francisco*
Michael Isaacson, *University of California, Santa Cruz*

Series Advisors
Bridget Carragher, *The Scripps Research Institute*
Wah Chiu, *Baylor College of Medicine*
Christian Colliex, *Université Paris Sud*
Ulrich Dahmen, *Lawrence Berkeley National Laboratory*
Mark Ellisman, *University of California, San Diego*
Peter Ingram, *Duke University Medical Center*
J. Richard McIntosh, *University of Colorado*
Giulio Pozzi, *University of Bologna*
John C. H. Spence, *Arizona State University*
Elmar Zeitler, *Fritz-Haber Institute*

Books in Series
Published
Heide Schatten, *Scanning Electron Microscopy for the Life Sciences*

Forthcoming
Nigel Browning et al., *Dynamic Transmission Electron Microscopy*
Michael Isaacson, *Microscopic Nanocharacterization of Materials*
Richard Leapman, *Energy Filtered Electron Microscopy and Electron Spectroscopy*
Eric Lifshin, *The Scanning Electron Microscope*
Chris Jacobsen and Janos Kirz, *X-Ray Microscopy*
Joel Kubby, Meng Cui and Sylvain Gigan, *Wavefront Shaping for Biomedical Imaging*
Thomas F. Kelly and Simon Ringer, *Atomic-Scale Tomography*

Liquid Cell Electron Microscopy

FRANCES M. ROSS
IBM Thomas J. Watson Research Center

CAMBRIDGE
UNIVERSITY PRESS

University Printing House, Cambridge CB2 8BS, United Kingdom

One Liberty Plaza, 20th Floor, New York, NY 10006, USA

477 Williamstown Road, Port Melbourne, VIC 3207, Australia

4843/24, 2nd Floor, Ansari Road, Daryaganj, Delhi – 110002, India

79 Anson Road, #06–04/06, Singapore 079906

Cambridge University Press is part of the University of Cambridge.

It furthers the University's mission by disseminating knowledge in the pursuit of education, learning, and research at the highest international levels of excellence.

www.cambridge.org
Information on this title: www.cambridge.org/9781107116573
10.1017/9781316337455

© Cambridge University Press 2017

This publication is in copyright. Subject to statutory exception
and to the provisions of relevant collective licensing agreements,
no reproduction of any part may take place without the written
permission of Cambridge University Press.

First published 2017

Printed in the United Kingdom by Clays, St Ives plc

A catalog record for this publication is available from the British Library.

Library of Congress Cataloging-in-Publication Data
Names: Ross, Frances M., 1964- author, editor.
Title: Liquid cell electron microscopy / Frances M. Ross, IBM T.J. Watson Research Center.
Description: New York, NY: Cambridge University Press, [2017]
Identifiers: LCCN 2016030576| ISBN 9781107116573 (Hardback) | ISBN 1107116570 (Hardback)
Subjects: LCSH: Electron microscopy–Technique. | Liquids–Microscopy. | Transmission electron microscopy.
Classification: LCC QH212.T7 R67 2016 | DDC 502.8/25–dc23 LC record available at https://lccn.loc.gov/2016030576

ISBN 978-1-107-11657-3 Hardback

Additional resources for this publication at www.cambridge.org/Ross

Cambridge University Press has no responsibility for the persistence or accuracy of URLs for external or third-party Internet Web sites referred to in this publication and does not guarantee that any content on such Web sites is, or will remain, accurate or appropriate.

Contents

List of Contributors		*page* xii
Preface and Acknowledgements		xvii

Part I Technique   1

1   **Past, Present, and Future Electron Microscopy of Liquid Specimens**   3
Niels de Jonge and Frances M. Ross

     1.1   Introduction   3
     1.2   The Rapidly Developing Liquid Cell Microscopy Technique   4
     1.3   Liquid Cell Microscopy for Materials Science, Biology, and Beyond   12
     1.4   Which Type of Microscopy Should I Use?   19
     1.5   Future Prospects   21
     1.6   Conclusions   25
     References   25

2   **Encapsulated Liquid Cells for Transmission Electron Microscopy**   35
Eric Jensen and Kristian Mølhave

     2.1   Introduction   35
     2.2   Microfabricated Chip Designs   37
     2.3   Other Encapsulation Methods   45
     2.4   What Happens When the Liquid Cell Fails?   46
     2.5   Membrane Bulging: Mitigation and Measurement   47
     2.6   Stimuli and Correlative Measurements: Biasing, Heating, Flow, and Spectroscopy   49
     2.7   Conclusions and Outlook   51
     References   52

3   **Imaging Liquid Processes Using Open Cells in the TEM, SEM, and Beyond**   56
Chongmin Wang

     3.1   Introduction   56
     3.2   Fundamental Concepts for Open Cell Experiments in S/TEM   57

	3.3 Open Cells for Imaging Droplets, Crystal Growth, Particle Motion, and Surface Passivation	59
	3.4 Open Cells for *In Situ* Battery Reactions	63
	3.5 Extension of the Open Cell Concept to Other Imaging and Spectroscopic Techniques	72
	3.6 Perspective	73
	Acknowledgements	74
	References	75
4	**Membrane-Based Environmental Cells for SEM in Liquids**	**78**
	Andrei Kolmakov	
	4.1 Introduction	78
	4.2 Basics of SEM through Membranes	80
	4.3 Examples of Environmental Cell Designs and Liquid SEM Applications	85
	4.4 Novel Two-Dimensional Materials as Electron-Transparent Membranes for Liquid SEM Cells	92
	4.5 Outlook	99
	Acknowledgements	101
	References	101
5	**Observations in Liquids Using an Inverted SEM**	**106**
	Chikara Sato and Mitsuo Suga	
	5.1 Introduction	106
	5.2 Instrument Design and Sample Geometry of the ASEM	106
	5.3 Applications of ASEM	109
	5.4 Correlative Microscopy (CLEM)	116
	5.5 Other SEM Techniques for Examining Liquids at Atmospheric Pressure	120
	5.6 Conclusions	123
	Acknowledgements	124
	References	124
6	**Temperature Control in Liquid Cells for TEM**	**127**
	Shen J. Dillon and Xin Chen	
	6.1 Introduction: Controlled Temperature Experiments	127
	6.2 Electron Beam-Induced Heating	131
	6.3 Temperature Measurements	132
	6.4 Applications	133
	6.5 Outlook	137
	References	138

7	**Electron Beam Effects in Liquid Cell TEM and STEM**	140
	Nicholas M. Schneider	
	7.1 Introduction	140
	7.2 Electron Energy Loss in Liquids	140
	7.3 Electron Beam Heating	144
	7.4 Introduction to the Radiation Chemistry of Water	147
	7.5 Homogeneous Irradiation	150
	7.6 Finite Beam Irradiation with Diffusion	152
	7.7 Practical Effects of Radiolysis	154
	7.8 Radiolysis beyond Neat Water	159
	7.9 Conclusions and Outlook	160
	References	161
8	**Resolution in Liquid Cell Experiments**	164
	Niels de Jonge, Nigel D. Browning, James E. Evans, See Wee Chee, and Frances M. Ross	
	8.1 Introduction	164
	8.2 Spatial Resolution in Liquid Cell TEM	165
	8.3 Spatial Resolution in Liquid Cell STEM	174
	8.4 Temporal Resolution in TEM and STEM	177
	8.5 Image Simulations in Liquid Cell TEM and STEM	179
	8.6 Some Practicalities and Pitfalls of Liquid Cell TEM and STEM	179
	8.7 Summary and Outlook	185
	Acknowledgements	185
	References	186

Part II Applications   189

9	**Nanostructure Growth, Interactions, and Assembly in the Liquid Phase**	191
	Hong-Gang Liao, Kai-Yang Niu, and Haimei Zheng	
	9.1 Introduction	191
	9.2 Formation of Nanoparticles in TEM	192
	9.3 Single Particle Growth Trajectories	194
	9.4 Important Factors in Nanoparticle Growth	197
	9.5 Growth of Materials Architectures	200
	9.6 Nanoparticle Diffusion and Assembly	201
	9.7 Etching and Corrosion	205
	9.8 Conclusions and Outlook	205
	Acknowledgements	206
	References	206

10	**Quantifying Electrochemical Processes Using Liquid Cell TEM**	210
	Frances M. Ross	
	10.1 Introduction	210
	10.2 Design of Liquid Cells for Quantitative Electrochemical Experiments	211
	10.3 Electrochemical Nucleation and Growth in Plan View	219
	10.4 Growth Front Propagation via Lateral Measurements	224
	10.5 Experimental Challenges	228
	10.6 Outlook	232
	References	233
11	**Application of Electrochemical Liquid Cells for Electrical Energy Storage and Conversion Studies**	237
	Raymond R. Unocic and Karren L. More	
	11.1 Introduction	237
	11.2 Electrical Energy Storage and Conversion Systems: Challenges and Opportunities	237
	11.3 Closed Cell Electrochemical-S/TEM for Energy Storage and Conversion Studies	238
	11.4 Electroanalytical Measurement Techniques	242
	11.5 Application of Electrochemical-S/TEM for Battery Research	244
	11.6 Application of ec-S/TEM for Fuel Cell Research	252
	11.7 Summary	254
	Acknowledgements	254
	References	255
12	**Applications of Liquid Cell TEM in Corrosion Science**	258
	See Wee Chee and M. Grace Burke	
	12.1 Introduction	258
	12.2 Studying Corrosion in Aqueous Environments	259
	12.3 Studies of Corrosion using Liquid Cell TEM	261
	12.4 Considerations Pertaining to Studying Corrosion with Liquid Cell TEM	266
	12.5 Microfluidic Cell Design for Electrochemical Corrosion Experiments	270
	12.6 Outlook	271
	Acknowledgements	272
	References and Notes	272
13	**Nanoscale Water Imaged by *In Situ* TEM**	276
	Utkur Mirsaidov and Paul Matsudaira	
	13.1 Introduction	276
	13.2 Interfacial Fluids	277
	13.3 Nanodroplet Condensation	280
	13.4 Fluids in Nanochannels	282

13.5	Voids and Nanobubbles in Liquid Films	285
13.6	Outlook	286
	References	287

14 Nanoscale Deposition and Etching of Materials Using Focused Electron Beams and Liquid Reactants — 291
Eugenii U. Donev, Matthew Bresin, and J. Todd Hastings

14.1	Overview of Gas-Phase Focused Electron Beam-Induced Processing (FEBIP)	291
14.2	Methods for LP-FEBIP	293
14.3	Survey of LP-FEBID of Transition Metals	299
14.4	Multi-element LP-FEBID	303
14.5	Liquid-Phase Focused Electron Beam-Induced Etching (LP-FEBIE)	306
14.6	Mechanisms for LP-FEBIP	309
14.7	Outlook	310
	Acknowledgements	310
	References	310

15 Liquid Cell TEM for Studying Environmental and Biological Mineral Systems — 316
Michael H. Nielsen and James J. De Yoreo

15.1	Introduction	316
15.2	Mechanisms of Mineral Formation	317
15.3	Liquid Holder Design	319
15.4	Calcium Carbonate Formation Pathways	321
15.5	Nucleation within an Organic Matrix	324
15.6	Particle-Based Crystallization	326
15.7	Conclusions and Future Applications	328
	Acknowledgements	330
	References	330

16 Liquid STEM for Studying Biological Function in Whole Cells — 334
Diana B. Peckys and Niels de Jonge

16.1	Introduction	334
16.2	Liquid STEM Technology	334
16.3	Studying Membrane Proteins in Whole Cells in Liquid	339
16.4	Live Cell Liquid STEM	344
16.5	Gold Nanoparticle Uptake Studied in Whole Cells	347
16.6	Comparison with Cryo-TEM	348
16.7	Conclusions and Outlook	350
	Acknowledgements	351
	References	351

17	**Visualizing Macromolecules in Liquid at the Nanoscale**	356

Andrew C. Demmert, Madeline J. Dukes, Elliot Pohlmann, Kaya Patel, A. Cameron Varano, Zhi Sheng, Sarah M. McDonald, Michael Spillman, Utkur Mirsaidov, Paul Matsudaira, and Deborah F. Kelly

    17.1 Introduction: The Critical Need for Imaging Dynamic Events in Life Sciences   356
    17.2 Recent Technical Advances: How Liquid Cell TEM Can Address This Critical Need   357
    17.3 The Affinity Capture Technique to Tether Unlabeled Biological Complexes onto Si_xN_y 357
    17.4 Correlative Nanoscale Imaging: What Information Can We Learn from Combining Liquid Cell TEM and Cryo-EM?   361
    17.5 New Directions: Use of Direct Electron CMOS Detectors to Acquire "Molecular Movies" of Fundamental Processes   368
    Acknowledgements   369
    References   369

18	**Application of Liquid Cell Microscopy to Study Function of Muscle Proteins**	371

Haruo Sugi, Shigeru Chaen, Tsuyoshi Akimoto, Masaru Tanokura, Takuya Miyakawa, and Hiroki Minoda

    18.1 Introduction: Our Motivation for Liquid Cell Microscopy of Muscle Contraction   371
    18.2 Experimental Methods for Recording Myosin Head Movement   375
    18.3 ATP-Induced Movement of Individual Myosin Heads   381
    18.4 Conclusions and Outlook   388
    Acknowledgements   389
    References   389

Part III Prospects   391

19	**High Resolution Imaging in the Graphene Liquid Cell**	393

Jungwon Park, Vivekananda P. Adiga, Alex Zettl, and A. Paul Alivisatos

    19.1 Introduction to Graphene Liquid Cells: Advantages, Opportunities, and Fabrication Methods   393
    19.2 Studying Growth Mechanisms in Atomic Detail by GLC-TEM   396
    19.3 Applications of GLC-TEM in Biological Studies   400
    19.4 Future Directions   404
    References   406

20	**Analytical Electron Microscopy during *In Situ* Liquid Cell Studies**	408

Megan E. Holtz, David A. Muller, and Nestor J. Zaluzec

    20.1 Introduction   408
    20.2 Electron Energy Loss Spectroscopy   413

	20.3	X-ray Energy Dispersive Spectroscopy	420
	20.4	Summary	431
	Acknowledgements	432
	References	432

21 Spherical and Chromatic Aberration Correction for Atomic-Resolution Liquid Cell Electron Microscopy — 434

Rafal E. Dunin-Borkowski and Lothar Houben

	21.1	Introduction	434
	21.2	Spherical Aberration Correction in the TEM	435
	21.3	Spherical Aberration Correction in STEM	444
	21.4	Chromatic Aberration Correction in the TEM	446
	21.5	Conclusions	452
	Acknowledgements	453
	References	453

22 The Potential for Imaging Dynamic Processes in Liquids with High Temporal Resolution — 456

Nigel D. Browning and James E. Evans

	22.1	Introduction	456
	22.2	Why Do We Need Better Temporal Resolution?	457
	22.3	Hardware/Software Developments for Fast Temporal Resolution	460
	22.4	Materials and Biological Examples	466
	22.5	Conclusions	471
	Acknowledgements	471
	References	472

23 Future Prospects for Biomolecular, Biomimetic, and Biomaterials Research Enabled by New Liquid Cell Electron Microscopy Techniques — 476

Taylor Woehl and Tanya Prozorov

	23.1	Introduction	476
	23.2	Visualizing Protein Structure in Liquid Water at High Resolution	476
	23.3	Elucidating Fundamental Biomineralization Mechanisms via *In Vivo* Imaging	479
	23.4	Visualizing Electromagnetic Fields and Nanoparticle Interactions in Biomolecular Systems	483
	23.5	Biomimetics	488
	23.6	Mesocrystal Formation	491
	23.7	Conclusions	495
	Acknowledgements	495
	References	495

Index — 501

Contributors

Vivekananda P. Adiga
University of California, Berkeley

Tsuyoshi Akimoto
Teikyo University

A. Paul Alivisatos
Lawrence Berkeley National Laboratory

Matthew Bresin
University of Illinois at Urbana-Champaign

Nigel D. Browning
Pacific Northwest National Laboratory

M. Grace Burke
The University of Manchester

Shigeru Chaen
Nihon University

See Wee Chee
National University of Singapore

Xin Chen
East China University of Science and Technology

Andrew C. Demmert
Virginia Tech Carilion Research Institute

Niels de Jonge
INM – Leibniz Institute for New Materials

James J. De Yoreo
Pacific Northwest National Laboratory

Shen J. Dillon
University of Illinois at Urbana-Champaign

Eugenii U. Donev
Sewanee: The University of the South

Madeline J. Dukes
Protochips, Inc.

Rafal E. Dunin-Borkowski
Forschungszentrum Jülich

James E. Evans
Pacific Northwest National Laboratory

J. Todd Hastings
University of Kentucky

Megan E. Holtz
Cornell University

Lothar Houben
Forschungszentrum Jülich

Eric Jensen
Technical University of Denmark

Deborah F. Kelly
Virginia Tech Carilion Research Institute

Andrei Kolmakov
National Institute of Standards and Technology

Hong-Gang Liao
Lawrence Berkeley National Laboratory

Paul Matsudaira
National University of Singapore

Sarah M. McDonald
Virginia Tech Carilion Research Institute

Hiroki Minoda
Tokyo University of Agriculture and Technology

Utkur Mirsaidov
National University of Singapore

Takuya Miyakawa
University of Tokyo

Kristian Mølhave
Technical University of Denmark

Karren L. More
Oak Ridge National Laboratory

David A. Muller
Cornell University

Michael H. Nielsen
Lawrence Livermore National Laboratory

Kai-Yang Niu
Lawrence Berkeley National Laboratory

Jungwon Park
Harvard University

Kaya Patel
Virginia Tech Carilion Research Institute

Diana B. Peckys
INM – Leibniz Institute for New Materials

Elliot Pohlmann
Virginia Tech Carilion Research Institute

Tanya Prozorov
Ames DOE Laboratory

Frances M. Ross
IBM T. J. Watson Research Center

Chikara Sato
National Institute of Advanced Industrial Science and Technology (AIST)

Nicholas M. Schneider
University of Pennsylvania

Zhi Sheng
Virginia Tech Carilion Research Institute

Michael Spillman
Direct Electron, LP

Mitsuo Suga
JEOL Ltd.

Haruo Sugi
Teikyo University

Masaru Tanokura
University of Tokyo

Raymond R. Unocic
Oak Ridge National Laboratory

A. Cameron Varano
Virginia Tech Carilion Research Institute

Chongmin Wang
Pacific Northwest National Laboratory

Taylor Woehl
Ames DOE Laboratory

Nestor J. Zaluzec
Argonne National Laboratory

Alex Zettl
University of California, Berkeley

Haimei Zheng
Lawrence Berkeley National Laboratory

Preface and Acknowledgements

Electron microscopy amazes and educates us with its images of atomic level structure, measurements of nanoscale chemistry, and movies showing otherwise invisible phenomena. However, electron microscopy is generally used only on solid samples, since liquids, particularly those with low vapor pressure such as water, are challenging to image within the microscope vacuum. The difficulty of dealing with water and other liquids was recognized early on in the development of electron microscopy, but practical solutions to the engineering challenges appeared only recently with the advent of microfabrication and micromanipulation techniques. We can now build liquid cells that allow us to apply the powerful capabilities of the electron microscope to imaging and analysis of liquid specimens. This has opened up wonderful opportunities ranging from nanoscale materials and processes in liquids to biological structures imaged at high resolution without freezing or drying. Physical and chemical changes taking place within batteries during operation, the attachment of atoms during the self-assembly of nanocrystals, and protein structures within whole cells in liquid water are examples in which a microscopic view is providing unique insights.

This book describes the practical aspects of liquid cell electron microscopy. We start with a historical perspective, then consider several different strategies for experimental design as well as resolution and electron beam effects. We next describe research results in fields as diverse as energy storage, biomineralization, fluid physics, and the structure of different types of biomaterials in their native, hydrated state. We finally look forward to future experiments with improved spatial and temporal resolution or analytical capabilities that probe increasingly complex materials and processes.

The motivation for bringing these discussions together is the exciting current state of liquid cell electron microscopy. Since the equipment is not too expensive and generally works in existing electron microscopes, many programs involving electron microscopy of liquids have developed around the world. These projects have provided unique, high quality scientific information that frequently appears in high impact journals. Within just the last few years, electron microscopy experiments involving liquids have become a regular component of materials science and microscopy conferences, generating lively and occasionally confrontational discussions. Liquid cell electron microscopy in all its different variations is well positioned to improve its reach still further. The ease of use of the equipment is improving continuously, our understanding is becoming more quantitative, and new capabilities are appearing that allow further exploration of frontiers in materials, environmental science, life sciences, and beyond.

Combining the expertise of many materials scientists, physicists, biologists, and microscopists in this edited volume enables us to address two aims. One is to help electron microscopists by providing guidance on how to obtain meaningful data from experiments on liquid samples. This is addressed through the chapters on experimental design, artifacts, and quantification. The other aim is to provide information that will help scientists who are not microscopists to decide for themselves the value of adopting or collaborating on electron microscopy techniques for imaging and analysis of liquids. This is addressed through the descriptions of scientific questions that have already been answered, as well as the discussions of the current status and future capabilities for obtaining information on liquid samples. By drawing on the experience of many of the leaders in this field we hope this book will fulfill these two objectives. While we cover a broad range of topics, our intention is that not too much specialized knowledge is required in any one of them.

I am grateful for all the help I have received during this project. I would especially like to thank my family for their patience and enthusiastic support and my colleagues at IBM and Cambridge University Press for their understanding and help as this book came together. It has been a great pleasure and a powerful learning experience to collaborate with colleagues around the world. Although there are omissions – a field moving this quickly cannot be captured in a book – I hope that we can nevertheless give a snapshot of the evolution, current position, and future of liquid cell electron microscopy, and preview the remarkable opportunities that I believe will be realized in the future.

Frances M. Ross
Yorktown Heights, New York

Part I

Technique

1 Past, Present, and Future Electron Microscopy of Liquid Specimens

Niels de Jonge and Frances M. Ross[1]

1.1 Introduction

This is an exciting era for electron microscopy. Over the last several years, the resolution of electron microscopes has increased dramatically through the successful implementation of aberration correction. Solving the decades-old lens aberration problem now allows astonishingly detailed and quantitative analysis of atomic level structures, as well as the ability to use lower voltage electrons for imaging while maintaining high resolution. At the same time, improvement in electron detectors has transformed the applications of electron microscopy. Images can be obtained with fewer electrons, reducing damage to the sample; higher speed imaging of dynamic processes is now possible; and more efficient collection of signals such as X-rays improves chemical analysis at the nanoscale. In this book, we describe an advance that is smaller in scale, but that is also transforming electron microscopy and its applications to materials science, life science, and beyond: the ability to image samples that contain liquids, particularly water. Electron microscopy of structures and processes in liquids is important over broad areas of science and technology. The changes that occur inside batteries during operation, the attachment of atoms during the self-assembly of nanocrystals, and the structures of biological materials in their native, liquid state are areas in which microscopy is essential. This wide applicability has driven a wave of interest in developing ways to perform electron microscopy on liquid samples. And as the technique of liquid cell electron microscopy has become more widespread, it promises exciting possibilities for solving grand challenges in materials, geology, biology, physics of fluids, and many other fields.

The difficulty of imaging liquids was recognized from the start of the development of electron microscopy. And somewhat like aberration correction (although on a vastly less complex scale), the solution was understood early on but its implementation required modern advances. The key issue is how to separate a liquid, in particular a high vapor pressure liquid such as water, from the vacuum in the electron microscope, while still obtaining reasonable images. In transmission electron microscopy (TEM), the image quality is generally reduced as the sample thickness increases, so this implies that the liquid must be in the form of a thin layer, in the range of nanometers to a few

---

[1] Note: Part of this chapter is adapted from F. M. Ross, Opportunities and Challenges in Liquid Cell Electron Microscopy, *Science*, **350** (2015), aaa9886, doi:10.1126/science.aaa9886.

micrometers. This goal was not achieved until modern microfabrication techniques could be used to build thin windows of silicon nitride with a controlled, sub-micrometer separation, between which the liquid could be confined. This "closed" liquid cell was rapidly developed to include electrodes and flow capabilities, and interfaced to the microscope with dedicated sample holders. Similarly, in scanning electron microscopy (SEM), the availability of thin and robust windows allowed closed cells to be developed for imaging liquids, complementing environmental SEM (ESEM), which is based on sample chambers capable of accommodating water vapor and cooling the sample.

The first TEM experiments to make use of modern liquid cell designs addressed questions in electrochemical deposition, nanomaterials synthesis, diffusion in liquids, and the structure of biological materials; liquid cell SEM enabled imaging of biological structures without traditional preparation procedures. The results demonstrated the ability of liquid cell imaging to probe areas that had traditionally been inaccessible to electron microscopy, and in doing so to achieve useful and unique information [1]. As interest in the technique increased, the quality and functionality of TEM and SEM closed liquid cells improved, and open cell TEM was devised for examination of electrochemical and other processes. These developments in turn expanded interest still further. Since the equipment for liquid cell TEM and SEM is not too expensive to purchase or can be made relatively easily in one's own laboratory, and is compatible with existing microscopes, programs were started by research groups around the world. Thus the impact of liquid cell electron microscopy has increased rapidly [2], in terms of both the scientific areas of applicability and the types of measurements that are possible. Materials and processes examined now include corrosion, biomolecular structure, battery operation, beam-induced nanofabrication, bubble dynamics, radiation effects, and biomineralization. The rich capabilities of modern microscopes have been applied to liquid cell experiments, including elemental analysis through energy loss or X-ray signals, aberration correction, and high speed image recording. Correlative techniques are starting to produce exciting information in which electron microscopy is combined with light microscopy, synchrotrons, or other probes. Liquid cell microscopy has the potential to extend into new areas, to adopt advances in instrumentation even more fully, and perhaps even solve "grand challenge" problems.

1.2 The Rapidly Developing Liquid Cell Microscopy Technique

1.2.1 Liquids in the TEM

The early pioneers of transmission electron microscopy were very interested in imaging hydrated samples and water itself for applications in both materials and life science [3, 4]. Remarkable progress was made, given the challenges of observing even solid materials with the microscopes of the time. Two techniques were developed for getting water into the electron microscope while still maintaining a good enough vacuum to operate the electron source. These have evolved directly into the techniques that are in use today.

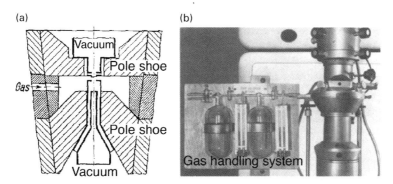

Figure 1.1. Transmission electron microscope with differentially pumped environmental chamber for imaging specimens at higher pressures. (a) The specimen in a gaseous environment is separated from the high vacuum of the electron column via pump-limiting apertures. The apertures are placed on the pole pieces (shoes) of the objective lens. (b) Photograph of electron microscope with gas handling system. From Ref. 5 with kind permission from Springer Science+Business Media.

One approach [5] is to use differential pumping to control the pressure at the sample region. Figure 1.1 shows how a relatively high pressure at the sample is maintained by placing pressure-limiting apertures on the pole pieces of the objective lens, enabling imaging in ambient air at up to 0.3 bar. Although not discussed in the original paper, this pressure is sufficient for water droplets to condense at room temperature. It would thus be fair to say that the first system for imaging specimens beneath a water layer was constructed in 1942! This "open cell" approach became highly successful in environmental SEM, as discussed below. The same open cell principle is now used in modern *in situ* TEMs to study gas phase reactions at moderate pressures – for example, imaging the growth of carbon nanotubes in a reactive gas atmosphere localized around the sample by differential pumping [6]. For reactions involving liquid water, the TEM community appeared less interested, perhaps because the geometry of the condensing water droplets was not controlled and the maximum pressure was limited. Nevertheless, open cell TEM continued to advance in diverse areas [3, 7], especially recently, when open cells using droplets of ionic liquids have helped visualize materials transformations during Li-ion battery operation [8–10]. The open cell TEM method and applications are described in more detail in Chapter 3.

The second approach is known as "closed cell" electron microscopy [11]. By enclosing the liquid between two electron-transparent windows, the limitation on maximum pressure imposed by the open cell could be circumvented. The earliest closed cells [11] made use of thin nitrocellulose films produced by floating a solution containing colloidal cotton on a mercury surface. The films were transferred to a perforated platinum support frame by raising the frame upward through the surface of the mercury. These windows were thick and provided poor resolution, and it was difficult to control the window separation, but interest continued over the next two decades. In the early 1960s, an environmental chamber [12, 13] consisting of two thin foils separated by a few micrometers (Figure 1.2a) enabled electron microscopy of condensed water

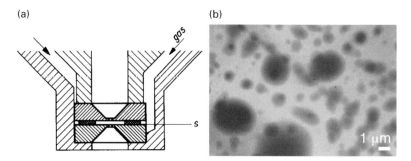

Figure 1.2. Environmental chamber closed with thin windows. (a) Two aperture holders support thin films and are placed face to face in the upper pole shoe of the objective lens. A thin film between the aperture holders (s) serves as a spacer of about 5 μm thickness. (b) Electron micrograph of water droplets condensed on the support film at a pressure of 0.13 bar. From Ref. 12 with kind permission from Springer Science+Business Media.

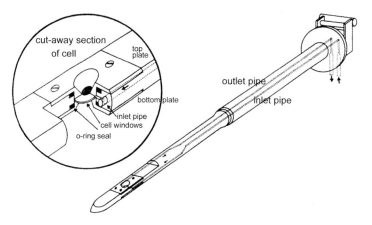

Figure 1.3. Early design of an environmental cell integrated in a side entry sample holder, with access pipes for gas flow from outside the electron microscope. Reprinted from Ref. 14, Copyright 1973, with permission from Elsevier.

droplets in a gas environment at 0.13 bar (Figure 1.2b). One could finally envisage imaging at atmospheric pressure, including samples enclosed in liquid, given foils of sufficient strength separated by spacers of sufficient narrowness. The next developments involved sample holders. The cell shown in Figure 1.2 is fixed in the microscope polepiece, making it difficult to modify the sample environment during imaging. Environmental cells integrated into side-entry sample rods allow simpler control, as shown in Figure 1.3, an early holder design that enables gas flow to and from the specimen [14]. With this type of equipment [4], the feasibility of imaging unstained biological materials such as bacteria in a wet environment was demonstrated, especially at higher accelerating voltages. Figure 1.4 shows a spatial resolution on the order of 100 nm achieved using ultra-high voltage TEM [4]. Similar systems developed for

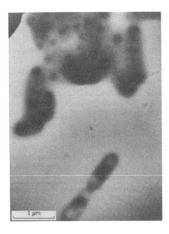

Figure 1.4. Coliform bacteria immersed in a layer of water in a closed chamber imaged with a 800 kV TEM. The environmental chamber was at room temperature with a water vapor pressure of 30 mbar. From Ref. 4. Reprinted with permission from AAAS.

standard voltage TEM were later used to inject liquid between windows for the study of a range of phenomena, including unstained bacteria [15], DNA [16], clays [17], catalytic reactions [18], and myosin head movement [19], described in more detail in Chapter 18.

These steps forward hinted at the opportunities possible with closed cell TEM, but the materials and technology available introduced severe practical difficulties. The windows were still thick and their separation, several micrometers, was large enough to reduce the image resolution to a value often not much better than that of light microscopy. Attempts to improve the resolution by filling the closed cell only partly with liquid resulted in poor control of the degree of water immersion of the sample. For biologists, the introduction of practical techniques to rapidly freeze biological specimens [20] superseded attempts to image in water with TEM. By rapidly freezing nanometer-scale biological objects, such as protein complexes and viruses, amorphous ice could be formed that preserved biological structure and prevented damage by ice crystals. Cryo-electron microscopy (cryo-EM) has become a key tool for structural biology [21–23], although the imaging of whole bacterial and eukaryotic cells, and of course dynamic processes, remain challenges.

The recent surge of interest in closed liquid cell TEM in both biology and materials science can be attributed to the use of modern microfabrication techniques to overcome the challenges of building closed liquid cells. Initially somewhat unreliable homemade cells that were sealed by hand with glue [24], modern closed liquid cells and their associated equipment have now developed to a point where they are readily available and simpler to use. Commercial systems can be purchased, and several groups have fabricated their own equipment [25–32]. The cells generally use windows made of silicon nitride supported on silicon chips. This highly practical solution, described further in Chapter 2, enables manufacture of strong windows of well-defined dimensions, strength, and electron transparency. Modern materials science also provides other choices for the window material, most importantly graphene [33], whose advantage in terms of image resolution is discussed further in Chapter 19.

Microfabrication brings another benefit to the fabrication of liquid cells for TEM: the integration of functionality into the cell. Liquid heating and cooling (Chapter 6), electrochemical reactions with customized electrode materials and geometries (Chapter 10), and patterned or modified silicon nitride surfaces (Chapter 17) enable a wider range of experiments. The sample holder has been developed in step with the liquid cell (Chapter 2). Its function extends beyond simply holding the cell securely. It carries the electrical connections between electrode or heater elements and their external controllers. It may also provide the vacuum seal by clamping the chips, and allow liquid flow via inlet and outlet tubes driven by a syringe pump. Integrated microfluidic systems for liquid flow [34] enable the possibility of replenishing or changing the solution chemistry rapidly while imaging.

One of the key ongoing developments in closed liquid cell TEM is its adaptation to the full range of TEM capabilities. Most liquid cell images are recorded using conventional bright field TEM or high angle annular dark field (HAADF) STEM imaging modes. Single images are recorded to minimize the dose for biological structures, while movies allow analysis of dynamic materials processes. Higher sensitivity detectors reduce the dose required (Chapter 8), or offer higher frame rates and hence better time resolution. Dark field and high resolution liquid cell TEM are available, and aberration-corrected imaging (Chapter 20) and ultrafast image acquisition (Chapter 22) are under development. Analytical microscopy is also making its mark on liquid cell experiments (Chapter 21), via both electron energy loss spectroscopy (EELS) [35, 36] and X-ray energy dispersive analysis (XEDS) [37, 38]. Improvements in liquid cell and holder designs, as well as higher performance microscopes and detectors, promise continuing development of liquid cell TEM.

1.2.2 Liquids in the SEM

For imaging liquids, SEM shares general features with TEM. Liquid droplets with a low vapor pressure, for example ionic liquids [39, 40], can be imaged without special precautions. For high vapor pressure liquids, particularly water, both open and closed cell concepts are useful and provide different types of information. Furthermore, innovative concepts for correlative light and electron microscopy (CLEM) have adapted the geometry of the microscope and sample to obtain data from the same sample area.

Figure 1.5 shows the first environmental system for a SEM, published in 1970 [41]. This was a closed cell system in which a supported 10 nm-thick carbon film sealed liquid inside an environmental chamber. We return to closed cell SEM below. In the following two decades an open chamber approach was also developed that enabled imaging at pressures high enough to permit liquid water [42, 43]. This concept, environmental SEM (ESEM), is based on a differential pumping aperture in the objective lens to maintain the sample chamber at an elevated pressure, similar to the open cell design described above for TEM [5].

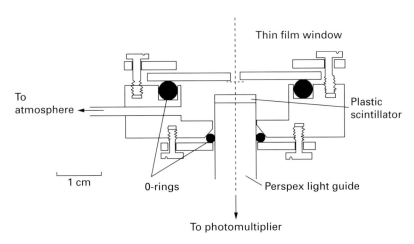

Figure 1.5. Environmental cell for a scanning electron microscope. From Ref. 41, © IOP Publishing. Reproduced with permission. All rights reserved.

Open chamber ESEM proved highly successful and is widely used in industry and academia to image a variety of specimens in a vapor environment with a spatial resolution in the few nanometer range [44]. Some microscopes can handle pressures in the sample chamber of up to 4000 Pa. Typical operation for optimal resolution is at a sample temperature of 4 °C and a pressure of 814 Pa, so that the sample is maintained at equilibrium between water and vapor. By fine-tuning the pressure it is possible to remove most of the bulk water over the sample to eventually leave a film of water thin enough to allow imaging. A key feature of ESEM is the variety of signals that can be collected. The standard backscatter detector generates contrast at objects somewhat below the surface of the liquid. The gaseous secondary electron detector provides surface information. It is mounted above the sample and held at a positive potential with respect to the sample, so that secondary electrons are accelerated towards it and cause an avalanche of ionized gas amplifying the signal. This enhancement in detection efficiency is an important benefit of the water vapor present in ESEM [43]. The spatial resolution achieved with these two detectors is typically around 10 nm for thin liquid layers. It is also possible to use a detector beneath the sample to record electrons transmitted through the sample. This approach, known as wet-STEM [45, 46], requires a very thin liquid layer (especially given the low accelerating voltage, compared to TEM) but is capable of detecting nanoscale objects of a high density within a liquid layer via atomic number (Z) STEM contrast, with minimal sample preparation. Resolution of 3 nm has been demonstrated for gold nanoparticles in a water layer [47]. The large sample chamber volume in SEM allows sample tilting, such that 3D ESEM is feasible [48]. ESEM applications involving liquid layers include imaging of biological cells [47], nanoparticle movement [49] and wettability studies [50–53]. In Chapter 16 we discuss ESEM of whole biological cells under thin water layers. The reader is referred elsewhere [44] for a more detailed description of ESEM and its broad range of applications.

When the sample involves a thicker liquid layer or complications such as electrodes or liquid flow, closed cell SEM is expected to have distinct advantages over ESEM. The closed cell concept was first revived [54] by enclosing the liquid in a capsule sealed with a thin polymer membrane, and forming images using a backscatter detector. The window thickness used, ~100 nm, limited the spatial resolution to ~20 nm for heavy materials. On the introduction of super resolution fluorescence microscopy techniques [55] with a similar resolution range, biological research adopted these revolutionary optical methods. But closed cell SEM continued to evolve to improve resolution and enable correlative or *in situ* capabilities.

Several strategies have been developed to increase the resolution in closed cell SEM, as described in Chapter 4. Reducing the window thickness using silicon nitride or graphene [56–58] is important. It is even possible to make a pore in a capsule window a few micrometers in diameter to allow imaging of a liquid meniscus without any membrane present [59]. The meniscus is stable over a narrow range of pressure differential, temperature, and surface tension. Resolution is also increased by using TEM-style closed cells with STEM detection through the liquid and windows [28]. Graphene can even be used to coat and seal samples on a thick substrate without mounting it in a capsule [60].

Correlative microscopy has been enabled through several imaginative concepts based on electron-transparent windows. Such designs are discussed in Chapter 5. A closed cell with an electron-transparent window above and a glass slide below allows SEM imaging from above and light microscopy from below, using an optical lens within the SEM vacuum system [61, 62]. Alternatively, the liquid can be open to the atmosphere, but placed in a dish that contains a silicon nitride viewing window. The sample is imaged from below with an "inverted SEM" using backscatter contrast, and from above with light microscopy [63]. The use of an open cell culture dish makes it easy to combine biological techniques, such as incubation and fixing, with SEM observation. A related inverted SEM concept involves focusing the electron beam in a small spot on a phosphor layer on the viewing window, thus creating a nanoscale light source for high resolution correlative fluorescence microscopy [64]. Finally, it is also possible to mount a silicon nitride window directly at the exit of an SEM column [65] to image samples in air. Unlike the simultaneous collection of the electron and light images with inverted SEM, this microscope utilizes a precise stage to move the sample between SEM and light microscopes. We anticipate exciting developments in these and other concepts for liquid cell CLEM.

1.2.3 Limitations of Liquid Cell Electron Microscopy

A series of developments is beginning to address the two key challenges of TEM and SEM of liquids: image resolution and electron beam effects.

For imaging purposes in TEM, liquids behave like amorphous materials and do not show distinctive contrast. However, when a liquid surrounds an object or covers a surface, the resulting images will generally show a lower resolution than if the liquid were absent. There are two main reasons for this (Chapter 8): loss of information due to multiple scattering of the electrons in the thick liquid and window layers, and blurring due to Brownian motion of suspended objects. Blurring can be mitigated by higher image acquisition speeds, though at the cost of higher dose (Chapter 22). Multiple scattering has to be controlled through both liquid geometry and window thickness. Thinner windows improve image quality without affecting the material or process under study. As described in Chapter 19, the development of graphene for windows has increased the spatial resolution for liquid cell TEM, with the challenge now to control the liquid geometry to include capabilities such as electrodes or liquid flow. Similarly, thinner liquid layers improve image quality. In silicon nitride microfabricated cells the liquid layer is usually thicker than intended, especially towards the center of the window, because the windows are deflected outwards by the pressure difference between the interior of the cell and the microscope vacuum. Chapter 2 describes several strategies for control of the window separation [28, 66, 67]. But it is worth noting that there are limits on how thin the liquid layer should be made. We cannot expect a liquid cell experiment to be a faithful representation of a phenomenon in bulk liquid if the liquid is too thin [10]. Examples of this are seen in electrochemical growth, Chapter 10 (where diffusion gradients control kinetics, making it necessary [68] to evaluate the effects of the limited electrolyte dimensions), and fluid physics, Chapter 17 (where Brownian motion appears different in ultra-thin compared to thicker layers [27], with particles slowing [69] near the window).

The resolution in closed cell SEM is discussed in Chapter 4. Here, the total liquid thickness is important only when imaging using STEM in the SEM, but optimizing the window materials is even more critical than for TEM, given the lower electron energy. Graphene and graphene oxide windows [57, 70] are more transparent to low energy electrons than silicon nitride, providing a practical solution for closed liquid capsules, and even allowing analytical techniques to be applied.

Electron beam effects are the other major consideration for liquid cell microscopy. This complex topic is discussed in more detail in Chapter 7. The energy deposited by the SEM or TEM beam does not significantly heat the liquid, but it does drive radiolytic processes that generate reactive species such as (in water) hydrated (solvated) electrons or hydrogen ions. Fortunately, water radiolysis has been well studied and the results can be applied to electron microscopy. For the conditions appropriate to TEM, radiolysis products reach equilibrium concentrations in the irradiated region within seconds, attaining concentrations that depend on dose rate, illuminated area, liquid thickness, and total liquid volume [71]. Radiolytically produced hydrogen gas can exceed its solubility limit and form bubbles [72]. Hydrogen ions can change the solution pH [71]. Hydrated electrons can alter the structure of biological materials, or drive the beam-induced growth of metallic nanoparticles by reducing metallic cations [73], allowing possibilities for beam writing [33, 72, 74]. A greater understanding is required for solutions with multiple dissolved species or non-aqueous materials such as ionic liquids. However, existing knowledge of radiation physics can guide us, and it has already been shown that beam effects can be mitigated using scavenging strategies [75]. As microscopists become increasingly familiar with beam effects in water, the low dose techniques developed for life science are becoming standard, and the benefits of high sensitivity detectors in reducing the dose required per image are appreciated even more.

1.3 Liquid Cell Microscopy for Materials Science, Biology, and Beyond

Modern microfabricated liquid cells were first used for electrochemical experiments (Chapter 10), imaging metal deposition onto electrode surfaces and correlating the movies with the electrochemical parameters of voltage and current. The synthesis of nanoparticles and nanostructures in TEM and SEM was another key materials area (Chapters 9 and 14), providing information on growth physics and Brownian motion, as well as fabricating useful structures. In life sciences (Chapter 16), it was quickly realized that, remarkably, gold labels on biological structures could be resolved even through several micrometers of water. Open cell TEM (Chapter 3) provided information on battery-relevant reactions. Liquid cell microscopy continues to produce insights and exciting new information in these areas, as well as in new fields such as biomineralization, the imaging of unlabeled biostructures, bubble dynamics, corrosion and phase transformations. Here we provide highlights of recent results to give a general overview of the applications of liquid cell TEM and SEM.

1.3.1 Electrochemistry

The importance of liquid cell microscopy to electrochemistry is that liquid cell experiments enable us to connect structural and electrochemical information: other techniques that provide electrochemical data do not possess the same combination of temporal and spatial resolution [10, 24, 76]. The outcome is a detailed probe of growth mechanisms for key processes such as corrosion or battery cycling. Sample and equipment developments have made open and closed electrochemical liquid cell experiments more powerful: these include new geometries and materials for closed cell electrodes, with the ability to interface arbitrary materials to standard electrodes; improvements in open cell sample design; wider use of glovebox handling systems for materials that are sensitive to air or moisture, such as the high vapor pressure organic liquids used in Li-ion batteries; and improved design for quantitative low current (nanoampere) measurement [76, 77].

1.3.1.1 Understanding Growth Morphology

Refining electrochemical growth models is an exciting opportunity for liquid cell microscopy. Closed cell experiments first followed nucleation of individual copper clusters on electrode surfaces [24, 68, 78, 79], aiming to understand the role of bulk and surface diffusion on nucleus size and density. Continuing growth allows the nuclei to coalesce into a continuous polycrystalline film which grows laterally away from the electrode [80]. Instabilities in the lateral growth front, such as dendrite formation, have been observed in systems relevant to battery operation [81–86]. The development of roughness can be tracked and local measures are accessible, such as the growth rate at each point on a dendrite [82]. Quantitative measurements could eventually help to understand and control growth front morphology through additives or pulse deposition [87]. Since diffusion fields play such a key role in electrochemical growth, an exciting development has been the demonstration that under certain circumstances one can image the distribution of ions in solution via their effect on transmitted electrons [81]. As discussed in Chapter 10, all these experiments suggest extensive possibilities for detailed interrogation of electrochemical growth physics.

1.3.1.2 Reactions in Battery Anodes, Cathodes, and Electrolytes

Interfacial electrochemical reactions control the transport of charge and mass in batteries and energy conversion systems. Liquid cell microscopy enables direct imaging of key phenomena during battery operation, and has already produced promising results for materials relevant to Li-ion batteries. In electrode materials, dramatic structural and chemical changes take place during cycling. Lithium incorporation changes the volume of anode materials by hundreds of percent, clearly visible in open [10] and closed [87] liquid cell TEM, as discussed in Chapter 3. Chemical changes in electrode materials, such as Li transport across a $LiFePO_4$ cathode material, have been resolved using EELS [36]. Battery electrolytes also undergo key changes during cycling, breaking down and forming a solid electrolyte interphase (SEI) layer composed of inorganic and organic electrolyte byproducts. Use of a three-electrode cell [88] allows correlation of the onset potential for the electrolyte solvent reduction and the SEI nucleation and growth. A detailed understanding of these phenomena will help to improve Li-ion battery safety and cycle life [89]. The excitement and opportunities of research in batteries and fuel cells are described in Chapter 11.

Future progress relies on how well the experiment simulates a "real" device, with arbitrary anode and cathode materials in an appropriate geometry, as well as a reference electrode, and realistic diffusion pathways through the liquid. The development of methods to place blocks or strips of materials onto liquid cell electrodes is therefore important [10, 77, 87], as is the availability of electrode materials such as glassy carbon [36]. Electron beam effects must also be considered, and may even indicate parameters of electrolyte stability [90].

1.3.1.3 Corrosion and Related Phenomena

In Chapter 12, the ability to relate nanoscale phenomena, such as pit formation, with macroscopic electrochemical parameters is shown to be an important motivation for using liquid cell microscopy to examine corrosion. Localized corrosion occurs when aggressive species such as Cl^- break down the protective oxide film on materials such as stainless steel and Al or Ti alloys. TEM movies can show the initiation of pitting when metals are exposed to salt solutions with or without an applied potential [91, 92], and can follow the kinetics and dependence of corrosion morphology on salt concentration. Corrosion mechanisms are often quantified through the dependence of the dissolution current on the applied voltage. Such "Tafel plots" can be obtained in a liquid cell by depositing a film of interest over the window and placing a second electrode nearby [92]. Materials such as steels require a different approach, where a lamella is cut and welded to the liquid cell electrode [93]. The challenges in corrosion experiments include beam effects [91] and reactions at dissimilar metal junctions, for example between the material of interest and the liquid cell electrode [92]. Indeed, liquid cell microscopy provides good opportunities for studying junctions and galvanic replacement reactions [75].

1.3.2 Nucleation, Growth, and Coalescence in Liquids

Liquid cell TEM and SEM can both be used to create solid phases when the electron beam irradiates certain liquids. Nucleation and growth of nanoparticles in TEM, described in Chapter 9, showcases the unique opportunities that high spatial and temporal resolution provide for understanding growth mechanisms, diffusion, and coalescence. Directed growth from liquids using a focused beam in SEM or STEM, described in Chapter 14, complements focused electron- and ion-beam assisted growth from vapor, and allows a wide range of materials to be patterned into nanostructures.

1.3.2.1 Growth of Individual Particles

Nanoparticles nucleate and grow on the liquid cell walls when ions in solution are reduced by radiolytically produced hydrated electrons. Although beam-induced, such growth is relevant to solution phase growth stimulated by other forms of irradiation, as used to form particles with a narrow size distribution and without using surfactants. Liquid cell movies provide quantitative measurements of individual particles as they grow, and a direct view of the mechanisms at work. Initial experiments [73] showed growth pathways for Pt nanocrystals. This key study stimulated experiments on different elements, core-shell and alloy particles, and particles in templates [33, 94–98], formation of extended, dendritic structures [99, 100], and dissolution of nanoparticles in corrosive environments [101, 102]. Measurements as a function of dose rate [94] can indicate whether diffusion or attachment controls growth. Calculations of the changes in chemistry induced by the beam are important in understanding the processes taking place [103]; in fact, the balance between radiolytic oxidizing and reducing species can lead to situations in which the beam intensity determines particle stability [71, 104]. With improved image resolution (Chapter 20), it even becomes possible to determine the nature and evolution of the facets on the nanoparticles. Surfactants can control the facets present [105], and it is possible to see whether facets develop according to the Wulff plot in the presence of surfactants [106].

The development of heating capabilities has allowed reactions to be driven by temperature as well as by the electron beam, probing other methods of nanoparticle synthesis or modification. Hydrothermal precipitation [107] and nucleation and growth in ionic liquids [108] can be observed during heating and cooling. Unusual particle growth modes [109] become visible. Oxidation at elevated temperature can lead to Kirkendall void formation [110], and laser heating can trigger nucleation [111].

In these nanoparticle growth experiments, the information obtained depends on the time resolution and the number of tracked particles. Thus, a key advance for this type of experiment is the development of data compression and automated video analysis techniques, discussed in Chapter 22.

1.3.2.2 Patterned Growth

Rather than irradiating a broad area and forming many small nanocrystals, one can irradiate a specific area to achieve liquid phase beam-induced growth [112]. Such patterned deposits have been formed in TEM closed cells [32, 72, 74], but innovative liquid cell SEM experiments have demonstrated a much broader range of materials and morphologies, described in Chapter 14.

1.3.2.3 Nanoparticle Coalescence

As particles coalesce into larger assemblies, or are pushed together by the beam [113], liquid cell TEM and SEM provide a remarkable view of the processes at work. The motions and rotations of particles as they approach yield direct information on interparticle forces [95, 114]. Coalescence takes place on preferred planes [33] and particles may approach multiple times until they rotate into registry and snap together [114]. Defect formation during coalescence can be imaged, as can structural rearrangements after coalescence. The packing of aggregates several micrometers in size can be studied directly in liquid using ESEM [49]. Superlattices or diffusion-limited aggregates may form, and aggregation parameters [115, 116] and orientational order parameters [117] can be measured. The effects of surfactants can be evaluated, including beam-induced changes [118]. A common theme, of great importance to synthesis and biomineralization, is the rich variety of phenomena at work; the final shape of a particle assembly provides only limited information on its formation pathway.

1.3.3 Phase Transformations in Liquids

Electron microscopy has a distinguished history in exploring phase transformations in solids. Such experiments require control of the key thermodynamic parameter, temperature. The development of strategies for liquid cell heating and cooling, discussed in Chapter 6, has allowed liquid cell electron microscopy to access phase transformations in liquids.

When water is heated, bubbles form, and their nucleation, growth, and stability can be measured and compared with thermodynamic models [119]. In this experiment, heating was achieved via Joule strips made of Pt lines, generating strong temperature gradients. But heating can also employ an external reservoir, resulting in minimal gradients, or a laser [111] through a fiber optic or port on the column, enabling rapid temperature changes. The variety of heating methods enables the experiment to be optimized for the phenomenon under study.

Cooling allows ice to form from liquid water, provided beam-induced heating is controlled. Solidification of saline solutions containing gold nanoparticles [120], using a cold finger in contact with the liquid cell, provides insights into the competition between hexagonal and cubic ice nucleation and growth as a function of temperature, as well as particle rejection and occlusion as the ice advances. These types of experiments are complementary to cold stage experiments in ESEM and TEM in which water [121] or ice [122] condense from vapor.

1.3.4 Physics of Fluids at the Nanoscale

The physics of fluids is another area where our understanding can benefit from observations at nanometer length scales and especially at improved temporal resolution. The experiments described in Chapter 13 show directly how water moves at small length scales, how nanoscale bubbles form and move, and how nanoscale objects move within thin water films. Such experiments are relevant to fields such as catalysis, cavitation, lubrication, degassing of fluids, and boiling, where a detailed understanding of the nucleation and motion of small droplets and bubbles in confined volumes is required.

Wetting experiments are carried out routinely and straightforwardly using ESEM [44], for example, to study the wetting of cellulose fibers [50, 51]. TEM provides the opportunity to examine more complex liquid behavior. It is easy (often unavoidable) to form bubbles by radiolysis. As a large bubble forms between silicon nitride windows, the receding water leaves a thin wetting layer. The formation and motion of voids and droplets in this layer [123, 124] can be quantified and related to the fluidity of nanoscale water. Small bubbles in thin liquid films drift up thickness gradients, and this motion can be modeled by considering the effects of the forces at the tri-junctions [125]. TEM also allows liquid dynamics to be imaged in volumes that are even more constrained than in a typical microfabricated liquid cell. Cylinders can be made from carbon nanotubes [126], or by drying water on a graphene stack that curls into scrolls around the liquid [127]. Bubble dynamics, condensation, and other processes can be visualized in these restricted volumes, in particular by imaging parallel to the liquid–solid interface.

Quantitative data on Brownian motion are readily obtained from liquid cell experiments, especially if high speed image recording is available. Nanoparticles in thin layers move much more slowly than expected for a bulk liquid, whether the window is made of silicon nitride [27, 115, 128–130] or graphene [33, 131]. The discrepancy can be several orders of magnitude! Such highly damped motion is helpful for image resolution, but unexpected based on calculations of water in pores [132]. In some cases motion may be constrained because the liquid is thin, but in other experiments the slow movement may hint at an interfacial layer of ordered liquid with higher viscosity [69, 133].

1.3.5 Environmental and Biological Mineralization

The mechanisms that control key biological and environmental processes such as mineral formation are not well understood, in part due to the difficulty of making observations during growth. Liquid cell microscopy is starting to make an impact in this area, with exciting results described in Chapter 15. The technique can be used to examine aggregation processes, for example in the geological material iron oxyhydroxide [114], and key biomineralization processes, such as the nucleation of iron oxide [134] or calcium carbonate [135–137] in the presence of organic additives. The latter experiments used a dual-inlet stage, diffusing carbon dioxide through one inlet into a calcium-bearing solution to continuously raise supersaturation, to show that the binding of calcium by the organic matrix biases the system towards formation and stabilization of amorphous rather than crystalline calcium carbonate.

Biomineralization can also be examined in whole, single-cell organisms. Magnetite magnetosomes have been imaged in their natural cellular environment using correlative liquid cell STEM and fluorescence microscopy [138]; such data could aid in elucidating the role of protein localization and templating in biomineralization. The demonstration of imaging both soft macromolecular matrices and hard mineral constituents suggests that liquid cell TEM will be applicable to many key processes in this area.

1.3.6 Liquid Cell Microscopy for Life Science

Microscopy has always been a driver for discovery in life science. Fluorescence microscopy can label specific proteins and image their dynamics and interactions in fixed or living cells, at several tens of nanometers resolution using super-resolution techniques. The resolution of electron microscopy can be higher. But this comes at a cost: the material must be encased in amorphous ice at cryogenic temperature or dried or embedded at room temperature. Preserving biological structure during dehydration or freezing is a challenge, and of course these techniques also remove the possibility of making dynamic observations or imaging living cells. The initial demonstrations that labeled biological structures could be resolved through micrometers of water in liquid STEM [1, 139], Chapter 16, and that biological processes can be stimulated by injecting nutrients [19], Chapter 18, showed that liquid cell electron microscopy can provide useful information while circumventing some of the sample preparation issues. Life science applications of liquid cell microscopy have developed in scope and complexity, and liquid cell S/TEM and SEM now provide a viable complement to conventional microscopy methods to address the complexity of biological materials.

1.3.6.1 Whole Cells and Live Cells

The ability to label specific parts of a cell with metal nanoparticles, then image these labels through thick liquid, provides a tool to study cell structure and function, as described in Chapter 16. Eukaryotic cells can be grown on liquid cell microchips, incubated with labels that tag specific proteins or are taken up by the cells, then enclosed by adding the top microchip. Liquid cell STEM or SEM can then visualize the tagged structures [63, 139, 140], measure particle uptake [141, 142], or evaluate interaction of cells with particles [143]. Such experiments have shown the distribution of epidermal growth factor receptor in cells and the uptake of nanoparticles into vesicles [139, 141]. New insights have been gained about the HER2 receptor, a membrane protein playing an important role in breast cancer [140]. To provide a complete picture of a cellular function, imaging can be repeated for cells incubated for different lengths of time and under varied conditions. By optimizing experimental procedures, it is possible to measure tens of cells within a few hours [144], an important advance in obtaining statistical information.

If the enclosure has suitable size and surface condition and nutrients are supplied, unfixed cultured cells can be kept alive in a liquid cell chamber at room temperature with liquid flow for several hours [141, 145, 146]. But is it possible for cells to remain alive with physiological functions intact while images are recorded? Flowing the liquid can provide nutrients and perhaps remove radiolysis products and heat during imaging, but the liquid surroundings may not "cage" the damaging radiolysis products, as is suspected to occur in cryo-EM where such caging may protect the biomaterial. Thus, the comparative benefits of liquid cell and cryo-EM for mitigating radiation damage remain under discussion [147]. In any case, unfixed cells appear not to be viable after even one image is recorded [146]. This has driven great ongoing interest in establishing whether tolerable doses exist for *in vivo* liquid cell microscopy [138, 146]. Doses that preserve enzyme function are substantially lower than doses that preserve structural information, and the dose delivered in a single image is typically above the lethal dose. The prospects for measuring biological processes using liquid cell S/TEM therefore need careful exploration.

SEM techniques provide a useful complement to liquid cell S/TEM for larger and thicker samples. Although many eukaryotic cells do not fit in the liquid cell TEM compartment, this limitation does not exist for SEM. In ESEM, the surface structure of cells grown on a substrate can be imaged with a backscatter detector [148]. In closed cell SEM, electrons of energies as low as 2 keV can be used if the cells are covered only with a thin graphene layer [60]. In SEM systems that are open to the atmosphere (Chapter 4), experiments can be performed in cell culture dishes and the cells can be fixed and imaged in SEM [63]. If the cells are thick but have a thin region at their edges, STEM detection in the SEM provides the highest resolution and can be used to examine the stoichiometry and distribution of membrane receptors or nanoparticle uptake in intact cells, over tens of cells [140, 144, 146]. Techniques for combining light microscopy with SEM are particularly promising for approaching the goal of live cell study, since the exposure of the sample to electrons is minimized. Light microscopy provides overview images, maps the regions where proteins tagged with a fluorescent label reside, and studies live cells. SEM then provides high resolution information at selected regions after fixation, or in a live state at a selected time point, after which the cellular structure decays.

1.3.6.2 Tracking Dynamics in Labeled Biological Systems

It is possible to visualize certain biological motions by attaching labels to the moving structures. As described in Chapter 18, myosin head movement can be measured in response to adenosine triphosphate (ATP) [19, 149, 150] using labeled synthetic myosin muscle filaments over which ATP is allowed to diffuse. Labels can also be attached to smaller biomolecules such as DNA to provide information on DNA configuration under the electron beam [129].

1.3.6.3 Imaging Unlabeled Macromolecules, Soft Materials, and Dynamic Phenomena

Under favorable conditions it is possible to image certain biomolecules directly, without the use of nanoparticle labels and at relatively low dose. Such conditions, described in Chapter 17, include the lowest possible liquid and window thickness and a strategy for minimizing motion blur during image acquisition. One approach is to closely encase the material with graphene, minimizing the membrane thickness and the volume of surrounding liquid [151, 152]. Another approach uses patterned silicon nitride substrates with microwells to control the liquid thickness [153], and tethers biostructures with a functionalized substrate [153–155].

Such improvements can result in high quality imaging of biomaterials and soft materials in the liquid environment. High contrast materials, such as ferritin molecules [156], micelles containing heavy metals [154], and the magnetosomes mentioned above [138], are readily imaged to show their overall structure. Analytical techniques can provide composition and bonding, as shown for ferritin using energy loss spectroscopy [156], and dynamics can be studied, as shown for Pt-containing micelles [157]. But even low atomic number materials are visible using liquid cell microscopy. Imaging of liposomes and polymers [158, 159] provides shape and size in water, a useful comparison with cryo-EM results. More complex biostructures can also be examined. For example, liquid cell imaging of the acrosomal process, a membrane that extends from sperm heads, gives results that are consistent with cryo-EM, and protein crystals have been imaged at good resolution [160]. The combination of TEM and fluorescence microscopy has yielded the spatial distributions and interactions of subcellular organelles, such as the cytoskeleton and its contact with adjacent cells [60]. And finally, a viral pathogen has been imaged at ~3 nm resolution in the process of transcribing RNA [161]. These data even show rearrangements in the internal structure of the virus during RNA synthesis. Various tests for viability are essential in evaluating experiments that attempt to probe dynamic biochemical processes [161], and the considerations described above for dose sensitivity are particularly urgent.

Although not a focus of this book, ESEM is routinely employed at the interface of biological molecules and soft/hard matter, for example, to examine drug release from polymeric enclosures of pharmaceutical materials upon wetting [52]. The composition and aggregation properties of nanoparticles used for drug targeting, radiation therapy, or diagnosis can readily be studied with STEM in the ESEM [162]. ESEM contributes strongly to research related to the pharmaceutical and food industries, polymers, and emulsions, and the reader is referred to books focusing on this area [44].

Liquid cell electron microscopy is a new technique for the life sciences, so we emphasize the importance of validating the results by comparing with cryo-EM and also establishing radiation damage criteria. However, these recent advances provide encouragement that liquid cell microscopy might complement cryo-EM, fluorescence, and diffraction techniques in providing static and possibly some types of dynamic imaging of biological systems in water, fulfilling the critical need to develop real-time, high resolution imaging tools for life sciences.

1.4 Which Type of Microscopy Should I Use?

In surveying the wide variety of experiments involving liquids that have become possible with TEM and SEM, we find that the imaging strategies can be broadly classified into several distinct system configurations. Table 1.1 provides an overview and the key features, benefits, and disadvantages of the different configurations.

Table 1.1 Summary of techniques used for imaging liquids using electron microscopy

System	Closed cell TEM		Closed cell SEM			Open cell/ETEM		ESEM	
	TEM	STEM	SEM capsule	SEM +STEM	SEM +CLEM	Open cell (with ionic liquids)	ETEM	WetSTEM	ESEM
Commercially available	y	y	y	n	y	y	y	y	y
Compatible with standard microscope	y	y	y	y	some versions	y	n	n	n
Spatial resolution on low Z material	+	+	−	+/−	+/−	+/−	+	+/−	+/−
Spatial resolution on high Z material	+	+	+/−	+	+	+/−	+	+/−	+/−
Speed in movies	+	−	−	−	−	+	+	−	−
Thickness of sample	10s of nm	up to few μm	bulk	up to few μm	10s of nm	large	10s of nm	up to few μm	bulk
Ease of sample loading	−	+/−	+	+	+	+/−	−	+	+
Control of liquid layer geometry	+	+	+	+	+	−	−	−	−
Temperature control	y	y	n	n	n	n	y	y	y
Correlative microscopy demonstrated	n	y	y	y	y	n	n	y	n
Analytical microscopy demonstrated	y	y	y	y		n	y	n	y
Technique described in chapter(s)	2	2	4	4	5	3	3	3	Ref. 44
Applications described in chapter(s)	9–13, 15, 18, 19	9–13, 15, 16, 19	14	14, 16	5, 16	3	3	3	Ref. 44

Optimal planning of an experiment requires the right choice of method and ideally a calculation or simulation of the expected contrast and spatial and temporal resolution, and the effect of the dose on the sample. As a first step towards designing the best experiment to solve a given problem, think about which technique will provide the most useful data while also remembering that it is helpful to try out different techniques rather than relying fully on any one method.

1.5 Future Prospects

Continuing improvements in equipment and experimental techniques will allow new materials to be studied and new capabilities and stimuli to be applied to samples in liquids. Correlative approaches where electron microscopy is combined with light microscopy or synchrotron sources promise a deeper probe of chemical, electrochemical, photochemical, and biological processes. Analytical microscopy will enable more detailed measurement of composition and chemical bonding in water, while high speed and aberration-corrected imaging will extend the range of phenomena that can be examined. The topics below provide some hints to the future directions we believe that liquid cell microscopy will go.

1.5.1 Geological and Atmospheric Materials

What can we learn about the kinetics and structures produced when water interacts with complex natural materials?

1.5.1.1 Clays

The hydration of clay minerals is important for soil properties, in developing building materials, and for mining activities. We expect that microscopy of clay minerals in a controlled environment that includes liquid water will provide new information on the pathways by which the key structural transformations occur. Early research in this area [3, 14, 17] – the hydration of Portland cement was a popular topic – was hampered by the lack of a reliable method for handling water *in situ*. Modern liquid cells have a greater chance of producing quantitative results, for example, in imaging the behavior of the swelling clays (montmorillonites or bentonites) in fluids of different compositions. Hydration behavior in confined volumes is well suited for liquid cell experiments and could provide information relevant to oil sand extraction. Many minerals have hydrated and anhydrous varieties; some have commercial value and others cause problems when they change state. There appears to be no shortage of interesting problems in this area.

1.5.1.2 Extreme Temperature and Pressure Geological Processes

Conventional liquid cells can withstand pressures of only a few atmospheres [28]. How high can we push this pressure limit? Optimizing the mechanical properties of the windows could open an interesting new frontier. Inorganic fluids confined in environmental cells at high pressures can perhaps provide nanoscale insights into geological processes such as nanostructure synthesis, fuel production, or volcanic activity. Liquid cell observations could provide new insights into hydrothermal reactions, relevant to crystal growth, or even stress corrosion cracking. At low temperatures, the structure and stability of methane clathrates would be an interesting field of study.

1.5.1.3 Atmospheric Aerosols

Environmental TEM has provided valuable information in understanding the behavior and properties of atmospheric particles (e.g. [163]). But its use is limited when the relative humidity is high, and in particular the extreme of complete saturation. Full control of the water environment will enable studies of cloud formation, where saturation and supersaturation values are high. We could imagine, for example, subjecting various types of common atmospheric particles to water and liquids with different salinities and observe phase changes during cooling at controlled rates.

1.5.1.4 Biominerals

The exciting recent progress in this area suggests that there will be further application to a wider range of mineral systems and organic matrices, particularly for nucleation studies that include the effects of surface modification. Given the multiple mineral phases and reaction pathways possible, a comparison of different systems would allow an understanding of the principles that determine which pathway will be followed, and how nucleation can be directed with control over phase, orientation, and location.

1.5.2 The Physics of Fluids: A Grand Challenge

A grand challenge for understanding interfacial fluids has been the difficulty of imaging liquid–solid interfaces with good time and space resolution [164]. Given the improvements in liquid cell microscopy resolution and detector sensitivity, we anticipate that studies over the next few years could illuminate features of the liquid–solid interface such as ordering and the hydration layer, and provide direct observations of phase transitions. Diffusion through liquids is important in reactions associated with catalysis, battery operation, biomaterials, and tribology. Analytical microscopy may provide an interesting insight here. And with further improvement in temperature control, it will be exciting to examine low temperature liquids such as nitrogen (or even helium?), and solvents such as liquid ammonia.

1.5.3 Electrochemistry: New Materials and a Grand Challenge

As we improve our understanding of the opportunities and limitations of liquid cell electron microscopy for electrochemical reactions, it becomes possible to imagine experiments of greater complexity and in a wider variety of materials systems.

1.5.3.1 Complex Processes and Materials

Processes such as co-deposition and electroless deposition are so far unexplored, and deposition studies have addressed only a few metals. Future experiments will broaden this range. A recent example is the imaging of a conducting polymer during deposition [165]; many such polymers are synthesized by electropolymerization [166]. Another key area is temperature-dependent electrochemistry, including low temperature Li-ion battery function. Extreme environments, requiring as-yet undeveloped capabilities for high temperature and pressure, could perhaps address molten salt electrochemical reactions relevant, say, to Al refining.

1.5.3.2 The Electrochemical Double Layer

A grand challenge for microscopy is to image directly the electrochemical double layer and measure its behavior during electrochemical processes. The length scale, a few nanometers, and the strong electric field associated with the double layer make this a challenge. However, recent technical developments may allow this within the next few years. The experiments may involve combinations of holography and liquid cell microscopy to probe the double layer at an atomically flat electrode surface.

1.5.3.3 Magnetic Materials

The holography/liquid cell combination provides other fascinating opportunities. Magnetic thin films such as the NiFe layers used in magnetic read heads and inductors are formed by electroless deposition at moderate temperatures. Motivated by the possibility of improved control of nucleation and domain structure, one could simultaneously measure the developing microstructure and magnetic field. Other areas of application include the interactions of magnetic nanoparticles with each other and with macromolecular matrices in fluid, possibly while being magnetized, and the behavior of the surfactant-coated particles within ferrofluids.

1.5.4 Catalysis

Environmental electron microscopy has provided detailed insights into catalysis from gases (e.g. [6]) but is relatively less explored for reactions that form liquids [18] or that are catalyzed by liquids [167]. As liquid geometry and flow become more controlled, we can now envisage imaging catalyst reactions involving water. One can imagine water splitting by catalysts driven electrochemically or with light fed through a fiber optic, as already demonstrated in gas phase catalysis [168], and more detailed examination of the changes in catalyst particles during operation, as already seen for fuel cell catalysts [169].

1.5.5 High Speed Phenomena

A universal frontier in liquid cell electron microscopy is the ability to record liquid cell data on a high-frame-rate detector. Better time resolution expands the type of phenomenon that can be addressed, and also improves spatial resolution by reducing motion-induced blur. However, to address even faster processes, Chapter 22 describes how the technique of dynamic TEM (DTEM) can be adapted for liquid cells [170]. A laser hits the sample, supplying heat or an optical stimulation, and also triggers a short burst of electrons to record an image, delayed by a specified time. The electron burst provides a time resolution in the micro- to nanosecond regime. Microanalysis is possible via energy-filtered TEM using an in-column filter, suggesting a broad range of applications. It is already feasible to employ ultrafast electron sources to record diffraction patterns, and this method could be applied to specimens in liquid as well [171].

1.5.6 Biological Materials and Processes

1.5.6.1 Whole Live Cells?

Imaging cells in their native liquid state already offers the possibility of high resolution information without freezing or drying. Imaging live cells is a grand challenge, and it is not yet clear whether it can be accomplished. It will require a better understanding of dose effects, say through correlative light and liquid cell electron microscopy to determine the extent to which biological functions are preserved during liquid cell imaging. Indeed, irradiation effects on cells (say with taken-up nanoparticles) could provide insights into cancer therapies. Because various parts of a cell show different dose tolerances, an interesting possibility is to develop liquid cells that allow dose-tolerant regions to be imaged with minimized overall dose [172]. The use of mixtures of labels could be explored in liquid cell TEM to distinguish different components within a cell, as is done for fluorescence microscopy.

1.5.6.2 Proteins and Other Complex Materials

The current progress in liquid cell microscopy of unlabeled biomaterials, combined with adaptation of cryo-EM techniques for low dose imaging, dose fractionation, and image analysis, suggests that liquid cell electron microscopy may provide insights into the structure of materials such as block copolymers, protein domains, macromolecule-mediated nanoparticle assembly, and even food materials (Chapter 23). The benefits and disadvantages of the liquid environment need to be better explored, but temperature control could prove useful in understanding interactions and processes in these types of materials.

1.5.6.3 Imaging Biological Dynamics

Movies of processes such as the dynamics of large proteins, changes in membrane geometry, or the assembly of microfilaments would provide fundamental and practical insights. However, prospects for real-time imaging ultimately come down to the dose required per image compared to the key limitation of dose tolerance. Increasing the material contrast reduces the dose required; achieving higher contrast (without using labels or defocus techniques) may require exploring the use of phase plates [173] or electron holography [174]. Temperature-dependent imaging of biomaterials could allow dynamics to be measured at optimum temperatures and biological processes to be explored under extreme environments. Processes may be triggered by introducing chemicals into the solution flow, or using a laser to heat or stimulate directly with light (Chapter 22). Since biological systems also respond to electromagnetic fields, externally applied fields could provide new information, with results measured using holography, if issues of dose and the holographic reference beam could be resolved. The constraints on these experiments are discussed in more detail in Chapter 23. Could far future experiments produce charge maps of proteins, show the self-assembly and folding of proteins, or the field-induced interactions between individual biomolecules and binding sites? Some of these speculations may appear far from reality, but the recent pace of progress encourages us to expect a strong impact of liquid cell microscopy for understanding biological structures and processes.

1.6 Conclusions

The examples we have described in this chapter, covering both current research and future ideas, suggest that liquid cell electron microscopy is well positioned to explore new frontiers in nanomaterial growth, fluid physics, radiation physics, corrosion and electrochemical processes, environmental science, and biomaterial structure and function. Based on continuing improvements in the experimental capabilities, data acquisition, and data interpretation, liquid cell observations can address key materials challenges and provide a unique and exciting view of liquid phase materials and processes. In the following chapters we will examine these areas in more detail. We hope you enjoy reading them.

References

1. N. de Jonge and F. M. Ross, Electron microscopy of specimens in liquid. *Nat. Nanotechnol.*, **6** (2011), 695–704.
2. H. G. Liao and H. Zheng, Liquid cell transmission electron microscopy, *Annu. Rev. Phys. Chem.*, **67** (2016), 719–747.
3. E. P. Butler and K. F. Hale, Chapter 6 in *Dynamic Experiments in the Electron Microscope* (Amsterdam: North-Holland, 1981).
4. D. F. Parsons, Structure of wet specimens in electron microscopy. *Science*, **186** (1974), 407–414.
5. E. Ruska, Beitrag zur uebermikroskopischen Abbildungen bei hoeheren Drucken. *Kolloid Z.*, **100** (1942), 212–219.
6. S. Helveg, C. López-Cartes, J. Sehested et al., Atomic-scale imaging of carbon nanofibre growth. *Nature*, **427** (2004), 426–429.
7. D. F. Parsons, V. R. Matricardi, R. C. Moretz and J. N. Turner, Electron microscopy and diffraction of wet unstained and unfixed biological objects. *Adv. Biol. Med. Phys.*, **15** (1974), 161–270.
8. J. Y. Huang, L. Zhong, C. M. Wang et al., In situ observation of the electrochemical lithiation of a single SnO_2 nanowire electrode. *Science*, **330** (2010), 1515–1520.
9. C. M. Wang, W. Xu, J. Liu et al., In situ transmission electron microscopy and spectroscopy studies of interfaces in Li ion batteries: challenges and opportunities. *J. Mater. Res.*, **25** (2010), 1541–1547.
10. C.-M. Wang, H.-G. Liao and F. M. Ross, Observation of materials processes in liquids by electron microscopy, *MRS Bulletin*, **40** (2015), 46–52.

11. I. M. Abrams and J. W. McBain, A closed cell for electron microscopy. *J. Appl. Phys.*, **15** (1944), 607–609.
12. H. G. Heide, Elektronenmikroskopie von Objekten unter Atmosphaerendruck oder unter Drucken, welche Austricknen verhindern. *Naturwissenschaften*, **47** (1960), 313–317.
13. H. G. Heide, Electron microscopic observation of specimens under controlled gas pressure. *J. Cell Biol.*, **13** (1962), 147–152.
14. D. D. Double, Some studies of the hydration of Portland cement using high voltage (1MV) electron microscopy. *Mater. Sci. Eng.*, **12** (1973), 29–34.
15. T. L. Daulton, B. J. Little, K. Lowe and J. Jones-Meehan, In situ environmental cell–transmission electron microscopy study of microbial reduction of chromium(VI) using electron energy loss spectroscopy. *Microsc. Microanal.*, **7** (2001), 470–485.
16. W.-A. Chiou et al., In situ TEM study of DNA/gold nanoparticles in liquid environment. *Microsc. Microanal.*, **5** (Suppl. 2) (1999), MSA.
17. A. Fukami, K. Fukushima and N. Kohyama, Observation technique for wet clay minerals using film-sealed environmental cell equipment attached to high-resolution electron microscope. In R. Bennett *et al.*, eds., *Microstructure of Fine-Grained Sediments* (New York: Springer, 1991) pp. 321–331.
18. P. L. Gai, Development of wet environment TEM (wet-ETEM) for in situ studies of liquid-catalyst reactions on the nanoscale. *Microsc Microanal.*, **8** (2002), 21–28.
19. H. T. Sugi, K. Akimoto, S. Sutoh *et al.*, Dynamic electron microscopy of ATP-induced myosin head movement in living muscle filaments. *Proc. Natl. Acad. Sci. USA*, **94** (1997), 4378–4392.
20. K. A. Taylor and R. M. Glaeser, Electron microscopy of frozen hydrated biological specimens. *J. Ultrastruct. Res.*, **55** (1976), 448–456.
21. J. Frank, *Three-Dimensional Electron Microscopy of Macromolecular Assemblies: Visualization of Biological Molecules in Their Native State* (Oxford: Oxford University Press, 2006).
22. V. Lucic, F. Foerster and W. Baumeister, Structural studies by electron tomography: from cells to molecules. *Annu. Rev. Biochem.*, **74** (2005), 833–865.
23. H. Stahlberg and T. Walz, Molecular electron microscopy: state of the art and current challenges. *ACS Chem. Biol.*, **3** (2008), 268–281.
24. M. J. Williamson, R. M. Tromp, P. M. Vereecken, R. Hull and F. M. Ross, Dynamic microscopy of nanoscale cluster growth at the solid-liquid interface. *Nat. Mater.*, **2** (2003), 532–536.
25. R. Franks, S. Morefield, J. Wen *et al.*, A study of nanomaterial dispersion in solution by wet-cell transmission electron microscopy. *J. Nanosci. Nanotechnol.*, **8** (2008), 4404–4407.
26. K.-L. Liu, C.-C. Wu, Y.-J. Huang *et al.*, Novel microchip for in situ TEM imaging of living organisms and bio-reactions in aqueous conditions. *Lab Chip*, **8** (2008), 1915–1921.
27. H. M. Zheng, S. A. Claridge, A. M. Minor, A. P. Alivisatos and U. Dahmen, Nanocrystal diffusion in a liquid thin film observed by in situ transmission electron microscopy. *Nano Lett.*, **9** (2009), 2460–2465.
28. J. M. Grogan and H. H. Bau, The Nanoaquarium: a platform for in situ transmission electron microscopy in liquid media. *J. Microelectromech. Syst.*, **19** (2010), 885–894.
29. A. J. Leenheer, J. P. Sullivan, M. J. Shaw and C. T. Harris, A sealed liquid cell for in situ transmission electron microscopy of controlled electrochemical processes. *J. Microelectromech. Syst.*, **24** (2015), 1061–1068.

30. M. Tanase, J. Winterstein, R. Sharma *et al.*, High-resolution imaging and spectroscopy at high pressure: a novel liquid cell for the TEM. *Microsc. Micranal.*, **21** (2015), 1629–1638.
31. C. Mueller, M. Harb, J. R. Dwyer and R. J. Dwayne Miller, Nanofluidic cells with controlled pathlength and liquid flow for rapid, high-resolution in situ imaging with electrons. *J. Phys. Chem. Lett.*, **4** (2013), 2339–2347.
32. M. den Heijer, I. Shao, A. Radisic, M. C. Reuter and F. M. Ross, Patterned electrochemical deposition of copper using an electron beam. *APL Materials*, **2** (2014), 022101.
33. J. M. Yuk, J. Park, P. Ercius *et al.*, High-resolution EM of colloidal nanocrystal growth using graphene liquid cells. *Science*, **336** (2012), 61–64.
34. E. A. Ring and N. de Jonge, Microfluidic system for transmission electron microscopy. *Microsc. Microanal.*, **16** (2010), 622–629.
35. K. L. Jungjohann, J. E. Evans, J. Aguiar, I. Arslan and N. D. Browning, Atomic-scale imaging and spectroscopy for in situ liquid scanning transmission electron microscopy. *Microsc. Microanal.*, **18** (2012), 621–627.
36. M. E. Holtz, Y. Yu, D. Gunceler *et al.*, Nanoscale imaging of lithium ion distribution during in situ operation of battery electrode and electrolyte. *Nano Lett.*, **14** (2014), 1453–1459.
37. N. J. Zaluzec, M. G. Burke, S. J. Haigh and M. A. Kulzick, X-ray energy-dispersive spectrometry during in situ liquid cell studies using an analytical electron microscope. *Microsc. Microanal.*, **20** (2014), 323–329.
38. E. A. Lewis, S. J. Haigh, T. J. A. Slater *et al.*, Real-time imaging and local elemental analysis of nanostructures in liquids. *Chem. Commun.*, **50** (2014), 10019–10022.
39. S. Kuwabata, A. Kongkanand, D. Oyamatsu and T. Torimoto, Observation of ionic liquid by scanning electron microscope. *Chem. Lett.*, **35** (2006), 600–601.
40. S. Arimoto, M. Sugimura, H. Kageyama, T. Torimoto and S. Kuwabata, Development of new techniques for scanning electron microscope observation using ionic liquid. *Electrochim. Acta*, **53** (2008), 6228–6234.
41. J. A. Swift and A. C. Brown, An environmental cell for the examination of wet biological specimens at atmospheric pressure by transmission scanning electron microscopy. *J. Phys. E*, **3** (1970), 924–926.
42. G. D. Danilatos, Review and outline of environmental SEM at present. *J. Microsc.*, **162** (1991), 391–402.
43. D. A. Moncrieff, P. R. Barker and V. N. E. Robinson, Electron scattering by gas in the scanning electron microscope. *J. Phys. D*, **12** (1979), 481–488.
44. D. J. Stokes, *Principles and Practice of Variable Pressure/Environmental Scanning Electron Microscopy (VP-ESEM)* (Chichester: John Wiley & Sons, 2008).
45. A. Bogner, G. Thollet, D. Basset, P. H. Jouneau and C. Gauthier, Wet STEM: a new development in environmental SEM for imaging nano-objects included in a liquid phase. *Ultramicroscopy*, **104** (2005), 290–301.
46. A. Bogner, P.-H. Jouneau, G. Thollet, D. Basset and C. Gauthier, A history of scanning electron microscopy developments: towards "wet-STEM" imaging. *Micron*, **38** (2007), 390–401.
47. D. B. Peckys, J. P. Baudoin, M. Eder, U. Werner and N. de Jonge, Epidermal growth factor receptor subunit locations determined in hydrated cells with environmental scanning electron microscopy. *Sci. Rep.*, **3** (2013), 2621–2626.
48. K. Masenelli-Varlot, A. Malchere, J. Ferreira *et al.*, Wet-STEM tomography: principles, potentialities and limitations. *Microsc. Microanal.*, **20** (2014), 366–375.

49. F. Novotny, P. Wandrol, J. Proska and M. Slouf, In situ wetSTEM observation of gold nanorod self-assembly dynamics in a drying colloidal droplet. *Microsc. Microanal.*, **20** (2014), 385–393.
50. A. Jansson, A. Nafari, A. Sanz-Velasco *et al.*, Novel method for controlled wetting of materials in the environmental scanning electron microscope. *Microsc. Microanal.*, **19** (2013), 30–37.
51. A. Jansson, C. Boissier, M. Marucci *et al.*, Novel method for visualizing water transport through phase-separated polymer films. *Microsc. Microanal.*, **20** (2014), 394–406.
52. Z. Barkay, Wettability study using transmitted electrons in environmental scanning electron microscope. *Appl. Phys. Lett.*, **96** (2010), 183109.
53. Z. Barkay, In situ imaging of nano-droplet condensation and coalescence on thin water films. *Microsc. Microanal.*, **20** (2014), 317–322.
54. S. Thiberge, A. Nechushtan, D. Sprinzak *et al.*, Scanning electron microscopy of cells and tissues under fully hydrated conditions. *Proc. Natl. Acad. Sci. USA*, **101** (2004), 3346–3351.
55. S. W. Hell, Far-field optical nanoscopy. *Science*, **316** (2007), 1153–1158.
56. E. Jensen, C. Kobler, P. S. Jensen and K. Molhave. In-situ SEM microchip setup for electrochemical experiments with water based solutions. *Ultramicroscopy*, **129** (2013), 63–69.
57. J. Kraus, R. Reichelt, S. Günther *et al.*, Photoelectron spectroscopy of wet and gaseous samples through graphene membranes. *Nanoscale*, **6** (2014), 14394–14403.
58. W. Yang, Y. Zhang, M. Hilke and W. Reisner, Dynamic imaging of Au-nanoparticles via scanning electron microscopy in a graphene wet cell. *Nanotechnology*, **26** (2015), 315703.
59. L. Yang, X.-Y. Yu, Z. Zhu, T. Thevuthasan and J. P. Cowin, Making a hybrid microfluidic platform compatible for in situ imaging by vacuum-based techniques. *J. Vac. Sci. Technol. A*, **29** (2011), 061101.
60. M. Wojcik, M. Hauser, W. Li, S. Moon and K. Xu, Graphene-enabled electron microscopy and correlated super-resolution microscopy of wet cells. *Nat. Commun.*, **6** (2015), 7384.
61. N. Liv, A. C. Zonnevylle, A. C. Narvaez *et al.*, Simultaneous correlative scanning electron and high-NA fluorescence microscopy. *PLOS One*, **8** (2013), e55707.
62. N. Liv, I. Lazić, P. Kruit and J. P. Hoogenboom, Scanning electron microscopy of individual nanoparticle bio-markers in liquid. *Ultramicroscopy*, **143** (2014), 93–99.
63. H. Nishiyama, M. Suga, T. Ogura *et al.*, Atmospheric scanning electron microscope observes cells and tissues in open medium through silicon nitride film. *J. Struct. Biol.*, **169** (2010), 438–449.
64. Y. Nawa, W. Inami, A. Chiba *et al.*, Dynamic and high-resolution live cell imaging by direct electron beam excitation. *Opt. Express*, **20** (2012), 5629–5635.
65. N. Vidavsky, S. Addadi, J. Mahamid *et al.*, Initial stages of calcium uptake and mineral deposition in sea urchin embryos. *Proc. Natl. Acad. Sci. USA*, **111** (2014), 39–44.
66. J. F. Creemer, S. Helveg, G. H. Hoveling *et al.*, Atomic-scale electron microscopy at ambient pressure. *Ultramicroscopy*, **108** (2008), 993–998.
67. E. Jensen and K. Molhave, Monolithic chip system with a microfluidic channel for in situ electron microscopy of liquids. *Microsc. Microanal.*, **20** (2014), 445–451.
68. A. Radisic, P. M. Vereecken, J. B. Hannon, P. C. Searson and F. M. Ross, Quantifying electrochemical nucleation and growth of nanoscale clusters using real-time kinetic data. *Nano Lett.*, **6** (2006), 238–242.
69. A. Verch, M. Pfaff and N. De Jonge, Exceptionally slow movement of gold nanoparticles at a solid:liquid interface investigated by scanning transmission electron microscopy. *Langmuir*, **31** (2015), 6956–6964.

70. M. Krueger, S. Berg, D. A. Stone *et al.*, Drop-casted self-assembling graphene oxide membranes for scanning electron microscopy on wet and dense gaseous samples. *ACS Nano*, **5** (2011), 10047–10054.
71. N. M. Schneider, M. M. Norton, B. J. Mendel *et al.*, Electron–water interactions and implications for liquid cell electron microscopy. *J. Phys. Chem. C*, **118** (2014), 22373–22382.
72. J. M. Grogan, N. M. Schneider, F. M. Ross and H. H. Bau, Bubble and pattern formation in liquid induced by an electron beam. *Nano Lett.*, **14** (2014), 359–364.
73. H. M. Zheng, R. K. Smith, Y. W. Jun *et al.*, Observation of single colloidal platinum nanocrystal growth trajectories. *Science*, **324** (2009), 1309–1312.
74. M. W. van de Put, C. C. Carcouet, P. H. Bomans *et al.*, Writing silica structures in liquid with scanning transmission electron microscopy. *Small*, **11** (2015), 585–590.
75. E. Sutter, K. Jungjohann, S. Bliznakov *et al.*, In situ liquid-cell electron microscopy of silver-palladium galvanic replacement reactions on silver nanoparticles. *Nat. Commun.*, **5** (2014), 4946.
76. R. R. Unocic, R. L. Sacci, G. M. Brown *et al.*, Quantitative electrochemical measurements using in situ ec-S/TEM devices. *Microsc. Microanal.*, **20** (2014), 452–461.
77. B. L. Mehdi, M. Gu, L. R. Parent *et al.*, In situ electrochemical transmission electron microscopy for battery research. *Microsc. Microanal.*, **20** (2014), 484–492.
78. A. Radisic, F. M. Ross and P. C. Searson, In situ study of the growth kinetics of individual islands during electrodeposition of copper. *J. Phys. Chem. B*, **110** (2006), 7862–7868.
79. A. Radisic, P. M. Vereecken, P. C. Searson and F. M. Ross, The morphology and nucleation kinetics of copper islands during electrodeposition. *Surf. Sci.*, **600** (2006), 1817–1826.
80. N. M. Schneider, J. H. Park and J. M. Grogan, Visualization of active and passive control of morphology during electrodeposition. *Microsc. Microanal.*, **20** (2014), 1530–1531.
81. E. R. White, S. B. Singer, V. Augustyn *et al.*, In situ transmission electron microscopy of lead dendrites and lead ions in aqueous solution. *ACS Nano*, **6** (2012), 6308–6317.
82. M. Sun, H.-G. Liao, K. Niu and H. Zheng, Structural and morphological evolution of lead dendrites during electrochemical migration. *Sci. Rep.*, **3** (2013), 2227.
83. Z. Zeng, W.-I. Liang, H.-G. Liao *et al.*, Visualization of electrode-electrolyte interfaces in $LiPF_6$/EC/DEC electrolyte for lithium ion batteries via in-situ TEM. *Nano Lett.*, **14** (2014), 1745–1750.
84. B. L. Mehdi, J. Qian, E. Nasybulin *et al.*, Observation and quantification of nanoscale processes in lithium batteries by *operando* electrochemical (S)TEM. *Nano Lett.*, **15** (2015), 2168–2173.
85. R. L. Sacci, J. M. Black, N. Balke *et al.*, Nanoscale imaging of fundamental Li battery chemistry: solid-electrolyte interphase formation and preferential growth of lithium metal nanoclusters. *Nano Lett.*, **15** (2015), 2011–2018.
86. A. J. Leenheer, K. L. Jungjohann, K. R. Zavadil, J. P. Sullivan and C. T. Harris, Lithium electrodeposition dynamics in aprotic electrolyte observed in situ via transmission electron microscopy, *ACS Nano*, **9** (2015), 4379–4389.
87. M. Gu, L. R. Parent, L. Mehdi *et al.*, Demonstration of an electrochemical liquid cell for *operando* transmission electron microscopy observation of the lithiation/delithiation behavior of Si nanowire battery anodes. *Nano Lett.*, **13** (2013), 6106–6112.
88. R. L. Sacci, N. J. Dudney, K. L. More *et al.*, Direct visualization of initial SEI morphology and growth kinetics during lithium deposition by in situ electrochemical transmission electron microscopy. *Chemical Commun.*, **50** (2013), 2104–2107.

89. F. Wu and N. Yao, Advances in sealed liquid cells for in-situ TEM electrochemical investigation of lithium-ion battery. *Nano Energy*, **11** (2015), 196–210.
90. P. Abellan Baeza, B. L. Mehdi, L. R. Parent *et al.*, Probing the degradation mechanisms in electrolyte solutions for Li-ion batteries by in-situ transmission electron microscopy. *Nano Lett.*, **14** (2014), 1293–1299.
91. S. W. Chee. D. Duquette, F. M. Ross and R. Hull, Metastable structures in Al thin films prior to the onset of corrosion pitting as observed using liquid cell transmission electron microscopy. *Microsc. Microanal.*, **20** (2014), 462–468.
92. S. W. Chee, S. H. Pratt, K. Hattar *et al.*, Studying localized corrosion using liquid cell transmission electron microscopy. *Chem. Commun.*, **51** (2015), 168–171.
93. X. Zhong, M. G. Burke, S. Schilling, S. J. Haigh and N. J. Zaluzec, Novel hybrid sample preparation method for in situ liquid cell TEM analysis. *Microsc. Microanal.*, **20** (S3) (2014), 1514–1515.
94. T. J. Woehl, J. E. Evans, I. Arslan, W. D. Ristenpart and N. D. Browning, Direct in situ determination of the mechanisms controlling nanoparticle nucleation and growth. *ACS Nano*, **6** (2012), 8599–8610.
95. H. G. Liao, L. K. Cui, S. Whitelam and H. M. Zheng, Real-time imaging of Pt_3Fe nanorod growth in solution. *Science*, **336** (2012), 1011–1014.
96. K. L. Jungjohann, S. Bliznakov, P. W. Sutter, E. A. Stach and E. A. Sutter, In situ liquid cell electron microscopy of the solution growth of Au–Pd core–shell nanostructures. *Nano Lett.*, **13** (2013), 2964–2970.
97. A. De Clercq, W. Dachraoui, O. Margeat *et al.*, Growth of Pt–Pd nanoparticles studied in situ by HRTEM in a liquid cell. *J. Phys. Chem. Lett.*, **5** (2014), 2126–2130.
98. L. R. Parent, D. R. Robinson, T. J. Woehl *et al.*, Direct in situ observation of nanoparticle synthesis in a liquid crystal surfactant template. *ACS Nano*, **6** (2012), 3589–3596.
99. G. Zhu, Y. Jiang, F. Lin *et al.*, In situ study of the growth of two-dimensional palladium dendritic nanostructures using liquid-cell electron microscopy. *Chem Commun.*, **50** (2014), 9447–9450.
100. T. Kraus and N. de Jonge, Dendritic gold nanowire growth observed in liquid with transmission electron microscopy. *Langmuir*, **29** (2013), 8427–8432.
101. Y. Jiang, G. Zhu, F. Lin *et al.*, In situ study of oxidative etching of palladium nanocrystals by liquid cell electron microscopy. *Nano Lett.*, **14** (2014), 3761–3765.
102. J. Wu, W. Gao, H. Yang and J.-M. Zuo, Imaging shape-dependent corrosion behavior of Pt nanoparticles over extended time using a liquid flow cell and TEM. *Microsc. Microanal.*, **20(S3)** (2014), 1508–1509.
103. J. H. Park, N. M. Schneider, J. M. Grogan *et al.*, Control of electron beam-induced Au nanocrystal growth kinetics through solution chemistry. *Nano Lett.*, **15** (2015), 5314–5320.
104. K. W. Noh, Y. Liu, L. Sun and S. J. Dillon, Challenges associated with in-situ TEM in environmental systems: the case of silver in aqueous solutions. *Ultramicroscopy*, **116** (2012), 34–38.
105. H.-G. Liao and H. Zheng, Liquid cell transmission electron microscopy study of platinum iron nanocrystal growth and shape evolution. *J. Am. Chem. Soc.*, **135** (2013), 5038–5043.
106. H. G. Liao, D. Zherebetskyy, H. Xin *et al.*, Facet development during platinum nanocube growth. *Science*, **345** (2014), 916–919.
107. Y. Liu, K. Tai and S. J. Dillon, Growth kinetics and morphological evolution of ZnO precipitated from solution. *Chem. Mater.*, **25** (2013), 2927–2933.

108. Y. Kimura, H. Niinomi, K. Tsukamoto and J. M. García-Ruiz, In situ live observation of nucleation and dissolution of sodium chlorate nanoparticles by transmission electron microscopy. *J. Am. Chem. Soc.*, **136** (2014), 1762–1765.
109. H. L. Xin and H. Zheng, In Situ observation of oscillatory growth of bismuth nanoparticles. *Nano Lett.*, **12** (2012), 1470–1474.
110. K.-Y. Niu, J. Park, H. Zheng and A.P. Alivisatos, Revealing bismuth oxide hollow nanoparticle formation by the Kirkendall effect. *Nano Lett.*, **13** (2013), 5715–5719.
111. J. E. Evans, K. L. Jungjohann, N. D. Browning and I. Arslan, Controlled growth of nanoparticles from solution with in situ liquid transmission electron microscopy. *Nano Lett.*, **11** (2011), 2809–2813.
112. E. U. Donev and J. T. Hastings, Electron-beam-induced deposition of platinum from a liquid precursor. *Nano Lett.*, **9** (2009), 2715–2718.
113. H. M. Zheng, U. M. Mirsaidov, L. W. Wang and P. Matsudaira, Electron beam manipulation of nanoparticles. *Nano Lett.*, **12** (2012), 5644–5648.
114. D. S. Li, M. H. Nielsen, J. R. I. Lee et al., Direction-specific interactions control crystal growth by oriented attachment. *Science*, **336** (2012), 1014–1018.
115. T. J. Woehl, C. Park, J. E. Evans et al., Direct observation of aggregative nanoparticle growth: kinetic modeling of the size distribution and growth rate. *Nano Lett.*, **14** (2013), 373–378.
116. J. M. Grogan, L. Rotkina and H. H. Bau, In situ liquid-cell electron microscopy of colloid aggregation and growth dynamics. *Phys. Rev. E*, **83** (2011), 061405.
117. J. Park, H. Zheng, W. C. Lee et al., Direct observation of nanoparticle superlattice formation by using liquid cell transmission electron microscopy. *ACS Nano*, **6** (2012), 2078–2085.
118. Y. Liu, X.-M. Lin, Y. Sun and T. Rajh, In situ visualization of self-assembly of charged gold nanoparticles. *J. Am. Chem. Soc.*, **135** (2013), 3764–3767.
119. E. R. White, M. Mecklenburg, S. B. Singer, S. Aloni and B. C. Regan, Imaging nanobubbles in water with scanning transmission electron microscopy. *Appl. Phys. Express*, **4** (2011), 055201.
120. K. Tai, Y. Liu and S. J. Dillon, In situ cryogenic transmission electron microscopy for characterizing the evolution of solidifying water ice in colloidal systems. *Microsc. Microanal.*, **20** (2014), 330–337.
121. D. Bhattacharya, M. Bosman, V. R. S. S. Mokkapati, F. Y. Leong and U. Mirsaidov, Nucleation dynamics of water nanodroplets. *Microsc. Microanal.*, **20** (2014), 407–415.
122. C.-Y. Ruan, V. A. Lobastov, F. Vigliotti, S. Chen and A. H. Zewail, Ultrafast electron crystallography of interfacial water. *Science*, **304** (2004), 80–84.
123. U. Mirsaidov, C.-D. Ohl and P. Matsudaira, A direct observation of nanometer-size void dynamics in an ultra-thin water film. *Soft Matter*, **8** (2012), 7108–7111.
124. U. M. Mirsaidov, H. Zheng, D. Bhattacharya, Y. Casana and P. Matsudaira, Direct observation of stick-slip movements of water nanodroplets induced by an electron beam. *Proc. Natl. Acad. Sci. USA*, **109** (2012), 7187–7190.
125. M. Norton, J. H. Park, S. Kodambaka, F. M. Ross and H. Bau, Dynamics of sub-micron bubbles growing in a wedge in the low capillary number regime. *Bull. Ameri. Phys. Soc.*, **59** (2014); and H. Bau, J. M. Grogan, M. Norton and F. M. Ross, On the surface tension of nanobubbles. APS Division of Fluid Dynamics Meeting (2013).
126. D. Mattia and Y. Gogotsi, Review: static and dynamic behavior of liquids inside carbon nanotubes. *Microfluidics and Nanofluidics*, **5** (2008), 289–305.

127. U. Mirsaidov, V. R. S. S. Mokkapati, D. Bhattacharya *et al.*, Scrolling graphene into nanofluidic channels. *Lab Chip*, **13** (2013), 2874–2878.
128. E. A. Ring and N. de Jonge, Video-frequency scanning transmission electron microscopy of moving gold nanoparticles in liquid. *Micron*, **43** (2012), 1078–1084.
129. E. R. White, M. Mecklenburg, B. Shevitski, S. B. Singer and B. C. Regan, Charged nanoparticle dynamics in water induced by scanning transmission electron microscopy. *Langmuir*, **28** (2012), 3695–3698.
130. J. Y. Lu, Z. Aabdin, N. D. Loh, D. Bhattacharya and U. Mirsaidov, Nanoparticle dynamics in a nanodroplet. *Nano Lett.*, **14** (2014), 2111–2115.
131. Q. Chen, J. M. Smith, J. Park *et al.*, 3D motion of DNA-Au nanoconjugates in graphene liquid cell electron microscopy. *Nano Lett.*, **13** (2013), 4556–4561.
132. P.-A. Cazade, R. Hartkamp and B. Coasne, Structure and dynamics of an electrolyte confined in charged nanopores. *J. Phys. Chem. C*, **118** (2014), 5061–5072.
133. H. I. Kim, J. G. Kushmerick, J. E. Houston and B. C. Bunker, Viscous "interphase" water adjacent to oligo(ethylene glycol)-terminated monolayers. *Langmuir*, **19** (2003), 9271–9275.
134. S. Kashyap, T. J. Woehl, X. Liu, S. K. Mallapragada and T. Prozorov, Nucleation of iron oxide nanoparticles mediated by Mms6 protein in situ, *ACS Nano*, **8** (2014), 9097–9106.
135. M. H. Nielsen, J. R. I. Lee, Q. N. Hu, T. Y. J. Han, and J. J. De Yoreo, Structural evolution, formation pathways and energetic controls during template-directed nucleation of $CaCO_3$. *Faraday Discuss.*, **159** (2012), 105–121.
136. M. H. Nielsen, S. Aloni and J. J. De Yoreo, In situ TEM imaging of $CaCO_3$ nucleation reveals coexistence of direct and indirect pathways. *Science*, **345** (2014), 1158–1162.
137. P. J. M. Smeets, K. R. Cho, R. G. E. Kempen, N. A. J. M. Sommerdijk and J. J. De Yoreo, In situ TEM shows ion binding is key to directing $CaCO_3$ nucleation in a biomimetic matrix. *Nat. Mater.*, **14** (2015), 394–399.
138. T. J. Woehl, S. Kashyap, E. Firlar *et al.*, Correlative electron and fluorescence microscopy of magnetotactic bacteria in liquid: toward in vivo imaging. *Sci. Rep.*, **4** (2014), 6854.
139. N. de Jonge, D. B. Peckys, G. J. Kremers and D. W. Piston, Electron microscopy of whole cells in liquid with nanometer resolution. *Proc. Natl. Acad. Sci. USA*, **106** (2009), 2159–2164.
140. D. B. Peckys, U. Korf and N. de Jonge, Local variations of HER2 dimerization in breast cancer cells discovered by correlative fluorescence and liquid electron microscopy. *Sci. Adv.*, **1** (2015), e1500165.
141. D. B. Peckys and N. de Jonge, Visualization of gold nanoparticle uptake in living cells with liquid scanning transmission electron microscopy. *Nano Lett.*, **11** (2011), 1733–1738.
142. M. J. Dukes, D. Peckys and N. de Jonge, Correlative fluorescence- and scanning transmission electron microscopy of quantum dot labeled proteins on whole cells in liquid. *ACS Nano*, **4** (2010), 4110–4116.
143. E. S. Pohlmann, K. Patel, S. Guo *et al.*, Real-time visualization of nanoparticles interacting with glioblastoma stem cells. *Nano Lett.*, **15** (2015), 2329–2335.
144. D. B. Peckys, J.-P. Baudoin, M. Eder, U. Werner and N. de Jonge, Epidermal growth factor receptor subunit locations determined in hydrated cells with environmental scanning electron microscopy. *Sci. Rep.*, **3** (2013), 2626.
145. D. B. Peckys, P. Mazur, K. L. Gould and N. de Jonge, Fully hydrated yeast cells imaged with electron microscopy. *Biophys. J.*, **100** (2011), 2522–2529.
146. D. B. Peckys and N. de Jonge, Liquid scanning transmission electron microscopy: imaging protein complexes in their native environment in whole eukaryotic cells. *Microsc. Microanal.*, **20** (2014), 189–198.

147. R. Glaeser, Comment on electron microscopy of biological specimens in liquid water. *Biophys. J.*, **103** (2012), 163–164.
148. S. E. Kirk, J. N. Skepper and A. M. Donald, Application of environmental scanning electron microscopy to determine biological surface structure. *J. Microsc.*, **233** (2009), 205–224.
149. H. Sugi, H. Minoda, Y. Inayoshi *et al.*, Direct demonstration of the cross-bridge recovery stroke in muscle thick filaments in aqueous solution by using the hydration chamber. *Proc. Natl. Acad. Sci. USA*, **105** (2008), 17396–17401.
150. H. Sugi, S. Chaen, T. Akimoto *et al.*, Electron microscopic recording of myosin head power stroke in hydrated myosin filaments. *Sci. Rep.*, **5** (2015), 15700.
151. N. Mohanty, M. Fahrenholtz, A. Nagaraja, D. Boyle and V. Berry, Impermeable graphenic encasement of bacteria. *Nano Lett.*, **11** (2011), 1270–1275.
152. J. Park, H. Park, P. Ercius *et al.*, Direct observation of wet biological samples by graphene liquid cell transmission electron microscopy. *Nano Lett.*, **15** (2015), 4737–4744.
153. M. J. Dukes, R. Thomas, J. Damiano *et al.*, Improved microchip design and application for in situ transmission electron microscopy of macromolecules. *Microsc. Microanal.*, **20** (2014), 338–345.
154. K. Degen, M. Dukes, J. R. Tanner and D. F. Kelly, The development of affinity capture devices: a nanoscale purification platform for biological in situ transmission electron microscopy. *RSC Adv.*, **2** (2012), 2408–2412.
155. B. L. Gilmore, S. P. Showalter, M. J. Dukes *et al.*, Visualizing viral assemblies in a nanoscale biosphere. *Lab Chip*, **13** (2013), 216–219.
156. C. Wang, Q. Qiao, T. Shokuhfar and R. F. Klie, High-resolution electron microscopy and spectroscopy of ferritin in biocompatible graphene liquid cells and graphene sandwiches. *Adv. Mater.*, **26** (2014), 3410–3414.
157. S. M. Hoppe, D. Y. Sasaki, A. N. Kinghorn and K. Hattar, In-situ transmission electron microscopy of liposomes in an aqueous environment. *Langmuir*, **29** (2013), 9958–9961.
158. M. T. Proetto, A. M. Rush, M.-P. Chien *et al.*, Dynamics of soft nanomaterials captured by transmission electron microscopy in liquid water. *J. Am. Chem. Soc.*, **136** (2014), 1162–1165.
159. F. A. Plamper, A. P. Gelissen, J. Timper *et al.*, Spontaneous assembly of miktoarm stars into vesicular interpolyelectrolyte complexes. *Macromol. Rapid Commun.*, **34** (2013), 855–860.
160. U. M. Mirsaidov, H. Zheng, Y. Casana and P. Matsudaira, Imaging protein structure in water at 2.7 nm resolution by transmission electron microscopy. *Biophys. J.*, **102** (2012), L15–17.
161. A. C. Varano, A. Rahimi, M. J. Dukes *et al.*, Visualizing virus particle mobility in liquid at the nanoscale, *Chem. Commun.*, **51** (2015), 16176–16179.
162. V. A. Maraloiu, M. Hamoudeh, H. Fessi and M. G. Blanchin, Study of magnetic nanovectors by Wet-STEM, a new ESEM mode in transmission. *J. Coll. Interf. Sci.*, **352** (2010), 386–392.
163. K. Adachi, E. J. Freney, P. R. Buseck, Shapes of internally mixed hygroscopic aerosol particles after deliquescence, and their effect on light scattering. *Geophys. Res. Lett.*, **38** (2011), L13804.
164. P. G. de Gennes, Wetting: statics and dynamics. *Rev. Mod. Phys.*, **57** (1985), 827–863.
165. J. Liu, B. Wei, J. D. Sloppy *et al.*, Direct imaging of electrochemical deposition of poly(3,4-ethylene dioxythiophene) (PEDOT) by transmission electron microscopy. *Macro Lett.*, **4** (2015), 897–900.

166. S. Sadki, P. Schottland, N. Brodie and G. Sabouraud, The mechanisms of pyrrole electropolymerization. *Chem. Soc. Rev.*, **29** (2000), 283–293.
167. F. M. Ross, Controlling nanowire structures through real time growth studies. *Rep. Prog. Phys.*, **73** (2010), 114501–114522.
168. L. Zhang, B. K. Miller and P. A. Crozier, Atomic level observation of surface amorphization in anatase nanocrystals during light irradiation in water vapor. *Nano Lett.*, **13** (2013), 679–684.
169. G.-Z. Zhu, S. Prabhudev, J. Yang *et al.*, In situ liquid cell TEM study of morphological evolution and degradation of Pt–Fe nanocatalysts during potential cycling. *J. Phys. Chem. C*, **118** (2014), 22111–22119.
170. N. D. Browning, M. A. Bonds, G. H. Campbell *et al.*, Recent developments in dynamic transmission electron microscopy. *Curr. Opin. Solid State Mater. Sci.*, **16** (2012), 23–30.
171. M. W. Mourik, W. J. van Engelen, E. J. D. Vredenbregt and O. J. Luiten, Ultrafast electron diffraction using an ultracold source. *Struct. Dynam.*, **1** (2014), 034302.
172. N. de Jonge, System and methods for live cell transmission electron microscopy. US Patent Application 13,299,241 (2011).
173. R. Danev and K. Nagayama, Transmission electron microscopy with Zernike phase plate. *Ultramicroscopy*, **88** (2001), 243–252.
174. P. Simon, H. Lichte, P. Formanek *et al.*, Electron holography of biological samples. *Micron*, **39** (2008), 229–256.

2 Encapsulated Liquid Cells for Transmission Electron Microscopy

Eric Jensen and Kristian Mølhave

2.1 Introduction

The evolution of microscopy has taken several quantum leaps from the first microscopes, using light and lenses made of glass, to modern, aberration-corrected electron microscopes. The trend in TEM today is to provide more than an image of a dry sample; in particular, to obtain information under process-relevant conditions, while external stimuli are being applied, or in a native state. This has driven advances in both open and closed cell microscopy to examine samples in complex environments. The open cell, using differential pumping to maintain a region of high pressure, can be used in specially modified TEMs or SEMs. The closed cell, where the liquid is fully encapsulated, allows an environment of high vapor pressure liquid or high pressure gases, and can be used in a standard TEM or SEM that does not have capabilities for differential pumping.

In this chapter we describe the key design considerations for closed liquid cells for the TEM. The use of volatile liquid samples in the TEM requires a careful design of both the cell and its holder. The closed cell must, of course, isolate the liquid from the vacuum of the TEM while simultaneously ensuring that the region of interest is electron transparent and also providing as large a field of view as possible. The encapsulating membrane must be thin and of a material made from light atoms, to be as electron transparent as possible. However, it must also be robust: the pressure difference between a liquid at atmospheric or higher pressure and the vacuum of the TEM will stress the membrane, forcing it to bulge outwards towards the vacuum, especially if the window area is large. Stress may cause fracture, resulting in a possible leak into the vacuum; bulging will increase the liquid thickness, thereby reducing resolution.

It is therefore clear that TEM liquid cell design requires a compromise in materials properties. The membrane must be thin and light for electron transparency, but also have a high Young's modulus to minimize bulging. The cell geometry must include relatively small window areas, and membranes that are close together and have a controlled separation. The mechanical property requirements lead us to materials such as silicon dioxide, silicon nitride, silicon-rich silicon nitride, hexagonal boron nitride, or graphene. Graphene is ideal in many respects. However, the desire to integrate features such as electrodes, flow capabilities, and heaters, as well as to control the spacing between the two membranes, implies the use of microfabrication techniques to mass-produce liquid cells that are generally based on pairs of silicon chips that each have a thin membrane and can each include a range of functionalities. Since current mainstream microfabrication techniques are best suited for creating high performance setups with silicon-based materials, silicon-rich nitride, Si_xN_y (see below) has become the most popular choice for TEM liquid cell membranes. A complete setup for liquid cell TEM also includes a sample holder. This may have liquid ports to enable flow, and electrical connections to electrodes or heaters that can be built into the liquid cell chips.

In the following sections we first provide a brief overview of the evolution of *in situ* electron microscopy of encapsulated liquids, then highlight various designs for encapsulation. Microfabrication techniques have been optimized to create several types of liquid cell, and we describe the processing steps involved. However, not all liquid cell systems are made by microfabrication, and new opportunities arise from other approaches to encapsulation. A single carbon nanotube can be a liquid container, and some graphene liquid cells are made by hand without lithographic processes. We discuss the benefits and limitations of the different approaches. We then discuss the competing issues of resolution and membrane bulging, and correlative imaging where multiple detection methods are utilized on the same sample.

Many of the design considerations for liquid cells in TEM are also relevant to encapsulation of liquids in SEM. SEM "capsules" with a single window have advantages for liquid cell microscopy [1], for example in allowing for a larger liquid volume. But TEM liquid cells with two windows encapsulating a thin liquid layer can also be used in the SEM, allowing for STEM in the SEM which can provide higher resolution information than conventional SEM (Chapter 1). Liquid cell SEM will not be discussed further here, as details are given in Chapter 4.

2.1.1 The Evolution of Liquid Cells: Electron Microscopy Meets Microfabrication

Liquid cells for electron microscopy were attempted in the 1940s [2], and *in situ* liquid electron microscopy was also carried out in the 1970s where the first cells were assembled [3]. However, these early cells lacked the ability to sustain atmospheric pressure [4] and also lacked the low mass thickness required for high resolution imaging. With the invention of environmental electron microscopes with differential pumping, liquids could be introduced in a limited way, at specific thermodynamic conditions of low pressure and low temperature in ESEM [5] and to some extent ETEM [6, 7]. Interest in closed liquid cells diminished.

However, when microfabrication technology was used to meet the demands of closed cell electron microscopy for room temperature, liquid water, TEM experiments were reintroduced [8]. The first of this new generation of liquid cells used two silicon chips, each with a window; the window separation was controlled with a spacer layer, and the chips were glued together. The importance of introducing external stimuli via electrodes was realized early on, and these first microfabricated liquid cells allowed for *in situ* two- and three-electrode electrochemical measurements [8, 9]. The cells were static, sealing in the liquid while allowing imaging with the electron microscope. As well as electrochemical processes, topics of interest included quantifying particle dispersion in liquid [10], examining viability of bacteria under irradiation [11], and investigating beam-induced nanoparticle synthesis [12]. Later, liquid flow [13] was introduced by adding an inlet and outlet to the liquid cell, and introducing the liquid by pumping it via tubes led through the sample holder.

To improve image resolution through the liquid, the high angle annular dark field (HAADF) STEM imaging mode proved to be particularly successful for detecting high Z nanoparticles in a low Z liquid sample [14, 15]. Gradual improvements in liquid cell design and the use of liquid cells in higher performance microscopes eventually allowed the resolution to reach the angstrom scale [16–18]. Commercial chip-based systems also started to appear widely, making use of microfabrication technology for mass production of chips with functionalities such as flow, electrodes, and heating (e.g. [16, 19]). Because these chip-based systems can be used with most conventional TEMs, closed liquid cell TEM is now available in a great number of microscopy laboratories and its long-term future as a key electron microscopy technique seems certain [20].

2.2 Microfabricated Chip Designs

There are several different ways to encapsulate liquids in a liquid cell in order to image them with TEM. In Sections 2.2.1–2.2.3 we discuss liquid cells that are based on microfabricated chips. These setups can be broadly divided into two types, as shown in Figure 2.1. The first type involves paired silicon chips with windows made of electron-transparent thin-film membranes. These chips are mounted together in what we here call a "sandwich" chip system. This is the most popular method in use today and is described further in Sections 2.2.1 and 2.2.2. The second is the monolithic chip system, where a channel that will contain the liquid is fabricated on a single chip. This is described further in Section 2.2.3. Each approach has benefits and disadvantages. For the sandwich-type

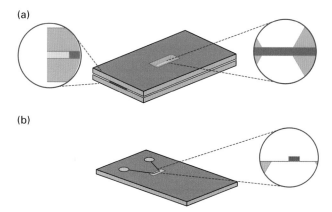

Figure 2.1. The two basic types of microfabricated *in situ* liquid TEM setups. The drawings are not to scale. (a) A cross-sectional view of a sandwich setup. Two silicon chips are placed together separated by a spacer material (yellow). Each chip has a hole etched in the center over which an electron-transparent membrane (brown) is suspended. Liquid (blue) fills the volume between the windows. (b) A monolithic chip with a fabricated channel that encloses the liquid. The membrane material also forms the channel walls. In both cases, the chips are typically each 200–300 μm thick, with the windows and liquid on the order of 50 nm thick.

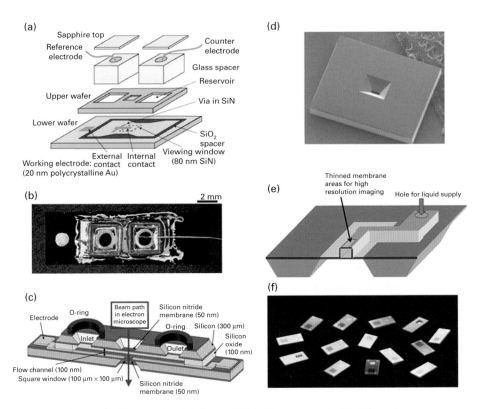

Figure 2.2. Overview of several different microfabricated TEM chip systems. (a, b) Schematic and photograph of the electrochemical liquid cell used in the seminal paper by Williamson *et al.* [8], where two chips are glued together, with an Au electrode lithographically defined on the bottom chip. (c) A bonded chip system with electrodes as well as inlets and outlets for liquid flow through a microfluidic channel defined by the spacer layer between the two chips [21]. (d) A single Si chip with a window, designed for use in a clamped setup [15]. The chip is viewed from the back to show the etched region that forms the window. (e, f) A cross section and photographs of a monolithic system. Copyright: (a) and (b) from Ref. 8 with copyright to the Nature Publishing Group, (c) from Ref. 21 with copyright to the *Journal of the Indian Institute of Science*, (d) Ref. 15, Creative Commons License, (f) photograph by Jesper Scheel.

setup, the challenges include ensuring alignment and controlling the distance between the windows. The monolithic configuration requires substantial microfabrication work compared to the sandwich chip, but can provide a more controlled geometry, for example by minimizing window size and hence bulging. Figure 2.2 shows examples of a few of the microfabricated chip systems that have been used in liquid cell experiments. Non-microfabricated liquid cells, such as the graphene liquid cell and nanotube-based liquid containers, are mentioned in Section 2.3.

2.2.1 Membrane Materials and Fabrication

The material of the membrane in all the different liquid cell TEM setups in Figure 2.2 must fulfill three parameters: (1) it must be electron transparent; (2) it must have a high stiffness to reduce bulging; and (3) it must be compatible with current cleanroom techniques. For most systems, those developed in academia as well as those commercially available, the amorphous ceramic silicon-rich silicon nitride, referred to hereafter as Si_xN_y, is the preferred choice for reasons discussed below. Other materials that have been considered include graphene and graphene oxide, hexagonal boron nitride (hBN), stoichiometric silicon nitride, and silicon dioxide, as well as polyimide, a polymer.

Of the materials listed above, graphene and hBN can be made extremely thin, hence showing minimal contrast in TEM. However, they are not yet fully compatible with current mainstream cleanroom techniques. While the materials can be bought or can be synthesized in some cleanrooms, creating reproducible wafer-scale devices with graphene, graphene oxide or hBN windows is not yet a standard process. Their use in liquid cells therefore requires a different approach [21–24], which is summarized in Section 2.3 and discussed in more detail in Chapter 19.

Stoichiometric silicon nitride, Si_3N_4, has a lower Young's modulus than graphene and hBN, but is used in thicker form than graphene or hBN in order to be sufficiently stiff. It is compatible with most standard cleanroom techniques and is thus an electron-transparent amorphous (or polycrystalline) film that is easily fabricated in the cleanroom. A standard technique for creating Si_3N_4 thin films on a silicon wafer is the use of low pressure chemical vapor deposition (LPCVD) in a furnace at elevated temperatures. However, during fabrication the silicon expands due to the high temperature, and on cooling the mismatch in the thermal expansion coefficients between the film and the wafer creates stress in the film. Non-stoichiometric silicon-rich silicon nitride with a better-matching thermal expansion coefficient is therefore preferred over stoichiometric Si_3N_4. The ratio of silicon to nitrogen is typically about 2:1 in Si_xN_y. It is a very widely used membrane material [8–10, 12–18, 21, 26]. Its amorphous structure significantly reduces the Young's modulus compared to stoichiometric Si_3N_4, to the point where it is comparable to silicon dioxide.

Silicon dioxide (SiO_2) has an even lower Young's modulus than the above materials, but also shows far lower residual stress than Si_3N_4. It has been used successfully in some liquid cell experiments [11]. But the main reason it is not as popular as silicon nitride is its relatively high etch rate in potassium hydroxide (KOH). This complicates wafer processing, as discussed below.

Another material in use is polyimide [27, 28]. Of the materials above, polyimide has the lowest Young's modulus and therefore bulges the most. It is also incompatible with many cleanroom techniques. Furthermore, standard thicknesses are about 150 nm, on the high side for designing TEM liquid cells for high resolution imaging. However it is used in SEM, including commercial SEM liquid cell capsules, where bulging is less important. The materials choices for liquid cell capsules for SEM are discussed in more detail in Chapter 4.

The microfabrication of a single membrane chip for a TEM liquid cell proceeds as illustrated in Figure 2.3. First a silicon wafer typically 200–400 μm thick is coated with an LPCVD Si_xN_y film that will both form the membrane and act as an etch mask for window formation. The window is formed using a KOH wet etch to remove the silicon in the region where the membrane is suspended, leaving the membrane unaffected.

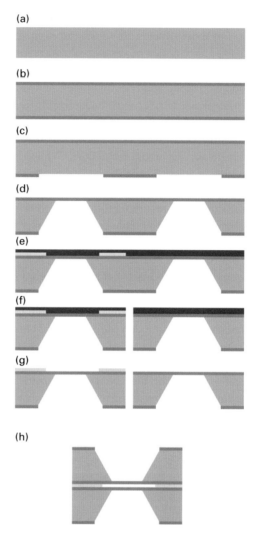

Figure 2.3. Basic microfabrication scheme for sandwich-type chips. Grey: Si, brown: Si$_x$N$_y$, yellow: spacer layer, red: photoresist. (a) Si wafer; (b) Si$_x$N$_y$ low stress LPCVD deposition; (c) back-side lithography and opening in SiN$_x$; (d) KOH wet etch of Si through the back-side hole masked by the Si$_x$N$_y$; (e) fabricate optional spacer layer, and add a layer of resist to protect membranes; (f) cut individual chips with a diamond saw; (g) user removes protective resist prior to use by solvent wash, and performs a careful clean of the chip surfaces; (h) mount in the sample holder and place or flow liquid in the regions that lack spacer.

Typical conditions are 80°C, 28 wt% KOH, with this etch step often taking several hours. For example, a 5 hour KOH etch will etch through 350 μm Si, while removing only ~6 nm Si$_x$N$_y$. The KOH etch leaves slanted Si(111) sidewalls determined by the crystal orientation of the wafer. This KOH etch step is what makes silicon dioxide membranes problematic. Although SiO$_2$ etches much more slowly than Si, a substantial amount will be removed, say 1.8 μm over 5 hours, so a thick initial oxide film would be required. A small uncertainty in the initial oxide thickness will then result in membranes whose thickness is difficult to control.

After window formation, a spacer layer is deposited on one surface, so that when a pair of chips is clamped together, the spacer layer defines the void into which the liquid can flow. This spacer layer may be oxide, the polymer SU8, or another material. At this stage in the processing, electrodes can also be patterned on one of the chips via lithography, metal deposition, and lift-off. A protective resist is added to strengthen the membrane and keep it clean when dicing, packaging, and sending chips to users. Finally, the individual chips are cut from the wafer with a diamond dicing saw. The user can remove the protective resist just before mounting the chips. Typically a plasma clean is also performed, which removes any remaining carbon and additionally renders the surface more hydrophilic for a few hours.

2.2.2 Strategies for Sandwich Setups

There are many different variations of what we have called the sandwich liquid cell setup for TEM. Some of the earlier designs were reviewed in Ref. 19, while the following sections attempt to describe the rich inventiveness of the current state of the art. Different strategies have been explored for inserting the liquid into the cell, for controlling the distance between the windows (spacer layers, spacer materials, posts, monolithic designs), and for attaching the chips together (gluing, bonding or clamping with o-rings) and preventing the liquid from escaping into the microscope vacuum. Design of flow and electrochemical capabilities is also important. Liquid flow is required if the liquid must be replenished or its composition changed during the experiment.

2.2.2.1 Glued Chips, Spacer Layers and Inlets

As shown in Figure 2.2, the earliest microfabricated liquid cells consisted of pairs of chips that were glued together [29–31]. Choosing the glue is particularly important since it must not out-gas in the TEM vacuum or interact chemically with the liquid sample. Each pair of chips was positioned in a clamp so that the windows were aligned, squeezed together using screws, then glue was applied around the edges. Thus, the glue provided part of the hermetic seal between the liquid and the microscope vacuum, but it did not control the spacing between the chips. The spacing was defined by a spacer layer of SiO_2 patterned on the lower chip.

Control of the liquid thickness via the patterned spacer layer was a key feature of this liquid cell. The liquid thickness of 500 nm to 1 μm was relatively large by current standards, but nevertheless provided sufficient resolution to image dynamic processes within the liquid. A second important feature was the method of introducing liquid into the cell. Figure 2.2a shows two vias in the upper chip [8]. These small openings (fabricated in the same manner as the windows, but with the membrane subsequently removed) allowed liquid to be delivered using a hypodermic syringe so that capillary forces could pull the liquid into the viewing area. Plates glued above the openings completed the hermetic seal. The final component of the entire liquid cell setup was a custom sample holder. As well as holding the sample, this provided electrical feedthroughs to the electrodes within the liquid.

There are numerous variations and improvements to this basic design, many of which are reviewed in Ref. 20. The spacer layer can be made from other materials such as gold or indium, or alternatively can be made by dispersing polystyrene spheres: when two chips are pushed together, the spheres ensure a minimum separation. Rather than using glue, the chips can be connected by bonding, as described in the next section, or clamping, described in the following one. The hermetic seal can be created using o-rings rather than glue. Finally, the sample holder should be designed to make the experiment as easy and reliable as possible. Glued or bonded liquid cells can sometimes fit into standard TEM holders, but otherwise dedicated holders are needed to accommodate these larger and thicker chip samples. Built-in feedthroughs for the liquid and for electrical connections can be included in the holder. Methods for delivering the liquid and electrical stimuli are discussed further in Section 2.6.

2.2.2.2 Bonded Chips

Wafer bonding is a high temperature and high pressure process that forms a hermetic seal between two wafers. It is a good alternative to gluing when fabricating liquid cells for TEM, as shown in Figure 2.2c. Such cells are fabricated by patterning the interior structures (for example the spacer layer and electrodes), bonding the top and bottom wafers, then processing the pair of wafers to form the windows by etching from top and bottom simultaneously. At the same time, inlets and outlets are also etched. Finally, the bonded and processed wafer is diced into individual liquid cells. Bonded chip systems were first made for imaging samples in gaseous environments [32, 33], and allow gas flow at relatively high pressure. Wafer bonding has benefits for both gas and liquid cell fabrication [21, 32], in that it produces a reliable vacuum seal, a high precision alignment between the top and bottom windows and a well-defined liquid chamber. The main disadvantage is that it is difficult to place certain samples inside the liquid cell; drop-casting or depositing a layer on the interior of a liquid cell is easier if clamped chips are used, as discussed below.

2.2.2.3 Clamped Chips: Sealing and Alignment

Simply squeezing two chips together with o-rings is a straightforward approach to forming a liquid cell. This is the approach that has been chosen by commercial manufacturers of liquid cell systems, where the TEM holder mechanically clamps the pair of chips using o-rings. The o-rings provide flexible support for the clamping so that the chips do not crack. Separation is ensured by spacer structures on the chips. However, a clamped pair of chips can leak liquid around its edges. Clamping is therefore combined with a design that preserves the seal between the microscope vacuum and the liquid. Clamped designs generally allow both static and flowing liquid to be imaged. Strategies for holder and clamping design are described in Figure 2.4 and examples shown in Figures 2.5, 2.6, and 2.7.

Fabricating the two chips separately has distinct advantages. It is possible to include complex functionality (such as electrodes or heaters) on only one chip, reducing costs

Encapsulated Liquid Cells for TEM 43

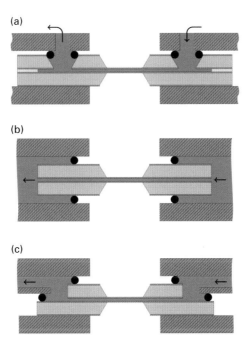

Figure 2.4. Schematic of several possible approaches for getting liquid into the volume between the windows. In (a) the two chips are sealed around their edges by gluing or by wafer bonding and the liquid enters through vias on one chip. In a static design, the cell is filled and the vias are sealed (as in Refs. 8, 21, 26, and 34), while in a flow design, as shown here, liquid enters and leaves through ports on the sample holder (as in Ref. 48). In (b, c) the liquid enters the cell from the side. A spacer material (not shown) keeps the chips at the desired separation. The vacuum seal and the pressure holding the chips together are provided by o-rings that are compressed by a plate on the holder that is held to the body of the holder with screws (not shown). Grey: Si, brown: Si_xN_y, blue: liquid, black: o-ring, orange: sample holder or plate, yellow: spacer layer/bonding agent.

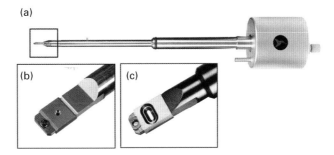

Figure 2.5. Example of a liquid flow TEM holder design by Hummingbird Scientific, illustrating the ability to use different chip/flow configurations in modular tips on a single TEM holder platform. (a) The holder; (b, c) sealing hardware for flow configurations (b) and (c) from Figure 2.4. The standard flow configuration is shown in (b); the electrochemical EDX-compatible design is shown in (c). Images copyright Hummingbird Scientific.

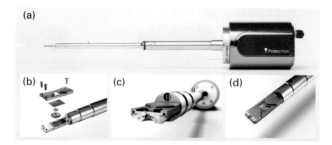

Figure 2.6. Example of a liquid cell design by Protochips Inc. (a) The holder; (b) exploded view showing the tip assembly with o-rings and contacts for electrochemistry or heating; (c) cross-sectional illustration showing the liquid pathway; (d) top view illustration showing the cutout that optimizes the design for EDX. Images copyright Protochips Inc.

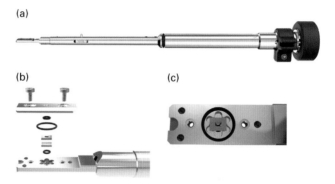

Figure 2.7. Example of a liquid flow holder design by DENSsolutions B.V. (a) Liquid holder; (b) holder tip with a precision slot for loading the *Nano-Cell* (dual chip sample carrier). The precision slot self-aligns the *Nano-Cell*'s top and bottom electron-transparent windows. (c) Top view showing the *Nano-Cell* loaded inside the slot, which acts as a bypass to prevent over-pressure due to fluidic resistances. Images copyright DENSsolutions B.V.

and perhaps increasing flexibility for design changes. It is possible to modify the surfaces of the chips before assembly, and in particular one can place specific samples (such as nanoparticles) onto the interior surface, by drop casting or evaporation, before assembling the liquid cell.

To assemble the clamped setup into its holder, the holder is built with a small pocket that fits the specific liquid cell chip. The chips are dropped in and the pockets align the windows on the chips automatically, provided the chips are diced accurately enough. Accurate alignment is important since misaligned windows reduce the viewing area; further, alignment minimizes the effects of window bulging around the edge of the viewing area, as described in Section 2.5 below. Because of the importance of window alignment, other strategies have also been developed to align chips that are placed on each other. These include alignment spheres in grooves [34], and a KOH etched self-aligning system [35] with a metal support on the membranes.

2.2.3 Monolithic Chips

The above systems generally have membranes with lateral dimensions larger than about 50 μm. This is due to a combination of the KOH etch process, where smaller membrane areas are difficult to make reproducibly, and the accuracy with which the two sandwiched chips can be aligned, whether clamped or glued. Given the 50 μm span of the membrane, its thickness is typically of the order 30 nm or more to ensure mechanical stability under 1 atm. pressure and an acceptable yield during microfabrication.

Creating microfluidic channels on a single chip's membrane is an alternative strategy. Alignment is no longer an issue, and the channel width can also be reduced to the ~2 μm limit of UV lithography. Such monolithic chips with channels hence provide smaller windows with less bulging and well-defined imaging regions where even thinner membranes can in principle be used.

One such system uses a microchannel of silicon nitride on top of a single silicon chip [36]. The silicon wafer is covered with a Si_xN_y membrane layer, as with previous setups. A sacrificial layer, e.g. of polycrystalline silicon, is then added on top of the Si_xN_y, patterned into the shape of the channel, and then covered with a second layer of Si_xN_y. By opening the channel at the inlets and etching away the sacrificial material all that is left is a Si_xN_y channel (see Figure 2.3c).

Such chip systems are self-contained, like some of the bonded chip setups [21, 32]. The main advantage is that the microchannels can reduce the window size and hence the bulging and allow for thinner films to be used. These systems can also be easier to use with conventional TEM holders since the total liquid cell is typically half the thickness of the sandwich systems. By sealing the liquid in the microchannel, monolithic chips can also act as disposable single use units if cross contamination is a cause for concern in the reusable clamped chip TEM holder systems.

2.3 Other Encapsulation Methods

Apart from the mainstream sandwiched chip setups discussed above, several other methods of encapsulating liquids (and gases) have been used for imaging in the TEM.

2.3.1 Graphene

Before the discovery of graphene, amorphous carbon was utilized as window material in liquid cells, in the form of thin membranes on metal supports in a holder system with widely spaced windows. This had successful bioimaging applications [37, 38]. However, amorphous carbon fell out of favor because it lacks the benefit of integration with standard microfabrication procedures. Graphene is also difficult to process, but its other properties make it a compelling choice for certain applications. Graphene has excellent mechanical properties, and is also conductive and hence avoids sample charging. It is also an effective barrier for water. Additionally, since it is a single atomic layer thick, there is near-zero contribution to beam scattering, improving image resolution.

Chapter 19 describes graphene and related materials in detail, including different methods of cell fabrication. Graphene liquid cells are made by non-lithographic approaches that form "blister" cells where a microscopic droplet of liquid is encapsulated between two adhering graphene sheets. Such cells provide exciting results such as sub-angstrom resolution of nanoparticle synthesis and sintering, and high resolution imaging of biological structures [22–25]. Currently, graphene microfabrication procedures are still non-standard and require experience for successful results. It is therefore difficult to integrate the benefits of graphene with the functionality of microfabricated liquid cells, such as flow and electrical connections. However, a new method for dropcasting graphene oxide membranes [39] might potentially provide a method for creating atomically thin windows that are integrated with microfabricated liquid cells for TEM. This will be an exciting future prospect for high resolution *in situ* experiments.

2.3.2 Carbon and Non-carbon Nanotubes

Some years prior to the "blister" graphene liquid cells, several groups successfully captured liquid inside carbon nanotubes and used these cells to image liquid in the TEM [40, 41]. Such experiments can allow for imaging liquid dynamics but it is difficult to achieve sufficient experimental control for reproducible results. The experiments are best suited for a limited range of liquids but not for studies requiring flow or application of electrical bias; also, insertion of the liquid can be difficult.

Other nanotubes, made of materials such as boron nitride, silicon dioxide, and magnesium oxide, or scrolls of various materials, have also been used for *in situ* liquid TEM imaging. A particular benefit is the opportunity to make observations with the liquid/substrate interface parallel to the electron beam. Some examples of the application of nanotubes in fluid physics are given in Chapter 13.

2.4 What Happens When the Liquid Cell Fails?

As is clear from the discussion above, the liquid cell has two basic functions: confining the liquid into a layer thin enough for transmission of electrons, and keeping the liquid separated from the microscope vacuum. If the first criterion is not met, the consequence is poor image quality, as discussed in Section 2.5 below. But what happens if the second criterion is not met? A "catastrophic" rupture of the membranes is a fairly common experience for users of microfabricated liquid cells. Silicon nitride membranes can crack, especially if the electron beam is focused in a small spot. Static liquid cells contain a finite, relatively small volume of liquid (perhaps 1 mm^3). If this leaks out, it may produce a pressure burst that closes the microscope column valves, but probably will not produce a long-lasting pressure rise. If the liquid cell is a flow cell connected to a larger reservoir of liquid, for example via tubing and a pump (see Section 2.6), a larger volume of liquid can in principle enter the microscope vacuum. However, the effects can still be surprisingly mild. Cracked membranes expose a relatively small liquid area to the vacuum, so the leak rate is low; furthermore, the drop in pressure can freeze the liquid and block the opening, also reducing the leak rate. This low leak rate, combined with the differential pumping that exists between the electron gun and the sample area, means that the pressure at the gun can remain low even during a membrane rupture. If the microscope has an anticontamination device (cold finger), this will also act to pump the liquid and mitigate the pressure rise, resulting in a faster recovery. If the liquid contains corrosive or toxic components, one clearly has other sources of concern. But for the majority of liquid cell experiments, a failure of the mechanical integrity of the liquid cell usually costs only the time and patience of the operator.

2.5 Membrane Bulging: Mitigation and Measurement

In Section 2.1 we described the bulging caused by the pressure difference between the interior of the liquid cell and the microscope vacuum. This outward deflection of the membranes increases the separation between the membranes, and hence the liquid thickness (assuming that the interior is completely filled with liquid). The liquid thickness will therefore be greater than the designed spacing between the two chips. As an example, a sandwich setup with two 50 nm thick and 50 μm wide Si_xN_y membranes can have liquid thickness up to ~3 μm in the center of the window [42]. Liquid thickness can be roughly estimated from the image contrast, but the most accurate measurement is through electron energy loss spectroscopy [43–45]. This gives t/λ, the total thickness divided by the mean free path of electrons through the liquid and membranes. More detail is given in Chapter 21.

Membrane bulging is an important factor in practical liquid cell TEM for several reasons. Most importantly, it affects the image resolution, since resolution is decreased by multiple scattering from the combined mass thickness of the membranes and the liquid. The effects of bulging on resolution are discussed quantitatively in Ref. 20 and Chapter 8. This means in practice that the highest resolution images are obtained in the narrow strip around the periphery of the window where the liquid is thinnest. Another consequence of membrane bulging is that the non-uniform liquid thickness can complicate modeling of a physical process under study, altering for example diffusion pathways. Membrane bulging can also trap any gas bubbles in the thickest part of the window, reducing mass–thickness there [46] but also creating a complicated liquid geometry around the sample.

Bulging of thin films due to pressure depends strongly on the shortest side-length of the membrane, as well as the material thickness and mechanical properties [42, 47]. Figure 2.8 shows the maximum bulging of a 50 nm thick membrane due to a 1 bar pressure difference as a function of the smallest membrane side-length, based on the formula from Ref. 42. A sandwich closed cell made by clamping two such chips would experience twice this bulging from the two membranes. It is clear that reducing the side-length of the window is a good way to decrease bulging – although of course this has to be balanced against the smaller field of view in the experiments. Reduced bulging is achieved practically in several liquid cells in which the window dimensions are small [34, 36], or where small regions of narrow membranes are built into larger but

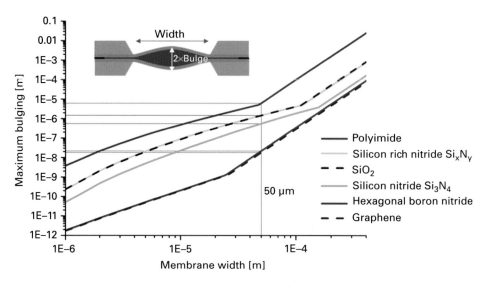

Figure 2.8. Bulging as a function of membrane width. The graph shows the maximum deflection of a 50 nm membrane for 10^5 Pa pressure difference. The material parameters used for this calculation are the Young's modulus and residual stress. A polyimide membrane would probably be over 150 nm thick and therefore bulge slightly less. A graphene membrane, a single atomic layer thick, would have a very different pressure response. The graph is therefore most useful for comparing the other materials and evaluating the balance of window size and liquid thickness. The black vertical and horizontal lines in the graph indicate the bulging for a 50 μm wide membrane, which is typical in most setups. Copyright Eric Jensen [64].

thicker membranes [32, 33, 36, 48]. Adding mechanical connections between the two membranes in the window area, such as posts or contact areas, also reduces bulging in both sandwich designs and monolithic designs of liquid cells [21, 29, 32, 35].

Although mechanical formulas can predict the deflection of a given membrane, they are not straightforward to apply in practical liquid cells. This is because small misalignments (even a few micrometers) between the position of the windows in the upper and lower chips can alter the spacing between the membranes, especially near the periphery of the imaging area. One way to mitigate this alignment issue is to use rectangular windows but place them orthogonally. This reduces the area available for observation, but ensures that the corners of the imaging area are at the edges of both membranes, minimizing the liquid thickness.

We finally note that loading the liquid at lower pressure should in principle mitigate bulging effects. It is also interesting to point out that, in principle, a hermetically sealed liquid cell, completely filled with liquid with no air space or dissolved gas, and perfectly rigid apart from the membranes, should not experience any membrane bulging: an incompressible liquid cannot permit the membranes to bulge as this would change the liquid volume. Microelectromechanical devices such as on-chip pumps may therefore be used to control membrane deflection. It will be interesting to explore the possibilities of pressure control in both optimizing liquid cell designs and investigating new phenomena associated with pressure and liquid thickness.

2.6 Stimuli and Correlative Measurements: Biasing, Heating, Flow, and Spectroscopy

The previous sections have focused on strategies for encapsulating the liquid in a suitably thin layer. But as more and more processes are measured in the liquid cell, different types of stimuli must be applied, and different types of information must be obtained. The following sections discuss setups in which electrical bias, heating, or liquid flow are used to trigger processes, and spectroscopy and fluorescence imaging are used to measure the properties of the liquid sample.

2.6.1 Flowing Liquids

Flowing liquids are important for a great range of liquid-based experiments. Liquid flow allows reactants to be replenished, products of radiolysis to be removed, solution chemistry to be changed during observation, and the physics of flow itself to be studied. Two liquids or a liquid plus gas can also be mixed at or near the viewing window.

Liquid flow is achieved in sandwich setups as well as monolithic setups. In a sandwich setup, the liquid has to enter the space between the membranes either from the side of the chips, or from vias patterned into the chips (Figure 2.4).

In Figure 2.4a, flowing the liquid in and out through vias minimizes the liquid volume in the system, and allows the flow rate to be directly controllable [48]. However, designing vias into the chips increases the complexity of the microfabrication. The sample holder needs to include small o-rings to direct the flow in and out of the vias. In Figure 2.4b, c, flowing the liquid in and out through the sides of the chips is simpler in terms of fabrication. The chips are placed in a pocket in the sample holder with excess volume on either side. Liquid enters this buffer region through tubes. It flows into the gap between the two chips, but can also flow in a bypass route around the outside of the pair of chips. The liquid between the windows is affected by the flow and gradually replaced. But, due to the geometry of such systems, most of the liquid flowing through the chip recess in the holder flows around the chips and not between them due to the hydraulic resistance in the narrow confined layer between the chips. Whichever way the liquid enters in Figure 2.4, its flow can be directed using channels in the spacer layer between the chips [8, 21, 48].

Both of these approaches have been successful in enabling numerous experiments that involve liquid flow. Recently, holders have also become commercially available that allow mixing of liquids as they enter the chip recess in the holder by having two inlet and one outlet capillary tubes connecting through the holder. Liquids and gases can also be mixed [50]. The possibilities of liquid flow are exciting and we anticipate further development in this area.

2.6.2 Electrodes for Electrochemistry

Electrochemical electrodes allow for inducing non-spontaneous chemical reactions and analyzing ongoing reactions by electrochemical methods. Such systems are useful for elucidating the complex heterogeneous electrochemical processes found in systems such as batteries and fuel cells. Generally, liquid TEM electrochemical experiments attempt to image the working electrode to see the interfacial chemistry driven by the experiment. Three or more electrodes can therefore be incorporated into the liquid cell to enable electrochemical measurements, by adding processing steps for patterning, metal deposition, and lift-off.

The material and placement of the electrodes are important design considerations. The simplest solution for microfabricated setups is to make the working electrode, the reference electrode, and the counter electrode out of the same material, and optimize their dimensions and relative position. Standard calibrated reference electrodes are difficult to implement in the liquid cell. Fabricating each electrode out of a material specifically relevant to the given experiment is optimal and can be achieved by strategies such as functionalization of each electrode by deposition of different materials. Furthermore, the electrodes must be connected to contact pads, generally at one end of the chip. It is important that the potential drop along each electrode is minimized, that the area of the counter electrode is larger than that of the working electrode, and that the reference electrode is placed near the working electrode.

These microfabricated electrodes are often quite large compared to the imaging region of the liquid cell. This allows for larger currents to flow, which are easier to measure, but also means that the reaction occurs where imaging is not possible. To avoid reactions outside the visible area the electrode can be masked using silicon dioxide or the resist SU-8, or an active component can be soldered to an electrode, often using focused ion beam welding. Further details of electrochemical electrode design and applications can be found in Chapters 10 and 11.

2.6.3 Heaters

Although some liquid cells can be placed directly into a conventional TEM heating holder with a resistively heated coil surrounding the sample [51], heating of liquids is now typically achieved through microfabricated heater coils, an approach that has proven successful for heating experiments in gases [32, 33, 52]. The heater is patterned onto or within membranes, and includes four-point measurements and resistive feedback to control the temperature. Such membrane heater systems deliver less power into the microscope than a conventional heater, reducing sample drift. The speed of temperature change, as well as the uniformity of the temperature across the window, are important considerations in designing the chip. Temperature control strategies and applications of heated (and cooled) liquid cells are discussed in detail in Chapter 6.

2.6.4 Correlative Spectroscopic Methods

In TEM, several spectroscopic measurement methods provide additional information about a sample, beyond what is available from bright field or high resolution imaging. Many of these methods have been applied to liquid cell TEM and here we briefly touch upon some examples.

Material composition can be obtained from electron energy loss (EELS) and energy dispersive X-ray (EDX) spectroscopy. These methods are useful for correlating structural and chemical information during or after chemical reactions such as nanoparticle synthesis or electroplating [53].

EDX provides an overview map of the chemical makeup, although it is less sensitive to the low atomic number elements than EELS and also has lower spatial resolution [54]. However, the EDX detector sits at a low angle relative to the specimen. In liquid TEM sandwich setups – although not for monolithic chips or graphene liquid cells – the sample is located inside the recess that forms the window and the o-ring seal. A liquid cell holder modification to optimize EDX analysis is described in Refs. 55 and 56. Results obtained from EDX in the liquid cell are shown in Chapter 21.

In liquid cell TEM, EELS has been used to measure the t/λ ratio, which relates the thickness of the sample to the mean free path of electrons in the material [57]. This is important for quantifying liquid thickness and is discussed further in Chapter 8. Estimating the optimal signal-to-noise ratio for specific materials [45] and even measuring the vibrational spectrum of molecules in the liquid [58] are possible. EELS in liquids is discussed in more detail in Chapter 21.

Light stimulation is an interesting and powerful addition to the liquid cell TEM technique. For biologists using SEM, correlative light (fluorescence) and electron microscopy (CLEM) already enables investigations of biological functions and processes in liquid (Chapter 5). TEM is less well integrated in this regard, since light and electron images have been correlated by physically moving the sample in its liquid cell from one microscope to the other [59,60] (see Chapter 16 for an example). However, it is possible to include optical fluorescence directly in the TEM column: one commercial system allows combined light and subsequent TEM imaging of selected areas, automatically overlaying acquired TEM images on the fluorescence image. Thus, biological function can be related to higher resolution structural information on a larger scale than is possible with TEM alone. For materials scientists, *in situ* optical measurements or the stimulation of a process can be carried out by directing a laser onto the sample with mirrors or fiber optics. Several examples exist for processes involving a gas environment [61, 62], but there has been relatively little work in liquids, with one interesting example being laser-stimulated particle growth [63]. There is plenty of scope for development in this area.

2.7 Conclusions and Outlook

In their 70-year history, setups for TEM of liquids have matured into complex, full-featured *in situ* laboratories. With the recent proliferation of commercially available systems, interest in liquid cell experiments has become widespread. The availability of such equipment in many microscopy research institutes allows for the exploration of problems in materials science, physics, chemistry, electrochemistry, and biology, with fruitful comparison of results across research groups.

The initial rush to obtain images of processes in liquid in TEM is over and now the results can be better characterized and validated. At present, much work in *in situ* liquid cell TEM focuses on including a greater range of stimuli, such as flow, bias, heating, or mixing. But quantifying beam effects and the effect of liquid thickness, flow, and even membrane material is necessary for improving the accuracy and usefulness of the results.

To make liquid cells into a robust and even more widely applicable imaging method, the current devices could be improved in many ways. The inclusion of heaters is a useful step forward, given the strong influence of temperature on many processes. For better control over electrochemical reactions, various ways to include and improve on-chip reference electrodes are likely to be tested, and there will be an ever-increasing development in electrode and membrane structures and functionalization. In the current clamped chip systems with capillary inlets and outlets, cross contamination is a concern, and single-use self-contained disposable bonded chip systems or monolithic designs could be an advantage. The devices that are designed to have a well-defined liquid layer thickness, such as the monolithic chip and those including pillars or other methods to reduce bulging, might also bring improvements in resolution and interpretability. Given the complexity of heterogeneous chemical processes and the need for direct images to verify what is actually occurring, we anticipate a strong emphasis on correlative methods that can resolve questions about chemical composition, structure, and biological function, and quantification of what previously has only been conjectured.

References

1. E. Jensen, C. Købler, P. S. Jensen and K. Mølhave, In-situ SEM microchip setup for electrochemical experiments with water based solutions. *Ultramicroscopy*, **129** (2013), 63–69.
2. I. M. Abrams and J. W. McBain, A closed cell for electron microscopy. *J. Appl. Phys.*, **100** (1944), 607–609.
3. D. D. Double, Some studies of the hydration of Portland cement using high voltage (1 MV) electron microscopy. *Mater. Sci. Eng.*, **12** (1973), 29–34.
4. D. J. Smith, Characterisation of nanomaterials using transmission electron microscopy. In J. Hutchison and A. Kirkland, eds., *Nanocharacterisation* (London: Royal Society of Chemistry, 2007) pp. 1–27.
5. G. D. Danilatos, Foundations of environmental scanning electron microscopy. *Adv. Electron. Electron Phys.*, **71** (1988), 109–250.
6. P. L. Gai, R. Sharma and F. M. Ross, Environmental (S)TEM studies of gas-liquid-solid interactions under reaction conditions. *MRS Bull.*, **33** (2008), 107–114.
7. C. M. Wang, H. G. Liao and F. M. Ross, Observation of materials processes in liquids by electron microscopy. *MRS Bull.*, **40** (2015), 46–52.
8. M. J. Williamson, R. M. Tromp, P. M. Vereecken, R. Hull and F. M. Ross, Dynamic microscopy of nanoscale cluster growth at the solid-liquid interface. *Nat. Mater.*, **2** (2003), 532–536.
9. A. Radisic, P. M. Vereecken, J. B. Hannon, P. C. Searson and F. M. Ross, Quantifying electrochemical nucleation and growth of nanoscale clusters using real-time kinetic data. *Nano Lett.*, **6** (2006), 238–242.

10. R. Franks, S. Morefield, J. Wen et al., A study of nanomaterial dispersion in solution by wet-cell transmission electron microscopy. *J. Nanosci. Nanotechnol.*, **8** (2008), 4404–4407.
11. K.-L. Liu, C.-C. Wu, Y.-J. Huang et al., Novel microchip for in situ TEM imaging of living organisms and bio-reactions in aqueous conditions. *Lab Chip*, **8** (2008), 1915–1921.
12. H. Zheng, S. A. Claridge, A. M. Minor, A. P. Alivisatos and U. Dahmen, Nanocrystal diffusion in a liquid thin film observed by in situ transmission electron microscopy. *Nano Lett.*, **9** (2009), 2460–2465.
13. N. de Jonge, D. B. Peckys, G. J. Kremers and D. W. Piston, Electron microscopy of whole cells in liquid with nanometer resolution. *Proc. Natl. Acad. Sci. USA*, **106** (2009), 2159–2164.
14. N. De Jonge, N. Poirier-Demers, H. Demers, D. B. Peckys and D. Drouin, Nanometer-resolution electron microscopy through micrometers-thick water layers. *Ultramicroscopy*, **110** (2010), 1114–1119.
15. D. B. Peckys, G. M. Veith, D. C. Joy and N. de Jonge, Nanoscale imaging of whole cells using a liquid enclosure and a scanning transmission electron microscope. *PLoS One*, **4** (2009), e8214.
16. K. L. Jungjohann, J. E. Evans, J. A. Aguiar, I. Arslan and N. D. Browning, Atomic-scale imaging and spectroscopy for in situ liquid scanning transmission electron microscopy. *Microsc. Microanal.*, **18** (2012), 621–627.
17. H.-G. Liao, D. Zherebetskyy, H. Xin et al., Facet development during platinum nanocube growth. *Science*, **345** (2014), 916–919.
18. D. Li, M. H. Nielsen, J. R. I. Lee et al., Direction-specific interactions control crystal growth by oriented attachment. *Science*, **336** (2012), 1014–1018.
19. R. R. Unocic, R. L. Sacci, G. M. Brown et al., Quantitative electrochemical measurements using in situ ec-S/TEM devices. *Microsc. Microanal.*, **20** (2014), 452–461.
20. F. M. Ross and N. de Jonge, Electron microscopy of specimens in liquid. *Nat. Nanotechnol.*, **6** (2011), 695–704.
21. J. M. Grogan, N. M. Schneider, F. M. Ross and H. H. Bau, The Nanoaquarium: a new paradigm in electron microscopy. *J. Indian Inst. Sci.*, **92** (2012), 295–308.
22. J. M. Yuk, J. Park, P. Ercius et al., High-resolution EM of colloidal nanocrystal growth using graphene liquid cells. *Science*, **336** (2012), 61–64.
23. J. M. Yuk, H. K. Seo, J. W. Choi and J. Y. Lee, Anisotropic lithiation onset in silicon nanoparticle anode revealed by in situ graphene liquid cell electron microscopy. *ACS Nano*, **8** (2014), 7478–7485.
24. C. Wang, Q. Qiao, R. F. Klie and T. Shokuhfar, High resolution in-situ study of reactions in graphene liquid cells. *Microsc. Microanal.*, **20** (2014), 1520–1521.
25. A. De Clercq, W. Dachraoui, O. Margeat et al., Growth of Pt–Pd nanoparticles studied in situ by HRTEM in a liquid cell. *J. Phys. Chem. Lett.*, **5** (2014), 2126–2130.
26. H. Zheng, R. K. Smith, Y.-W. Jun et al., Observation of single colloidal platinum nanocrystal growth trajectories. *Science*, **324** (2009), 1309–1312.
27. S. Thiberge, A. Nechushtan, D. Sprinzak et al., Scanning electron microscopy of cells and tissues under fully hydrated conditions. *Proc. Natl. Acad. Sci. USA*, **101** (2004), 3346–3351.
28. S. Thiberge, O. Zik and E. Moses, An apparatus for imaging liquids, cells, and other wet samples in the scanning electron microscopy. *Rev. Sci. Instrum.*, **75** (2004), 2280–2289.
29. M. J. Williamson, Investigations of materials issues in advanced interconnect structures, Ph.D. Thesis, University of Virginia (2002).
30. A. Radisic, Electrochemical nucleation and growth of copper, Ph.D. Thesis, The Johns Hopkins University (2005).

31. M. den Heijer, In-situ transmission electron microscopy of electrodeposition: technical development, beam effects and lithography, M.Sc. Thesis, Leiden University (2008).
32. J. F. Creemer, S. Helveg, G. H. Hoveling *et al.*, Atomic-scale electron microscopy at ambient pressure. *Ultramicroscopy*, **108** (2008), 993–998.
33. J. F. Creemer, S. Helveg, P. J. Kooyman *et al.*, A MEMS reactor for atomic-scale microscopy of nanomaterials under industrially relevant conditions. *J. Microelectromech. Syst.*, **19** (2010), 254–264.
34. A. J. Leenheer, J. P. Sullivan, M. J. Shaw and C. T. Harris, A sealed liquid cell for in situ transmission electron microscopy of controlled electrochemical processes. *J. Microelectromech. Syst.*, **24** (2015), 1061–1068.
35. T.-W. Huang, S.-Y. Liu, Y.-J. Chuang et al., Self-aligned wet-cell for hydrated microbiology observation in TEM. *Lab Chip*, **12** (2012), 340–347.
36. E. Jensen, A. Burrows and K. Mølhave, Monolithic chip system with a microfluidic channel for in situ electron microscopy of liquids. *Microsc. Microanal.*, **20** (2014), 445–451.
37. T. L. Daulton, B. J. Little, K. Lowe and J. Jones-Meehan, In situ environmental cell–transmission electron microscopy study of microbial reduction of chromium(VI) using electron energy loss spectroscopy. *Microsc. Microanal.*, **7** (2001), 470–485.
38. T. L. Daulton, B. J. Little, K. Lowe and J. Jones-Meehan, Electron energy loss spectroscopy techniques for the study of microbial chromium(VI) reduction. *J. Microbiol. Methods*, **50** (2002), 39–54.
39. M. Krueger, S. Berg, D. Stone *et al.*, Drop-casted self-assembling graphene oxide membranes for scanning electron microscopy on wet and dense gaseous samples. *ACS Nano*, **5** (2011), 10047–10054.
40. Y. Gao and Y. Bando, Nanotechnology: carbon nanothermometer containing gallium. *Nature*, **415** (2002), 599.
41. A. L. Yarin, A. G. Yazicioglu, C. M. Megaridis, M. P. Rossi and Y. Gogotsi, Theoretical and experimental investigation of aqueous liquids contained in carbon nanotubes. *J. Appl. Phys.*, **97** (2005), 124309.
42. J. Yang and O. Paul, Fracture properties of LPCVD silicon nitride thin films from the load-deflection of long membranes. *Sens. Actuators A Phys.*, **97–98** (2002), 520–526.
43. P. Abellan, T. J. Woehl, R. G. Tonkyn *et al.*, Implementing in situ experiments in liquids in the (scanning) transmission electron microscope ((S)TEM) and dynamic TEM (DTEM). *Microsc. Microanal.*, **20** (2014), 1648–1649.
44. K. L. Klein and I. M. Anderson, Current challenges of TEM imaging with a liquid flow cell. *Microsc. Microanal.*, **18** (2012), 1154–1155.
45. M. E. Holtz, Y. Yu, J. Gao, H. D. Abruña and D. A. Muller, In situ electron energy-loss spectroscopy in liquids. *Microsc. Microanal.*, **19** (2013), 1027–1035.
46. B. C. Regan, M. Mecklenburg, E. R. White, S. B. Singer and S. Aloni, Imaging nanobubbles in water with scanning transmission electron microscopy. *Appl. Phys. Express*, **4** (2011), 055201.
47. J. Yang, J. Gaspar and O. Paul, Fracture properties of LPCVD silicon nitride and thermally grown silicon oxide thin films from the load-deflection of long Si_3N_4 and SiO_2/Si_3N_4 diaphragms. *J. Microelectromech. Syst.*, **17** (2008), 1120–1134.
48. C. Mueller, M. Harb, J. R. Dwyer and R. J. D. Miller, Nanofluidic cells with controlled pathlength and liquid flow for rapid, high-resolution in situ imaging with electrons. *J. Phys. Chem. Lett.*, **4** (2013), 2339–2347.
49. M. Tanase, J. Winterstein, R. Sharma *et al.*, High-resolution imaging and spectroscopy at high pressure: a novel liquid cell for the TEM. *Microsc. Micranal.*, **21** (2015), 1629–1638.

50. M. H. Nielsen, S. Aloni and J. J. De Yoreo, In situ TEM imaging of $CaCO_3$ nucleation reveals coexistence of direct and indirect pathways. *Science*, **345** (2014), 1158–1162.
51. H. L. Xin and H. Zheng, In situ observation of oscillatory growth of bismuth nanoparticles. *Nano Lett.*, **12** (2012), 1470.
52. H. L. Xin, K. Niu, D. H. Alsem and H. Zheng, In-situ TEM study of catalytic nanoparticle reactions in atmospheric pressure gas environment, *Microsc. Microanal.*, **19** (2013), 1558–1568.
53. A. E. Goode, A. E. Porter, M. P. Ryan and D. W. McComb, Correlative electron and X-ray microscopy: probing chemistry and bonding with high spatial resolution. *Nanoscale*, **7** (2015), 1534–1548.
54. R. F. Egerton, Electron energy-loss spectroscopy in the TEM. *Rep. Prog. Phys.*, **72** (2009), 016502.
55. E. A. Lewis, S. J. Haigh, T. J. A. Slater *et al.*, Real-time imaging and local elemental analysis of nanostructures in liquids. *Chem. Commun.*, **50** (2014), 10019–10022.
56. N. J. Zaluzec, M. G. Burke, S. J. Haigh and M. A. Kulzick, X-ray energy-dispersive spectrometry during in situ liquid cell studies using an analytical electron microscope. *Microsc. Microanal.*, **20** (2014), 323–329.
57. K. Iakoubovskii, K. Mitsuishi, Y. Nakayama and K. Furuya, Thickness measurements with electron energy loss spectroscopy. *Microsc. Res. Tech.*, **71** (2008), 626–631.
58. T. Miyata, M. Fukuyama, A. Hibara et al., Measurement of vibrational spectrum of liquid using monochromated scanning transmission electron microscopy-electron energy loss spectroscopy. *Microscopy*, **63** (2014), 377–382.
59. N. De Jonge, in P. W. Hawkes, ed., *Advances in Imaging and Electron Physics* Volume **190** (Elsevier, 2015) pp. 1–102.
60. M. J. Dukes, D. B. Peckys and N. de Jonge, Correlative fluorescence microscopy and scanning transmission electron microscopy of quantum-dot-labeled proteins in whole cells in liquid. *ACS Nano*, **4** (2010), 4110–4116.
61. F. Cavalca, T. W. Hansen, J. B. Wagner *et al.*, In situ light spectroscopy in the environmental transmission electron microscope (ETEM). *Microsc. Microanal.*, **18** (2012), 1184–1185.
62. L. Zhang, B. K. Miller and P. A. Crozier, Atomic level observation of surface amorphization in anatase nanocrystals during light irradiation in water vapor. *Nano Lett.*, **13** (2013), 679–684.
63. J. E. Evans, K. L. Jungjohann, N. D. Browning and I. Arslan, Controlled growth of nanoparticles from solution with in situ liquid transmission electron microscopy. *Nano Lett.*, **11** (2011), 2809–2813.
64. E. Jensen, Engineering electrochemical setups for electron microscopy of liquid processes. Ph.D. Thesis, Denmark Technical University (2012).

3 Imaging Liquid Processes Using Open Cells in the TEM, SEM, and Beyond

Chongmin Wang

3.1 Introduction

With the development of liquid stages, imaging liquids in the electron microscope has become a routine technique, and the field of *in situ* electron microscopy of liquids has already evolved to the point that quantitative information can be obtained. Two general methods have been developed for imaging processes under a liquid environment in TEM (or STEM) and SEM. One method, described in Chapter 2, is to enclose the liquid within a membrane-based closed cell. The other method, described here, is the open cell concept, which makes use of low vapor pressure liquids. In particular, solvents based on ionic liquids have a low enough vapor pressure that they can be loaded directly into a vacuum chamber. Imaging can be carried out as reactions take place within a liquid, or as changes occur in solid materials adjacent to the liquid. The absence of the membranes needed for closed cell microscopy can provide improved resolution for imaging and spectroscopy in open cell experiments.

A versatile range of imaging modes is possible using open cells. In the TEM or STEM, processes taking place within the liquid are examined by transmitting the electron beam through either a small droplet of the liquid, a thin layer of liquid on a thin solid substrate, or a free-standing liquid layer supported, for example, on a holey grid. Alternatively, the process of interest may take place in a solid material that is adjacent to the liquid. An exciting class of such experiments involves ion transport reactions relevant to battery operation, such as Li-ion transport. These reactions can be made to take place in a solid structure, such as a nanowire, that has one end penetrating into a droplet that acts as a source of Li or other ions. Since the reaction of interest is in the solid material while the liquid is not imaged, the liquid does not need to be electron transparent and a large droplet can be used. These experiments provide a compelling view of the reactions taking place during battery cycling, and have driven a broad interest in open cell TEM for various electrochemical reactions.

In the SEM, open cell experiments can be used to examine the structure or dynamics of low vapor pressure liquid droplets on a bulk substrate using backscattered electrons. It is also possible, as in TEM, to image ionic liquids as free-standing thin layers supported on a holey grid. In this case the images are acquired using a STEM detector. STEM in the SEM can produce higher resolution images of structures within liquids, compared to standard SEM imaging below the surface of thick liquid droplets.

These open cell experiments involving ionic liquids can be carried out using standard S/TEM or SEM instruments; only the sample holder and the sample preparation procedure need to be adapted. Thus, open cell experiments on ionic liquids can provide unique information with relatively little investment in non-standard equipment.

Applying the open cell concept to other types of liquids, particularly water, is naturally of great interest too. Water can be imaged in spite of its higher vapor pressure if a dedicated environmental SEM (ESEM) or TEM (ETEM) with cooling stage is available, so that a partial pressure of water vapor can be maintained in the microscope to stabilize water droplets [1]. Under these circumstances, the geometries described above for ionic liquids also work for water: droplets on a bulk substrate in ESEM, thin layers or small droplets in ETEM, or freestanding thin films of water supported on a grid and imaged in ETEM or ESEM with transmitted electrons. Although the focus of this chapter is on open cells involving ionic liquids, we briefly touch upon related experiments using water in environmental electron microscopes. The reader is referred to Chapter 16 for one example of particular importance, the imaging of thin biological materials in a hydrated environment using STEM in the ESEM.

We first discuss basic concepts for open cell imaging, and show examples of structures and reactions in thin liquid films. We also discuss a related use of ionic liquids as a conformal coating layer. We then describe more complex applications of the open cell concept, where a battery-relevant reaction takes place in a solid material that is in contact with a liquid droplet. These experiments require an electrical bias and therefore a somewhat more complex sample geometry, but provide a remarkable view of some important liquid-mediated reactions. We conclude with a discussion of the use of open cell SEM for battery reactions. We foresee exciting opportunities for open cell TEM and SEM in studying structure, dynamics, and reactions in other liquid systems, and in correlative experiments where a range of other signals are collected from the same sample.

3.2 Fundamental Concepts for Open Cell Experiments in S/TEM

Implementation of open cell experiments for studies of liquid phase reactions relies on a special category of solvent: room-temperature ionic liquids. These make up a new class of green solvents composed of bulky organic cations and smaller anions, and they exist in the liquid state at room temperature [2]. Ionic liquids have unique properties compared with other types of solvent: typically a high chemical and thermal stability, high viscosity, high ionic conductivity, wide voltage windows for electrochemical processes, and negligible vapor pressure. Their vapor pressure and thermal stability make ionic liquids suitable for vacuum compatibility. Ionic liquids can be loaded directly into a high vacuum chamber for *in situ* observation of liquid droplets and liquid-based processes.

In SEM, droplets of ionic liquid on a surface may be imaged directly [4]. But for TEM and for imaging in transmission mode in the SEM (STEM in the SEM), the liquid thickness must be small enough for transmission of the electron beam. This can be

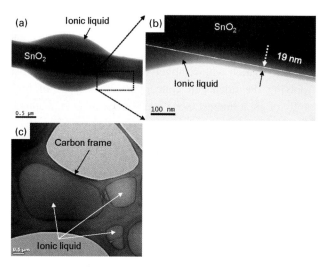

Figure 3.1. TEM images showing: (a) ionic liquid droplet formed on a tin oxide (SnO$_2$) nanowire; (b) wetting of an ionic liquid on SnO$_2$ to form a surface layer of ~19 nm; and (c) stretching of an ionic liquid film over a hole on a lacey carbon TEM grid. The ionic liquid illustrated is 10% LiTFSI in P$_{14}$TFSI. (a, b) Reproduced with permission from Ref. 3. © 2010 Materials Research Society.

achieved in different ways. Taking advantage of the wetting characteristics of ionic liquids with nanofibers, Figure 3.1a illustrates formation of a liquid bead on a nanowire in TEM. The liquid shown here is an electrolyte that contains 10% of lithium bis(trifluoromethane sulfonyl)imide (LiTFSI) dissolved in a hydrophobic ionic liquid, 1-butyl-1-methylpyrrolidium (P$_{14}$) TFSI (P$_{14}$TFSI). This type of material is useful for studies of battery-relevant reactions, as discussed below. In Figure 3.1b, the end of the liquid bead and the wetting layer on the surface of the nanofiber are thin enough to allow electron beam transmission. Figure 3.1c illustrates a different approach for forming a thin liquid layer across a small hole created by the carbon frame of a typical lacey carbon TEM grid. This liquid geometry is useful for both TEM and for STEM in the SEM. It enables direct observation of droplet dynamics and nucleation and growth phenomena in ionic liquid solvents, as well as in water under certain circumstances, as discussed below. The liquid thickness depends on the size and shape of the hole as well as the viscosity of the ionic liquid. Compared to closed cell experiments, in which the liquid thickness is determined by the spacer thickness coupled with the bulging of the membranes, these open cell experiments can be carried out with very thin liquid films and consequently higher spatial resolution. A final useful geometry for the liquid, discussed in Section 3.4, involves a droplet that is too thick for imaging, but that is adjacent to a solid material such as a nanowire where a reaction of interest takes place under controlled stimulus.

We mention here another use of the open cell/ionic liquid combination, which is the passivation of samples that are sensitive to high vacuum. Soaking such samples in an ionic liquid forms a thin conformal coating layer on the surface. It has been found that

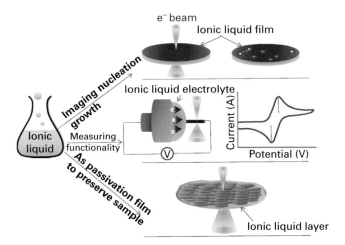

Figure 3.2. Ionic liquids used for three different purposes in the electron microscope: imaging processes in thin films or droplets, measuring functionality, and as a passivation film to preserve vacuum-sensitive samples.

this layer can prevent dehydration of samples in vacuum. Further, the good ionic conductivity of the ionic liquid reduces charging effects for insulating samples, such as biological materials and ceramics.

Figure 3.2 illustrates schematically the applications of ionic liquids for TEM and SEM based on the open cell concept.

3.3 Open Cells for Imaging Droplets, Crystal Growth, Particle Motion and Surface Passivation

The first open cell experiments we describe are relatively straightforward: an ionic liquid is placed in the microscope on a conventional sample holder. The liquid is used for dynamic studies such as electron beam-induced particle growth (Section 3.3.1) or for coating a sample (Section 3.3.3). These examples illustrate some of the phenomena and structures that can be examined with the help of ionic liquids and simple open cell samples. With the use of ESEM, analogous observations can be carried out in thin layers of water (Section 3.3.2). More complex experiments involving application of an electrical bias are described in Section 3.4.

3.3.1 Nanoparticle Nucleation and Growth in Ionic Liquid Films

Nanocrystals are fundamental to modern science and technology, and receive great attention because of their unique optical, electrical, mechanical, and chemical properties [5–7]. Since these properties are closely related to their size, shape, and structure, synthesis techniques that enable control over size and shape have been extensively studied, especially in solution. It is possible to make nanocrystals with a variety of shapes, a narrow size distribution, and high crystal quality [8–10]. The surface structure can also be controlled by adjusting growth parameters such as the concentration of reactants, presence of surfactants, temperature, or environment [11,12]. Experiments that probe the formation and morphology evolution of colloidal nanocrystals are essential for developing ways to synthesize materials with desired properties. Such experiments can be carried out in the TEM or SEM, since in either case the electron beam can reduce metal ions to solid metal via radiolysis (Chapter 7) and thereby create colloidal nanocrystals whose growth can be imaged directly. Closed cell experiments on colloidal growth in TEM are described in Chapter 9, while larger structures formed by electron beam-induced deposition in closed cell SEM are discussed in Chapter 14. These experiments have provided a detailed view of growth pathways, particularly in aqueous solutions.

Ionic liquid-based solvents provide some advantages over aqueous solutions for nanoparticle growth since their electrostatic and steric properties allow for the stabilization of nanocrystals without additional surfactants or capping ligands [2]. As with aqueous solutions, the electron beam can induce metal nanoparticle formation by reducing metal ions in the solution. For example, Ag nanoparticles nucleate when the beam irradiates the ionic liquid solution $Ag(CH_3CH_2CH_2NH_2)_2N(SO_2CF_3)_2$, and reduces Ag ions to solid Ag. Using open cell microscopy for imaging particle growth allows observations without the presence of surrounding membranes, as in a closed cell. The liquid is drop cast onto a 200 mesh lacey carbon TEM grid (Figure 3.3). As expected

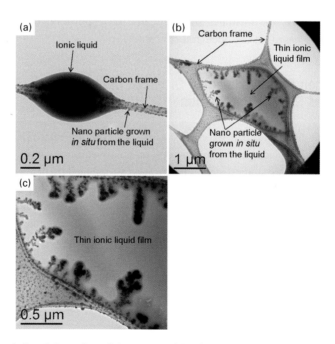

Figure 3.3. Electron beam-induced formation of Ag nanoparticles from an ionic liquid solution: (a) Ag particle growth from a liquid film on a carbon frame, (b) Ag particle growth from a free-standing liquid film. Note the heterogeneous nucleation from the carbon frame. (c) Contrast differences within the liquid around the particle, which may indicate a chemical composition or thickness difference. Chongmin Wang, unpublished data.

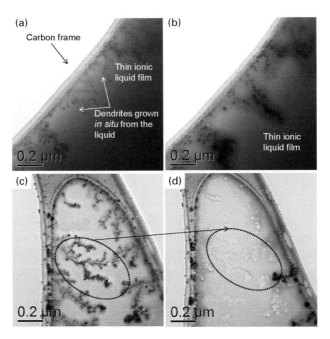

Figure 3.4. (a, b) Image pair showing the coarsening of Ag nanoparticles grown *in situ* under TEM irradiation. (c, d) Image pair showing Ag dissolution at higher dose in the ionic liquid. Chongmin Wang, unpublished data.

from Figure 3.1, two types of thin liquid film form: beads with a wetting layer on the carbon frame, Figure 3.3a, and a freestanding liquid film stretched across a hole, Figure 3.3b, c. The experiments show that Ag preferentially nucleates at the carbon frame and forms relatively large structures in the freestanding liquid film, Figure 3.3b, c, or smaller particles in the wetting layer, Figure 3.3a. The dendritic growth pattern in the freestanding film is further illustrated in Figure 3.4, which also shows the effects of electron dose. Continued dosing leads to coarsening, Figure 3.4a, b, or even dissolution, Figure 3.4c, d. This example hints at the opportunities open cell TEM can offer for convenient observation of nucleation and growth in ionic liquid-based solutions, promising a detailed understanding of single nanoparticle growth kinetics and other reactions in ionic liquids.

Although the thin nature of the freestanding liquid film and the absence of membranes are beneficial for the spatial resolution, there exist challenges associated with this type of open cell experiment. High electron doses can polymerize ionic liquids into solids, so controlling the dose is essential. For this particular beam-stimulated growth process, it is important to calibrate the dose and also, since nucleation and growth are fast, use a detector with high frame rate. Finally, the thin nature of the film can lead to depletion effects as solute atoms are removed from the liquid. Modeling the diffusion process that resupplies the ions is an important component of quantitative data interpretation. Of course, this is equally true of closed cell experiments.

3.3.2 Freestanding Films of Water: WetSTEM

The crystal growth in Figures 3.3 and 3.4 was conveniently imaged in vacuum in a thin liquid layer supported on a holey carbon grid. Thin water layers can also be formed on the same type of grid. However, because of their vapor pressure, water films are only stable under certain conditions of temperature and water vapor pressure. Environmental SEM is therefore necessary, and the resulting imaging mode is known as "WetSTEM" [13, 14]. A droplet of water containing the material of interest is pipetted onto a holey carbon-coated TEM copper grid. The carbon layer faces down so that the copper grid bars form small wells that retain the water. Once in the ESEM, a careful choice of pressure and temperature (controlled with a Peltier cooler on the sample holder) allows the microscopist to maintain a stable water layer, and evaporate or condense controlled amounts of water to optimize its thickness. Typical conditions may be 2 °C and 5.4 Torr water vapor. Imaging takes place with transmitted electrons using a STEM detector in the SEM. WetSTEM is a versatile technique that has been used to image materials such as carbon nanotubes, metal nanoparticles, polymers, and bacteria in water, soot in hexane, and processes such as aggregation of particles in water [13, 14], and wetting of materials by condensing water [15]. WetSTEM is also useful for imaging structures within biological cells, as discussed in more detail in Chapter 16. In principle, it is also possible to create the same thin water layer in ETEM to examine processes in liquid water. ETEM can be used for ice nucleation or for experiments in a water vapor atmosphere (e.g. [16]), but we are not aware of experiments involving liquid water films; this is a promising area for research.

3.3.3 Vacuum Sensitive Samples in the Open Cell

The unique characteristics of ionic liquid-based solvents lead to an interesting application of ionic liquids in SEM and TEM. It has been discovered that coating certain samples with a thin layer of ionic liquid can have a dual effect of reducing charging (for example in ceramics and biological samples) and preventing the rapid dehydration of biological samples when loaded into a high vacuum chamber for observation [17]. For reducing sample charging, the typical ionic liquid includes, among others [18], 1-butyl-3-methylimidazolium tetrafluoroborate (BMI-BF_4); 1-butyl-3-methylimidazolium hexafluorophosphate (BMI-PF_6); and trimethyl-n-propylammonium bis(trifluoromethanesulfonyl) imide (TMPA-TFSI). Coating can be done conveniently by simply soaking the sample in the ionic liquid, causing a thin layer of liquid to cover the sample surface. As illustrated in Figure 3.1, this surface coating layer may be ~20 nm thick, which is thin enough to allow even low energy electrons to escape, in the case of SEM imaging. The layer is, however, thicker than the sputter-coated carbon films (several nanometers thick) also used for SEM imaging. It has been reported that even at voltages as low as 1 kV, a clear image still can be captured through the ionic liquid, suggesting that even this low an energy is sufficient for injecting electrons into the ionic liquid [4].

3.4 Open Cells for *In Situ* Battery Reactions

We now consider a class of experiments that is unique to open cell TEM. By passing current through a sample consisting of electrodes plus a Li-containing ionic liquid, reactions can be imaged directly in the electrodes during transport of Li ions. This provides a remarkable view of the processes taking place during charge and discharge of batteries, complementing *ex situ* and electrical characterization techniques, and has stimulated immense interest in open cell electron microscopy.

These experiments are motivated by the indispensable nature of energy storage technologies such as Li-ion batteries for portable electronics, electric vehicles, and renewable energy technologies [19]. Improving energy storage technology critically depends on a fundamental understanding of the behavior of materials during the cyclic charge and discharge of the storage device. One of the fundamental challenges in battery research is therefore the direct observation of the structural and chemical evolution of the components and how this evolution is correlated with the battery performance. *Ex situ* or *post-mortem* studies of structure and chemistry provide key insights into the evolution of battery materials, but lack dynamic information. The liquid electrolyte used for most commercial rechargeable batteries, including Li-ion batteries, is incompatible with the TEM vacuum, so a fundamental challenge is the operation of the liquid electrolyte plus electrode system *in situ*. Inspired by the progress of *in situ* TEM techniques for imaging in a liquid environment, as outlined elsewhere in this book, an objective became the design and fabrication of a miniature Li-ion battery to allow *in situ* TEM observation during operation. Three strategies have been explored to make such observations viable. (1) Liquid electrolyte-based batteries may be sealed within a closed cell to prevent electrolyte evaporation, as described in Chapter 11. (2) "Nanobatteries" in which a solid state electrolyte is used may be fabricated by focused ion beam (FIB) cutting [20, 21] and directly placed in the TEM column [22]. (3) Batteries using alternative electrolytes based on low vapor pressure ionic liquids [23, 24] can be placed directly in the TEM column in the open cell configuration [3, 25, 26]. Although the electrolyte is not the same as that used in commercial batteries, these open cell experiments provide key information on materials evolution during operation, as described in the following sections.

3.4.1 Open Cell Design for Battery Reactions with Ionic Liquid-Based Electrolytes

The basic concept of the open cell for battery studies is illustrated in Figure 3.5. The essential components of the cell are a single nanowire, forming the observable electrode, a vacuum-compatible ionic liquid or solid Li_2O as the electrolyte, and $LiCoO_2$ or Li metal as the counter electrode. Because the open cell configuration avoids the use of membranes, it offers the possibility of imaging the battery operation with atomic-level spatial resolution and with the analytical capability needed to study Li ion insertion mechanisms into electrode materials during the charge/discharge cycle.

A typical example is a miniature prototype battery built using $LiCoO_2$ as the cathode, an ionic liquid electrolyte, and a SnO_2 nanowire as the anode. The cathode material is

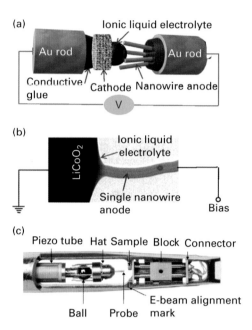

Figure 3.5. (a) Schematic drawing showing the fundamental concept of an open cell for *in situ* operation of a rechargeable battery using an ionic liquid-based electrolyte. (b) TEM image showing a nanobattery with LiCoO$_2$ as cathode, an ionic liquid-based electrolyte, and silicon coated on a carbon nanowire as anode. (c) A photograph showing the details of a nanomanipulator that can be used to implement the nanobattery concept in (a). (a) and (b) Reproduced with permission from Ref. 27. © 2015 Materials Research Society.

actually a mixture of LiCoO$_2$ particles and carbon, a standard choice in coin cell batteries. This mixture is coated as a thin film on an Al foil. The electrolyte is air stable. The salt, LiTFSI (see Figure 3.1), is dissolved to 10% in the hydrophobic ionic liquid P$_{14}$TFSI. This ionic liquid-based electrolyte is first verified to be stable in the column of the TEM at a typical vacuum of 10^{-7} Torr. It shows no significant evaporation after a week in the TEM column, therefore allowing a systematic cyclic charging and discharging of the battery *in situ*. The SnO$_2$ nanowire anode gives the advantage of electron transparency and therefore allows *in situ* observation of both chemical and structural changes during operation. It is attached to a W wire electrode, which connects to the outside circuit through a Cu film. To ensure a good electrical connection between the SnO$_2$ and the W, the contact region is welded by deposition of Pt at the joint using a focused ion beam in a dual beam FIB/SEM. Once the battery is assembled, the ionic liquid tends to spread along the nanowire surface to form a thin coating layer. This is helpful in allowing this configuration to mimic in some degree the real battery configuration where the liquid electrolyte completely surrounds the active component in the electrode. This point will be discussed in more detail below. The open cell configuration just described is greatly enhanced by using a sample holder that includes a nanomanipulator, making it easier to study single nanowire anodes. Since its initial development, this type of setup has been widely used for studying lithiation behavior in a range of materials.

3.4.2 Open Cell Design for Battery Reactions Using Native Metal Oxide Electrolytes

A modification of the open cell plus ionic liquid electrolyte experiments described above leads to a different type of open cell for *in situ* TEM study of lithiation. In this configuration, all components are solid [28, 29]. As an example, suppose Li metal is used as the anode. The nanobattery must be assembled in an Ar-filled glove box and transferred to the TEM goniometer in a glove bag. During insertion of the sample holder into the TEM column, the Li metal surface is instantaneously oxidized to form a surface layer of Li_2O. This layer of Li_2O then serves as a solid electrolyte. In a typical experiment (Figure 3.6), a Si nanowire forms one electrode, Li_2O on Li is a solid electrolyte, and bulk Li metal is the counter electrode. In principle, the battery assembled in this way is a charged battery and completing the circuit will lead to the discharging process, which is alloying of Li with Si. However, experimentally this reaction does not take place due to the low conductivity of Li ions in Li_2O at room temperature [30]. Instead, a negative potential of typically 2–4 V has to be applied between Si and Li to drive Li into Si. This experiment therefore does not represent a real rechargeable battery, but still provides insights into the structural response of Si on Li-ion insertion. More generally, this type of open cell configuration provides a quick and convenient way for probing the intrinsic response of a given material to Li-ion insertion or extraction. A similar principle can be used for other metallic ions to examine ionic insertion and extraction behavior. Examples include Na- and Ca-ion batteries using SnO_2 cathodes [31].

3.4.3 Ion Insertion and Transport in Electrodes During Battery Cycling

Battery operation relies critically on electron and ionic transfer across solid–solid and solid–liquid interfaces and within each of the constituent phases. Repeated charging and

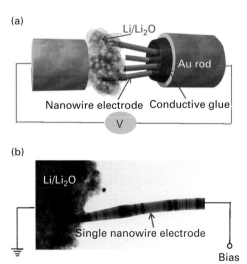

Figure 3.6. (a) Schematic drawing showing the fundamental concept of an open cell for *in situ* TEM study of a charged battery using Li metal as the anode and Li_2O as the electrolyte. (b) TEM image showing the nanobattery with a Si nanowire as a cathode and Li_2O as the electrolyte. Reproduced with permission from Ref. 27. © 2015 Materials Research Society.

discharging induces ionic migration which changes the microstructure both at the electrolyte–electrode interface and within the electrode (the active materials). Although it has been established that this structural evolution of active materials is responsible for battery failure, the mechanisms as a function of charge/discharge cycling are not well understood. Overall, this imposes the fundamental question of how the microstructures evolve and affect the battery's properties. Open cell TEM has helped to reveal many of the details of lithiation mechanisms and structural evolution in a range of materials, especially anode materials including Si [32, 33], Ge [34], Al_2O_3 [35], SnO_2 [26], ZnO [36], graphene [37], and carbon nanotubes [38]. These studies provide a fascinating and often surprising view of ion insertion into host materials and the corresponding structural and chemical evolution.

As a classic example of the use of open cell TEM for ion insertion reactions, Figure 3.7 shows the structural evolution of SnO_2 when Li ions are inserted under electrochemical bias [25]. Tin oxide represents one of the most promising anode

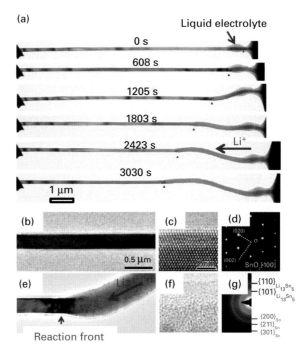

Figure 3.7. Charging behavior of SnO_2 nanowires. (a) Morphology evolution of a SnO_2 nanowire during charging. As the reaction front (marked by red triangles) passes, the originally straight nanowire swells and expands in both radial and axial directions. The reaction fronts are heavily strained. (b–d) Microstructure of a pristine SnO_2 nanowire showing its single crystal nature. (e–g) Microstructure of the same nanowire after lithiation. Isotropic expansion in all directions is clear. A strained region with visible dislocation cloud is present in the reaction front marked in (e). In (g), diffraction confirms coexistence of amorphous Li_xSn, Sn, and Li_2O after lithiation. The predominant features in the diffraction pattern are the two broad bands centered at the positions corresponding to the $\{101\}$ ($d = 3.959$ Å) and $\{110\}$ ($d = 2.349$ Å) planes of hexagonal $Li_{13}Sn_5$, JCPDS# 29-0838. The diffraction rings from the Li_2O phase are usually weak and thus are not marked. Reproduced with permission from Ref. 26. © 2011 American Chemical Society.

materials to replace the carbon-based anodes described above for Li-ion batteries, with a high theoretical capacity of 781 mA h/g and a demonstrated reversible capacity exceeding 500 mA h/g. Intercalation anodes such as graphite experience a small volume change (usually less than 10%) as Li is inserted, but Sn- or Si-based alloying anodes with higher capacity show much larger volume changes. This volume change generates large lithiation-induced stress, up to 10 GPa, with consequent fracture and pulverization of the electrodes. Open cell TEM shows how this volume change takes place. The initially straight and uniform pristine SnO_2 nanowire swells and bends as the reaction front passes (Figure 3.7a). Overall, there is a measured total volume expansion of ∼240% with ∼45% radial and ∼60% axial elongation. The reaction front migrates at ∼0.6 nm/s, forming a defective zone leaving behind it a mixture of amorphous Li_2O, Sn, and Li_xSn phases. Lithiation is expected to proceed in two steps: (1) $4Li^+ + SnO_2 + 4e^- \rightarrow 2Li_2O + Sn$; and (2) $Sn + xLi + xe^- \rightarrow Li_xSn$ ($0 < x < 4.4$), and the phases seen are consistent with this reaction sequence.

Non-lithium metals such as sodium are alternative charge-carrying ions for rechargeable batteries. Since Na and Li have the same +1 charge, it is assumed that what has been learnt about Li-ion batteries can be transferred directly to Na-ion batteries [31]. Open cell TEM allows the insertion and transport of Na ions into SnO_2 to be observed

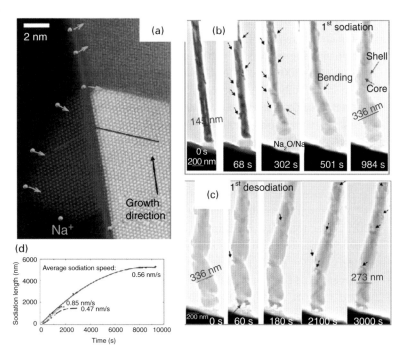

Figure 3.8. (a) STEM Z-contrast image showing the atomic structure of a twinned SnO_2 nanowire (marked by the red lines) with grain boundary structures (arrow). (b, c) TEM images showing the first sodiation and the first desodiation, respectively. The red arrow in the image at 302 s in (b) shows the location of the reaction front, the red arrow in the 60 s image in (c) shows a crack formed at the end of the nanowire during desodiation, and black arrows in (c) mark pores formed during desodiation. (d) Sodiation kinetics for different nanowires with Na_2O electrolyte. Reproduced with permission from Ref. 31. © 2013 American Chemical Society.

(Figure 3.8), and has established that the reaction leads to the formation of amorphous Na_xSn particles covered by a shell of crystalline Na_2O. This suggests that upon Na insertion into SnO_2, a displacement reaction occurs: $(4+x)Na^+ + (4+x)e + SnO_2 \rightarrow Na_xSn(amorphous) + 2Na_2O(crystalline)$. As x reaches the critical value of 3.75, the amorphous Na_xSn transforms to crystalline $Na_{15}Sn_4$. Insertion of Na can chemo-mechanically soften the reaction product to a greater extent than lithiation. Therefore, unlike lithiation of SnO_2, no dislocation plasticity is seen ahead of the sodiation front. The sodiation speed is approximately 1/20 of that of lithiation, in good agreement with the calculated larger diffusion barrier of Na^+ in SnO_2. The reverse reaction, extraction of Na (desodiation), transforms the Na_xSn to Sn nanoparticles. As the volume shrinks, nanopores form and metallic Sn particles appear, confined in hollow shells of Na_2O, mimicking a peapod structure. These pores greatly increase electrical impedance, explaining the poor cyclability of SnO_2. This direct comparison of the results from Na and Li highlights the critical role of ionic size and electronic structure on the charge/discharge rate and failure mechanisms in these batteries.

3.4.4 Structural and Chemical Evolution of Electrode Materials

Silicon has attracted significant attention for use as the negative electrode material in Li-ion batteries [39], in part because it has a specific capacity about ten times that of commercial graphite. In contrast to traditional intercalation electrodes, Si and Li react via an alloying process, which results in both solid state amorphization of Li_xSi and an enormous (~300%) volume expansion due to the uptake of up to 4.4 Li atoms per Si atom in the fully alloyed material. The large volume change upon cycling is a fundamental issue that has prevented the widespread use of Si in batteries. This is because a thin, passivating solid electrolyte interphase (SEI) is vital for long-term cycling, but the expansion and contraction of Si structures promotes unstable growth of the (SEI) surface film. Very thick SEI films grow due to the continual exposure of fresh Si surfaces to the electrolyte [40], resulting in high ionic resistance and poor cycling performance. Fracture also compounds this issue: the SEI grows on the new surfaces of a fractured electrode particle, which could also promote electrical isolation since the SEI is electronically insulating [41]. In short, the volume changes during Li alloying/dealloying are the cause of many problems associated with this system, so it is vital to understand the nature of these processes.

Open cell TEM of the lithiation of crystalline Si nanoparticles provides an unprecedented view of behavior upon Li-ion insertion and extraction (Figure 3.9) including volume expansion, self-limited lithiation, and pulverization for particles larger than a critical dimension. During lithiation, volume expansion of each particle occurs via the growth of the Li_xSi phase at the expense of the crystalline Si. Fast surface diffusion allows for relatively uniform spherical lithiation in each particle, so that a growing Li_xSi shell forms around a shrinking crystalline Si core. The lithiated shells on many of the particles in these images are observed to crystallize into $Li_{15}Si_4$ during lithiation while the Si cores are still present. The contrast between the crystalline core and the lithiated shell allows for precise measurements of lithiation kinetics. There is a strong size

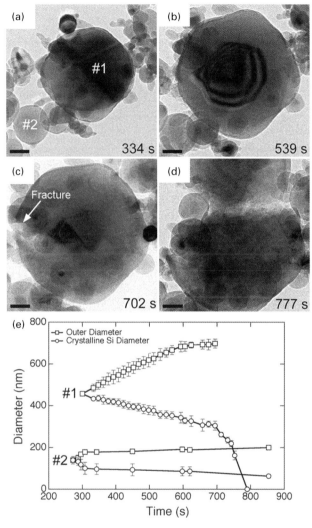

Figure 3.9. Lithiation of a group of pristine Si nanoparticles attached to a lithiated nanowire that acts as the Li source (to the left of the frame). The largest particle has ∼450 nm initial diameter. (a–d) Time series showing lithiation and fracture after applying a bias of −3.8 V: (a) Lithiation has just started to occur, with the Li_xSi phase visible around the lower edge of particle #1. Particle #2 is already partially lithiated. (b) After 3.5 min, the crystalline Si core is clearly visible as the darker, faceted region in the center of particle #1. (c) Fracture initiates at the left surface of particle #1. (d) After fracture, the interior Si region of the particle is quickly lithiated and the crack grows, tearing the particle apart. Scale bars 100 nm. (e) Measured outer and crystalline core diameters for particles #1 and #2 as a function of time. The reaction front does not slow in particle #1, presumably because the a/b ratio never decreases to the point where large enough stresses develop. Fracture occurs after ∼695 s of lithiation, after which the crystalline core of particle #1 disappears rapidly. The reaction front in particle #2 slows, and particle #1 is fully reacted before #2. Reproduced with permission from Ref. 42. © 2012 Wiley-VCH.

dependence of fracture in Figure 3.9, as particles less than ~150 nm in diameter generally do not fracture [43]. Figure 3.9 shows an example of a particle that fractures as Li flows into it through the smaller particles to the left [42]. In a real battery electrode, this type of mechanical failure could possibly result in electrical isolation of Si fragments. Figure 3.9e shows lithiation kinetics. The fast lithiation after fracture seen in the larger particle was also seen in particles of intermediate size in which reaction front slowing first occurred. It is interesting to note that the fracture and subsequent rapid lithiation cause the larger particle to become fully lithiated before the smaller particle in this experiment.

These and other open cell experiments provide a picture of the microstructure evolution of Si nanoparticles upon lithiation via the following five stages [44]: (1) Li ions diffuse along the nanoparticle surface, leading to a high concentration of Li at the Si surface layer. (2) Lithiation initiates at the surface and proceeds towards the core, via propagation of a boundary between a-Li_xSi and crystalline Si. (3) Accompanying the conversion of Si to a-Li_xSi is a significant volume increase, which is accommodated by radial expansion. (4) As the reaction progresses, its rate decreases, and nanoparticles retain a residual core even following prolonged lithiation. Detailed experimental and theoretical analysis has indicated that Si lithiation is an interface reaction-controlled process, which normally yields a constant lithiation rate. However, the stress due to the volume expansion reduces the reaction rate in nanoparticles. (5) Continued lithiation leads to an increase in the Li concentration in the a-Li_xSi, eventually transforming at a critical value, $x = 3.75$, to crystalline $Li_{15}Si_4$ (c-$Li_{15}Si_4$). Open cell observations show that a-$Li_{3.75}Si$ spontaneously and congruently transforms to c-$Li_{15}Si_4$ by a process that is solely controlled by the Li concentration in the a-Li_xSi. This does not involve large-scale atomic migration or phase separation, but does involve local bonding configuration changes. The grain size of the crystalline phase provides a characteristic length scale at which the atomic bonding configuration has changed to nucleate the crystalline phase, as measured in bright field STEM images and diffraction (Figure 3.10).

3.4.5 The Deficiency of the Open Cell: A Comparison of Closed and Open Cell and the Critical Open Cell Issues

It is clear from the above discussion that open cell TEM has provided unique insights on the structural and chemical evolution of electrodes during lithiation/delithiation and related reactions. However, in assessing the relevance of this information to real life battery operation, we need to be aware of three typical deficiencies associated with the open cell configuration as used in these types of electrochemical experiments. First, the electrolyte is only in point contact with the electrode. This may modify the diffusion pathways of Li ions in the electrode, such that the result is not representative of the real battery where the electrode is fully immersed in the liquid electrolyte. Second, in experiments where Li_2O is used as the electrolyte, a large overpotential has to be applied in order to drive the Li ions into the electrode. This may change the kinetics and phase behavior during solid state electrode lithiation. Thirdly,

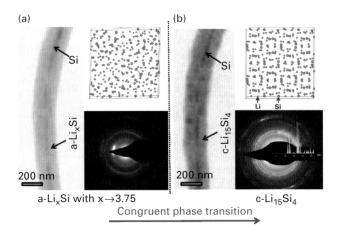

Figure 3.10. STEM bright field image sequence revealing the spontaneous crystallization of a-Li_xSi to c-$Li_{15}Si_4$ when x reaches 3.75. (a) STEM bright field image of a-Li_xSi nanowire with residual silicon core in some sections; atomic arrangement in a-$Li_{3.75}Si$; and diffraction pattern of a-Li_xSi. (b) STEM bright field image showing polycrystalline c-$Li_{15}Si_4$; atomic arrangement in c-$Li_{15}Si_4$; and diffraction pattern of c-$Li_{15}Si_4$. The $Li_{15}Si_4$ is polycrystalline with a typical grain size of several tens of nanometers. Images recorded in an open cell with Li_2O as electrolyte. Reproduced with permission from Ref. 44. © 2013 American Chemical Society.

the experimental restriction to ionic liquid or Li_2O electrolytes precludes study of some of the fundamental processes that only occur in real electrolytes and battery operating conditions, such as the interaction between electrolyte and electrode and the SEI layer formation. To provide data that can be compared with open cell results, recent work is focused on developing *in situ* or more precisely *operando* TEM studies of Li-ion batteries using a battery-relevant liquid electrolyte and a Li metal counter electrode [45]. These experiments require closed cells, and are discussed in more detail in Chapter 11.

To gain insight into the effect of the geometry of the open cell, it is worthwhile to compare the same reaction in closed and open liquid cells [46]. Lithiation of a single crystalline Si nanowire is illustrated in Figure 3.11. In the closed cell, Figure 3.11a, b, the nanowire is fully immersed in the liquid electrolyte. The insertion of Li ions into Si occurs from all possible directions and proceeds in a core-shell mode with a uniform shell thickness along the whole nanowire. The closed cell observations show the response of the whole nanowire to Li insertion. In the open cell, Figure 3.11c–e, the Li ion source is only in contact with the end of the nanowire, resulting in a sequential lithiation process. Despite the point contact, it is apparent that the Si nanowire is lithiated in a core-shell fashion, but the lithiation front shows tapering features. The fact that lithiation is still a core-shell process in the open cell is related to the fast diffusion of Li on the nanowire surface, replicating to some extent the immersion of the nanowire in a Li ion source in the closed cell. The open cell has an advantage in terms of imaging spatial resolution and spectroscopy capability compared to the closed cell, as illustrated in Figure 3.11f, allowing the open cell experiments to reveal formation of the Li_xSi phase and distribution of the inserted Li ions.

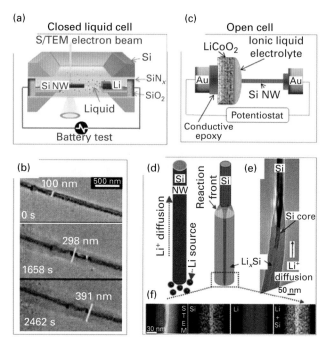

Figure 3.11. Comparison of closed and open cells for battery studies. (a) Closed cell with two electrodes, on each of which is attached one component of a battery. The cathode is a Si nanowire (NW) and the anode is metallic Li. (b) Series of images during the lithiation of a crystalline Si nanowire (in this case coated with Cu to increase conductivity), performed by holding the Cu–Si voltage at ~0.03 V. (c) Open cell with an ionic liquid electrolyte. (d) Schematics and (e) an image recorded at a constant voltage of −2 V versus Li metal during lithiation of an (uncoated) Si nanowire showing the sequential nature of the process. (f) STEM image and electron energy loss spectroscopy map of element distribution of the process depicted in (d) and (e). (a, b, d–f) Reproduced with permission from Ref. 45. © 2013 American Chemical Society; (c) Reproduced with permission from Ref. 46. © 2015 Materials Research Society.

This comparison therefore indicates a consistent overall reaction, but differences in the detailed structural evolution related to diffusion pathways. Both experiments are worthwhile: the closed cell for the global dynamics of the whole structure, plus the SEI formation; and the open cell for exploring the details of phase sequences at high spatial and spectroscopic resolution. Both are essential for understanding battery performance.

We finally note that the examples given here are mostly based on anode materials, typified by large volume changes. In principle, both open and closed cells are applicable for cathode materials, and broadly for other types of ionic transport measurement.

3.5 Extension of the Open Cell Concept to Other Imaging and Spectroscopic Techniques

The low vapor pressure of ionic liquids implies that the open cell concept can essentially be transplanted to any analytical instrument that requires high vacuum. This includes SEM

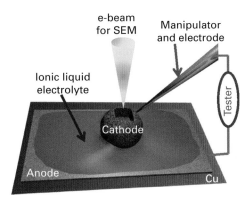

Figure 3.12. Schematic drawing showing the concept of open cell SEM. A nanomanipulator probe was used to immerse or partially immerse an attached cathode oxide particle into a liquid droplet covering an anode material. Reproduced with permission from Ref. 27. © 2015 Materials Research Society.

and XPS [47]. As an example, based on the open cell concept using an ionic liquid electrolyte, single Li-ion battery cathode particles can be imaged during electrochemical cycling in SEM [48]. The experimental setup is shown in Figure 3.12. The electrode materials are partially dipped into the ionic liquid electrolyte. To see the internal structure of the electrode during operation of the battery, a cross section is cut using FIB. The experiment provides information on large-scale deformation of the electrode, here Li$(Ni_{0.8}Co_{0.15}Al_{0.05})O_2$-based cathode material. It shows that significant voids develop between grains even during the first charge (oxide delithiation), and the electrolyte penetrates through the crack network into the particle interior. Comparing these results to the microstructure seen in oxide particles subjected to cycling confirms the occurrence of these processes in practical cells and suggests that the physical separation and isolation of grains may contribute to performance degradation in Li-ion cells. This SEM open cell has the advantage of providing a large-scale overview at the electrode level, even though (as for the open cell TEM experiments described above) some processes, such as SEI formation, are different in the ionic liquid compared to the electrolytes used in real batteries.

A similar open cell concept in ESEM can be used for all-solid-state batteries. For a lithium–oxygen battery [47], a nanowire-geometry electrode (a carbon nanotube) makes contact with a lithium oxide layer formed on Li metal. Such experiments show the structural evolution of discharge products on cycling the battery, and show that the growth and decomposition of Li_2O_2 are related to the uneven distribution of ionic and electronic conductivities. This information is difficult to capture by other methods and presents new insights into understanding the reaction mechanism of the all-solid-state lithium–oxygen battery.

3.6 Perspective

We have outlined the fundamental concepts for probing processes in liquids based on the open cell concept. Essential to open cell operation is the use of a low vapor pressure ionic liquid as a solvent. Chemical and physical processes that involve such liquids can be imaged directly, without the presence of membranes as in the closed cell, hence with potentially higher achievable resolution for imaging and spectroscopy. The open cell concept has created a new pathway for analysis of complex processes occurring within these liquid environments, potentially fostering a paradigm shift in materials characterization by probing the real-time transformations of active materials, nucleation and growth processes, and side reactions such as the electrochemical degradation of electrolytes. To obtain quantitative information in any electron microscopy experiment, the electron dose effects must be understood. Minimizing such effects requires adjusting the accelerating voltage, reducing the dose, or a combination of both, as described in Chapter 7.

The application of open cell TEM has so far been most notable for probing ionic transport in anode materials, which is normally associated with large volume and structural changes. In cathode materials, ion insertion and extraction produce a less dramatic structural change. Open cell experiments may thus be more limited for cathode applications until the technique is used with higher performance microscopes, proper orientation of the crystal (i.e. double tilt sample holders), extreme stability of the system under imaging conditions, and ultimately detection of mobile ions such as Li, Mg, Na, and Ca. This requires temporal resolution of faster than the standard 30 frames per second, implying the use of modern detectors. Tremendous scope exists in this field of research. Open cell SEM promises complementary information on ionic liquid-based processes, as compared to TEM. Device-level imaging and analytical microscopy are possible during electrochemical cycling of "bulk" materials. For samples that can be examined in thin layer form using STEM detection, higher resolution information is available. The scope of open cell SEM for ionic liquid experiments appears just as broad as that of open cell TEM. And finally, the open cell techniques described for ionic liquids can be adapted for water or other high vapor pressure liquids by carrying out the experiments in ESEM or ETEM. For all the variants of open cell microscopy, exciting opportunities are certain to arise from the combination of information at different length scales from TEM and SEM and from *in situ* or *operando* XPS, light microscopy, or other experiments.

Acknowledgements

The author would like to acknowledge the support of the Assistant Secretary for Energy Efficiency and Renewable Energy, Office of Vehicle Technologies of the U.S. Department of Energy under Contract No. DE-AC02-05CH11231, Subcontract No. 6951379 under the Battery Materials Research (BMR) Program. The development of the electrochemical liquid cell is supported by the Chemical Imaging Initiative at Pacific Northwest National Laboratory (PNNL). The work was conducted in the William R. Wiley Environmental Molecular Sciences Laboratory, a national scientific user facility sponsored by DOE's Office of Biological and

Environmental Research and located at PNNL. PNNL is operated by Battelle for the Department of Energy under Contract DE-AC05-76RLO1830.

References

1. D. J. Stokes, *Principles and Practice of Variable Pressure/Environmental Scanning Electron Microscopy (VP-ESEM)* (Chichester: John Wiley & Sons, 2008).
2. C. Janiak, Ionic liquids for the synthesis and stabilization of metal nanoparticles. *Z. Naturforsch.*, **68B** (2013), 1059–1089.
3. C. M. Wang, W. Xu, J. Liu, *et al.*, In situ transmission electron microscopy and spectroscopy studies of interfaces in Li ion batteries: challenges and opportunities. *J. Mater. Res.*, **25** (2010), 1541–1547.
4. S. Kuwabata, A. Kongkanand, D. Oyamatsu and T. Torimoto, Observation of ionic liquid by scanning electron microscope. *Chem. Lett.*, **35** (2006), 600–603.
5. Y. Xia, Y. J. Xiong, B. Lim and S. E. Skrabalak, Shape-controlled synthesis of metal nanocrystals: simple chemistry meets complex physics? *Angew. Chem. Int. Ed.*, **48** (2009), 60–103.
6. M. Law, J. Goldberger and P. D. Yang, Semiconductor nanowires and nanotubes. *Annu. Rev. Mater. Res.*, **34** (2004), 83–122.
7. V. I. Klimov, A. A. Mikhailovsky, S. Xu *et al.*, Optical gain and stimulated emission in nanocrystal quantum dots. *Science*, **290** (2000), 314–317.
8. T. S. Ahmadi, Z. L. Wang, T. C. Green, A. Henglein and M. A. ElSayed, Shape-controlled synthesis of colloidal platinum nanoparticles. *Science*, **272** (1996), 1924–1926.
9. L. J. Lauhon, M. S. Gudiksen, C. L. Wang and C. M. Lieber, Epitaxial core-shell and core-multishell nanowire heterostructures. *Nature*, **420** (2002), 57–61.
10. Y. Yin and A. P. Alivisatos, Colloidal nanocrystal synthesis and the organic-inorganic interface. *Nature*, **437** (2005), 664–670.
11. N. Tian, Z. Y. Zhou, S. G. Sun, Y. Ding and Z. L. Wang, Synthesis of tetrahexahedral platinum nanocrystals with high-index facets and high electro-oxidation activity. *Science*, **316** (2007), 732–735.
12. H. G. Liao, Y. X. Jiang, Z. Y. Zhou, S. P. Chen and S. G. Sun, Shape-controlled synthesis of gold nanoparticles in deep eutectic solvents for studies of structure-functionality relationships in electrocatalysis. *Angew. Chem. Int. Ed.*, **47** (2008), 9100–9103.
13. A. Bogner, G. Thollet, D. Basset, P. H. Jouneau and C. Gauthier, Wet STEM: a new development in environmental SEM for imaging nano-objects included in a liquid phase. *Ultramicroscopy*, **104** (2005), 290–301.
14. A. Bogner, P. H. Jouneau, G. Thollet, D. Basset and C. Gauthier. A history of scanning electron microscopy developments: towards "Wet-STEM" imaging. *Micron*, **38** (2007), 390–401.
15. Z. Barkay, Wettability study using transmitted electrons in environmental scanning electron microscope. *Appl. Phys. Lett.*, **96** (2010), 183109.
16. K. Yoshida, A.N. Bright, M.R. Ward *et al.*, Dynamic wet-ETEM observation of Pt/C electrode catalysts in a moisturized cathode atmosphere. *Nanotechnology*, **25** (2014), 425702.

17. M. Sakaue, M. Shiono, M. Konomi et al., New preparation method using ionic liquid for fast and reliable SEM observation of biological specimens. *Microsc. Microanal.*, **20 (Suppl. 3)** (2014), 1012–1013.
18. N. Brodusch, H. Demers and R. Gauvin, Ionic liquid used for charge compensation for high resolution imaging and analysis in the FE-SEM. *Microsc. Microanal.*, **20 (Suppl 3)** (2014), 38–39.
19. J. M. Tarascon and M. Armand, Issues and challenges facing rechargeable lithium batteries. *Nature*, **414** (2001), 359–367.
20. K. Yamamoto, Y. Iriyama, T. Asaka et al., Dynamic visualization of the electric potential in an all-solid-state rechargeable lithium battery. *Angew. Chem. Int. Ed.*, **49** (2010), 4414–4417.
21. A. Brazier, L. Dupont, L. Dantras-Laffont et al., First cross-section observation of an all solid-state lithium-ion "nanobattery" by transmission electron microscopy. *Chem.Mater.*, **20** (2008), 2352–2359.
22. C. M. Wang, W. Xu, J. Liu et al., In situ transmission electron microscopy observation of microstructure and phase evolution in a SnO_2 nanowire during lithium intercalation. *Nano Lett.*, **11** (2011), 1874–1880.
23. S. F. Lux, M. Schmuck, B. Rupp et al., Mixtures of ionic liquids in combination with graphite electrodes: the role of Li-salt. *ECS Trans.*, **16** (2009), 45–49.
24. A. Lewandowski and A. Świderska-Mocek, Properties of the graphite-lithium anode in N-methyl-N-propylpiperidinium bis(trifluoromethanesulfonyl)imide as an electrolyte. *J. Power Sources*, **171** (2007), 938–943.
25. J. Y. Huang, L. Zhong, C. M. Wang et al., In situ observation of the electrochemical lithiation of a single SnO_2 nanowire electrode. *Science*, **330** (2010), 1515–1520.
26. L. Q. Zhang, X. H. Liu, Y. Liu et al., Controlling the lithiation-induced strain and charging rate in nanowire electrodes by coating. *ACS Nano*, **5** (2011), 4800–4809.
27. C. M. Wang, In situ transmission electron microscopy and spectroscopy studies of rechargeable batteries under dynamic operating conditions: a retrospective and perspective view. *J. Mater. Res.*, **30** (2015), 326–339.
28. X. H. Liu, H. Zheng, L. Zhong et al., Anisotropic swelling and fracture of silicon nanowires during lithiation. *Nano Lett.*, **11** (2011), 3312–3318.
29. F. Wang, H.-C. Yu, M.-H. Chen et al., Tracking lithium transport and electrochemical reactions in nanoparticles. *Nat. Commun.*, **3** (2012), 1201.
30. M. M. Islam, and T. Bredow, Density functional theory study for the stability and ionic conductivity of Li_2O surfaces. *J. Phys. Chem. C*, **113** (2009), 672–676.
31. M. Gu, A. Kushima, Y. Shao et al., Probing the failure mechanism of SnO_2 nanowires for sodium-ion batteries. *Nano Lett.*, **13** (2013), 5203–5211.
32. X. H. Liu, L. Q. Zhang, L. Zhong et al., Ultrafast electrochemical lithiation of individual Si nanowire anodes. *Nano Lett.*, **11** (2011), 2251–2258.
33. C.-M. Wang, X. Li, Z. Wang et al., In situ TEM investigation of congruent phase transition and structural evolution of nanostructured silicon/carbon anode for lithium ion batteries. *Nano Lett.*, **12** (2012), 1624–1632.
34. X. H. Liu, S. Huang, S. T. Picraux et al., Reversible nanopore formation in Ge nanowires during lithiation–delithiation cycling: an in situ transmission electron microscopy study. *Nano Lett.*, **11** (2011), 3991–3997.
35. Y. Liu, N. S. Hudak, D. L. Huber et al., In situ transmission electron microscopy observation of pulverization of aluminum nanowires and evolution of the thin surface Al_2O_3 layers during lithiation–delithiation cycles. *Nano Lett.*, **11** (2011), 4188–4194.

36. A. Kushima, X. H. Liu, G. Zhu et al., Leapfrog cracking and nanoamorphization of ZnO nanowires during in situ electrochemical lithiation. *Nano Lett.*, **11** (2011), 4535–4541.
37. X. H. Liu, J. W. Wang, Y. Liu et al., In situ transmission electron microscopy of electrochemical lithiation, delithiation and deformation of individual graphene nanoribbons. *Carbon*, **50** (2012), 3836–3844.
38. Y. Liu, H. Zheng, X. H. Liu et al., Lithiation-induced embrittlement of multiwalled carbon nanotubes. *ACS Nano*, **5** (2011), 7245–7253.
39. M. S. Whittingham, Materials challenges facing electrical energy storage. *MRS Bull.*, **33** (2008), 411–419.
40. H. Wu, G. Chan, J. W. Choi et al., Stable cycling of double-walled silicon nanotube battery anodes through solid-electrolyte interphase control. *Nat. Nanotechnol.*, **7** (2012), 310–315.
41. J. Christensen and J. Newman, Stress generation and fracture in lithium insertion materials. *J. Solid State Electrochem.*, **10** (2006), 293–319.
42. M. T. McDowell, I. Ryu, S. W. Lee et al., Studying the kinetics of crystalline silicon nanoparticle lithiation with in situ transmission electron microscopy. *Adv. Mater.*, **24** (2012), 6034–6041.
43. X. H. Liu, L. Zhong, S. Huang et al., Size dependent fracture of silicon nanoparticles during lithiation. *ACS Nano*, **6** (2012), 1522–1531.
44. M. Gu, Z. G. Wang, J. G. Connell et al., Electronic origin for the phase transition from amorphous Li_xSi to crystalline $Li_{15}Si_4$. *ACS Nano*, **7** (2013), 6303–6309.
45. M. Gu, L. R. Parent, B. L. Mehdi et al., Demonstration of an electrochemical liquid cell for operando transmission electron microscopy observation of the lithiation/delithiation behavior of Si nanowire battery anodes. *Nano Lett.*, **13** (2013), 6106–6112.
46. C. M. Wang, H. G. Liao, F. M. Ross, Observation of materials processes in liquids by electron microscopy. *MRS Bull.*, **40** (2015), 46–52.
47. H. Zheng, D. D. Xiao, X. Li et al., New insight in understanding oxygen reduction and evolution in solid-state lithium–oxygen batteries using an in situ environmental scanning electron microscope. *Nano Lett.*, **14** (2014), 4245–4249.
48. D. J. Miller, C. Proff, J. G. Wen, D. P. Abraham, J. Bareño, Observation of microstructural evolution in Li battery cathode oxide particles by in situ electron microscopy. *Adv. Energy Mater.*, **3** (2013), 1098–1103.

4 Membrane-Based Environmental Cells for SEM in Liquids

Andrei Kolmakov

4.1 Introduction

4.1.1 Motivation and Terminology

Environmental scanning electron microscopy (ESEM) and SEM in liquids are among the most active research areas in modern electron microscopy and spectroscopy. Research interest in these techniques is broad as they enable nanoscale research into dynamic systems in fields as diverse as materials, biological, medical, and environmental sciences.

Complementing conventional high vacuum SEM (Figure 4.1a), three schemes have been demonstrated for SEM in liquids and gases. The first approach is based on SEM microscopes that employ differentially pumped environmental cells (also often called "open cells"). In such systems, high vapor pressure or fluidic regions of interest are separated from the high vacuum of the microscope and detectors by multi-stage differential pumping through small apertures (Figure 4.1b). Placement of the pressure-limiting apertures into the column of the microscope allows for high pressure inside the entire SEM sample chamber. Modern commercial environmental SEMs employ this methodology, enabling instrument operation at elevated pressure sufficient to maintain up to a few kPa of gas or vapor near the sample [1]. This class of instrumentation is referred to as ESEM, variable pressure SEM (VPSEM), or low vacuum SEM (LVSEM). By cooling water vapor within the instrument, nanoscale imaging of liquid water is possible using ESEM or scanning transmission electron microscopy (STEM) within the microscope chamber [2]. These open cell modes of imaging liquids are discussed in Chapter 3. The second approach uses molecular-impermeable electron-transparent membranes to isolate samples that are at atmospheric pressure, or that include liquids, from the high vacuum of the microscope [3]. Such environmental cells (also often called "closed cells") can be filled with liquid or gases and placed inside the microscope chamber to enable liquid SEM (LSEM) or wet SEM (WetSEM) [4] (Figure 4.1c). Finally, the third method for imaging under high pressure or in liquid specimens is based on sealing the entrance of the objective lens of the SEM with an electron-transparent membrane, thus isolating the column and the electron detectors from the ambient atmosphere. This technique, known in various forms as atmospheric pressure SEM (ASEM) or AirSEM, is discussed in

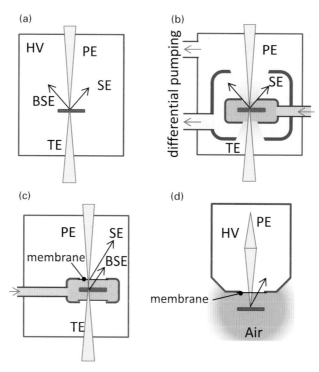

Figure 4.1. (a) Standard high vacuum SEM/TEM setup; (b) open cell environmental SEM/TEM; (c) closed cell environmental SEM/TEM; (d) atmospheric pressure SEM. HV, PE, SE, BSE, and TE stand for high vacuum, primary electrons, secondary electrons, backscattered electrons, and transmitted electrons, respectively.

Chapter 5. It enables probing of a proximal object at ambient atmosphere conditions [5–7], as shown in Figure 4.1d.

The historical development of membrane-based transmission electron microscopy is intriguing and enlightening. More than seven decades of development of such liquid cells, from early work on collodion membranes by Abrams and McBain [3] up to the recent state-of-the-art monolithic silicon-based chips for microfluidic electron microscopy [8], are described comprehensively in Chapters 1 and 2 of this book. This chapter concentrates on the development, capabilities, and applications of membrane-based SEM in liquids, describing the advantages and limitations of the technique. This is an increasingly active research area and, at this point in time, no versatile commercial product exists to meet the diverse experimental needs of the research community. The chapter thus discusses ongoing research efforts based on custom-fabricated instrumentation that improves the capabilities of the membrane-based environmental cells (E-cells). We use the terms liquid SEM, membrane SEM, WetSEM, or ambient pressure scanning electron microscopy (APSEM) throughout the chapter to refer to this technique unless specified otherwise, implying that both dense gaseous and liquid media may be used with this technique.

4.1.2 Why SEM?

The vast majority of modern electron microscopy research in liquids and gases is based upon the use of either (S)TEM or ESEM. In spite of somewhat lower resolution, membrane-based liquid SEM has its specific niche with the following advantageous features:

(a) Surface and/or near-surface sensitivity, enabling high resolution analysis of interfacial processes.
(b) In contrast to ESEM, which relies on vapor condensation, higher ranges of pressure and temperature can be achieved independently inside the closed cells. The pressure is limited only by the mechanical stability of the membranes. A pressure differential of a few hundred kPa can routinely be achieved.
(c) Closed E-cells enable research into chemically reactive, toxic, or even radioactive media without contamination of the main chamber of the microscope.
(d) Unique or precious samples can be studied without losing them due to sublimation or evaporation.
(e) Compared to closed cell TEM or ETEM, there are fewer limitations for the imaged sample size and/or sample preparation. Membrane-based SEM techniques enable both large lateral fields of view (FOV), up to a few mm^2, as well as straightforward incorporation of optical, electrical, micro-mechanical, fluidic, and other connectors. As a result, SEM-based membrane techniques meet the requirements for correlative optical microscopy and spectroscopy.
(f) Finally, this methodology can be adapted by a large number of scientists using standard SEM instruments.

4.2 Basics of SEM through Membranes

4.2.1 Image Formation Mechanism and Detectors

The processes occurring during electron beam interactions with objects inside a liquid cell can be best illustrated by Monte Carlo simulations of electron trajectories [9] (Figure 4.2a, b). When a high energy electron beam with an energy E_b impinges upon the membrane and enclosed liquid specimen, the primary electrons dissipate their energy and change their directions via multiple inelastic and elastic processes. As a result, the electrons, ions, radicals, and electronic excitations become distributed within an interaction volume (defined by the colored trajectories in Figure 4.2a) whose dimensions depend on the primary energy of the beam, the properties of the membrane and the medium behind it, and the probe–surface interaction angle. One of the commonly used estimates of the dimensions of the interaction volume is the electron range R_{KO} [nm] given by the parametric equation:

$$R_{KO} = \frac{28\,W}{Z^{0.89}\rho}(E_b)^{1.64}, \qquad (4.1)$$

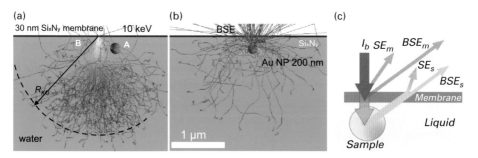

Figure 4.2. (a) Electron trajectories of a 10 keV electron beam in water after the beam passes through a 30 nm Si$_x$N$_y$ membrane (blue). The electron range R_{KO} is marked by the dashed line. (b) The drastic change of electron trajectories after the probe beam interacts with a 200 nm Au nanoparticle immersed in water 100 nm below the membrane; backscattered electrons (BSE) form an image of the object in the liquid. (c) The major electron signals in membrane-based SEM in liquids including secondary electrons (SE) and BSE due to membrane (m) and sample (s) interaction.

given by Kanaya and Okayama [10]. Here, E_b [keV] is the electron beam energy, W [g/mol] is the atomic weight, Z is the atomic number and ρ [g/cm^3] is the density of the media. For example, liquid water enclosed behind a 30 nm Si$_x$N$_y$ membrane yields R_{KO} that ranges from ~50 nm to ~2000 nm for 1 keV and 10 keV primary electrons, respectively (see Figure 4.2a). While the secondary electrons (SE) originate from the first few nanometers of the membrane, it is important to note that ~90% of backscattered electrons (BSE) originate within ~0.1R_{KO} (for high Z materials) to 0.3R_{KO} (for light elements) [11]. Thus, for liquid SEM imaging the R_{KO} parameter (and its energy dependence) has crucial importance, since it determines the ultimate "probing depth" in the liquid and therefore the contrast and resolution in the resultant images.

As an example, we consider a high Z object A (such as an Au nanoparticle) immersed in a low Z liquid background B which is separated from the high vacuum of the electron source and detectors by a thin membrane (Figure 4.2a, b). Assuming a standard Si$_x$N$_y$ membrane thickness of a few tens of nanometers, low energy secondary electrons (SE) from the object behind the membrane are completely attenuated. The signal (S) detected by a standard cumulative SE/BSE electron detector from such an object has the following contributions:

$$S = \text{SE}_m + \text{BSE}_m + \text{BSE}_s + \text{SE}_s \tag{4.2}$$

(Figure 4.2c), which includes SE and BSE electrons from the membrane (subscript m) and sample (subscript s) behind the membrane. Note that the SE$_s$ component (often called SE type 2 electrons or SE$_2$) originates from the backscattered electrons, BSE$_s$, scattered from the sample in the direction of incidence of the probe. These BSE$_s$ excite a detectable SE$_s$ from the membrane. The total yield of the backscattered electrons η (also called the backscattered electron coefficient) will be:

$$\eta = \eta_m + \eta_s. \tag{4.3}$$

Similarly, the secondary electron coefficient is:

$$\delta = \delta_m + \Delta \cdot \eta_s. \tag{4.4}$$

Here subscripts s and m again correspond to sample and membrane and Δ is the efficiency of the secondary electron emission from the membrane by outgoing BS electrons emitted from the sample. Using electron detector efficiencies ε_{SE}, ε_{BSE}, the cumulative electron signal can be written as:

$$S = \varepsilon_{SE}(\delta_m + \Delta \cdot \eta_s) + \varepsilon_{BSE}(\eta_m + \eta_s). \tag{4.5}$$

The contrast C between two points A and B in an SEM image (Figure 4.2) is defined [4] as:

$$C = (S_A - S_B)/S_A, \text{ which can be reduced to } C = (\eta_{sA} - \eta_{sB})/\eta_{sA}, \tag{4.6a, b}$$

assuming that the primary electron beam has sufficiently high energy (few keV) to penetrate through the membrane and that there is no contrast from the membrane and liquid backgrounds. Under these conditions, SEM contrast is predominantly determined by the ratio of the backscattered electron coefficients of the object and the liquid. This result emphasizes the importance of backscattered detectors for membrane-based SEM.

The validity of these considerations has been experimentally verified by Thiberge et al. using liquid cells equipped with 145 nm thick polyimide membranes [4]. One of their informative experimental results describes the energy dependence of the ratio:

$$T = \left(\eta_s(1+\Delta)\right)/\eta_{s,\text{theor}}, \tag{4.7}$$

as shown in Figure 4.3. The numerator in this equation denotes the sample signal (without terms from the featureless membrane) experimentally recorded through 145 nm polyimide membrane, while the denominator describes the theoretical value, $\eta_{s,\text{theor}}$, for the membrane material. Since this is a ratio of the amount of BSE transmitted through the membrane to the theoretically expected total, this dimensionless parameter has minimal material dependence and defines the electron transparency of the particular membrane. Several important conclusions can be deduced from these experimental dependences:

(a) The experimental electron transmittance curves of the membrane have a characteristic *transparency threshold* indicating, for example, that 145 nm thick polyimide membranes are practically opaque for electrons with energies below 4 keV, as shown in Figure 4.3. This transparency threshold is not a constant value but can be reduced towards lower primary energy by decreasing the thickness and/or the density of the membrane material.

(b) The transmittance curve shape also depends on the medium behind the membrane. For a water sample (low Z) the electron transmittance of the polyimide membrane is systematically lower compared to Au (high Z) (Figure 4.3). The latter can be a consequence of the so-called *Z-filtering* effect of the membrane.

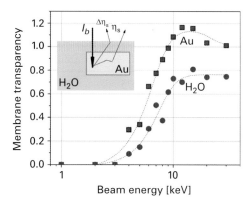

Figure 4.3. Energy dependence of the electron transparency of 145 nm thick polyimide membrane as a function of primary electron energy and material behind the membrane. Notice the existence of the characteristic transparency threshold and abnormal transparency around 10 keV for high Z material. Adapted from Ref. 4.

It is actually more transparent with a high Z sample beneath it [4]. To understand this, note that for any given primary energy E_b, the maximum of the energy spectrum for backscattered electrons from high Z targets is centered closer to E_b; for light materials it is closer to $E_b/2$. Thus, electrons from low Z targets become more attenuated by the membrane on their way to the detector.

(c) The presence of the apparent maximum in the transmittance curve for Au and its value in excess of 1 can be attributed to the significant contribution of SE_2 electrons to the total detectable signal from the sample. Further increase in the energy of the BSE_s reduces the cross section for SE_2 generation, implying that an optimal condition exists for signal generation and contrast that enables detection of both BSE_s and SE_2 from objects beneath the membrane.

4.2.2 Signal Quality and Spatial Resolution

The attenuation of the primary beam electrons and BSE signal by the membrane raises the question: what probe beam current is required to generate sufficient contrast from objects immersed in liquid behind the membrane? This is also an important question from the point of view of the sample's chemical stability, since higher beam currents can generate image artifacts, degrade samples, and complicate the interpretation of experimental results (see Chapter 7, and Section 4.3 for an example). A measure of image contrast is based on the Rose criterion, stating that the signal difference between points (subscript A and B) should be $(S_A - S_B) > 5n^{1/2}$ to yield reliable imaging of object A behind the membrane over the background B (Figure 4.2a). Here, n is the number of electrons reaching the detector [11]. The criterion can be rewritten [4] as a so-called *threshold equation*:

$$I_B > \frac{16}{\eta_s T C^2}, \quad (4.8)$$

Table 4.1 Calculated minimal SEM beam current needed to image some characteristic objects immersed in water using the threshold equation

Material	Z	η	Contrast in water	I_B (pA)
Water	7.22	0.075		
Cells	7.07	0.073	0.027	2900
Oil	5.8	0.055	0.267	30
Gold	79	0.78	0.9	0.25

Scanning time $T = 100$ s and η values for 20 keV were used; adapted from Ref. 4.

where I_B is current [pA], η_S is the backscattering coefficient of the sample under the membrane, T [s] is the time to complete the image scan, and $C = (\eta_{sA} - \eta_{sB})/\eta_{sA}$ is the contrast between sample and background. Using Equation (4.8), Thiberge et al. [4] calculated the minimum SEM beam current required to image characteristic objects immersed in water (Table 4.1). The data show that reliable imaging of samples, such as living cells with effective Z numbers and backscattering coefficients close to water, require more than four orders of magnitude larger beam current compared to materials with larger Z, such as gold. Potentially, this is the limiting factor for imaging unstained living cells in liquid SEM, since such radiation doses would lead to direct radiation damage of the cells or their indirect degradation due to radiolysis, or the chemical reactions induced in the surrounding water (see also Section 3.2.3 and Chapters 7, 16, 18, and 23).

The diameter of the primary electron beam also broadens due to scattering by the membrane and the enclosed liquid media (Figure 4.4a). Thus, the probe beam diameter increases with the distance between an immersed sample and the membrane. This has been analyzed both theoretically and experimentally [4, 12]. The effective electron beam diameter in the liquid is described by:

$$d_{\text{eff}} = \sqrt{d_b^2 + d_m^2 + d_{\text{water}}^2(t)}. \quad (4.9)$$

Here, d_b is the incident beam diameter, d_m is beam broadening due to scattering inside the membrane, and $d_{\text{water}}(t)$ is beam broadening from the water layer of thickness t. Monte Carlo electron trajectory simulations for $d_{\text{water}}(t)$ yielded the beam diameter vs. object depth curve in Figure 4.4b. The beam diameter can be as large as 100 nm at ~400 nm below the surface of standard 145 nm polyimide membrane. Two objects separated by 200 nm would barely be resolved under these conditions.

In practice, the experimentally observed resolution for objects with $Z > Z_{\text{water}}$ was found to be much better than the simulated value. This is attributed to the much higher electron scattering from high Z objects in water, compared to the water background. In this case, the resolution is not defined by the incident probe beam diameter in water (as would be the case for objects with comparable Z) but mainly by the size of the object and by a dramatic increase of the BSE signal when the beam hits the object (compare BSE yield in Figures 4.2a and b). To illustrate this point, Figure 4.4c shows simulated

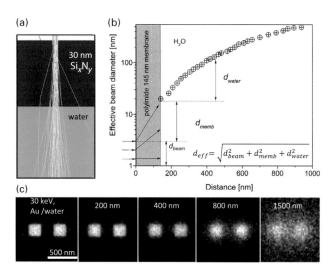

Figure 4.4. (a) Monte Carlo simulations of 30 keV beam broadening while passing through a 30 nm Si_xN_y membrane and impinging on water. (b) Effective electron beam diameter d_{eff} in water (after passing through 145 nm polyimide membrane) as a function of the probing depth (E_b = 20 keV; the data points were adapted from Ref. 5). (c) Simulated SEM images of two 200 nm × 200 nm × 1 μm Au rods separated by 200 nm as a function of the water layer thickness between the 30 nm Si_xN_y membrane and the sample.

SEM images of two rods, as a function of water thickness. The simulation demonstrates that these strongly absorbing high Z objects can be resolved up to 2 μm from the membrane using 30 keV primary electrons. In general, the broadening of the electron beam in liquid leads to a rule of thumb (see Ref. 4) that high Z objects can be detected if their size is comparable to the local size of the beam in liquid. Objects smaller than the broadened beam show lower signal-to-noise ratios (SNR). This effect is also illustrated in Figure 4.4c, where resolution and contrast deteriorate quickly as the beam diameter exceeds the size of the rod. This result implies a limitation on the maximum depth that particles of given size can be detected and resolved. It is important to note that increasing the primary beam energy improves the depth and the resolving power of liquid cell microscopy but does not necessarily lead to optimal imaging conditions. This is because the electron scattering cross section becomes reduced at higher energy so that small particles would not have enough material to provide a measurable BSE signal.

4.3 Examples of Environmental Cell Designs and Liquid SEM Applications

Recent progress in liquid electron microscopy is a direct result of success in high yield microfabrication technology of ultra-thin membranes that were initially developed for MEMS devices, ultrafiltration, and X-ray (electron) transparent windows. A number of vendors offer an array of Si, SiO_2, Si_xN_y, SiC, polyimide, and other membranes ranging from 5 nm to 200 nm in thickness, which can be used for assembling custom made liquid cells. To date, the majority of liquid cell SEM results have been obtained using commercial wet cells [13], though alternative membrane-based liquid cells have become available recently [14]. In this section we describe the commercially available wet cells and the related research that extends this instrumentation to new applications.

4.3.1 Basic Liquid SEM Capsule Design

The development of wet SEM technology [13] in the early 2000s addressed a growing demand for *in vivo* imaging of biological tissues and live cells with a better spatial resolution compared to conventional light microscopy. Detailed designs, capabilities, and applications of this approach can be found in the literature [4, 13, 15]. The cell design is depicted in Figure 4.5. It consists of two parts: the top sample dish with an electron-transparent membrane covering the sample compartment, and the stub. The cell can be vacuum sealed, with a rubber membrane used to join the two elements. The sample compartment can be filled with ~15 μL of liquid. Several modifications to the conventional cell body design have been proposed [13]. One such modification is the insertion of fiber optics for correlative SEM and cathodoluminescence imaging of hydrated samples [15] (see Figure 4.5b–d). The window of this cell is a robust 145 nm thick polyimide electron-transparent membrane supported by a metal grid. The grid provides sufficient mechanical stability for the membrane to withstand the atmospheric pressure differential and yet offers a large segmented ~4 mm^2 FOV. The rubber membrane that joins the two elements of the liquid cell is a thin elastomer material that expands in vacuum and functions as a pressure relief element that reduces the risk of membrane rupture during cell closing and SEM vacuum pumping.

The 145 nm thickness of the polyimide membrane implies electron beam energies above 10 keV and backscattered electron detection for SEM imaging of wet samples

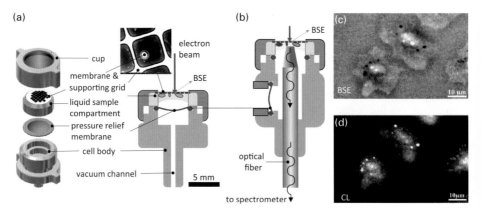

Figure 4.5. (a) The basics of commercial liquid SEM capsule design. (b) Simultaneous SEM and light microscopy in liquid SEM capsule via introduction of an optical fiber. (c) BSE image of fully hydrated untreated CHO cells, visualized simultaneously (d) by cathodoluminescence (CL) signal. Adapted from Ref. 15; Copyright (2004) National Academy of Sciences, USA.

(see Figure 4.3). Due to the strong attenuation of the BSE signal by the membrane and the water, it is crucial to have the sample in close proximity or adhered to the inner surface of the membrane. Thus, protocols have been developed that functionalize the membrane surface to promote nanoparticle adhesion as well as cell culture. To work with wet biological tissue, biopsies, plants sections, and other soft samples, an upgraded capsule [13] is equipped with a spring-loaded piston that gently pushes the tissue against the membrane.

4.3.2 Application Examples

Numerous studies in a variety of fields have been performed using the wet SEM capsule [16–21]. We summarize several results below, to provide a sense of the scope of applications that this technique can offer.

4.3.2.1 Catalysis for Water Remediation

Figure 4.6 shows the use of wet SEM for *in situ* observation of the removal of heavy metal ions from water by TiO_2-based photocatalysis [22]. Electron-hole pairs generated in catalyst particles by UV radiation from the Sun can diffuse to the particle surface and reduce or oxidize ionic and molecular pollutants in water such as heavy metal ions [23]. An example of such a photocatalytic reaction is the UV photon-induced reduction of

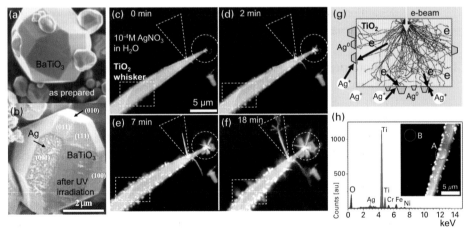

Figure 4.6. (a) SEM image of as-prepared bare $BaTiO_3$ photocatalyst; (b) similar particle after UV light exposure in 10^{-1} mol m^{-3} aqueous solution of $AgNO_3$. Silver reduction can be seen on (001) facet. Adapted from Ref. 24. (c)–(f) *In situ* sequential liquid SEM images of a titania whisker immersed in 10^{-1} mol m^{-3} aqueous solution of $AgNO_3$. Areas marked with circles, triangles, and squares show growing Ag deposits as a result of electron beam-induced reduction of Ag ions. Imaging conditions: 30 kV acceleration voltage and BSE detector. (g) The mechanism of beam-induced reduction of Ag ions at the surface of TiO_2. (h) XEDS spectra recorded from $AgNO_3$ solution (blue) and Ag-decorated TiO_3 nanowire (red). Adapted with permission from Ref. 22, Copyright (2010) American Chemical Society.

ionic Ag to metallic nanoparticles at the surface of BaTiO$_3$ in an aqueous AgNO$_3$ solution [24]. Figure 4.6a, b shows the process *ex situ*. BaTiO$_3$ is ferroelectric and its spontaneous polarization affects the chemical reactivity of different crystal facets. Silver is deposited only on (100) facets, implying that the internal polarization promotes the photocatalytic action at this interface [24]. *In situ* observation of such a process is possible in wet SEM capsules by adhering TiO$_2$ single crystal whiskers to the inside of the membrane then dropping ~10 μL of water or AgNO$_3$ aqueous solution into the E-cell compartment. No changes in morphology and shape of the titania nanostructure occur as a result of exposure to the 30 keV electron beam in pure water, while TiO$_2$ becomes progressively decorated with metal particles in the AgNO$_3$ solution (Figure 4.6c–f). This phenomenon is analogous to photocatalysis and is due to electron-induced reduction of Ag ions at the TiO$_2$ surface, but with the electron-hole pairs created in TiO$_2$ by the primary beam rather than UV light. Assuming that the stoichiometry of the beam-induced reaction is analogous to the classical photocatalytic reaction (photoreduction) [25] one can write:

$$4Ag^+ + 2H_2O \xrightarrow[\text{e-beam}]{\text{TiO}_2} 4Ag^0 + O_2 + 4H^+, \qquad (4.10)$$

where the first half-reaction is a transfer of an excited electron to Ag$^+$ bound to the TiO$_2$ surface, leading to the growth of Ag^0 domains with high quantum yield. The hole transfer to the surface-bound water or hydroxyl groups completes the second half-reaction (Figure 4.6g). Energy dispersive spectroscopy (XEDS) demonstrates Ag deposition on the nanowire (Figure 4.6h) and the feasibility of recording XEDS spectra on objects in liquid cells.

4.3.2.2 Phytotoxicity

The increasing production of nano-enabled materials such as engineered nanoparticles (ENP) raises concerns regarding waste byproducts and potentially negative ENP impacts on human health and ecosystems. Plants are consumed directly or indirectly by humans and it is therefore important to understand the uptake and accumulation of ENP in roots and leaves. Liquid SEM can be a powerful tool in exploration of the effects of ENP phytotoxicity on living plants. We have tested the capabilities and limitations of this imaging mode to probe the *in vivo* uptake and accumulation of nanoparticles and metal ions by plants at a cellular level. A recent example is the study of cellular accumulation and subcellular transport of Ag nanoparticles in *Arabidopsis thaliana* [26]. *Arabidopsis* root anatomy is well understood (Figure 4.7a) and is often used as a model plant system. In this experiment, various sizes and concentrations of Ag colloids were added to a hydroponic nutrient solution (Hoagland solution) in which seeds of *Arabidopsis thaliana* were germinated. After growth, the tips of the plant roots were sectioned in a wet state and adhered to the back side of the electron-transparent membrane of a wet SEM capsule, and the capsule was filled with water. Laser scanning confocal microscopy optical sections can be compared with the corresponding liquid cell SEM images (Figure 4.7b–e). Significantly improved resolving power and signal-to-noise ratio in the SEM images revealed that accumulation takes place mainly at

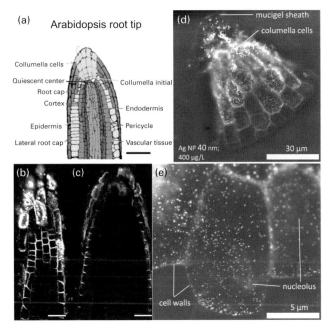

Figure 4.7. (a) The anatomy of *Arabidopsis* root tip. (b, c) Laser scanning confocal microscopy of *Arabidopsis* root tip exposed to 270 µg/L of 40 nm Ag ENP colloid in Hoagland solution for 4 weeks; with (b) a surface scan (8 µm below the root surface) and (c) a semi-median scan (~36 µm below the root surface) showing the distribution of Ag ENPs in larger root cap cells, epidermis, and columella initials. (d) ENP-decorated *Arabidopsis* roots imaged by SEM in water. (e) Higher resolution SEM images of fully hydrated *Arabidopsis* root cap cells with visible interior (nucleolus) and Ag nanoparticles decorating cellular walls. Adapted from Ref. 26.

the root cap, columella cells, and border cells [26]. Even at low concentrations and exposure times the cell walls in these locations are coated with individual nanoparticles at surface densities sufficient to impede incoming solute transport. Detailed SEM image analysis of the intercellular regions (supported by *ex situ* TEM studies) indicated the aggregation of Ag ENPs at plasmodesmata connecting adjacent cells. Silver ENP aggregation at plasmodesmata may block symplastic transport between cells and thus degrade the system viability.

4.3.2.3 High Radiation Dose Effects

In spite of the demonstrated advantages of liquid cell SEM for high resolution analysis of ENP fate in plants, there are several limitations of this technique for biological objects. Inelastic interaction of the probe beam with water and the biological material generates ionized and excited molecular species in the interaction volume, as discussed in Chapter 7. As a result, steady state concentrations of strongly reducing and oxidizing species such as hydrated electrons (e_h^-), hydroxyl radicals ($OH^\bullet$), hydrogen peroxide (H_2O_2), hydrogen radicals ($H^\bullet$), H_3O^+, molecular hydrogen and oxygen (H_2, O_2) are formed in and around the interaction volume [27]. Assuming that the radiation dose rate

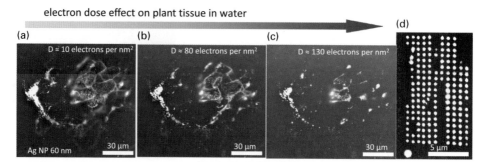

Figure 4.8. (a) Liquid SEM image of *Arabidopsis thaliana* with cellular walls decorated with 60 nm Ag nanoparticles. (b, c) Electron beam-induced Ag ENP dissolution and plant tissue degradation in water with increasing radiation dose (in electrons per nm^2). (d) Two orders of magnitude increase of the beam intensity at specified locations reverses the process and results in Ag^0 redeposition on the membrane inner surface.

in standard SEM measurements is somewhat larger (10^8–10^9 Gy s^{-1}) than for TEM ($\sim 10^8$ Gy s^{-1}), and following the scaling laws deduced in Ref. 27, one can expect the steady state molar concentrations of $OH^\bullet$ and H_2O_2 species to be in excess of 10^{-1} mol m^{-3}. These radiolysis byproducts contribute significantly to the chemical stability of nanoscale objects in solution and the degradation of biological samples. An example is shown in Figure 4.8, where degradation of cell walls can be observed after doses as low as 10 nm^{-2} (e/nm^2). Continued exposure leads to complete decay of the cell walls, as shown by reduced image contrast as well as release and dissolution of the decorating 60 nm Ag nanoparticles.

It is interesting to note that a dose two orders of magnitude higher, achieved by increasing dwell times at specific locations (bright dots in Figure 4.8d), redeposits Ag on the membrane surface. This interplay between Ag dissolution and growth as function of electron dose rate can be understood via the relative concentrations of reducing (e_h^-) and oxidizing ($OH^\bullet$) agents, which change with dose rate as discussed in Chapter 7. Such results are useful in controlling beam-induced nucleation and growth of nanoparticles as well as electron beam-induced deposition (EBID) in liquids, described in Chapter 14 [28]. Coupled with electrochemistry, EBID can promise greatly increased selectivity and deposition yield compared to its vacuum analog, along with a greater variety of deposited species. Flushing of gas or liquid media in the E-cell is easily achieved without compromising the SEM vacuum, potentially enabling sequential nanoscale deposition of different materials.

4.3.3 Upgrading the Basic Capsule Design

Commercial wet SEM capsules are mainly used by the biomedical research community at present, with less usage for materials research. This is partly due to the limited capabilities of these capsules to vary temperature or the chemical and electrical (electrochemical) environment. Development of such capabilities would broaden the utility

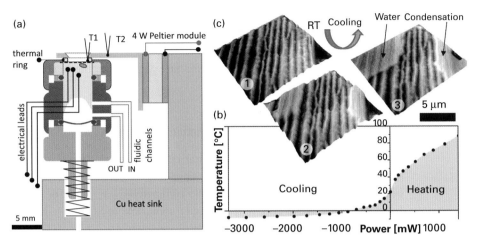

Figure 4.9. (a) Customization of commercial wet SEM capsule with addition of heating/cooling, electrical, and fluidic capabilities. (b) Temperature of the sample-supporting membrane in (a) as a function of power of the 4 W thermoelectric element. (c) Liquid SEM images of water condensation on *Morpho* butterfly scale on cooling inside water vapor-filled wet SEM capsule: see details in the text and in Ref. 29.

of wet SEM for materials science applications. Figure 4.9 shows our initial effort to develop these capabilities by modifying the central stub of a commercial capsule to include electrical leads and tubing for gas or fluid delivery. The metal coating of the capsule was removed to prevent short-circuits, and UV curable glue was used to vacuum seal the electrical leads and tubing. Temperature control is performed by a simple heating and cooling system based on a miniature 4 W Peltier element and copper heat sink (Figure 4.9a) [29]. In this system, the heat from the thermoelectric module is delivered to the electron-transparent membrane using a copper bar connecting the module to the membrane-supporting metal grid. This bar is spring loaded against the membrane supporting mesh to maintain thermal contact while enabling capsule assembly and disassembly. Temperatures from $-10\,°C$ to $100\,°C$ can be achieved at the membrane (Figure 4.9b). More powerful thermoelectric elements enable an even larger range.

The ability to cool or to heat samples in their native gaseous or liquid environments at different pressures opens new possibilities in SEM that cannot be achieved using commercially available cryo- or heating stages. For example, microscopic studies of water condensation, wetting, and flow in 3D photonic structures improve understanding of fundamental processes controlling opto-fluidics [30, 31] and sensors [32]. Figure 4.9c shows details of the initial stages of water condensation at the surface of butterfly wings, a natural photonic structure. The butterfly scales have characteristic mesoscopic ridge structures with stacked periodic layers of cuticle separated by air gaps when dry. They were adhered to the back side of the electron-transparent membrane. The capsule contained a water droplet separated from the membrane, so that dew could be formed or evaporated by cooling or heating the membrane. Water first condensed at the top of the ridges, then the condensation front spread along the ridges. As recently discovered, this condensation pattern is due to a gradient of surface polarity from the polar ridge tops to the non-polar bases [32]. The water therefore preferably condenses and spreads along the top part of the ridges leaving air pockets deeper inside the scales. This results in a hydrophobic wing structure that allows the butterfly to shed water droplets.

4.3.4 A Custom E-Cell Design for Electrochemical Studies in Liquids

Increased functionality of environmental cells is a current trend in TEM and STEM studies. Several types of fluidic, heating, and electrochemical cells have recently become available [33, 34]. Many of the commercially available or custom-developed E-cells for S/TEM are in principle directly applicable to SEM [14, 35]. However, they are rarely adopted by the SEM community, since the main advantages of the SEM, i.e. its large field of view, fast exchange, and ease of sample preparation, cannot be fully exploited in thin TEM flow cells with submicrometer fluidic channels. As a result, several E-cells have been specifically designed for complex SEM studies [36–38]. One such instrument is described below, designed to study the morphology and composition of nanoscale devices and interfaces during operation in realistic conditions.

These requirements imply complex circuitry on the membrane and parallel indexing of multiple electrodes. A solution is to employ a Si_xN_y membrane window on a silicon chip with a lithographically defined electrode array. The chip is wire bonded to standard vacuum-compatible ceramic chip carrier (Figure 4.10a, b), requiring only some modification such as a millimeter size hole through the center of the device [36]. For electrochemical studies, the electrical leads and other metals are coated with a SiO_2 insulating layer (Figure 4.10b). The assembly can be used in two different setups. For correlative light and electron microscopy, the cavity is filled with the electrolyte and covered with a glass lid sealed to the device with UV-curable adhesive. The lid acts as a viewing port making the cell usable with a light microscope as well as SEM for correlative microscopy. Alternatively, the fluidic chamber can be attached to the chip carrier (Figure 4.10c, d) to enable exchange of the solution within the device. Figure 4.10e shows a cyclic voltammogram of Ag in $AgNO_3$ solution on a SnO_2 nanowire electrode. The corresponding liquid SEM images demonstrate the morphology of deposited Ag^0 and its stripping at reverse potential. XEDS analysis was also carried out (not shown here) *in situ* to confirm the deposition of pure Ag on to the SnO_2 surface.

4.4 Novel Two-Dimensional Materials as Electron-Transparent Membranes for Liquid SEM Cells

4.4.1 Why Do We Need Them?

In spite of the progress in Si_xN_y, SiO_2, and polyimide membrane development for liquid SEM and STEM cells, the effect of the membrane itself on imaging is the most

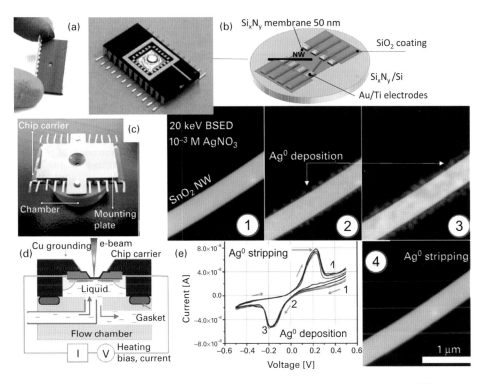

Figure 4.10. Design and results of multi-electrode cell for electrochemical studies in liquid using SEM and light microscopy: (a) standard ceramic chip carrier equipped with electron-transparent Si_xN_y window; (b) the electron-transparent window with multiple electrodes; (c) and (d) the corresponding fluidic cell; (e) a voltammogram and images recorded during Ag plating and stripping at a SnO_2 whisker immersed in 1 mol m^{-3} AgNO$_3$ solution. Adapted from Ref. 36.

significant problem to solve in order to improve SEM image quality and analytical capability, and reduce electron exposure time (dose) to the sample. This involves reducing the membrane thickness or atomic number. Commercially available Si_xN_y- and SiO_2-based membranes can be as thin as 5 nm, but as discussed in Chapter 2, such membranes can be produced reliably only with window sizes ~10 μm or larger, and are therefore prone to mechanical collapse due to the pressure differential. The solution is to fabricate larger, perforated supporting membranes with micrometer size orifices covered with thinner membranes. Such structures form the basis for modern designs for ambient pressure TEM and can in principle be adapted for ambient pressure SEM.

An alternative approach for reducing atomic number and membrane thickness is to use novel free-standing membranes [39, 40] made of single atomic layer two-dimensional (2D) materials such as graphene (G), graphene oxide (GO) and boron nitride (BN). These membranes are impermeable to liquids and gases. Compared to standard Si_xN_y or SiO_2 membranes, they have extremely high breaking strength [41] and high electron transparency: for graphene, the inelastic electron mean free path is larger than the thickness of a single layer [42]. Thus, incoming and outgoing electrons can pass without significant attenuation. Opportunities are even more intriguing for very low energy electrons (<5 eV). For such electrons, electron–electron scattering is further reduced and electron–phonon scattering is inefficient, so the membranes can be almost totally transparent. Thicker membranes (1–2 nm thick) made of multilayered 2D materials have even higher strength and more robust mechanical performance and are thus practical for liquid SEM measurements. As a result of these properties, a range of new opportunities for SEM emerge:

(a) True SEs can be used for imaging the surfaces of objects covered with such a membrane. Therefore, data can be acquired from specimens at ambient pressure, in fully hydrated conditions, with SEM resolution and surface sensitivity.

(b) An array of powerful surface sensitive analytical techniques such as X-ray photoelectron spectroscopy (XPS) and Auger electron spectroscopy (AES) can be applied to such objects.

(c) Lower primary beam energies can be used enabling smaller excitation volumes and reduced beam-induced damage.

(d) Charging can be negligible due to fast neutralization of the accumulated charge either by conducting electrons (in G, GO) or, for dielectric membranes (e.g. BN), via tunneling from the conducting liquid medium.

(e) Better stability of graphene membranes, since SEM electron beam energy is well below the energy threshold, ~80 keV, of the knock-on process for graphene [43].

The electron transparency of 2D free-standing G, BN, or GO membranes or coatings has been evaluated in TEM [40, 44, 45], SEM [46], and point projection microscopy [47, 48] for a range of electron energies. Quantitative measures of transparency can make use of the widely used overlayer thickness determination technique. In such measurements the electron attenuation length (EAL) parameter is used as an equivalent to inelastic mean free path assuming that elastic scattering is negligible for electrons on their way to the detector [49]. EAL relates the initial electron flux of secondary and backscattered electrons (I_0) from objects under the membrane to the flux (I) measured by the detector:

$$I/I_0 = \exp\left[-\left(\frac{L}{\lambda_M} + \frac{d}{\lambda_G}\right)\right]. \quad (4.11)$$

Here λ_G, λ_M are EALs for the graphene membrane and the medium between object and membrane, while d and L are the thicknesses of these layers, as shown in the inset of Figure 4.11. This method is widely used to determine the thickness of overlayers for materials with known EALs. Using this calculation, EALs for graphene and graphene oxide have been measured by collecting XPS [42, 50], AES [51], and secondary electron [52] signals from a substrate attenuated by a known thickness of graphene or GO. Figure 4.11 compiles some of these experimental data and compares them with theoretical predictions for graphite. Reasonable agreement is obtained for voltages above 70 eV. EALs for energies below this value are yet to be explored.

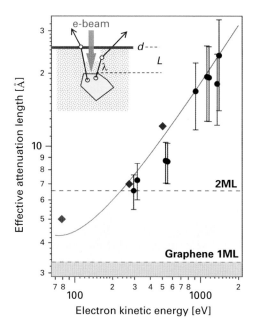

Figure 4.11. Effective attenuation length (EAL) for graphene monolayer: experimental data from Refs. 42 (circles) and 51 (diamonds). For comparison, horizontal dashed lines denote the thicknesses of single and bilayer graphene. The solid line is the predicted inelastic mean free path in graphite as calculated using the TPP-2M formula [49]. Adapted from Ref. 42.

4.4.2 Preparation of Graphene and Graphene Oxide Membranes

Graphene oxide was the first 2D material tested as an electron-transparent window for SEM E-cells [50, 53]. Its attractiveness stems from well-developed, high yield GO production protocols discussed in Ref. 54 and references therein. Conveniently, chemically exfoliated GO flakes have hydrophilic edges due to terminating COOH groups, and a hydrophobic basal plane due to the presence of domains of intact graphene. Being amphiphilic, GO flakes in water solutions segregate at air–water and water–solid interfaces to form membranes [55, 56], enabling suspended membranes to be fabricated via Langmuir–Blodgett (LB) or simple drop casting methods. When dried, such membranes have excellent tensile strength, ~100 MPa [57], which can be further enhanced by GO functionalization with divalent ions such as Mg^{2+} and Ca^{2+} [58].

Several approaches for fabricating GO suspended membranes have been reported [40, 50, 53, 59]. We have used diluted GO water or methanol solutions to cover micrometer-wide orifices with a single flake of GO using LB deposition [50], repeating multiple times to fabricate multilayer membranes. Such membranes are impermeable and mechanically stable if their diameter is below ~5 µm. Larger and thicker membranes can be fabricated easily by drying GO solution over small orifices [53]. Alternatively, GO solution can be dried on a support and then suspended with a selective backside chemical or electrochemical etch [53, 59].

GO membranes made of interlocked flakes have selective permeation properties [59] that favor the capillary uptake and diffusion of water between the stacked GO flakes. This permeation results from a flake separation distance (d) of 0.7–1.1 nm (depending on the degree of GO reduction) that can accommodate 1–2 layers of water. Interestingly, such GO membranes are vacuum tight for many other gases and liquids, such as atmospheric gases and alcohols [59]. The permeation makes GO excellent for selective filtering but poor for liquid water SEM experiments. For liquid SEM, one can either completely cover an orifice with an *individual* single layer or multilayer GO flake using the LB approach; or, for larger orifices, use thicker GO membranes that have less than one percolation channel inside the orifice. The total length l and surface density s of the percolation channels in a GO membrane of thickness h scale as $l \sim hL/d$ and $s \sim 1/L^2$ correspondingly, where L is the average flake size [59]. Thus, the number N of water-conducting channels in an orifice of size D is $N \sim (D/L)^2$. Assuming the average flake size $L \sim 10$ μm, water-impermeable multilayered GO membranes can be fabricated for $D < \sim L$, or 10 μm [53]. Furthermore, a mild anneal to 120–150 °C eliminates the percolating water layers, decreasing d without causing irreversible thermal reduction.

Graphene, on the other hand, lacks the aforementioned GO drawbacks. It can be fabricated as a single layer carbon sheet with low atomic number, small scattering cross section, and is chemically inert, ultra-thin and mechanically robust. Bulge and indentation tests demonstrate that suspended membranes are gas impermeable and yet have unprecedented robustness with stiffness in excess of 1 Tpa [60], sustaining a few hundred kPa pressure differential at $D < \sim 5$ μm. High quality, large-scale single and multilayer graphene is routinely grown by CVD on Cu or Ni substrates and can easily be transferred to a desired substrate by chemical etching of the metal using PMMA as a sacrificial supporting layer. Many other wet and dry transfer protocols have been developed in recent years and are reviewed in Refs. 61 and 62. Like GO, graphene can be transferred to micrometer-size apertures made by FIB, micro-patterning or laser drilling in Si_xN_y, SiO_2, or metal supports. The interfacial liquid layer and contaminants remaining after graphene growth and release can be removed via slow annealing in vacuum or under reducing (Ar/H$_2$ mixture) ambient [63]. This anneal also improves adhesion of graphene to the substrate.

4.4.3 Graphene and GO-Based E-cell Design

Figure 4.12 shows two different closed cell designs based on the graphene or GO membranes discussed above. The single-use Si-based cell in Figure 4.12a consists of a Si_xN_y primary membrane with FIB drilled micro-orifices, LB covered with one or more graphene or GO layers. The chip is filled with the liquid sample, placed on a supporting Si plate and sealed with UV-curable glue. A silver paste patch is used to make a conductive ground for the device. A liquid cell with exchangeable graphene or GO windows is depicted in Figure 4.12b. The core of this device is a metal plate or Si_xN_y membrane with a laser or FIB drilled aperture 1–3 μm in diameter. For graphene, a standard PMMA-based graphene transfer protocol is used to cover the orifice, followed by annealing in air and dissolving PMMA using warm acetone [64]. For GO, LB

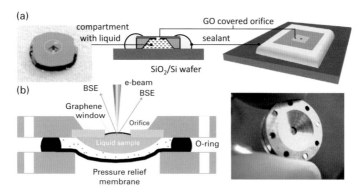

Figure 4.12. A closed graphene or GO E-cell: (a) single use cell; (b) refillable cell with exchangeable graphene or GO membranes. (a) Adapted from Ref. 50 and (b) with permission from Ref. 53, Copyright (2011) American Chemical Society.

or drop casting protocols are used to construct the membrane. An exchangeable disc with a droplet of the liquid sample is placed between two metal plates and vacuum sealed using a pressure relief rubber membrane gasket. This design allows isolation of about 20 µL of sample and reduction of the pressure differential onto the graphene or GO membrane in cases where the liquid contains an excess of trapped gas, or in vacuum.

4.4.4 SEM Imaging with GO and Graphene Membranes

The performance of a liquid cell with a ~40 nm GO membrane is shown in Figure 4.13a, b. Images are shown using both the BSED and Everhart–Thornley (ETD) detectors [53]. Since the ETD is sensitive to low energy secondary electrons, it reveals more details of the membrane surface morphology, such as multiple GO wrinkles. Electron scattering in this low Z membrane does not degrade the image resolution significantly (Figure 4.13c), and XEDS analysis can routinely be performed through the membrane (Figure 4.13d).

Even very low electron beam energies, such as 1 keV, appear to form clear images of high Z nanoparticles through the GO membrane. This may be surprising, since the membrane thickness is an order of magnitude larger than the inelastic mean free path for secondary electrons in GO [50]. It is plausible that the images are formed by backscattered along with the type S2 and S3 electrons that compose the SEM signal [11].

4.4.5 Radiation and Chemical Stability of Graphene-Based Membranes

Electron beam-induced water radiolysis is significant in liquid cell electron microscopy because of its impact on the material under study. But for liquid SEM, beam-induced damage to the membrane is also important since it limits the lifetime of the window and may contribute artifacts.

Direct formation of defects in the membrane can be caused by electron-induced atomic displacement (the "knock-on" process) or as a result of relaxation of an

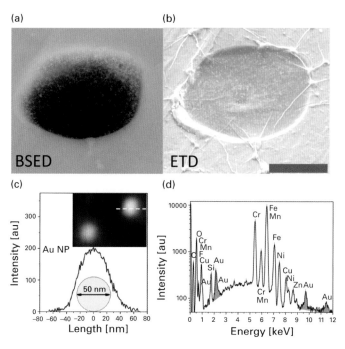

Figure 4.13. (a) SEM image of Au nanoparticles in solution through ~40 nm thick GO with beam energy 30 keV and BSE detector at 39° tilt. (b) SEM taken under the same conditions but using the Everhart–Thornley detector. The scale bar corresponds to 5 µm. (c) A high resolution SEM image and line profile of 50 nm Au nanoparticles in water attached to the back side of the membrane. (d) XEDS spectrum taken from Au nanoparticle solution through the GO membrane. Au is marked in orange; the other peaks originate from the surrounding stainless steel frame. Adapted with permission from Ref. 53, Copyright (2011) American Chemical Society.

electronic excitation inside the membrane. For graphene, the former mechanism requires energies above ~80 keV [43], so is predicted to be inefficient in SEM. On the other hand, the cross section for inelastic electron excitation processes increases at lower electron beam energies, and becomes significant at low keV. This damage mechanism can result in chemical bond breaking and local heating. The thermal conductivity of graphene at room temperature is high ($c.$ 5000 W m^{-1} K^{-1}) [65] and no noticeable damage is expected due to electron beam heating, even more so when in contact with gaseous or liquid media.

For practical SEM imaging through 2D membranes it is important to minimize all effects by reducing the electron dose, and know the threshold dose for defect formation. Beam effects are difficult to define as there is a significant variation in published experimental data and experimental conditions; often the dose is not recorded in SEM experiments. This issue is further complicated by a variety of energy dissipation mechanisms during electron beam irradiation of hydrated samples. GO membranes were found to be prone to photothermal reduction under flash light [66, 67] and laser light illumination. The measured threshold of ~10^2 mJ cm^{-2} for a 1 µm thick GO membrane [66] corresponds to a dose of only ~1 electron/nm^2 at 10 keV, assuming full absorption of the beam energy by the film. Alternative results indicate that GO films did not suffer reduction at doses as high as 10^3 electrons/nm^2 at 20 keV [68]. For graphene, electronic properties (such as mobility) and Raman spectroscopy can indicate the onset of damage. Graphene field effect transistors (GFETs) show electronic effects after doses as low as 20 electrons/nm^2 [69], but suspended graphene FETs require larger doses (10^2 electrons/nm^2) to show such effects [70]. Both types of device, however, show changes in Raman spectra at similar doses [70], implying that this may be due to effects such as local charging of the supporting substrate or carbonization of the graphene layer [71] rather than massive defect formation in the graphene itself. Bilayer graphene appears to be significantly more stable against radiation damage.

The presence of dense gaseous or liquid media in contact with graphene or GO can catalyze the formation of new defects or develop pre-existing ones indirectly, via generation of ionic or radical species that attack the membrane chemically. Scanning tunneling microscopy studies of graphene exposed to water show that this is particularly relevant to linear defects and grain boundaries in the 2D material [72]. Another example is the study of reduction of GO by plasmas of inert (Ar) and reducing gases (CH$_4$), or their mixtures [73]. Similar processes occur in water during beam-induced radiolysis (e.g. Ref. 74). Chemically reactive products such as molecular hydrogen (H$_2$), hydrogen peroxide (H$_2$O$_2$), and hydroxyl radicals (·OH) produced in water by radiolysis can accumulate [74] and reduce or oxidize the 2D membrane, weakening or dissociating C–C bonds at defect sites. In addition, under prolonged exposure, volatile products of radiolysis can eventually segregate as a separate phase forming microbubbles under the membrane. This potentially leads to a pressure buildup inside the E-cell and to a capillary-induced collapse of the membrane. We estimate that bubbles are formed in water under a graphene membrane at ~10^4 electrons/nm^2 at 20 keV [42, 75]. These effects of electron dose can be partly mitigated in fluidic cells with replenishable solution where liquid flow can remove radiolysis products from the scan area.

4.5 Outlook

Nanoscale imaging in liquids and at solid–gas–liquid interfaces is in demand in fields such as materials science, biomedics, forensics, chemistry, and environmental research. We expect membrane-based (closed cell) liquid SEM to be a growing field of research in the years to come. Based on the current trends, one can envision three directions for future activities and developments in this area of microscopy.

1. ***Theory, modeling, and fundamental research*** on chemical and physical aspects of the electron beam interaction with liquid matter and immersed objects. This will include the understanding of beam-induced local chemistry changes in liquids and dense gases, spatial and temporal distributions of reactive species, and their possible influence on the immersed objects. The probing of different systems such as operational electrochemical and fuel cells as well as living biological objects will continue along with efforts to minimize beam-induced

artifacts by improving the sensitivity of the detectors and by implementing modern image processing and data mining algorithms. There is also a clear trend towards correlative microscopy by combining different imaging modalities such as electron, ion, X-ray, and optical microscopies.

2. *Liquid cell designs*. Electron-transparent windows 10–50 nm thick based on modern Si_xN_y, SiO_2, or SiC microfabrication processes will dominate the technological platform for liquid SEM in the near future. The diversity of experimental tasks and objects makes it almost impossible to create a universal liquid SEM cell, and therefore, similar to the trends in liquid TEM, some degree of specialization is expected in the cell design. For example, fabrication of Si_xN_y windows dedicated to electrochemical research, variable temperature studies, in situ micromanipulation, and electrical measurements can be envisioned. Future liquid cells will be single use monolithic chips with on-board fluidics, where the source liquid is driven by electrophoretic, electro-osmotic, or micromechanical pumps that are integral parts of the "lab-on-chip" for liquid SEM technology.

3. *Membrane development*. Novel 2D materials with ultimately high electron transparency and mechanical strength will be an active area for research and technical developments, driven by their unique ability to provide powerful surface characterization techniques such as XPS for liquid interfaces. The requirement to have a large FOV conflicts with the mechanical stability of large-area graphene membranes. This will be resolved by development of platforms where the membrane is composed of multiple orifices. One possible design for a graphene-based liquid cell is proposed in Figure 4.14. The core of the cell is a multichannel or microporous matrix that contains a high density of isolated or interconnected microfluidic channels. The channel openings are covered with

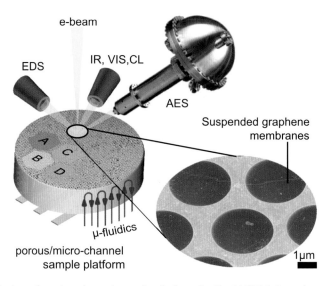

Figure 4.14. Prospective design of a microchannel sample platform for liquid SEM through an array of graphene-covered micro-orifices. Adapted from Ref. 42.

graphene membranes. Such a cell can be equipped with heaters and electrodes, and channels can be impregnated with different liquids (as in A–D in Figure 4.14). Such design is advantageous for combinatorial SEM studies including correlative electron (e.g. XPS, AES) or optical (cathodoluminescence, IR, or Vis) spectroscopies.

Acknowledgements

AK thanks his former students Shannon Berg, Mark Krueger, Joshua Cothren, Joshua Stoll, and Alexander Yulaev, whose work became the basis of this chapter. Technical discussions with Christopher Brown, Alex Liddle, Renu Sharma, Douglas C. Meier, and Nikolai Zhitenev of NIST are greatly appreciated.

Disclaimer: Official contribution of the National Institute of Standards and Technology; not subject to copyright in the United States.

References

1. G. D. Danilatos, Review and outline of environmental SEM at present. *J. Microsc. Oxford*, **162** (1991), 391–402.
2. A. Bogner, G. Thollet, D. Basset, P.-H. Jouneau and C. Gauthier, Wet STEM: a new development in environmental SEM for imaging nano-objects included in a liquid phase. *Ultramicroscopy*, **104** (2005), 290–301.
3. I. Abrams and J. McBain, A closed cell for electron microscopy. *J. Appl. Phys.*, **1** (1944), 607–609.
4. S. Thiberge, O. Zik and E. Moses, An apparatus for imaging liquids, cells, and other wet samples in the scanning electron microscope. *Rev. Sci. Instrum.*, **75** (2004), 2280–2289.
5. M Suga, H. Nishiyama, Y. Konyuba *et al.*, The atmospheric scanning electron microscope with open sample space observes dynamic phenomena in liquid or gas. *Ultramicroscopy*, **111** (2011), 1650–1658.
6. N. Vidavsky, S. Addadi, J. Mahamid *et al.*, Initial stages of calcium uptake and mineral deposition in sea urchin embryos. *Proc. Natl. Acad. Sci. USA*, **111** (2014), 39–44.
7. Y. Ominami, S. Kawanishi, T. Ushiki and S. Ito, A novel approach to scanning electron microscopy at ambient atmospheric pressure. *Microscopy*, **64** (2015), 97–104.
8. E. Jensen, A. Burrows and K. Mølhave, Monolithic chip system with a microfluidic channel for in situ electron microscopy of liquids. *Microsc. Microanal.*, **20** (2014), 445–451.
9. H. Demers, N. Poirier-Demers, A. Réal Couture *et al.*, Three-dimensional electron microscopy simulation with the CASINO Monte Carlo software. *Scanning*, **33** (2011), 135–146.
10. K. Kanaya and S. Okayama, Penetration and energy-loss theory of electrons in solid targets. *J. Phys. D: Appl. Phys.*, **5** (1972), 43–58.
11. J. Goldstein, D. E. Newbury, D. C. Joy *et al.*, *Scanning Electron Microscopy and X-ray Microanalysis* (New York: Springer, 2003).
12. N. Liv, I. Lazić, P. Kruit and J. P. Hoogenboom, Scanning electron microscopy of individual nanoparticle bio-markers in liquid. *Ultramicroscopy*, **143** (2014), 93–99.

13. V. Behar, A. Nechushtan, Y. Kliger *et al.*, Methods for SEM inspection of fluid containing samples. US Patent 7230242 B2 (2007).
14. D. A. Fischer, D. H. Alsem, B. Simon, T. Prozorov and N. Salmon, Development of an integrated platform for cross-correlative imaging of biological specimens in liquid using light and electron microscopies. *Microsc. Microanal.*, **19** (2013), 476–477.
15. S. Thiberge, A. Nechushtan, D. Sprinzak *et al.*, Scanning electron microscopy of cells and tissues under fully hydrated conditions. *Proc. Natl. Acad. Sci. USA*, **101** (2004), 3346–3351.
16. G. Venkiteela and Z. H. Sun, In situ observation of cement particle growth during setting. *Cement Concrete Comp.*, **32** (2010), 211–218.
17. K. Tiede, S. P. Tear, H. David and A. B. A. Boxall, Imaging of engineered nanoparticles and their aggregates under fully liquid conditions in environmental matrices. *Water Res.*, **43** (2009), 3335–3343.
18. C. Lorenz, K. Tiede, S. Tear *et al.*, Imaging and characterization of engineered nanoparticles in sunscreens by electron microscopy, under wet and dry conditions. *Int. J. Occup. Environ. Health*, **16** (2010), 406–428.
19. D. C. Joy and C. S. Joy, Scanning electron microscope imaging in liquids: some data on electron interactions in water. *J. Microsc. Oxford*, **221** (2006), 84–88.
20. A. K. F. Dyab and V. N. Paunov, Particle stabilised emulsions studied by WETSEM technique. *Soft Matter*, **6** (2010), 2613–2615.
21. O. Cohen, R. Beery, S. Levit *et al.*, Scanning electron microscopy of thyroid cells under fully hydrated conditions – A novel technique for a seasoned procedure: a brief observation. *Thyroid*, **16** (2006), 997–1001.
22. N. Kolmakova and A. Kolmakov, Scanning electron microscopy for in situ monitoring of semiconductor–liquid interfacial processes: electron assisted reduction of Ag ions from aqueous solution on the surface of TiO_2 rutile nanowire. *J. Phys. Chem. C*, **114** (2010), 17233–17237.
23. C. Wei, W. Y. Lin, Z. Zainal *et al.*, Bactericidal activity of TiO_2 photocatalyst in aqueous media: toward a solar-assisted water disinfection system. *Environ. Sci. Technol.*, **28** (1994), 934–938.
24. J. L. Giocondi and G. S. Rohrer, The influence of the dipolar field effect on the photochemical reactivity of $Sr_2Nb_2O_7$ and $BaTiO_3$ microcrystals. *Top. Catal.*, **49** (2008), 18–23.
25. J. M. Herrmann, Heterogeneous photocatalysis: fundamentals and applications to the removal of various types of aqueous pollutants. *Catal. Today*, **53** (1999), 115–129.
26. J. Geisler-Lee, Q. Wang, Y. Yao *et al.*, Phytotoxicity, accumulation and transport of silver nanoparticles by *Arabidopsis thaliana*. *Nanotoxicology*, **7** (2012), 323–337.
27. N. M. Schneider, M. M. Norton, B. J. Mendel *et al.*, Electron–water interactions and implications for liquid cell electron microscopy. *J. Phys. Chem. C*, **118** (2014), 22373–22382.
28. E. U. Donev and J. T. Hastings, Electron-beam-induced deposition of platinum from a liquid precursor. *Nano Lett.*, **9** (2009), 2715–2718.
29. A. S. Al-Asadi, J. Zhang, J. Li, R. A. Potyrailo and A. Kolmakov, Design and application of variable temperature setup for scanning electron microscopy in gases and liquids at ambient conditions. *Microsc. Microanal.*, **21** (2015), 765–770.
30. C. Monat, P. Domachuk and B. Eggleton, Integrated optofluidics: a new river of light. *Nat. Photonics*, **1** (2007), 106–114.
31. D. Erickson, D. Sinton and D. Psaltis, Optofluidics for energy applications. *Nat. Photonics*, **5** (2011), 583–590.

32. R. A. Potyrailo, T. A. Starkey, P. Vukusicb et al., Discovery of the surface polarity gradient on iridescent Morpho butterfly scales reveals a mechanism of their selective vapor response. *Proc. Natl. Acad. Sci. USA*, **110** (2013), 15567–15572.
33. R. R. Unocic, X. G. Sun, R. L. Sacci et al., Direct visualization of solid electrolyte interphase formation in lithium-ion batteries with in situ electrochemical transmission electron microscopy. *Microsc. Microanal.*, **20** (2014), 1029–1037.
34. K. Klein, I, Anderson and N. De Jonge, Transmission electron microscopy with a liquid flow cell. *J. Microsc.*, **242** (2011), 117–123.
35. K. Mølhave, C. Kallesøe, C. Y. Wen et al., Microfabricated systems for electron microscopy of nanoscale processes: in-situ TEM creation of Si nanowire devices and in-situ SEM electrochemistry. *Microsc. Microanal.*, **16** (2010), 322–323.
36. J. E. Cothren, Development of techniques and instrumentation for *in situ* imaging and spectroscopy of working nanodevices using ultrathin membrane based environmental cells. M.Sc. Thesis, Southern Illinois University at Carbondale (2011).
37. Y. Liu, Scanning electron microscopy to probe working nanowire gas sensors. M.Sc. Thesis, Southern Illinois University at Carbondale (2013).
38. S. Ueda, Y. Kobayashi, S. Koizumi and Y. Tsutsumi, In situ observation of water in a fuel cell catalyst using scanning electron microscopy. *Microscopy*, **64** (2015), 87–96.
39. J. C. Meyer, A. K. Geim, M. Katsnelson et al., The structure of suspended graphene sheets. *Nature*, **446** (2007), 60–63.
40. N. R. Wilson, P. A. Pandey, R. Beanland et al., Graphene oxide: structural analysis and application as a highly transparent support for electron microscopy. *ACS Nano*, **3** (2009), 2547–2556.
41. C. Lee, X. Wei, J. W. Kysar and J. Hone, Measurement of the elastic properties and intrinsic strength of monolayer graphene. *Science*, **321** (2008), 385–388.
42. J. Kraus, R. Reichelt, S. Günther et al., Photoelectron spectroscopy of wet and gaseous samples through graphene membranes. *Nanoscale*, **6** (2014), 14394–14403.
43. J. C. Meyer, F. Eder, S. Kurasch et al., Accurate measurement of electron beam induced displacement cross sections for single-layer graphene. *Phys. Rev. Lett.*, **108** (2012), 196102.
44. J. C. Meyer, C. O. Girit, M. Crommie and A. Zettl, Imaging and dynamics of light atoms and molecules on graphene. *Nature*, **454** (2008), 319–322.
45. R. S. Pantelic, J. C. Meyer, U. Kaiser and H. Stahlberg, The application of graphene as a sample support in transmission electron microscopy. *Solid State Commun.*, **152** (2012), 1375–1382.
46. L. Frank, E. Mikmeková, I. Müllerová and M. Lejeune, Counting graphene layers with very slow electrons. *Appl. Phys. Lett.*, **106** (2015), 013117.
47. J. Mutus, L. Livadaru, J. T. Robinson et al., Low-energy electron point projection microscopy of suspended graphene, the ultimate 'microscope slide'. *New J. Phys.*, **13** (2011), 063011.
48. J.-N. Longchamp, C. Escher, T. Latychevskaia and H.-W. Fink, Low-energy electron holographic imaging of gold nanorods supported by ultraclean graphene. *Ultramicroscopy*, **145** (2014), 80–84.
49. A. Jablonski and C. Powell, Practical expressions for the mean escape depth, the information depth, and the effective attenuation length in Auger-electron spectroscopy and x-ray photoelectron spectroscopy. *J. Vac. Sci. Technol. A*, **27** (2009), 253–261.
50. A. Kolmakov, D. A. Dikin, L. J. Cote et al., Graphene oxide windows for in situ environmental cell photoelectron spectroscopy. *Nat. Nanotechnol.*, **6** (2011), 651–657.

51. M. Xu, D. Fujita, J. Gao and N. Hanagata, Auger electron spectroscopy: a rational method for determining thickness of graphene films. *ACS Nano*, **4** (2010), 2937–2945.
52. V. Kochat, A. N. Pal, E. S. Sneha *et al.*, High contrast imaging and thickness determination of graphene with in-column secondary electron microscopy. *J. Appl. Phys.*, **110** (2011), 014315.
53. M. Krueger, S. Berg, D. Stone *et al.*, Drop-casted self-assembling graphene oxide membranes for scanning electron microscopy on wet and dense gaseous samples. *ACS Nano*, **5** (2011), 10047–10054.
54. S. Park and R. S. Ruoff, Chemical methods for the production of graphenes. *Nat. Nanotechnol.*, **4** (2009), 217–224.
55. D. Li, M. B. Müller, S. Gilje, R. B. Kaner and G. G. Wallace, Processable aqueous dispersions of graphene nanosheets. *Nat. Nanotechnol.*, **3** (2008), 101–105.
56. L. J. Cote, J. Kim, V. C. Tung *et al.*, Graphene oxide as surfactant sheets. *Pure Appl. Chem.*, **83** (2010), 95–110.
57. D. A. Dikin, S. Stankovich, E. J. Zimney *et al.*, Preparation and characterization of graphene oxide paper. *Nature*, **448** (2007), 457–460.
58. S. Park, K.-S. Lee, G. Bozoklu *et al.*, Graphene oxide papers modified by divalent ions: enhancing mechanical properties via chemical cross-linking. *ACS Nano*, **2** (2008), 572–578.
59. R. Nair, H. Wu, P. Jayaram, I. Grigorieva and A. Geim, Unimpeded permeation of water through helium-leak–tight graphene-based membranes. *Science*, **335** (2012), 442–444.
60. J. S. Bunch, S. S. Verbridge, J. S. Alden *et al.*, Impermeable atomic membranes from graphene sheets. *Nano Lett.*, **8** (2008), 2458–2462.
61. M. Xu, T. Liang, M. Shi and H. Chen, Graphene-like two-dimensional materials. *Chemical Rev.*, **113** (2013), 3766–3798.
62. M. Büttner and P. Oelhafen, XPS study on the evaporation of gold submonolayers on carbon surfaces. *Surf. Sci.*, **600** (2006), 1170–1177.
63. Y.-C. Lin, C. C. Lu, C. H. Yeh *et al.*, Graphene annealing: how clean can it be? *Nano Lett.*, **12** (2011), 414–419.
64. J. W. Suk, A. Kitt, C. W. Magnuson *et al.*, Transfer of CVD-grown monolayer graphene onto arbitrary substrates. *ACS Nano*, **5** (2011), 6916–6924.
65. A. A. Balandin, S. Ghosh, W. Bao *et al.*, Superior thermal conductivity of single-layer graphene. *Nano Lett.*, **8** (2008), 902–907.
66. L. J. Cote, R. Cruz-Silva and J. Huang, Flash reduction and patterning of graphite oxide and its polymer composite. *J. Am. Chem. Soc.*, **131** (2009), 11027–11032.
67. S. Gilje, J. Farrar, S. Dubin *et al.*, Photothermal deoxygenation of graphene oxide for patterning and distributed ignition applications. *Adv. Mater.*, **22** (2010), 419–423.
68. P. Kumar, K. Subrahmanyam and C. Rao, Graphene patterning and lithography employing laser/electron-beam reduced graphene oxide and hydrogenated graphene. *Mater. Express*, **1** (2011), 252–256.
69. G. Liu, D. Teweldebrhan and A. A. Balandin, Tuning of graphene properties via controlled exposure to electron beams. *IEEE Trans. Nanotechnol.*, **10** (2011), 865–870.
70. I. Childres, L. A. Jauregui, M. Foxe *et al.*, Effect of electron-beam irradiation on graphene field effect devices. *Appl. Phys. Lett.*, **97** (2010), 173109.
71. L. Tao, C. Qiu, F. Yu *et al.*, Modification on single-layer graphene induced by low-energy electron-beam irradiation. *J. Phys. Chem. C*, **117** (2013), 10079–10085.
72. X. Feng, S. Maier and M. Salmeron, Water splits epitaxial graphene and intercalates. *J. Am. Chem. Soc.*, **134** (2012), 5662–5668.

73. M. Baraket, S. G. Walton, Z. We *et al.*, Reduction of graphene oxide by electron beam generated plasmas produced in methane/argon mixtures. *Carbon*, **48** (2010), 3382–3390.
74. C. Royall, B. Thiel and A. Donald, Radiation damage of water in environmental scanning electron microscopy. *J. Microsc.*, **204** (2001), 185–195.
75. J. D. Stoll and A. Kolmakov, Electron transparent graphene windows for environmental scanning electron microscopy in liquids and dense gases. *Nanotechnology*, **23** (2012), 505704–505711.

5 Observations in Liquids Using an Inverted SEM

Chikara Sato and Mitsuo Suga

5.1 Introduction

In closed cell or environmental capsule electron microscopy, a liquid sample is placed in a capsule sealed by electron-transparent windows, and directly imaged *in situ* by TEM [1–3] and STEM [5] (Chapter 2) or SEM [4] (Chapter 4). Exploiting the possibilities of closed cell EM has led to significant breakthroughs in various fields, including ligand- and affinity-labeling studies in cell biology [5, 6] as well as electrochemistry [7]. However, the shape and small size of the closed cells, with capacity typically less than 20 µl, limit the types of samples that can be examined. The small dimensions also limit the culturable cell types and probably the labeling and washing efficiency achievable during experiments. We developed atmospheric scanning electron microscopy (ASEM) to allow the use of an open sample container at atmospheric pressure, overcoming these limitations [8]. ASEM has already been applied in bioscience and materials science and is expected to find applications in additional research fields, including clinical diagnosis. Particular benefits are the minimal sample preparation required, the ability to image samples immersed in liquid that are open to the air, and the opportunity to carry out correlative light and SEM (CLEM) on the sample.

In this chapter, we discuss the design considerations for ASEM and describe some of its existing applications. We also describe alternative SEM designs that share the same aims, enabling SEM of liquid samples in air with minimal preparation, or CLEM on samples in liquids. The microscopes we describe here provide an approach to imaging liquids that is very different from the closed cell (or environmental capsule) SEM discussed in Chapter 4. The information that they provide is complementary to that obtained from other liquid cell imaging modes such as closed cell TEM or STEM (Chapter 2) or open cell SEM (Chapter 3).

5.2 Instrument Design and Sample Geometry of the ASEM

5.2.1 Configuration of the ASEM

Our atmospheric SEM is referred to as the ClairScope™ (JEOL Ltd.) and is shown in Figure 5.1. It is based on an inverted microscope column that permits SEM of samples that remain at atmospheric pressure in a readily accessible, open container called the

Observations in Liquids Using an Inverted SEM   107

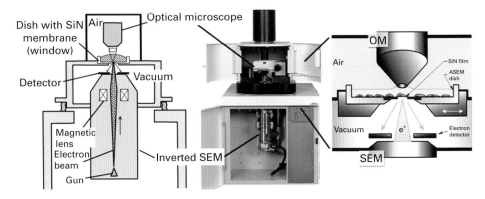

Figure 5.1. Configuration of atmospheric SEM. In the Jeol Clairscope™, the SEM is completely inverted, the electron gun being at the bottom. A removable dish with a Si_xN_y window in its base seals the column at the top. The samples are immersed in liquid of up to 2 ml volume. A light microscope is positioned above the dish as indicated, directly opposite the inverted SEM [8]. Biological samples are fixed, stained, and immersed in radical scavenger solution (10 mg/ml glucose) during SEM observation. Left and right panels from Refs. 8 and 12, respectively.

ASEM dish [8]. The ASEM dish contains a Si_xN_y thin film window which completes the vacuum seal at the top of the inverted SEM column. An optical microscope (OM) positioned above the ASEM dish can be used to observe wide areas of the sample from above; the SEM images smaller regions from below through the Si_xN_y membrane in the base of the dish. The optical axes of both microscopes are aligned and fixed to ensure that correlative images are recorded, and the specimen stage can be moved laterally in two dimensions for selecting the area to be imaged.

5.2.2 The ASEM Dish

The ASEM dish is made of polystyrene and is disposable [9]. In the standard version (Figure 5.2a, b) there is one 0.25 mm × 0.25 mm, 100 nm thick Si_xN_y window; dishes with eight such windows have also been developed [9, 10]. The inside of the dish base can be coated with various kinds of proteins or poly-L-lysine [11–13]. The dish can hold a few milliliters of culture medium, and allows various types of cells to be cultured in the stable environment provided by a CO_2 incubator. The cell types include primary cells obtained directly from organs. For example, neurons from mammalian brain can be cultured [12, 14], critical for the study of neural network formation under close-to-native conditions. The dish can also be modified for materials science applications. Electrodes on the Si_xN_y window enable electrochemical measurements (see Section 5.3.6), and a temperature-regulated titanium ASEM dish (Figure 5.2c) permits observation of temperature-dependent phenomena (Section 5.3.7).

5.2.3 Column Protection in Case of Accidental Membrane Breakage

In the ASEM, the sample is open to the atmosphere and separated from the vacuum of the inverted SEM by the 100 nm thick Si_xN_y window or windows in the base of the

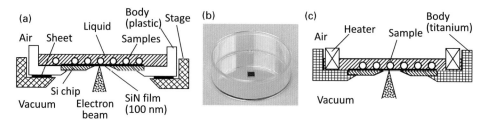

Figure 5.2. Configuration of the ASEM dish. (a) Diagram and (b) photograph of the standard ASEM dish, which is 35mm in diameter and is made of polystyrene. It has the shape of a Petri dish and is suitable for cell culture in a CO_2 incubator. The Si_xN_y membrane is manufactured by etching the Si side of a Si_xN_y-coated Si chip. The chip is glued to the windowed base of the dish with the Si_xN_y side facing up so that the inner surface of the dish base is flat [9]. (c) Temperature-regulated ASEM dish fabricated by building a ceramic heater and thermocouple into the body of a titanium dish. The windowed Si chip is glued to the base, as for the standard ASEM dish. From Ref. 19.

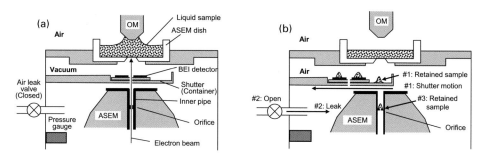

Figure 5.3. Fail-safe system that protects the ASEM column, showing the pressure gauge, a shutter equipped with a BSE detector and liquid container (#1), an air leak valve (#2), and an inner pipe with an orifice (#3). (a) Normal operation. (b) Response to membrane breakage. From Ref. 9.

sample dish. Although the membrane has sufficient resistance to withstand the pressure differential, it can be injured mechanically, for example penetrated by a pipette tip. A three-component protection system was therefore developed to prevent contamination of the ASEM column and electron gun by the liquid in case of accidental film breakage [9]. The protection system consists of a specialized shutter, an inner pipe, and an air-leak valve in the chamber beneath the ASEM dish (Figure 5.3a). The shutter is situated below the ASEM dish, at the top of the ASEM column, and has a container with a capacity of a few milliliters on its upper surface. When the pressure gauge detects a pressure increase due to film breakage, the shutter closes. This positions the container under the ASEM dish to collect any material lost and prevent liquid from entering the ASEM column (Figure 5.3b, #1). At the same time the high voltage turns off and the air-leak process activates (#2) to bring the column up to atmospheric pressure. The inner protection pipe lines the inside of the ASEM column. Any sample that passes through the central orifice of the BSE (backscattered electron) imaging detector before the shutter closes is retained by the orifice of the inner pipe (#3) and prevented from entering the gun region. The ASEM system can be recovered by exchanging the BSE detector, shutter, and inner pipe.

5.3 Applications of ASEM

Most of the applications outlined below were only possible because the ASEM dish is open to the atmosphere and easily accessible, allowing monitoring of the sample by OM, the addition of reagents, sample manipulation, and the investigation of phenomena that involve volume change. Applications involving correlative light and electron microscopy are considered separately in Section 5.4.

5.3.1 Observation of a Large Area of Tissue by Exploiting the Open Configuration of the ASEM Sample Holder

Film-sealed environmental cell SEM has successfully been used to observe wet tissue blocks excised from various organs such as heart and kidney [6], kidney [15], or brain tumor [16], avoiding dehydration artifacts. However, the samples are not readily accessible for manipulation or additional staining. Using the ASEM, large areas of tissues can be observed by manually sliding a tissue block across the ASEM dish (Figure 5.4; [10]). The tissues were fixed with glutaraldehyde, stained with metal solutions, and immersed in radical scavenger, 10 g/ml glucose solution, during SEM observation.

5.3.2 Observation of Lung Tissue Metastasized by Breast Cancer, Aiming at Intro-operative Cancer Diagnosis

The lungs are known to be organs easily metastasized by cancer. Indeed, according to World Health Organization statistics, lung cancer led to the most deaths from cancer

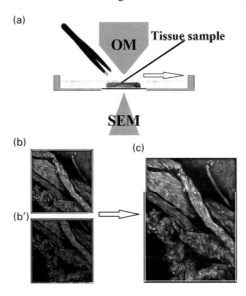

Figure 5.4. Large-area observation by sliding an excised tissue slab of spinal cord across the ASEM dish. (a) Schematic representation. (b', b) Tissue imaged before and after being moved, respectively, and (c) the combined overlapping images. From Ref. 10.

in 2012 [17]. At present, light microscopy is usually used to make the diagnosis. Comparison of lung tissue excised from normal mice and from mice with tumors by ASEM indicates that this instrument could become an important, less time-consuming diagnostic tool because the cryo thin-sectioning to a few μm, which usually takes about 15–30 minutes for each sample for light microscopy, is not required for ASEM [10]. Consistent with light microscopy observations of lung thin sections from mice (Figure 5.5a, e), ASEM of normal lung tissue revealed typical thin-wall structures with alveoli, alveolar ducts, a vein system, and trachea (Figure 5.5b–d), while only faint traces of the regular alveoli and alveolar ducts were discernable for metastasized

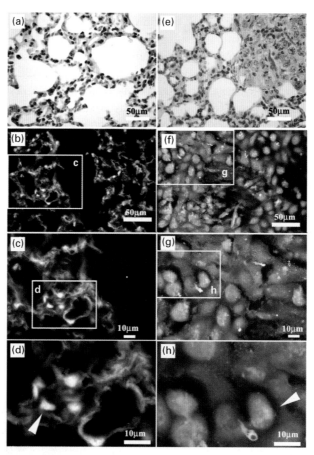

Figure 5.5. Comparison of normal lung and lung metastasized by breast cancer cells. (a) Light microscopy of a thin section of normal lung: hematoxylin-eosin (HE)-stained; nuclei are blue, cytoplasm is red. (b) Low magnification ASEM image of an independently prepared slab of normal lung tissue stained with platinum-blue (Pt-blue) and phosphotungstic acid. Alveoli with alveolar ducts, a vein system and trachea can be discerned. (c, d) Higher magnification images of the indicated fields. Arrowhead: normal sized nuclei. (e–h) Comparative observation of tissue excised from a lung metastasized with breast cancer cells: (e) light microscopy; (f–h) ASEM. The cells are a different shape and their nuclei are larger (arrowhead), showing they are cancer cells. From Ref. 10.

lung (Figure 5.5f h). Nuclei close to the surface of the tissue slabs appeared as bright cores and were larger in the breast cancer cells, which were also a different shape. Spinal-cord metastasis of a small number of breast cancer cells was also distinguished [10].

5.3.3 Platelet Generation by Megakaryocytes Cultured on the ASEM Dish

Platelets are indispensable cellular components to stop bleeding, and highly related to both myocardial and cerebral infarction under pathological conditions. Mature megakaryocytes (MKs) generate beaded cell projections called proplatelets and shed off platelets [18]. Monitoring the growth of MKs cultured in the open ASEM dish by light microscopy allowed these to be fixed at an appropriate time point, stained with heavy metal solutions *in situ* and observed using SEM. The pseudopodia extended beaded strings (Figure 5.6a, b), and the proplatelets contained vesicles (Figure 5.6c). After fixation and perforation of mature MKs, immunolabeling of P-selectin indicated that some of the vesicles might be α-granules, which express this adhesion protein on their surface. Additional labeling of α-tubulin suggests that the vesicles are transported on the microtubules of proplatelets (Figure 5.6d, e) [14].

5.3.4 Microparticle Movements in Water and the Effect of *In Situ* Administration of Chemicals

The configuration of the Clairscope allows reagent addition and monitoring of processes over time. The large capacity of the ASEM dish enables prolonged observation in a stable environment, minimizing the influence of evaporation.

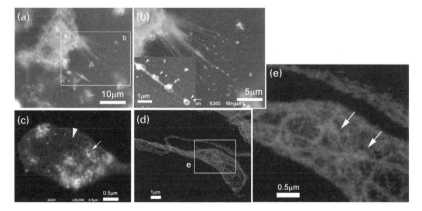

Figure 5.6. ASEM of platelet generation by megakaryocytes (MKs). MKs with proplatelet formation cultured on an ASEM dish. (a, b) Fixed and stained with Ti-blue. Beaded proplatelets extend from pseudopodia. (c) A proplatelet bead fixed and Au-tagged for P-selectin and counter-stained with heavy metal solutions. The arrow and arrowhead indicate putative α-granule and dense granule, respectively. (d, e) A proplatelet bead fixed, Au-tagged for P-selectin and further for α-tubulin to detect microtubules. Arrows indicate putative α-granules. From Ref. 14.

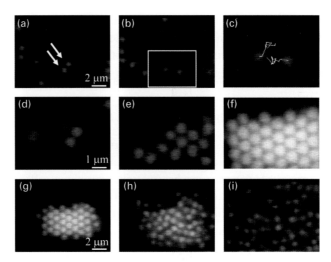

Figure 5.7. Movement of silica particles in water. (a, b) Magnification 5000×; sequential frames recorded at the same region at 0.5 s intervals. (c) The movement of the two particles in the white square in (b). (d–f) Magnification 10 000×; time between (d) and (e), 1 s; between (e) and (f), 24 s. (g–i) Magnification 5000×; sequential frames recorded at 2 s intervals. Particles moved randomly at 5000× magnification, but clustered at 10 000×. Modified from Ref. 19.

An initial experiment is shown in Figure 5.7, where colloidal silica particles suspended in water were repeatedly imaged at a magnification of 5000×. The resulting ASEM video showed constant random Brownian motion (Figure 5.7a–c). However, when the magnification was increased, the particles clustered and ordered into a close-packed structure (Figure 5.7d–f). When the magnification was reduced, the cluster dispersed again (Figure 5.7g–i). This "radiation-induced self-organization" [19] occurs because silica particles are positively charged in aqueous solution and are attracted by the negative charges injected on imaging. Since the beam intensity increases with magnification, packing is enhanced at high magnification. The repulsive force between the positively charged particles became dominant again when the current density was reduced.

Chemicals can be administered to a sample *in situ* using a long tube positioned above the center of the ASEM dish and connected to a syringe situated outside the shield door. Figure 5.8 shows the effect of adding several tens of microliters of saturated NaCl solution drop-wise to a silica particle sample. The particle speed of motion gradually decreased to almost zero 15 s after the initial salt addition, due to aggregation and attachment to the Si_xN_y window [9]. Various physical and/or chemical phenomena occurring when reagent is added to a sample have been monitored *in situ* using closed cell microscopy (see [20] for an example). Since the sample space of the ClairScope is open to the atmosphere, remote control of the pipette is possible, making it easy to control the addition of reagents and the amounts administered.

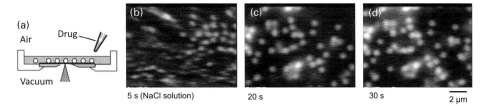

Figure 5.8. Effect of NaCl on the movement of silica particles. A schematic representation is shown in (a). Saturated NaCl solution was added during observation (b). The speed at which the silica particles moved gradually decreased, and almost all of the particles were stationary 15 s after the initial salt addition (c), and were at the same position 10 s later (d). From Ref. 19 (a) and Ref. 9 (b–d).

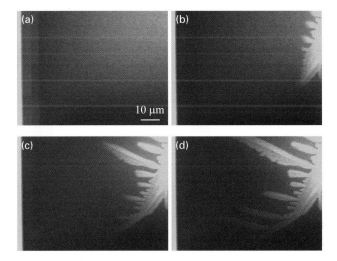

Figure 5.9. *In situ* ASEM images of the evaporation-induced crystallization of PBS solution. (a) The initial PBS droplet. (b–d) Progressive growth of dendrite-like structures. The interval between images is 2 s. From Ref. 19.

5.3.5 Volume Change: Evaporation of PBS Solution

The open sample dish makes it straightforward to study evaporation processes by ASEM. Figure 5.9 shows evaporation of phosphate-buffered saline solution (PBS: 154 mM NaCl, 3.0 mM Na_2HPO_4, 1.1 mM KH_2PO_4, pH 7.2) placed on the window. At first, the PBS droplet showed no significant contrast (Figure 5.9a) then, as evaporation proceeded, a dendrite-like structure suddenly emerged (Figure 5.9b), grew further (Figure 5.9c), and finally matured into a large dendrite (Figure 5.9d) [19]. The ability of ASEM to observe evaporation could be important for various fields, including the painting and printing industries. It might be possible to improve drying processes by observing what happens *in situ* in various solvents at different temperatures.

5.3.6 Electrochemical Reactions

The dynamic observation of electrochemical phenomena in aqueous electrolytes is important in developing batteries and fuel cells. A specially designed ASEM dish with a gold cathode and anode (Figure 5.10a, b) [19] allows changes occurring at the cathode to be monitored over time by ASEM. In Figure 5.10c, the deposition of Au on the cathode (dissolved from the anode) is visible on applying a large voltage between the electrodes in the presence of an electrolyte [19]. ASEM observations provide a rapid tool for monitoring electrochemical processes, and the data obtained are complementary to electrochemical observations made using open cell (Chapter 3) or closed cell TEM (Chapters 10 and 11).

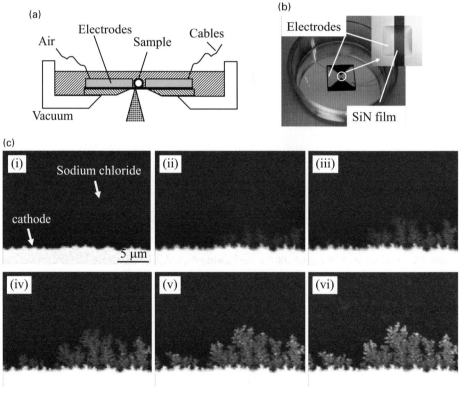

Figure 5.10. ASEM images of the electrochemical deposition of Au. Gold electrodes were immersed in NaCl solution and a high voltage (2.1 V) was applied between them causing Au dissolution at the anode and deposition at the cathode. (a, b) The electrochemical dish with two electrodes consisting of a 30 nm Ti adhesion layer and a 100 nm thick Au layer on the silicon nitride film, patterned using conventional photolithography to extend over the window. (c): (i) Area near the cathode, immediately after application of the voltage; (ii–vi) dendritic structures appeared and grew towards the anode. Images (ii–vi) were recorded 2.4, 3.0, 4.0, 6.0, and 8.0 s, respectively, after image (i). From Ref. 19.

5.3.7 Temperature-Dependent Phenomena

A metal nanoparticle paste is becoming popular as a new wiring material for electrical circuits. Small droplets containing the particles are deposited directly on the substrate by an inkjet machine. The liquid is then evaporated by raising the temperature and the particles are sintered. *In situ* observation using ASEM can help optimize these processes. To demonstrate the possibilities, Figure 5.11 shows water-dispersed silver paste placed on the window of a temperature-controlled ASEM dish (Section 5.2.2) and heated [9]. The sedimentation and sintering process can be monitored using the ASEM.

A second example is the temperature-dependent change that occurs in solder. The phase change occurring when solder is heated has been observed at low air pressure

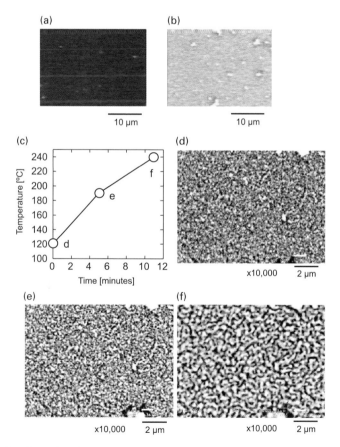

Figure 5.11. Silver paste sintering using the heated ASEM dish. A drop of silver paste was placed on the window. Initially, only a few 100 nm particles were observed (a). When the temperature rose to 100 °C, the silver particles started to stick to the surface of the window due to water evaporation, giving a bright ASEM image (b). At higher temperatures (c), the size of the silver particles gradually increased due to sintering (d–f). From Ref. 9.

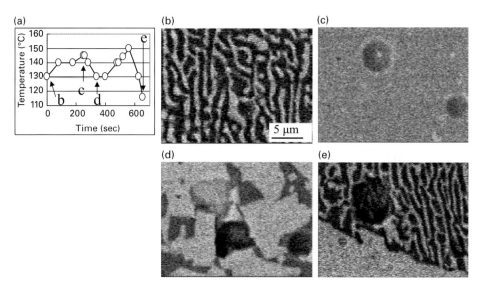

Figure 5.12. Temperature-dependent changes occurring in solder (Sn: 42 wt%, Bi: 58 wt%) during continuous observation. (a) Temperature profile; the bulk eutectic temperature of 138 °C is indicated in purple. (b–e) Representative images captured from a video documenting two sequential heating/cooling cycles; temperatures indicated in (a). (b) Solidified solder at 130 °C, revealing the presence of a Bi-rich and an Sn-rich solid solution. (c) After heating to 145 °C; the solder is molten and the contrast is almost uniform. (d) After cooling to 130 °C. (e) After heating to 150 °C and then cooling to 115 °C. The difference between (d) and (e) suggests that the morphology depends on the cooling conditions. From Ref. 19.

using environmental SEM (ESEM) [21]. However, because the fluxes in solder are volatile, the process at atmospheric pressure may be different from that at low pressure. In Figure 5.12, the process is continuously monitored at atmospheric pressure by ASEM. High contrast patterns are observed below the eutectic temperature (b, d, e) with an almost uniform contrast above the eutectic temperature. The morphology of the contrast depended on the cooling conditions.

The maximum temperature attainable with the ASEM dish is approximately 250 °C. Up to this limit, not only solders but also various kinds of temperature-dependent physical and chemical reactions, including thermosetting resins, should be observable at atmospheric pressure by ASEM.

5.4 Correlative Microscopy (CLEM)

Starting with the lateral (landscape) TEM (LEM 2000, Akashi Seisakusho Ltd., Tokyo, Japan) specifically developed for CLEM in the 1980s, the recent trend has been to use standard electron microscopes, cryo-electron microscopy, or closed cell TEM together with fluorescence microscopy (FM) for correlative imaging [22–24]. Biological samples usually need to be specifically labeled for CLEM, and a range of dual-labeling methods has been developed to allow comparison of FM mapping and the higher resolution EM images. Methods available include fluorescence-nanogold (FluoroNanogold™)-labeled antibodies and probes [25, 26], fluorescent semiconductor particles (quantum dots) [27, 28], and fluorescence conversion protocols [29].

The ability of the ASEM to image liquid samples at atmospheric pressure and the direct connection between the light microscope above the sample dish and the SEM below make ASEM well suited for CLEM observations. CLEM of a range of biological samples achieved using ASEM is the main focus of this section. Several of the other SEM technologies developed for the observation of wet samples in air, which will be discussed below in Section 5.5, have also been exploited for CLEM.

5.4.1 Structure of the Endoplasmic Reticulum

CLEM was used to find a metal stain that labels the endoplasmic reticulum (ER) for subsequent ASEM imaging to precisely localize this physiologically important tubular organelle [8]. First, for fluorescence microscopy, the protein disulfide isomerase (PDI) present in the ER of COS7 cells cultured on an ASEM dish was immunolabeled with a fluorescent tag. After fluorescence microscopy, the cells were counterstained *in situ* with Pt-blue solution and the same area imaged by ASEM. Comparison of the two staining patterns confirmed that Pt-blue solution stained the ER. The ASEM gave a more detailed view, revealing fine tubes bound with even finer structures (Figure 5.13).

5.4.2 Formation of a Supermolecular Complex by Sensing of Ca^{2+}

Ca^{2+} storage in the ER of cells plays an essential role in various physiological functions, including neuronal transmission and embryogenesis. The membrane protein STromal Interaction Molecule (STIM) 1 functions as a calcium sensor in the ER. It is known to change its spatial distribution in response to calcium concentration. When the cell's Ca^{2+}

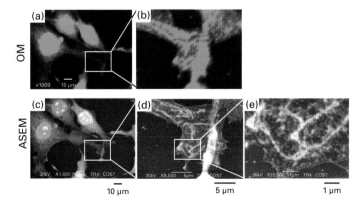

Figure 5.13. CLEM of the endoplasmic reticulum (ER). The ER of COS7 cells was labeled with Alexa Fluor® 488 for fluorescence microscopy and then *in situ* with Pt-blue for ASEM. Light microscopy shows the gross structure of the ER (top) while ASEM of the same area reveals the presence of many fine structures (bottom). From Ref. 8.

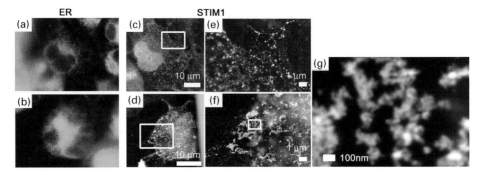

Figure 5.14. Rearrangement of STIM1 in the ER in response to Ca^{2+} store depletion. COS7 cells that express STIM1 were cultured on an ASEM dish and treated with thapsigargin to deplete their Ca^{2+} store, or left untreated, fixed and perforated. STIM1 was labeled with FluoroNanogold for ASEM. The ER was labeled with anti-PDI antibody (green). (a–c) Without store depletion: STIM1 was distributed on the ER. (d–g) With store depletion: STIM1 proteins formed strings on the ER near the cell surface. Images (c), (f), and (g) are enlargements of the boxes in (b), (e), and (f) respectively. Modified from Ref. 12.

store is depleted, a dynamic complex forms between STIM1 and another protein, Orai1, to create a calcium channel in the plasma membrane. In this process, STIM1 is believed to gather into puncta near the cell surface. It is difficult to observe the spatial distribution of STIM1 with sufficient resolution using light microscopy. However, Figure 5.14 shows changes in STIM1 distribution using a combination of light microscopy, FluoroNanogold immunolabeling and ASEM. STIM1 expressed in COS7 cells (Figure 5.14; white) is usually distributed throughout the ER (Figure 5.14; green). But in response to Ca^{2+} store depletion, ASEM showed a dynamic string-like gathering of STIM1 on the ER near puncta [12]. The immunolabeling required for this experiment was possible because of the preservation of the protein properties (antigenicity) of cells in aqueous solution.

5.4.3 Other Applications of CLEM in Cell Biology

CLEM using the ASEM has been applied to examine several other key questions regarding cell structure and function. As an example, axonal segmentation is observed in primary cultures of neurons from vertebrates and invertebrates and is believed to have important functions in embryonic development, especially in the early stages. Isolated neurons can be grown on the ASEM dish, and CLEM gives a detailed picture of the localization of different proteins in the axial fibers [13]. CLEM also helps to correlate axonal segmentation with specific cytoskeletal structures such as contacts between microtubule bundles [13]. Since microtubules work as a rail for cell trafficking, the contacts could influence cell trafficking.

CLEM can also provide further information on cell trafficking, which is performed by combinations of proteins in complex machineries, but with most of the mechanisms still unknown. Transfected cells can be immunolabeled, so that information on the spatial distribution of proteins can be identified both by fluorescence microscopy and by

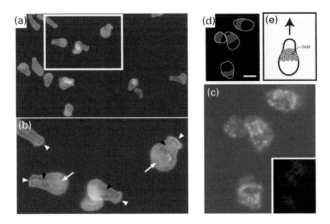

Figure 5.15. ASEM observation of *Mycoplasma* in solution. (a, b) *M. mobile* cells after fixation and staining with five different heavy metals. DNA (arrow) and a ring-like structure (arrowhead) were observed at opposite ends, with another structure of variable shape (black arrowhead) between; (b) enlargement of the box in (a), field of view 4 μm. (c) Localization of the leg protein Gli349 labeled with dually tagged secondary antibody, with inset the fluorescence image recorded by CLEM showing chromatin (blue) and Gli349 (red). (d) Diagram indicating the location of Gli349 in the ASEM image. (e) Diagram indicating the antibody labeling the cell "neck". Scale bar 0.5 μm. Modified from Ref. 11.

ASEM. Such experiments help to identify how knockdown of certain proteins in cargoes affects trafficking of enzymes in specific types of cells [9].

5.4.4 Imaging Bacteria in Water

ASEM provides characteristic easily recognizable views of bacteria such as *Mycoplasma mobile*. Figure 5.15 shows ASEM images of this bacterium after fixing and staining with heavy metals [11]. Each cell is seen to have a bulb-shaped body with a protrusion. A strongly stained area in the bulb can be attributed to DNA. It is separated from a cap-like density at the other end by a delicate, variably shaped structure that might be the cytoskeleton or some other structure [30]. Immunolabeling with a monoclonal antibody and subsequent ASEM shows that this structure may, at least partly, colocalize with the so-called "leg" structure. The ease with which *M. mobile* can be recognized makes ASEM a potential diagnosis tool for mycoplasma-related diseases such as pneumonia, which is not easily diagnosed by chest X-ray imaging. This could be particularly important as a complementary diagnostic tool in cases where the usual diagnosis method, the polymerase chain reaction, is precluded by the blockers included in nasopharyngeal aspirates [11].

5.4.5 Cathodoluminescence via Correlative Microscopy

It is possible to detect cathodoluminescence using the light microscope simultaneously as a sample is scanned with the electron beam [9, 31]. The BSE detector of the ASEM is sensitive to light, so the light microscopy has to be performed using a mercury lamp to

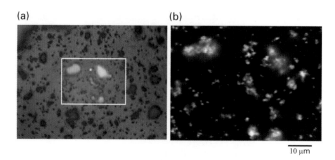

Figure 5.16. CLEM of cathodoluminescence. (a) Bright field light microscope image of ZnO including a field (white rectangle) continuously irradiated through the Si_xN_y window of the sample dish by the electron beam of the ASEM. (b) ASEM image of the rectangle in (a). Due to cathodoluminescence, only particles within the electron beam-irradiated area were luminescent. From Ref. 9.

provide faint bright field illumination of the whole area. A sample consisting of ZnO particles dispersed in water is shown in Figure 5.16. As the sample is continuously irradiated by the electron beam from below, it produces a bright bluish-white cathodoluminescence signal in the bright field image recorded from above by the light microscope. In these types of experiments, the detection efficiency for light can be improved by use of a photomultiplier. We suggest that new fields of CLEM can be opened up for analysis of complex biological samples by using immunolabeling [12] in combination with inorganic fluorescent materials [32], rather than gold particles, to realize multicolor staining for cathodoluminescence microscopy. A different approach is to coat the membrane with a luminescent thin film. This allows the SEM beam to be used to form a nanoscale light source [33] which can then interact with a sample placed on the luminescent film. By choice of imaging and window parameters, it is possible to arrange for the primary beam to be scattered mainly in the membrane and luminescent film, avoiding direct interaction and damage of the sample. The sample can be kept in air or liquid, and the emitted light with which it interacts can be tuned by choice of luminescent material to provide information about the sample.

5.5 Other SEM Techniques for Examining Liquids at Atmospheric Pressure

We have attempted to show in the preceding section the broad range of applicability of the ASEM in biology and materials science, emphasizing the ability of the ASEM to carry out experiments requiring CLEM. In this section we discuss and compare other developments in instrumentation that share the same aims: to examine liquids at atmospheric pressure and to enable CLEM of samples in water.

5.5.1 Closed Cells for CLEM: Bulk and Thin Liquids

As described in Chapter 4, a closed cell or environmental capsule for SEM can be placed in almost any SEM sample chamber. It completely isolates the sample from the microscope vacuum allowing bulk, wet samples to be imaged through a thin window. Cathodoluminescence microscopy is possible in closed cells by introducing a light guide at the bottom of the capsule [6]. Details of capsule fabrication, sample imaging, and applications are described in Chapter 4. Compared to ASEM as described above, the specialized equipment required is only the capsule, rather than a dedicated microscope. However, preparing a sample in a capsule is more time consuming than loading into the ASEM sample dish, and it is more difficult to modify the sample during observation, for example by addition of reagents.

The closed cells developed for TEM include two windows separated by a small distance, less than about 1 µm, so that electrons can be transmitted through the thin liquid layer. It was recognized early on that thin closed liquid cells can also be used in SEM, if the image is acquired with a STEM detector [34]. Thin liquid cells promise higher resolution than capsules, due to the reduced interaction volume for the primary electrons. Large objects, including many of the biological samples described in previous sections, cannot easily be placed in the restricted volume, but for materials science, STEM detection in the SEM successfully provides observations of smaller structures in water [35].

CLEM can be achieved by using a variation of the thin closed cell concept where the top window is electron transparent and the lower one is a glass slide [36]. It is then possible to image with a high numerical aperture epifluorescence microscope from below, where the objective lens of the microscope is positioned within the vacuum system of the SEM. Simultaneously, SEM imaging from above achieves resolution similar to that in a closed cell through a membrane. The liquid layer does not have to be thin, since electrons are not transmitted, making it more straightforward to load biological samples. The ability to record CLEM simultaneously will be advantageous in real-time studies of cellular processes.

5.5.2 *Air*SEM™

For some time it has been recognized that sealing an SEM column with an electron-transparent window can allow imaging of samples in air if they can be brought close enough to the window to reduce the path length in air [37]. *Air*SEM™ (b-Nano, Israel) is a modern implementation of this idea that enables SEM of liquids in the open atmosphere (Figure 5.17a, b; [38–40]), while also allowing CLEM. It is based around a field emission SEM column that is upright, with electrons exiting at the bottom. The exit is sealed by a fixed thin-film window, which is similar to that used in SEM capsules or liquid cell TEM as described in Chapter 2. Samples are positioned 50–400 µm below the window and are therefore physically separate from the microscope. The alignment between sample and SEM is computer controlled through a light microscope mounted parallel to the SEM. A rail beneath the two microscopes allows the sample to be moved from one to the other to identify the location of interest and perform correlative microscopy.

This microscope geometry offers several advantages. Sample preparation is straightforward because the material of interest can be placed on a conventional glass slide. Samples with large volume can be imaged, including those involving complex *in situ*

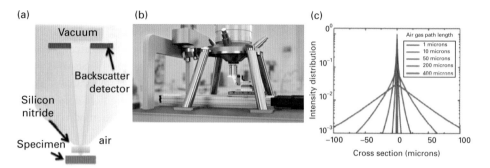

Figure 5.17. Design of the *air*SEM™. (a) Schematic; (b) photo. (c) Monte Carlo simulation of 30 keV electron beam profiles in air. (a) and (c) from Refs. 39 and 40, respectively; (b) from www.b-nano.com.

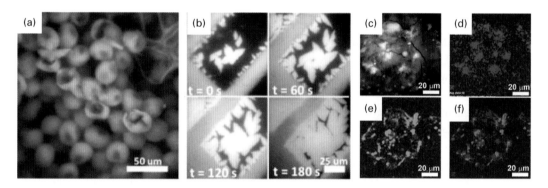

Figure 5.18. Applications of the *air*SEM. (a) Wet moss imaged on a glass side, illustrating the minimal preparation needed. (b) A time sequence from the growth of Pb dendrites on the surface of a liquid. (c–e) Correlative *air*SEM image of a ≤100 μm slice of a fixed sea urchin embryo at ambient conditions showing the BSE signal (c), calcium EDS map (d), and calcein fluorescence (e). (f) Superposition of (d) and (e) showing colocalized signals in yellow. Slight deformation, resulting from electron beam damage, may have decreased the colocalization area. (a, b) from Ref. 40 and (c–f) from Ref. 38.

experiments or detectors. In addition to backscattered electrons, other signals can be collected easily if required, including secondary ions, X-ray mapping and cathodoluminescence, as well as CLEM. A series of samples can be imaged rapidly since the column vacuum is not broken. Additional imaging tools can be included, such as atomic force microscopy, by adding a simple transfer. Finally, because the sample and window are decoupled, the window does not need to be replaced for each sample.

Typical operating conditions include 30 kV imaging electrons, a 30 nm Si_xN_y window, and an atmospheric gas path length between window and sample of up to 400 μm. The resolution can be high under these conditions, if a usable fraction of the incident beam remains unscattered, while the scattered electrons provide a structureless background (Figure 5.17c; [39]). This is achieved, for example, when imaging labeled biological samples. The window and atmospheric gas then simply reduce the contrast of features according to Beer's law. At 30 keV, the mean free path for the imaging electrons is ~30 nm in the window material and ~75 μm in air [40]. Scattering in the window therefore dominates the resolution, provided that the gas path length is not too long and the region of interest is near the top surface of the sample. Monte Carlo simulations (an example is given in Figure 5.17c; [39]) show that, under these conditions, nanoscale edge resolution is possible at several hundred μm path length in an air–He mixture, and suggest that the contrast can be increased by 2.5× by reducing the membrane thickness from 30 nm to 5 nm.

Both SEM and STEM operation are possible, and the technique has been applied to biological and materials specimens [38–40]. But because only sample regions close to the upper liquid surface are imaged without loss of resolution, the water layer above "wet" specimens should be thin with images recorded before drying [38]. We anticipate that this relatively new type of microscopy will gain additional applications in the future.

5.5.3 Hitachi Atmospheric Scanning Electron Microscope

The Hitachi atmospheric scanning electron microscope is a tungsten-filament SEM with upright configuration. Below the column is a thin window, through which the electron beam exits and scans the wet sample underneath to form an image using backscattered electrons [41]. The sample is not directly in air, but is imaged while in a smaller chamber that forms part of the SEM sample chamber and includes a stage that can bring the sample close to the window. The use of the microscope's sample chamber is beneficial in allowing observations at atmospheric pressure, at lower pressure if the sample chamber is pumped, or in a controlled environment if gases are flowed. This provides increased opportunities for experiments that may, for example, compare a material's structure at high and low pressure. However, the arrangement does not permit CLEM or easy manipulation of the sample by addition of reagents.

This microscope has successfully imaged cellulose fibers while drying, red blood cells, renal glomerulus, mildew fungus, and a leaf surface in which stomata and the surrounding epidermis were clearly resolved [41, 42]. The rapid turnaround time of this microscope suggests that useful information will come from many other types of sample.

5.6 Conclusions

We have described several applications of ASEM, as well as other techniques for observing open samples in atmospheric conditions. The benefits of simple and rapid sample preparation, and the opportunities available from correlative measurements, suggest that there will be many areas in which these techniques can make an important contribution and solve critical problems.

The cancer cell images and the characteristic images of mycoplasma we have shown suggest the applicability of ASEM to medical diagnosis. ASEM can locate structures in a several micrometer-thick observable sample depth, and has provided information on the three-dimensional nature of biological structures such as microtubules in axons. During such experiments, correlative light microscopy can be used to select areas for SEM inspection, to verify biological events such as the transfection of a gene, and to determine the distribution of proteins. As well as static structures, real-time monitoring of a sample during chemical administration or temperature change is possible. Already studied are phenomena such as random motion and self-organization of particles, electrochemical deposition, and phase changes such as solder melting. This range of time-resolved applications could be expanded into areas such as the aggregation, sedimentation, and chemical reactions of microparticles, and testing the quality delivered by industrial production systems. As an important future aim, ASEM promises to aid drug development using its capabilities of delivering a drug to a system in the open sample dish. Here, correlative microscopy will allow the overall effect of drugs on cell growth and morphology to be monitored by light microscopy, observing their effect on particular organelles as a function of time.

Improving the information obtainable from ASEM and related techniques requires instrumental developments. One area in which we expect to see progress is in the electron-transparent window. Larger area windows will increase observation efficiency, which is especially critical for diagnosis. Coated windows, perhaps including luminescent material or functionalized to enhance biological properties, can enhance the range of samples that can be observed. The use of thinner materials, perhaps based on graphene as discussed in Chapter 4, can improve the imaging resolution but need to be designed to preserve the ease and speed of sample preparation. With future targets in nanotechnology, nanomaterials, and soft materials including polymers, and studies of microbiota and micro-environments involved in ecology, SEM of samples in water under atmospheric conditions has many promising future opportunities.

Acknowledgements

We thank Dr. Toshihiko Ogura at the National Institute of Advanced Industrial Science and Technology (AIST) for valuable discussions concerning the development of the ClairScope, and also Dr. Hidetoshi Nishiyama at JEOL for his important research and discussions. The work was supported by Grant-in-Aid for Scientific Research on Innovative Areas, Structural Cell Biology (to CS), by CREST (to CS), by grants from the Ministry of Education, Culture, Sports, Science, and Technology (MEXT) (to CS), by matching funds between AIST and JEOL, and by AIST.

References

1. I. M. Abrams and J. W. McBrain, A closed cell for electron microscopy. *J. Appl. Phys.*, **15** (1944), 607–609.
2. T. L. Daulton, B. J. Little, K. Lowe and J. Jones-Meehan, *In situ* environmental cell-transmission electron microscopy study of microbial reduction of chromium(VI) using electron energy loss spectroscopy. *Microsc. Microanal.*, **7** (2001), 470–485.
3. N. de Jonge and F. M. Ross, Electron microscopy of specimens in liquid. *Nat. Nanotechnol.*, **6** (2011), 695–704.

4. S. Thiberge, O. Zik and E. Mosesa, An apparatus for imaging liquids, cells, and other wet samples in the scanning electron microscopy. *Rev. Sci. Instrum.*, **75** (2004), 2280–2289.
5. N. de Jonge, D. B. Peckys, G. J. Kremers, D. W. Piston, Electron microscopy of whole cells in liquid with nanometer resolution. *Proc. Natl. Acad. Sci. USA*, **106** (2009), 2159–2164.
6. S. Thiberge, A. Nechushtan and D. Sprinzak *et al.*, Scanning electron microscopy of cells and tissues under fully hydrated conditions. *Proc. Natl. Acad. Sci. USA*, **101** (2004), 3346–3351.
7. F. M. Ross, In Situ Transmission Electron Microscopy, in *Science of Microscopy*, Ed. P. W. Hawkes and J. C. H. Spence, pp. 445–534. (New York: Springer, 2007).
8. H. Nishiyama, M. Suga and T. Ogura *et al.*, Atmospheric scanning electron microscope observes cells and tissues in open medium through silicon nitride film. *J. Struct. Biol.*, **172** (2010), 191–202.
9. H. Nishiyama, M. Koizumi, K. Ogawa *et al.*, Atmospheric scanning electron microscope system with an open sample chamber: configuration and applications. *Ultramicroscopy*, **147** (2014), 86–97.
10. N. Memtily, T. Okada, T. Ebihara *et al.*, Observation of tissues in open aqueous solution by atmospheric scanning electron microscopy: applicability to intraoperative cancer diagnosis. *Int. J. Oncol.*, **46** (2015), 1872–1882
11. C. Sato, S. Manaka, D. Nakane *et al.*, Rapid imaging of mycoplasma in solution using atmospheric scanning electron microscopy (ASEM). *Biochem. Biophys. Res. Commun.*, **417** (2012), 1213–1218.
12. Y. Maruyama, T. Ebihara, H. Nishiyama, M. Suga and C. Sato, Immuno EM-OM correlative microscopy in solution by atmospheric scanning electron microscopy (ASEM). *J. Struct. Biol.*, **180** (2012), 259–270.
13. T. Kinoshita, Y. Mori, K. Hirano *et al.*, Immuno-electron microscopy of primary cell cultures from genetically modified animals in liquid by atmospheric scanning electron microscopy. *Microsc. Microanal.*, **20** (2014), 469–483.
14. K. Hirano, T. Kinoshita, T. Uemura *et al.*, Electron microscopy of primary cell cultures in solution and correlative optical microscopy using ASEM. *Ultramicroscopy*, **143** (2014), 52–66.
15. A. Nyska, C. A. Cummings, A. Vainshtein *et al.*, Electron microscopy of wet tissues: a case study in renal pathology. *Toxicol. Pathol.*, **32** (2004), 357–363.
16. I. Barshack, S. Polak-Charcon, V. Behar *et al.*, Wet SEM: a novel method for rapid diagnosis of brain tumors. *Ultrastruct. Pathol.*, **28** (2004), 255–260.
17. http://www.who.int/mediacentre/factsheets/fs297/en/.
18. T. Junt, H. Schulze, Z. Chen *et al.*, Dynamic visualization of thrombopoiesis within bone marrow. *Science*, **317** (2007), 1767–1770.
19. M. Suga, H. Nishiyama, Y. Konyuba *et al.*, The atmospheric scanning electron microscope with open sample space observes dynamic phenomena in liquid or gas. *Ultramicroscopy*, **111** (2011), 1650–1658.
20. K. Fukushima, A. Ishikawa and A. Fukami, Injection of liquid into environmental cell for in situ observations. *J. Electron Microsc.*, **34** (1985), 47–51.
21. N. Koopman, Application of ESEM to fluxless soldering. *Microsc. Res. Tech.*, **25** (1993), 493–502.
22. A. V. Agronskaia, J. A. Valentijn, L. F. van Driel *et al.*, Integrated fluorescence and transmission electron microscopy. *J. Struct. Biol.*, **164** (2008), 183–189.
23. A. Sartori, R. Gatz, F. Beck *et al.*, Correlative microscopy: bridging the gap between fluorescence light microscopy and cryo-electron tomography. *J. Struct. Biol.*, **160** (2007), 135–145.

24. M. J. Dukes, D. B. Peckys and N. de Jonge, Correlative fluorescence microscopy and scanning transmission electron microscopy of quantum-dot-labeled proteins in whole cells in liquid. *ACS Nano*, **4** (2010), 4110–4116.
25. R. D. Powell, C. M. Halsey, D. L. Spector *et al.*, A covalent fluorescent-gold immunoprobe: simultaneous detection of a pre-mRNA splicing factor by light and electron microscopy. *J. Histochem. Cytochem.*, **45** (1997), 947–956.
26. J. M. Robinson and D. D. Vandre, Efficient immunocytochemical labeling of leukocyte microtubules with FluoroNanogold: an important tool for correlative microscopy. *J. Histochem. Cytochem.*, **45** (1997), 631–642.
27. B. N. Giepmans, T. J. Deerinck, B. L. Smarr, Y. Z. Jones and M. H. Ellisman, Correlated light and electron microscopic imaging of multiple endogenous proteins using quantum dots. *Nat. Methods*, **2** (2005), 743–749.
28. A. M. Smith and S. Nie, Next-generation quantum dots. *Nat. Biotechnol.*, **27** (2009) 732–733.
29. G. Gaietta, T. J. Deerinck, S. R. Adams *et al.*, Multicolor and electron microscopic imaging of connexin trafficking. *Science*, **296** (2002), 503–507.
30. D. Nakane and M. Miyata, Cytoskeletal "jellyfish" structure of Mycoplasma mobile. *Proc. Natl. Acad. Sci. USA*, **104** (2007), 19518–19523.
31. Y. Nawa, W. Inami, A. Miyake *et al.*, Dynamic autofluorescence imaging of intracellular components inside living cells using direct electron beam excitation. *Biomed. Opt. Express*, **5** (2014), 378–386.
32. D. R. Glenn, H. Zhang, N. Kasthuri *et al.*, Correlative light and electron microscopy using cathodoluminescence from nanoparticles with distinguishable colours. *Sci. Rep.*, **2** (2012), 865.
33. W. Inami, K. Nakajima, A. Miyakawa and Y. Kawata, Electron beam excitation assisted optical microscope with ultra-high resolution. *Opt. Express*, **18** (2010), 12897–12902.
34. J. A. Swift and A. Brown, An environmental cell for the examination of wet biological specimens at atmospheric pressure by transmission scanning electron microscopy. *J. Phys. E: Sci. Instrum.*, **3** (1970), 924–926.
35. J. M. Grogan and H. H. Bau, The Nanoaquarium: a platform for in situ transmission electron microscopy in liquid media. *J. Microelectromech. Syst.*, **19** (2010), 885–894.
36. N. Liv, A. C. Zonnevylle, A. C. Narvaez *et al.*, Simultaneous correlative scanning electron and high-NA fluorescence microscopy. *PLoS One*, **8** (2013), e55707.
37. E. D. Green and G. S. Kino, Atmospheric scanning electron-microscopy using silicon-nitride thin-film windows. *J. Vac. Sci. Technol. B*, **9** (1991), 1557–1558.
38. N. Vidavsky, S. Addadi, J. Mahamid *et al.*, Initial stages of calcium uptake and mineral deposition in sea urchin embryos. *Proc. Natl. Acad. Sci. USA*, **111** (2014), 39–44.
39. K. Nguyen, M. Holtz and D. Muller, *Air*SEM: electron microscopy in air, without a specimen chamber. *Microsc. Microanal.*, **19 (Suppl. 2)** (2013), 428–429.
40. K. Nguyen, J. D. Richmond-Decker, M. Holtz, Y. Milstein and D. A. Muller, Spatial resolution of scanning electron microscopy without a vacuum chamber. *Microsc. Microanal.*, **20** (2014), 26–27.
41. Y. Ominami, S. Kawanishi, T. Ushiki and S. Ito, Observation of wet samples using a novel atmospheric scanning electron microscope. *Microsc. Microanal.*, **20** (2014), 1154–1155.
42. Y. Ominami, S. Kawanishi, T. Ushiki and S. Ito, A novel approach to scanning electron microscopy at ambient atmospheric pressure. *Microscopy*, **64** (2015), 97–104.

6 Temperature Control in Liquid Cells for TEM

Shen J. Dillon and Xin Chen

6.1 Introduction: Controlled Temperature Experiments

In situ and model *operando* EM experiments ultimately strive to reproduce the environmental conditions relevant to the corresponding laboratory experiment, industrial process, or physical phenomenon. Since temperature arguably represents the most broadly important thermodynamic control variable, *in situ* liquid cell electron microscopy must incorporate temperature control and measurement to effectively characterize a broad range of materials, chemical, physical, biological, and geological phenomena. This chapter describes the progress made to date in performing experiments in closed liquid cells in TEM, of the type described in Chapters 1 and 2, above and below ambient temperature. We discuss associated technical challenges and describe potential future opportunities for heating and cooling in other applications of liquid cell microscopy.

6.1.1 Strategies for Heating and Cooling

Processes that do work evolve heat, and thus an inordinate variety of methods can be employed to heat samples. Among those with the greatest potential applicability are electrical, optical, magnetic, fission, and mechanical heating, or the use of thermal sources or sinks that initially exist at the desired temperature. While mechanical work, nuclear reactions, and electromagnetic induction are common thermal sources for many applications and cannot be discounted completely for specialized *in situ* experiments, their use will not be considered in detail here. This section will, therefore, focus on the most promising and previously demonstrated approaches. These include Joule heating, equilibration with a remote thermal source, photon heating (e.g. laser, microwave, or infrared), and thermal fluid circulation. Key experimental considerations include the rate of thermal cycling, the precision of temperature control and calibration, thermal stability, and temperature range. These factors depend on the type of thermal source, its distance from the region of interest, and the design of the liquid cell. Of particular importance is the spatial distance between the heat source and region of observation. This may vary from nanometers to decimeters, and is a major factor in determining the thermal cycling rate, the magnitude of thermal gradients, and the requisite temperature control and calibration methods. Figure 6.1 depicts several general schemes. The benefits and disadvantages of each are described below.

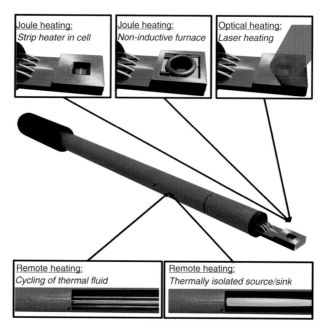

Figure 6.1. Schematic of a TEM specimen rod and liquid cell depicting several potential modifications to enable heating or cooling of a liquid sample.

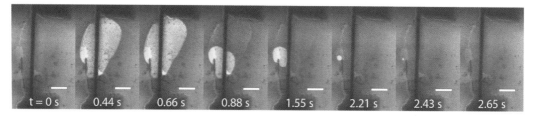

Figure 6.2. Time-lapse sequence of STEM images showing the growth and collapse of a nanoscale bubble (bright feature) during thermal cycling of a Pt strip (dark vertical line) Joule heated for 740 ms at 0.5 V and 1 mA. The scale bar is 1 μm. Reprinted with permission from Ref. 2.

6.1.2 Joule Heating

The majority of high temperature TEM experiments in both liquids and gases have utilized Joule heating. A detailed review of early work on the subject can be found in Ref. 1. Joule heaters can be fabricated directly on substrates so that the heat source is placed very close to the region of interest. For example, Figure 6.2 shows thermally induced bubble evolution [2] that utilized a 200 nm wide and 200 nm thick FIB-deposited Pt strip. Heating of the aqueous solution to $T > 100\,°C$ required 0.5 V and ~1 mA. In this experiment, the thermal source is localized to the region that is imaged. This allows for accurate local temperature control and rapid thermal cycling, but also produces large local thermal gradients.

A Joule heater strip or wire may be utilized simultaneously to both heat and measure temperature, through calibrated resistance measurements where the conductor functions as a resistance temperature detector (RTD). Knowledge of the temperature at any position within the sample then requires the application of thermal diffusion models. Fortunately, the local geometry may be characterized reasonably well *in situ*, enabling accurate modeling. The application of complementary techniques such as the 3ω method [3] could allow one to characterize the local thermal conductivity. This approach measures the frequency dependence of a voltage drop in the wire. This voltage drop dependence results from the temperature dependence of electrical resistance, providing a measure of the system's thermal impedance which may be quantified with appropriate models of the system geometry. With a calibrated Joule heater/RTD, the temperature within the strip can be accurate to better than 1 °C, but the temperature of the surrounding environment must be calculated. Based on a 500 nm thick layer of pure water transporting heat via diffusion only, the time constant for transport is on the order of 10^{-6} s.

Similar local Joule heating approaches have been employed broadly for *in situ* studies in vacuum, low pressure gases in open cell designs, and high pressure gases in closed cell designs [1, 4, 5]. Such designs enable heating to temperatures in excess of 1200 °C. A major benefit of local heating is the minimization of spatial drift resulting from thermal expansion, which can be a limiting factor for high resolution imaging. However, materials selection is critical if the Joule heater is going to be exposed to the liquid environment and the electron beam. The choice of material depends on the desired temperature range, accuracy, and the corrosion chemistry of the overall system. As well as the Pt shown above, other local Joule heater materials, such as W, SiC, and polycrystalline Si, have been used in heating stages for vacuum and gas operation, and can be obtained from several commercial vendors. If the liquid cell window is made of sufficiently thin material, such heating lines could be placed on the external surface of the cell to avoid chemical interactions between the liquid and the heater, limit its effect on the internal geometry of the liquid cell, and prevent potential enhanced electron beam-induced interactions at the heater–liquid interface. Examples include Pt [2] and Au [6]. Encapsulation of the heater between inert membranes, such as atomic layer or chemical vapor deposited or sputtered films, could also limit chemical interactions. This is the approach that is starting to be introduced in commercial heating liquid cells, where an encapsulated heater is patterned in the form of a broad strip that covers part of the window area or is adjacent to the window area. It is important to calibrate the internal temperature when using heaters on one window to understand the effects of thermal gradients vertically as well as laterally, and the effects of liquid flow across the heated area.

6.1.3 Equilibration with a Remote Heat Source

Commercial hot stages typically work by using a small non-inductive resistance furnace built into the holder that surrounds a standard 3 mm TEM sample. If the closed liquid cell is small enough, without for example bulky electrical or liquid feedthroughs, it can be placed directly into such a commercial hot stage. This is the case for certain liquid cells [7]: the resulting setup has been utilized to characterize particle nucleation and growth at elevated temperatures. Specific results are described in more detail in Chapter 9. Assuming good thermal contact at all the interfaces in the system, with thermal diffusion through either the liquid water or a silicon support substrate, the time constant for thermal transport in this type of setup is on the order of 0.1 s. In such an arrangement, thermocouples are typically placed adjacent to the heat source, approximately 1.5 mm from the center of the sample. This introduces a temperature drop that could be large enough to require temperature calibration at the region of interest. The error in the temperature measurement is most pronounced during rapid thermal cycling and least pronounced at steady state. Commercial embodiments of hot stages can access temperatures ranging from $-150\,°C$ to $1000\,°C$, but to date do not integrate liquid cells. A benefit of using a furnace hot stage for liquid cell experiments is the absence of any chemical interactions between the heater and the liquid in the cell.

An even more remote heat source, for example located within the specimen rod or external to and insulated from the microscope, may be utilized in situations where both heating and cooling are desirable. The liquid cell is placed on a thermally conducting stage, which is itself placed in contact with a thermal reservoir or thermoelectric device, enabling flexibility in both the temperature range accessible and the ability to design additional functionality into the liquid cell by not sacrificing valuable space on or near the cell to accommodate the heating elements. This approach tends to have long-scale thermal gradients, requiring accurate calibration at the sample position, but will produce small local gradients in the region of interest. One disadvantage is the slow thermal cycling, with a time constant on the order of 10^1–10^2 s. There is also a large spatial drift during thermal cycling, because a relatively large volume of material has to be heated, and a low maximum temperature, which depends on how well the heat source is thermally insulated from the microscope. For cooling experiments, it may be valuable to utilize a cold reservoir in concert with a heat source, such as Joule heating or optical heating, in order to tune the temperature to values above that of the reservoir. Microscale investigations of ice crystallization have often utilized thermoelectric heating and cooling in concert with a cold reservoir for precise control of temperature [8].

6.1.4 Photon Heating

Laser heating has been utilized for *in situ* studies of solid state phase transformations, primarily in the context of Dynamic TEM [9]. In liquid cells, laser heating has also been utilized to control nucleation and growth of particles from solution [10]. Laser heating may be implemented through fiber optics in the specimen rod or optical ports in the microscope column. Generally, laser heating provides rapid thermal cycling that results in larger thermal gradients, but the amount of photon energy converted to heat is dependent on the sample. Techniques such as thermoreflectance may be used to accurately calibrate temperature, but in many situations such approaches may not be practically viable. Nanoscale thermocouples or RTD measurements may also be utilized in concert with laser heating, but they introduce significant complexity in accurately modeling the temporal and spatial distributions of heat and its transport.

6.1.5 Thermal Fluid Circulation

Heat exchange from thermal fluids is a common general approach to heating and cooling. However, it is yet to be applied to liquid cell TEM, despite the fact that micro- and nanofluidic cells have been designed and utilized for *in situ* and *operando* experiments [11–13]. Thermal fluids pumped from an external reservoir provide high and low temperature flexibility and could be integrated in a manner that introduces limited spatial drift and relatively rapid thermal cycling. We anticipate a greater use of this approach in future liquid cell designs.

6.2 Electron Beam-Induced Heating

Typical TEM experiments are performed at a high power density at the sample, on the order of 10^3–10^6 W cm^{-2}. This could raise concerns regarding beam-induced heating. Indeed, experiments performed in aqueous solutions typically observe bubble formation above a critical current density [2, 14], which can be composition dependent [15], leading to speculation that these bubbles could relate to boiling or thermal degassing. However, most studies have concluded that such bubbles result from beam-induced decomposition of the solution, rather than heating. This is supported by the relative stability of the beam-induced bubbles, which suggests that the gas present is not water vapor and has low solubility in water [2, 14]. Bubble formation during liquid cell TEM experiments is discussed in Chapter 7, where it is shown that radiolysis of water by the intense electron beam can readily produce sufficient hydrogen to exceed its solubility limit and nucleate bubbles.

The electron beam does indeed heat the sample, but beam-induced heating can be shown to have a minor effect on the temperature of the solution in a liquid cell. Calculations of beam-induced heating are based on electron energy loss spectroscopy measurements or mean free path-based calculations of energy transferred to the sample during inelastic collisions, coupled with a thermal diffusion model describing the cell [14–16]. Two separate calculations have shown that beam-induced heating should be small: less than 2.5 °C in typical experiments [16] or less than 4 °C at 300 kV and ~1 A cm^{-2} [14]. The dependence of temperature rise on the key parameters of beam current and irradiated area for TEM and STEM are shown in Chapter 7. Note that beam-induced heating and melting is readily observed in solid TEM samples that consist of particles distributed on a grid. It has been suggested [16] that such melting results from point contacts with the substrate leading to large interfacial thermal resistance, which limits their ability to effectively dissipate energy.

Overall, the results thus far in the literature seem to confirm the expectation that beam-induced heating in liquids is negligible. However, the degree of beam-induced heating will be sensitive to geometry and the nature of thermal impedances within the liquid cell. For example, the formation of bubbles by radiolysis can change heat transfer in the closed liquid cell. Future work using liquid cells could see the use of 0D and 1D geometries such as nanoconfined liquids in graphene or other 2D materials, or in nanofluidic channels. Beam-induced heating in such geometries could depend significantly on thermal transport in the surrounding materials.

6.3 Temperature Measurements

Accurate temperature measurement and control is the basis for performing meaningful *in situ* or *operando* experimentation. Approaches to temperature measurement share challenges and solutions with related technologies such as microfluidics. As was the case for heating and cooling sources, temperature measurements may be made at positions varying from localized in the region of interest to remote – millimeters to centimeters away. Local measurements are always desirable, but can introduce experimental design challenges associated with integrating additional electrical, optical, mechanical, or phase change elements into the cell. Remote measurements are most appropriate in cases where thermal gradients are small and predictable, such that it is possible to reasonably calibrate the temperature in the region of interest.

The most promising methods for temperature determination include thermocouples, RTDs, optical-based measurements, and phase change references. Microscale thermocouples provide accurate measurements and good temporal resolution and are well suited for remote temperature measurements. Nanoscale thermocouples, more appropriate for local measurements, have been investigated for a variety of applications [17] and could be integrated into liquid cell design. Thermocouples are based on the intrinsic Seebeck voltage arising at a junction between dissimilar metals, and can function at both micro- and nanoscale when appropriate interfaces are manufactured sufficiently well [18]. RTDs have typically been favored for nanoscale thermography due to their accuracy and sensitivity at the nanoscale, along with their good temporal resolution and simplicity of design. RTDs can serve dual roles as heaters and measurement devices and have been utilized as nanocalorimeters [19] and for measurement of local thermal conductivity [3]. Optical methods such as infrared thermography, liquid crystal thermography, thermoreflectance, fluorescence thermography, or optical interference thermography could provide accurate temperature measurements at or below micrometer-scale resolution. However, they require the additional complication of integrating appropriate optics into the sample stage or microscope column. Similarly, scan probe-based techniques have been popular for high spatial resolution thermography [20]. While scan probes have been integrated with TEM stages, such experimental platforms have so far not been supplemented by environmental control. Phase change materials can provide a useful reference for a single temperature and provide localized measurements, and have been useful in evaluating the degree of electron beam-induced heating [21].

It should be noted that the Gibbs–Thomson effect will influence calibration in nanoscale systems, relative to the bulk reference [22]. In one example relevant to environmental cells, Ag nanoparticles imaged in 1 atm. of air and at ambient temperature were observed to oxidize in the presence of the electron beam [23]. At sufficiently high beam current density, the AgO unexpectedly vaporized due to beam-induced heating that resulted from the reaction enthalpy between Ag and atomic and ionized oxygen produced by the beam. Bulk AgO vaporizes at ~280 °C, suggesting the possibility of significant beam-induced heating in this system. In this case, the heating occurred because the beam altered the reaction pathway of the system, and predicting the temperature change resulting from the enthalpy of reaction would have been difficult by other means.

6.4 Applications

6.4.1 High Temperature Applications

High temperature experimentation is central to numerous scientific and engineering problems. This section highlights several example applications that have been reported as well as potential future opportunities.

In the growth of nanoscale crystals from liquid phases, controlled temperature has an important role in enabling the study of a broader variety of phenomena compared to the typical room temperature experiments. Two examples of phenomena seen at elevated temperature are an oscillatory growth of bismuth nanoparticles [7] and surface oxidation-induced internal Kirkendall void formation in Bi [24]. These phenomena were observed using a custom-fabricated liquid cell that fits within the 3 mm dimensions of a commercial heating stage. The details of this work are discussed in Chapter 9.

High temperature is of course required for studies of phase transitions such as boiling. Figure 6.1 showed an example where thermally induced nucleation, growth, and stability of H_2O vapor bubbles was characterized [2] using Pt lines to induce local heating above the boiling point of water. On cooling back to ambient temperature, some bubbles were found to be more stable than would be anticipated from simple thermodynamic treatments. This could relate to dual contributions of thermally induced boiling and beam-induced H_2 and O_2 that could form the bubble and subsequently stabilize it.

Hydrothermal precipitation makes up another important class of reactions that may be visualized using temperature-controlled liquid cells. The precipitation of ZnO from $Zn(NO_3)_2$-hexamethylenetetramine solutions at 95 °C has been characterized [15] by making use of a thermal reservoir outside the microscope to heat the samples. This allowed the growth kinetics associated with *ex situ* thermally induced precipitation to be compared quantitatively with *in situ* thermally induced precipitation and *in situ* electron beam-induced precipitation, all under conditions of diffusion-limited kinetics. The results are interesting in the context of liquid cell heating experiments in general. It was concluded that the kinetics were not ideally correlated between any of the three approaches. The *in situ* and *ex situ* thermally induced reaction kinetics differed due to the effects of sample confinement in the liquid cell on diffusion and particle coalescence. The kinetics associated with the *in situ* electron beam-induced reaction and the *in situ*

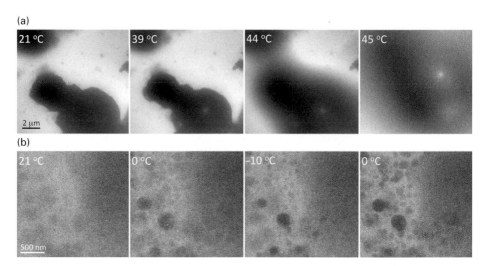

Figure 6.3. Time-lapse sequences of images showing (a) the melting of commercial chocolate above room temperature and (b) the solidification of a commercial ice cream formulation below room temperature.

thermally induced reaction differed due to the magnitude of the driving force for nucleation and growth, despite both reactions being performed in a diffusion-limited kinetic regime. In spite of these effects, the *in situ* study still provided useful insights into nanoscale interactions such as coarsening and coalescence mechanisms, as well as morphological evolution. Since the magnitude of the driving force for a reaction can affect its reaction pathway, as noted above for the case of AgO nucleation, in many chemical reactions it may be ideal to study a system under conditions where the thermodynamic driving force resulting from temperature (and pressure) control exceeds that induced by the electron beam. Temperature is typically the critical variable that controls the driving force for a chemical reaction, so the ability to vary the temperature creates opportunities to study representative reaction pathways associated with many phenomena that might not be accessible by electron beam-induced effects.

While only a few notable examples currently exist in the literature, the capability to perform *in situ* heating experiments in liquids generates a host of novel characterization opportunities. For example, Figure 6.3a follows the thermally induced melting of small particles of chocolate. A range of nanoscale food science problems could be addressed in liquid cells with temperature control, and the results would complement data from the relatively widespread use of environmental scanning electron microscopy in the study of food science [25–27]. Similarly, while *in situ* EM work on cellular or viral systems has been carried out at room temperature [28, 29], temperature-dependent studies in this field would significantly broaden the range of potential applications. High pressure and high temperature experiments could more broadly impact fields such as geology, where inorganic fluids might be confined in liquid cells at high pressures to provide nanoscale insights into geological processes, or high temperature electrochemistry, which is critical in refining and energy conversion.

6.4.2 Low Temperature Applications

Cryo-holders for electron microscopes have long been exploited to characterize samples at low temperatures. Applications have primarily focused on pre-frozen solutions, low damage imaging, or electrical measurements that require thermal noise to be minimized, such as electron beam-induced current measurements [30, 31]. However, cooling liquid cells allow a range of important phase transformations and other liquid phase phenomena to be visualized. In Figure 6.4, we show the solidification of saline solutions containing nanoparticles during *in situ* cooling [32]. The platform employed a liquid cell resting on a cold finger that was thermally isolated from the specimen rod. This experiment provides insights into the competition between hexagonal and cubic ice nucleation and growth at different temperatures. The associated length scales and morphology governing interactions between Au nanoparticles and the ice phase during

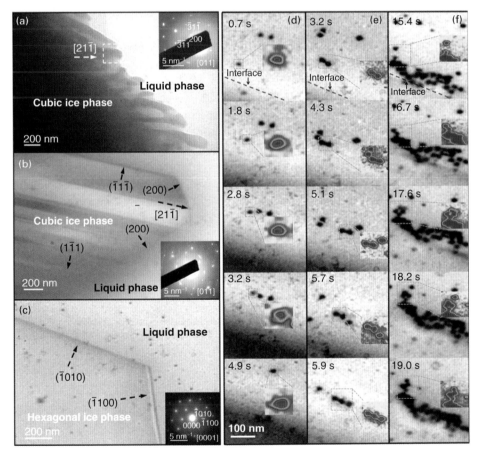

Figure 6.4. (a–c) Ice crystallizing from saline solution at 220, 245, and 260 K, respectively. (d–f) Three time-lapse TEM image series showing interactions between nanoparticles (black) and a crystallizing ice–water interface. As the ice (darker region) interface propagates, the Au nanoparticles in the water (lighter region) assemble and align with the crystallization front. The insets show the particle contrast in false color for clarity.

particle rejection and occlusion could be imaged. Another phase transformation is shown in Figure 6.3: crystallization in a commercial ice cream formulation, where coarsening and the final size of ice crystals is critical in determining the texture of the product. Further reports of *in situ* and *operando* experimentation in liquids and gases at low temperatures are few in the literature. However, a variety of potential applications can be envisioned. Temperature-dependent electrochemistry, including low temperature Li-ion battery function, the structure, evolution, and stability of methane clathrates, reactions in low temperature solvents such as liquid NH_3, and the behavior of low temperature liquids (N_2 or even H_2) would all be valuable areas where a low temperature liquid cell could provide unique information.

6.4.3 Limitations

Key limitations on the proliferation of *in situ* liquid cell TEM performed in controlled temperature environments have included a lack of commercially available holders, the difficulty of integrating multiple stimuli and probes into an experimentally flexible holder, the extent to which temperature can be known in the region of interest, and the ability of liquid cell construction materials to function in harsh environments. As these limitations are addressed, we anticipate a much greater range of experiments. Figure 6.5 shows a rudimentary custom-built specimen rod that integrates a Cu rod for external heating or cooling along with a control thermocouple and electrical feedthroughs for performing *in situ* temperature-controlled electrochemistry. While the design could be improved to be more compact, the size of the holder demonstrates the challenges of integration. One might, for example, wish to perform temperature-controlled photoelectrochemistry, and such a platform could additionally require a fiber optic. The plethora of potential applications of liquid cell TEM in functional systems necessitates the construction of more flexible commercially available experimental platforms.

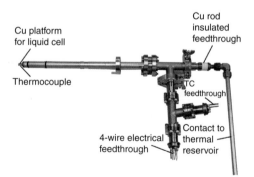

Figure 6.5. Image of a rudimentary temperature-controlled electrochemical specimen rod outfitted with a Cu "thermal feedthrough", which conducts between an external thermal reservoir and the sample stage and is insulated from the specimen rod. The platform also contains a control thermocouple, attached where the liquid cell mounts, and electrical feedthroughs. While relatively simplistic, the image highlights the challenge of fitting multiple probes, stimuli, and controls in a single specimen rod. A similar apparatus was used for the experiments in Figures 6.3 and 6.4.

As highlighted in Section 6.1, thermal gradients in the liquid cell can be large or small depending on the nature of the heating source, and temperature measurements can be made locally or more remotely with calibration. Nevertheless, knowing the exact temperature in the region of interest remains challenging, especially when accounting for beam-induced heating and temperature changes associated with the enthalpies of reactions. For the current state of the art, *in situ* liquid cell TEM is more suitable for characterizing generalities of reaction mechanisms and pathways than as a sensitive probe of critical temperatures or a tool for systematic kinetic measurements that correlate exactly with *ex situ* experiments at nominally the same temperature. These challenges can be overcome, but require significant effort and investment in next generation microscope and holder designs.

Liquid cell construction materials may be a limiting factor in moving towards more extreme liquid cell environments. For example, Si_xN_y corrodes under hydrothermal conditions or extremes in pH. Such windows are typically constructed on exposed Si substrates and the Si can corrode and contaminate the solution. Metallic enclosures, polymer o-rings, or adhesives commonly utilized to assemble and seal liquid cells are susceptible to thermal and chemical degradation. Since liquid cell TEM experiments are typically performed on small volumes, contamination from the enclosure could be particularly insidious in producing spurious results.

6.5 Outlook

A great variety of temperature controlled liquid cell TEM experiments will undoubtedly be applied to examine chemical reactions, electrochemical reactions, photochemical reactions, and others, by exploiting existing capabilities and retrofitting methods demonstrated in the published literature. New opportunities also exist for extending controlled temperature and environment experiments to more extreme conditions through integration of disparate types of experiments and the utilization of new liquid cell materials and designs. ETEMs have long been integrated with sources of additional stimuli, such as ion or particle irradiation. Recent efforts have been made to extend such extreme environments to experiments in liquids [33], which should enable characterization of phenomena such as irradiation-induced reactions and stress corrosion cracking in extreme environments. Mechanically robust closed liquid cells designed to operate at high pressures could be utilized to study hydrothermal reactions such as crystal growth, nanostructure synthesis, fuel production, or volcanic activity. The nature of hydrothermal chemical reactions remains enigmatic in many cases and nanoscale *in situ* observations could provide important new insights. Similarly, molten salt chemical and electrochemical reactions relevant to aluminum refining, crystal growth, and materials synthesis could be performed *in situ* provided that appropriate chemically and mechanically robust electrochemical cells are developed. Related experiments have been performed within nanotubes [34, 35], but greater flexibility is required to generalize the experiments to arbitrary systems. Finally, a variety of platforms have been developed for nanomechanical testing within the TEM. Integration of, for example, a

MEMS nanomechanical testing apparatus with high temperature and high pressure liquid cells will enable access to the study of a rich variety of mechanics problems in extreme environments.

References

1. E. P. Butler, In situ experiments in the transmission electron microscope. *Rep. Prog. Phys.*, **42** (1979), 833–896.
2. E. R. White, M. Mecklenburg, S. B. Singer, S. Aloni and B. C. Regan, Imaging nanobubbles in water with scanning transmission electron microscopy. *Appl. Phys. Express.*, **4** (2011), 055201.
3. D. G. Cahill, Thermal conductivity measurement from 30 to 750 K: the 3ω method. *Rev. Sci. Instrum.*, **61** (1990), 802–808.
4. T. Alan, T. Yokosawa, J. Gaspar *et al.*, Micro-fabricated channel with ultra-thin yet ultra-strong windows enables electron microscopy under 4-bar pressure. *Appl. Phys. Lett.*, **100** (2012), 081903.
5. T. Yokosawa, T. Alan, G. Pandraud, B. Dam and H. Zandbergen, In-situ TEM on (de)hydrogenation of Pd at 0.5–4.5 bar hydrogen pressure and 20–400 °C. *Ultramicroscopy*, **112** (2012), 47–52.
6. Y. Liu, X. Chen, K. W. Noh and S. J. Dillon, Electron beam induced deposition of silicon nanostructures from a liquid phase precursor. *Nanotechnology*, **23** (2012), 385302/1.
7. H. L. Xin and H. Zheng, In situ observation of oscillatory growth of bismuth nanoparticles. *Nano Lett.*, **12** (2012), 1470–1474.
8. C. A. Knight, A. L. DeVries and L. D. Oolman, Fish antifreeze protein and the freezing and recrystallization of ice. *Nature*, **308** (1984), 295–296.
9. N. D. Browning, M. A. Bonds, G. H. Campbell *et al.*, Recent developments in dynamic transmission electron microscopy. *Curr. Opin. Solid State Mater. Sci.*, **16** (2012), 23–30.
10. J. E. Evans, K. L. Jungjohann, N. D. Browning and I. Arslan, Controlled growth of nanoparticles from solution with in situ liquid transmission electron microscopy. *Nano Lett.*, **11** (2011), 2809–2813.
11. C. Mueller, M. Harb, J. R. Dwyer and R. J. D. Miller, Nanofluidic cells with controlled pathlength and liquid flow for rapid, high-resolution in situ imaging with electrons. *J. Phys. Chem. Lett.*, **4** (2013), 2339–2347.
12. K. L. Klein, I. M. Anderson and J. N. de Jonge, Transmission electron microscopy with a liquid flow cell, *J. Microsc.*, **242** (2011), 117–123.
13. R. R. Unocic, R. L. Sacci, G. M. Brown *et al.*, Quantitative electrochemical measurements using in situ ec-S/TEM devices. *Microsc. Microanal.*, **20** (2014), 452–461.
14. J. M. Grogan, N. M. Schneider, F. M. Ross and H. H. Bau, Bubble and pattern formation in liquid induced by an electron beam. *Nano Lett.*, **14** (2014), 359–364.
15. Y. Liu, K. Tai and S. J. Dillon, Growth kinetics and morphological evolution of ZnO precipitated from solution. *Chem. Mater.*, **25** (2013), 2927–2933.
16. H. Zheng, S. A. Claridge, A. M. Minor, A. P. Alivisatos and U. Dahmen, Nanocrystal diffusion in a liquid thin film observed by in situ transmission electron microscopy. *Nano Lett.*, **9** (2009), 2460–2465.
17. J. Christofferson, K. Maize, Y. Ezzahri *et al.*, Microscale and nanoscale thermal characterization techniques. *J. Electron. Packaging*, **130** (2008), 041101.

18. E. Shapira, D. Marchak, A. Tsukernik and Y. Selzer, Segmented metal nanowires as nanoscale thermocouples. *Nanotechnology*, **19** (2008), 125501.
19. S. L. Lai, G. Ramanath, L. H. Allen and P. Infante, Heat capacity measurements of Sn nanostructures using a thin-film differential scanning calorimeter with 0.2 nJ sensitivity. *Appl. Phys. Lett.*, **70** (1997), 43–45.
20. L. Shi and A. Majumdar, Thermal transport mechanisms at nanoscale point contacts. *J. Heat Transfer*, **124** (2002), 329–337.
21. T. Yokota, M. Murayama and J. M. Howe, In situ transmission-electron-microscopy investigation of melting in submicron Al-Si alloy particles under electron-beam irradiation. *Phys. Rev. Lett.*, **91** (2003), 265504/1.
22. S. L. Lai, J. Y. Guo, V. Petrova, G. Ramanath and L. H. Allen, Size-dependent melting properties of small tin particles: nanocalorimetric measurements. *Phys. Rev. Lett.*, **77** (1996), 99–102.
23. L. Sun, K. W. Noh, J.-G. Wen and S. J. Dillon, In situ transmission electron microscopy observation of silver oxidation in ionized/atomic gas. *Langmuir*, **27** (2011), 14201–14206.
24. K.-Y. Niu, J. Park, H. Zheng and A. P. Alivisatos, Revealing bismuth oxide hollow nanoparticle formation by the Kirkendall effect. *Nano Lett.*, **13** (2013), 5715–5719.
25. D. J. Stokes and A. M. Donald. In situ mechanical testing of dry and hydrated breadcrumb in the environmental scanning electron microscope (ESEM). *J. Mater. Sci.*, **35** (2000), 599–607.
26. E. H. C. Bromley, M. R. H. Krebs and A. M. Donald, Aggregation across the length-scales in β-lactoglobulin. *Faraday Discuss.*, **128** (2004), 13–27.
27. A. Blennow, M. Hansen, A. Schulz *et al.*, The molecular deposition of transgenically modified starch in the starch granule as imaged by functional microscopy. *J. Struct. Biol.*, **143** (2003), 229–241.
28. K.-L. Liu, C.-C. Wu, and Y.-J. Huang, Novel microchip for in situ TEM imaging of living organisms and bio-reactions in aqueous conditions *Lab Chip*, **8** (2008), 1915–1921.
29. K. L. Klein, I. M. Anderson, N. de Jonge *et al.*, Transmission electron microscopy with a liquid flow cell. *J. Microsc. Oxford*, **242** (2011), 117–123.
30. S. Tanaka, H. Tanaka, T. Kawasaki *et al.*, EBIC imaging using scanning transmission electron microscopy: experiment and analysis. *J. Mater. Sci.: Mater. Electron.*, **19** (2008), S324–327.
31. C. Cabanel, J. L. Maurice and J. Y. Laval, Scanning transmission electron beam induced current in polycrystalline silicon. *Mater. Sci. Forum.*, **10–12** (1986), 545–550.
32. K. Tai, Y. Liu and S. J. Dillon, In situ cryogenic transmission electron microscopy for characterizing the evolution of solidifying water ice in colloidal systems. *Microsc. Microanal.*, **20** (2014), 330–337.
33. K. Hattar, D. C. Bufford and D. L. Buller, Concurrent in situ ion irradiation transmission electron microscope. *Nucl. Instrum. Methods Phys. Res., Sect. B.*, **338** (2014), 56–65.
34. Y. Gogotsi, J. A. Libera, A. Güvenç-Yazicioglu and C. M. Megaridis, In situ multiphase fluid experiments in hydrothermal carbon nanotubes. *Appl. Phys. Lett.*, **79** (2001), 1021–1023.
35. N. Naguib, H. Ye, Y. Gogotsi *et al.*, Observation of water confined in nanometer channels of closed carbon nanotubes. *Nano Lett.*, **4** (2004), 2237–2243.

7 Electron Beam Effects in Liquid Cell TEM and STEM

Nicholas M. Schneider

7.1 Introduction

The advent of liquid cells for TEM, as described in Chapters 1, 2, and 6, has made possible the direct imaging of processes in liquids with the nanometer resolution of the electron microscope. The samples are, however, subject to high dose rate radiation in the form of the electron beam. A detailed understanding of the fundamental interactions between the radiation and the suspending medium is necessary to interpret, suppress, or exploit beam-mediated phenomena in liquid cell electron microscopy [1, 2]. The study of chemical effects due to irradiation is called radiation chemistry, and this field has received intensive attention due to its importance in medical imaging, nuclear energy, atmospheric science, and food preservation [3–8].

Ionizing radiation, such as photons, γ-rays, X-rays, and electrons, readily transfers energy to the irradiated medium with effects that are relatively independent of the type of radiation. Deposited energy excites and dislodges orbital electrons, resulting in both generation of heat and changes in chemistry. In this chapter, we discuss the theoretical background necessary to understand the effects of radiolysis, heating, changes in liquid chemistry with and without diffusion, and their implications for experimentally observed phenomena in liquid cell microscopy. The model was developed for standard TEM imaging conditions, but we also discuss its relevance to STEM. SEM experiments take place under different conditions, in particular accelerating voltage and energy absorption, and irradiation effects are discussed in Chapter 4.

7.2 Electron Energy Loss in Liquids

Dose is the measure of the average *amount of energy absorbed by a mass* exposed to ionizing radiation. The SI derived unit for dose is the gray (Gy) and is defined as the absorption of one joule per kilogram of matter. The *dose rate* is, therefore, the dose per unit time (Gy/s). It is important to distinguish this measure from the oft-reported electron flux (e^-/nm^2 s), which is *a measure of the ionizing radiation itself*, not the energy transferred to the sample. Here, we consider dose rate due to an electron beam incident upon a thin layer of water in order to understand radiation effects in liquid cell EM. Water is used as the normative example, but other liquids can easily be considered by using their appropriate material properties in the equations below.

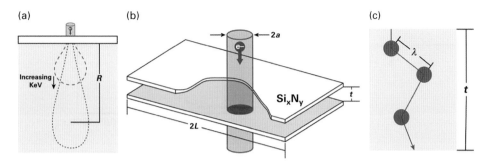

Figure 7.1. (a) Illustration of the shape taken by electron trajectories in the sample. R is the mean range. (b) Schematic of a cylindrical irradiated volume of radius a in the center of a window of dimension L enclosing a liquid of thickness t. (c) Schematic of an electron track through a liquid of thickness t. Each inelastic collision (separated by average distance of λ, the mean free path) produces excited species within a spur (red circle). The total track length s exceeds the thickness t.

Electrons passing through matter lose energy due to their interactions with nuclei and atoms of the irradiated medium. Electrons elastically scattered by nuclei may be scattered to large angles, but lose little energy. However, inelastic scattering causes both excitation and ionization as well as direction change. As the electrons pass through the medium, the average rate of energy loss is given by dE/ds, which is called the *(mean) stopping power*. An electron will come to rest once it has lost all of its energy after a path length of

$$R = \int \left(\frac{dE}{ds}\right)^{-1} dE, \qquad (7.1)$$

where R is called the *(mean) range* [9]. In reality, energy loss is a stochastic process and energy is deposited at about the average rate dE/ds with some range of variation, known as *straggle*. These variations in energy are also manifested in the range, causing actual particles to travel more or less than the average distance (Figure 7.1a). It is convenient, however, to consider only the mean, and this is called the *continuous slow down approximation* (CSDA). The straggle about the CSDA range is only about 10–15% [10]. As an example, the mean range for electrons in liquid water under atmospheric conditions at 20 kV, 30 kV, and 300 kV is 8.6 μm, 18 μm, and 840 μm, respectively [11]. In the following discussion, we consider only average values.

Electrons lose energy in two predominant ways: via Bremsstrahlung and collisions. Figure 7.2 shows how these two processes depend on energy [11]. Bremsstrahlung is radiation released while a charged particle accelerates or slows down. The *radiative stopping power*, S_R, is the energy lost by an electron by this means. Bremsstrahlung becomes the dominant form of energy loss above a *critical energy*, which is 90.8 MeV for water, as seen in Figure 7.2. The *collisional stopping power*, S_C, is the energy lost by electrons due to collisions causing excitation and ionization processes as they pass through the medium. This is the dominant form of energy loss for the acceleration

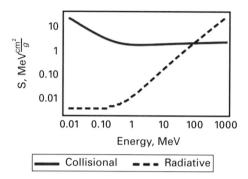

Figure 7.2. Collisional and radiative stopping power for electrons in water. From NIST [11].

voltages used in electron microscopy. The *total stopping power* is the sum of all forms of stopping power:

$$S_{\text{Total}} = \frac{dE}{ds} = S_R + S_C = \rho S, \quad (7.2)$$

and is often reported as the *density-normalized stopping power*, S (MeV cm^2 g^{-1}), which we refer to simply as stopping power from here on.

In liquid cell electron microscopy we need to estimate the total energy deposited in the medium, i.e. water. We use the stopping power to perform this estimation. The average energy deposited in the medium by a single electron will be the product of the average energy loss per track length and the average distance travelled:

$$dE_{e^-} = \frac{dE}{ds} s \, (\text{MeV}), \quad (7.3)$$

where s is the average track length. The dose absorbed by the water will be the energy deposited per mass, or:

$$\text{dose}_{e^-} = \frac{\frac{dE}{ds} s}{\rho V_I} \, (\text{MeV/g electron}), \quad (7.4)$$

where ρ is density and V_I is the irradiated volume (Figure 7.1b). Notice that this is really the stopping power divided by the volume. By substituting Equation (7.2) we get the dose due to a single electron:

$$\text{dose}_{e^-} = \frac{Ss}{V_I} \, (\text{MeV/g electron}). \quad (7.5)$$

To retrieve the dose rate for an electron beam, we break the volume up into the product of irradiated area, A_I (cm^2), and sample thickness, t, and multiply by the number of electrons per second, J, to obtain the dose rate, ψ:

$$\psi = \frac{Ss}{A_I t} J \, (\text{MeV/g s}). \quad (7.6)$$

Next, we convert into the correct units for dose rate, Gy/s, replace the irradiated area with an assumed circular beam geometry measured in m² instead of cm², and convert the electron flow into beam current, I (C/s). This results in a simple multiplicative factor of 10^5 (m² electron Gy g/cm² MeV C). The dose rate given by microscope parameters is then:

$$\psi = \frac{10^5 Ss}{\pi a^2 t} I \, (\text{Gy/s}), \tag{7.7}$$

where a is the beam radius. Finally, we assume that the average length over which each electron loses energy is equal to the sample thickness. This leads to the expression given in Ref. 1:

$$\psi = \frac{10^5 SI}{\pi a^2} \, (\text{Gy/s}). \tag{7.8}$$

For typical TEM conditions, a 300 keV beam of 1 μm radius and 1 nA current, Equation (7.8) gives a dose rate of 7.5×10^7 (Gy/s). It is important to realize that this value is many orders of magnitude higher than ranges studied in medical physics and nuclear reactor design (10^{-2} to 10^3 Gy/s).

In Equation (7.8), the assumption that the track length travelled by each electron is equal to the sample thickness is only correct for a thin sample, i.e. one where the sample thickness t is less than the mean free path between collisions, λ. The mean free path of electrons in water is shown in Figure 7.3 [12]. It can be seen that λ ranges from tens to hundreds of nanometers (λ = 45 nm, 63 nm, and 267 nm at 20 kV, 30 kV, and 300 kV, respectively). Compare these with the range, R, in the tens to hundreds of μm: at a microscopic level, we can think of the electron as undergoing an inelastic collision event every few tens of nm over a total distance of tens of μm (until it slows to zero) or less (if it already exited the sample). The locations of the inelastic events are called *spurs* and they form along the electron trajectory like beads on a string (Figure 7.1c).

To increase the accuracy of Equation (7.8), we need to calculate how far the electron actually travels during the time it is passing through the sample. Especially at low acceleration voltages, there may be significant lengthening of the track due to multiple

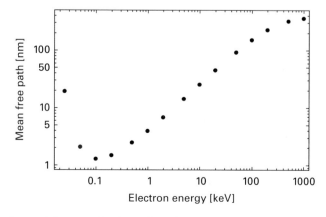

Figure 7.3. Electron mean free path in water. Produced from data presented in Ref. 12.

scattering. Multiple scattering in thin foils has been addressed in detail [13, 14]. In the limit where the total range, R, of the electron is much greater than the sample thickness, t, the average track length, s, can be accurately estimated by the first order approximation [13]:

$$s = t\left(1 + \frac{t}{\lambda}\right) \text{ (m)}. \tag{7.9}$$

Higher order corrections to Equation (7.9) are not required for typical liquid cell thicknesses in the few μm range.

For convenience, we can divide Equation (7.9) by the sample thickness to get the factor by which the electron track is extended due to multiple scattering events:

$$\frac{s}{t} = \left(1 + \frac{t}{\lambda}\right). \tag{7.10}$$

By substituting Equation (7.10) into Equation (7.7), the first order estimate of dose rate becomes:

$$\psi = \frac{10^5 SI}{\pi a^2}\left(1 + \frac{t}{\lambda}\right) \text{ (Gy/s)}, \tag{7.11}$$

which now includes the effect of multiple scattering on the average path taken by an electron (and the consequentially greater deposited energy). For a typical TEM experiment at 300 kV and a liquid thickness of 100nm, Equation (7.11) would estimate the dose rate to be 1.4 times greater than the zeroth order estimate of Equation (7.8). For a 1μm thick liquid layer, the first order dose rate would now be 4.8 times greater. In other words, the thickness of the sample plays a strong role in the energy deposition process.

So far, we have considered only the liquid in these energy deposition calculations. But all materials and interfaces within the liquid cell can locally alter the dose rate. The window material can cause scattering events and the emission of secondary electrons, which increases the local dose rate. The secondary electrons have low energy and hence very short range, so their effect is limited to a region within a few nanometers of the surface itself. Structures within the liquid cell, such as electrodes or even colloids in the liquid itself, will also modify the local energy deposition rate. The extent of these effects can be investigated via electron trajectory simulation software such as CASINO [15].

7.3 Electron Beam Heating

Having determined the rate at which energy is deposited into the sample, we now consider the changes that this energy causes to the sample. We discuss heating here and chemical effects in Section 7.4.

7.3.1 Calculation of Heating Profiles

As mentioned in Chapter 6, it is known experimentally that there is relatively little heating of liquid cell samples during typical TEM experiments [1, 16], even though thermally isolated samples can experience significant beam heating. To explain this observation, we start with the heat generation, calculated from the volumetric rate of energy deposited in the sample, then balance it with conduction away from the beam region [3, 18, 19].

The rate of energy deposition is:

$$q = 10^3 \rho \psi \ (\text{W/m}^3), \tag{7.12}$$

where ρ is the density, 10^3 (cm^3 J /g m^3 Gy) accounts for unit conversions, and ψ is the dose rate as given by Equation (7.8) or (7.11). The heat generation given by Equation (7.12) is assumed to take place uniformly over a cylinder (Figure 7.1b).

To calculate transfer of heat away from this volume, we take conduction as the dominant means of heat transfer. Furthermore, we assume that the surfaces in contact with the vacuum environment of the microscope act as insulators and that there are no significant temperature variations in the vertical direction (conduction/convection and radiation heat transfer through the vacuum are neglected). By considering the case where the beam is centered in the window region we retrieve the maximum temperature rise, since the center of the window is the most thermally isolated point in the system. The silicon chip used in most liquid cells acts as an effective heat sink due to the high thermal conductivity of silicon, its relatively large thermal mass, and its thermal connection to the holder. This leads to a constant temperature assumption for the boundary condition at the window edge (T_0 at $r = L$).

The temperature rise can be found by solving the dimensionless heat equation:

$$\frac{\partial \tilde{\theta}}{\partial \tilde{t}} - \tilde{\nabla}^2 \tilde{\theta} = 1 \qquad \text{(inside the beam, } 0 \leq \tilde{r} \leq 1\text{)}, \tag{7.13}$$

$$\frac{\partial \tilde{\theta}}{\partial \tilde{t}} - \tilde{\nabla}^2 \tilde{\theta} = 0 \qquad \text{(outside the beam, } 1 < \tilde{r}\text{)}. \tag{7.14}$$

Quantities with a tilde are dimensionless while plain letters represent dimensional quantities. In Equations (7.13) and (7.14), the temperature rise ($T - T_0$) is scaled by $a^2 q \, 10^{-6}/(\alpha_{\text{th}} C_p \rho)$ to get a dimensionless temperature rise $\tilde{\theta}$, where C_p is the specific heat at constant pressure, α_{th} is the thermal diffusivity, and 10^{-6} (m^3/cm^3) accounts for unit conversions to express ρ in g/cm^3. Length is scaled by the beam radius a and time is normalized by the diffusive time a^2/α_{th}. The boundary conditions are symmetry at the origin, constant temperature at the window edge, $\tilde{\theta}(\tilde{L}, \tilde{t}) = 0$, $\tilde{L} = L/a$, and initial temperature equal to the ambient, $\tilde{\theta}(\tilde{r}, 0) = 0$. An analytical expression for the steady state temperature profile can readily be obtained in dimensionless form both inside and outside the irradiated region [1]:

$$\tilde{\theta}_{\text{ss}}(\tilde{r}) = \begin{cases} \frac{1}{4}(1 - \tilde{r}^2) + \frac{1}{2}\ln[\tilde{L}] & (0 \leq \tilde{r} \leq 1) \\ \frac{1}{2}\ln\left[\frac{\tilde{L}}{\tilde{r}}\right] & (1 < \tilde{r} \leq \tilde{L}) \end{cases}. \tag{7.15}$$

The maximum temperature rise is at the beam center. It is given by substituting Equation (7.12) into the dimensional form of Equation (7.15) at $r = 0$. As expected, it is proportional to the dose rate:

$$\Delta T_{max} = \frac{a^2 \, 10^{-3}}{\alpha_{th} \, C_p} \psi \left(\frac{1}{4} + \frac{1}{2} Ln \left[\frac{L}{a} \right] \right). \quad (7.16)$$

7.3.2 Effect of Beam Current and Sample Thickness

By substituting Equation (7.8) or (7.11) for the dose rate, we can restate the temperature rise as a function of physical and microscope parameters:

$$\Delta T_{max} = \frac{S \, 10^2}{\pi \alpha_{th} C_p} I \left(1 + \frac{t}{\lambda} \right) \left(\frac{1}{4} + \frac{1}{2} Ln \left[\frac{L}{a} \right] \right), \quad (7.17)$$

where 10^2 (m^2 electron J/cm^2 MeV C) accounts for unit conversions. Equation (7.17) is a linear function of beam current, I, and thickness, t. The temperature rise, however, will have a non-linear relationship with beam energy (via stopping power, S) and radius, a. Figure 7.4 shows the maximum temperature rise, ΔT_{max}, as a function of beam current and radius in the case of a thin sample ($t/\lambda \to 0$) at voltages typical for TEM and SEM and other parameters given in the caption. Under typical TEM conditions, the temperature rise is only a few degrees Celsius. It can be much greater for low energy focused beams, such as a stationary STEM beam at 30 kV with beam currents on the order of a few tens of nanoamperes. The values are overestimates as they neglect heat transfer through the membrane which, for silicon nitride, has a thermal conductivity greater than water. Overall, we conclude that the beam does not raise the sample temperature much in a liquid cell, mainly because of the excellent thermal conductivity of the relatively thick sample. However, as we discuss next, non-thermal beam effects can cause more dramatic changes to liquid samples during observation.

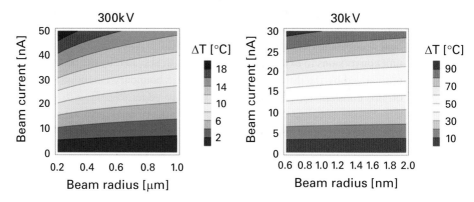

Figure 7.4. Estimated temperature rise in a water layer as a function of beam radius and current due to electron beam-induced heating for typical TEM (left) and lower voltage STEM (right). The calculations use window size $L = (2/\sqrt{\pi})50 \times 10^{-6}$ m, $\alpha_{th} = 1.4 \times 10^{-7}$ m^2/s, and $C_p = 4.18$ J/g K. Adapted with permission from Ref. 1. Copyright 2013 American Chemical Society.

7.4 Introduction to the Radiation Chemistry of Water

7.4.1 Energy Transfer and Short and Long Time Kinetics

The energy that is deposited by each collision can be explicitly estimated from the stopping power (S_{Total}), the inelastic cross section (σ), and the atomic number density (N),

$$E_{\text{collision}} = \frac{S_{\text{Total}}}{N\sigma}. \quad (7.18)$$

This energy is typically ~20 eV, enough to form highly reactive radicals. Within say 10 ps of the initial energy transfer, individual ionized water molecules and their ejected electrons begin to decompose the nearby water molecules into radical and molecular species. These include hydrated (solvated) electrons e_h^-, hydrogen radicals H$^{\bullet}$, hydroxyl radicals OH$^{\bullet}$, and H_2. The most reactive species, e_h^-, is a free electron surrounded by a cage of water molecules.

These reaction products, known as the *initial yield*, are concentrated inhomogeneously along the track of the ionizing radiation in the discrete volumes, called *spurs*, shown in Figure 7.1c [1, 3–7, 17–19]. In reality, spurs have Gaussian concentration profiles of the radiolytic products with variance σ_{spur}^2. For most species in water, $\sigma_{\text{spur}} \sim 1.2$ nm [20]. Within <1 μs, the spurs evolve as the species diffuse and participate in further reactions to yield the *primary products* [5–7, 19]:

$$H_2O \rightarrow e_h^-, H^{\bullet}, OH^{\bullet}, H_2, H_2O_2, H_3O^+, HO_2^{\bullet}. \quad (7.19)$$

By ~1 μs after the energy transfer event, the spurs have dissipated and the primary products are assumed to be uniformly distributed within the irradiated region [6]. The rate of generation (or destruction) of the primary products at this time is given as a constant called the *G-value*, first introduced by Burton in 1947 [22]. This is the number of molecules created or destroyed per 100 eV of energy deposited. The G-values for a particular product and medium depend on the type and energy of the radiation (electrons, heavy ions, γ-rays, etc.) and the solution chemistry. G-values have been computed and measured for a variety of systems [4, 6, 19]. Experimentally, G-values are measured by monitoring the absorption bands associated with a particular species generated during irradiation. However, since the production of all species is coupled, further information is obtained by adding scavengers to the solution. These are additives known to quickly consume a particular species, and hence provide a way to distinguish the production of individual components.

As long as the dose rate is sufficiently low, the rate of energy transfer is linear and the reported G-values for LET (linear energy transfer) radiation are a good measure of the production rates [1, 6]. At higher dose rates, the spurs start to overlap along the electrons' tracks. The overlap leads to an increase in the probability that radical species will react to form the more stable molecular species (H_2, O_2, H_2O_2, and recombined H_2O). Under these circumstances, the rate equations become non-linear with dose rate, meaning that the concept of a G-value breaks down. We can estimate the dose rate at which this occurs by first calculating the number of spurs produced during time T:

$$n_{\text{spur}} = \frac{tIT}{\lambda e}, \qquad (7.20)$$

where e is the fundamental charge. Then, the average volume containing a single spur is simply the irradiated volume divided by the number of spurs, or $\pi a^2 \lambda e / IT$. Assuming each volume is spherical, the average distance between spur centers is given by [1]

$$\overline{d}_{\text{spur}} = \left(\frac{6e\lambda a^2}{IT}\right)^{1/3}. \qquad (7.21)$$

In order to ensure there is no spur overlap, the average distance between spurs should be greater than $5\sigma_{\text{spur}}$, or $\overline{d}_{\text{spur}} \geq 6$ nm [1]. Given the lifetime of a spur of ~1 μs, Equation (7.21) shows when spur overlap is likely. For example, 300 keV electrons in a 10 nA beam with a 100 nm radius would give $\overline{d}_{\text{spur}} = 6$ nm, and greater beam current or smaller radius would imply spur overlap is likely to occur. In the case of a rastering beam, such as in STEM, the average distance between spur centers is given by [1]

$$\overline{d}_{\text{spur, Rastering}} = \left(\frac{12e\lambda av}{\pi I}\right)^{1/3}, \qquad (7.22)$$

where v is the beam velocity. For 300 keV electrons in a 1 nA rastering beam with a 1 nm radius, spur overlap would occur for rastering velocities less than 1.4 m/s ($\overline{d}_{\text{spur}} \approx 6$ nm when $v = 1.4$ m/s). Spur overlap should be considered when modeling radiolysis during liquid cell electron microscopy, as it will have strong effects on the G-values.

7.4.2 Formation, Diffusion, and Consumption Rates of Radiolysis Species

After the first 1 μs of irradiation, the system enters a state in which the primary products participate in a cascade of slower chemical reactions, creating additional species such as O_2. The time scales associated with liquid cell EM observations and experiments are seconds to hours, and therefore fall within this regime. To describe the evolving solution chemistry at times beyond 1 μs, it is necessary to consider large numbers of interrelated kinetic equations, one for each of the key reaction pathways between species [4, 8, 9]. For neat water alone, there are 16 relevant species and 79 reactions are needed to model the interactions between them. These are listed in Table 7.1.

The evolution of each species is simulated by obtaining a reaction-diffusion equation for that species, constructing rate equations from the kinetic model and implementing them assuming conservation of mass. As long as the dominant reactions and their associated rates are known for a given liquid (such as water with a particular mix of ions), a balanced kinetic model can be formed, making simulation possible. The coupled equations, when solved, give the concentration field of each species i, $C_i(\mathbf{x},t)$ as functions of space $(\mathbf{x})$ and time (t),

$$\frac{\partial C_i}{\partial t} = D_i \nabla^2 C_i - \sum_j k_{ij} C_i C_j + \sum_{j,k \neq i} k_{jk} C_j C_k + R_i. \qquad (7.23)$$

Table 7.1 Steady state concentration power law parameters

	Homogeneous			Heterogeneous	
Species	α_i	β_i	a_i	α_i	β_i
e_h^-	2.49E-10	0.51	3.18	1.02E-09	0.50
$H^\bullet$	6.31E-09	0.38	5.92	2.61E-08	0.33
H_2	8.88E-08	0.50	3.55	1.88E-07	0.52
H_2O_2	2.20E-07	0.44	3.16	4.43E-07	0.45
$HO_2^\bullet$	2.39E-11	0.71	7.04	8.90E-12	0.85
HO_2^-	4.34E-12	0.72	2.30	2.91E-12	0.85
$HO_3^\bullet$	2.32E-18	1.21	4.43	4.53E-19	1.50
H^+	2.07E-09	0.51	6.52	6.83E-09	0.48
O_2	2.11E-10	0.67	6.99	2.54E-10	0.77
O_2^-	8.21E-10	0.51	6.49	5.55E-10	0.63
$O_3^\bullet$	7.96E-16	0.84	6.42	4.45E-17	1.17
O_3^-	2.10E-15	0.83	5.72	1.77E-16	1.13
$OH^\bullet$	1.45E-08	0.46	0.40	4.28E-08	0.46
OH^-	1.20E-08	0.37	4.12	1.87E-08	0.43
O^-	4.56E-13	0.69	4.67	1.87E-13	0.84

Simple approximate estimates of the steady state concentrations of radiolysis products as a function of dose rate, valid for neat water irradiated at dose rates ranging from 10^6 to 10^{10} Gy/s. The coefficients are for use in Equations (7.27) and (7.28). α_i and β_i provide the steady state concentration $C_{ss,i} \approx \alpha_i \psi^{\beta_i}$ and a_i provides the time to steady state $t_{ss,i} \approx a_i \psi^m$ of species i. For the heterogeneous model, the concentrations are at the beam center for a beam radius 1 μm with the liquid cell sidewalls 50 μm distant. Reproduced from Ref. 4.

Here, the first term on the RHS represents diffusion. Each species will diffuse down concentration gradients, such as those between the irradiated and unirradiated volumes; D_i is the diffusivity of species i in water. The second and third terms represent destruction and production of species i through chemical reactions with all other species j present. k_{ij} is the associated reaction rate constant. The last term,

$$R_i = \frac{\rho \psi G_i}{F} \text{ (M/s)}, \tag{7.24}$$

is the volumetric production rate of species i by the irradiation. It includes the dose rate ψ given by Equation (7.8) or (7.15) and the G-value described in Section 7.4.1. G_i is non-zero only for the primary products within the irradiated region; ρ is the density and F is Faraday's constant.

To complete the model, initial and boundary conditions must be prescribed. Prior to irradiation, the initial concentration of any species is uniform throughout the domain,

$$C_i(\mathbf{x}, 0) = C_{i,0} \text{ (M)}. \tag{7.25}$$

For example, for neat water, $C_i(\mathbf{x},0) = 0$ for all species apart from H_2O, H^+, and OH^-; $C_{H_2O}(0) = 55.56$ (M) and $C_{H^+} = C_{OH^-}(0) = 10^{-7}$ M. Furthermore, for aerated solutions, we also have $C_{O_2}(\mathbf{x},0) = 255$ μM to represent the initial, saturation concentration of oxygen as calculated with a Henry's law constant of 0.0013 mol/kg bar [23]. As boundary conditions, we assume that all confining surfaces are impermeable:

$$\nabla C_i \cdot \hat{n}|_{\text{wall}} = 0, \quad (7.26)$$

where $\hat{n}$ is a unit vector normal to the surface. We are now ready to calculate the evolution of each species.

7.5 Homogeneous Irradiation

Suppose the entire liquid is irradiated homogeneously. This may occur, for example, in graphene cells (Chapters 1 and 19) when imaging isolated pockets of liquid that are smaller than the beam diameter. There will be no spatial concentration gradients, and the diffusion term of Equation (7.23) will drop out. For neat water, this leaves a set of 16 coupled ordinary differential equations, one for each species. This can readily be implemented in numerical solvers in order to obtain the temporal evolution of the solution chemistry. This implementation has been performed in Mathematica and MATLAB [2] and released as open source code available for download on GitHub [24]. The G-values used are for high energy electrons in neat water [19, 25] which are assumed to first order to be independent of solution composition.

Figure 7.5 shows the concentrations of e_h^-, $H^\bullet$, H_2, H_2O_2, $OH^\bullet$, and O_2 as functions of time. The simulation was carried out for the dose rate of 7.5×10^7 (Gy/s) associated with a 300 keV beam of 1 μm radius and 1 nA current according to Equation (7.8). At time $t = 0$ irradiation begins and is maintained continuously thereafter. Concentrations of H_2, H_2O_2, OH, and O_2 increase rapidly from zero to achieve steady state values within about 1 ms after the start of the irradiation. The concentrations of e_h^- and $H^\bullet$ initially increase, peak, and then decline to their steady state values within the same time frame. A key result is thus that the system rapidly reaches a steady state, and the concentrations of radiolysis products do not grow without bound. Instead, reverse reactions convert radiolysis products back to water, enabling the establishment of this steady state.

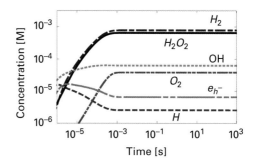

Figure 7.5. The concentrations of e_h^-, $H^\bullet$, H_2, H_2O_2, $OH^\bullet$, and O_2 as functions of time. Neat, de-aerated water is irradiated continuously at a dose rate of 7.5×10^7 Gy/s. Although only a subset of species is shown, the concentrations of all species can be obtained with the programs in Ref. 24. Reproduced with permission from Ref. 2. Copyright 2014 American Chemical Society.

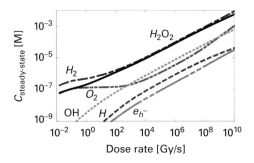

Figure 7.6. Steady state concentrations of e_h^-, $H^{\bullet}$, H_2, H_2O_2, $OH^{\bullet}$, and O_2 as functions of dose rate. Deaerated, neat water subjected to continuous irradiation. Reproduced with permission from Ref. 2. Copyright 2014 American Chemical Society.

The calculation can be repeated for the range of dose rates typical in liquid cells. A steady state is always observed (Figure 7.6), but both the magnitude of the steady state concentration and the time that it takes to reach steady state depend on the dose rate.

Close examination of Figures 7.5 and 7.6 suggests that the concentrations of radiolysis products as functions of time and dose rate appear, empirically, to behave according to power laws. This provides a shortcut for estimating irradiation effects without performing the full simulation.

We can approximate the steady state concentration vs. dose rate relationship by [2]

$$C_{ss,i} \approx a_i \psi^{\beta_i} \,(\text{M}) \, (10^6 < \psi < 10^{10} \, \text{Gy/s}). \tag{7.27}$$

The values of α_i and β_i ($0.37 < \beta_i < 1.21$) obtained by fitting Equation (7.27) to the numerical results are listed in Table 7.1.

We can approximate the time to steady state t_{ss} vs. dose rate relationship, where (ignoring initial fluctuations) t_{ss} is defined as the time needed for the concentration to achieve 95% of its steady state value, i.e.

$$\left[\frac{C_i(t_{ss}) - C_i(\infty)}{C_i(\infty)}\right] = 0.95, \tag{7.28}$$

by [2]

$$t_{ss,i} \approx a_i \psi^{-m} \,(\text{s}) \, (10^6 < \psi < 10^{10} \, \text{Gy/s}). \tag{7.29}$$

Here, the exponent $m = 0.50 \pm 0.01$ for all species. The coefficients a_i are of order 1 and are listed in Table 7.1.

Several interesting findings are evident from these simulations. First, note that the time it takes to establish a steady state in a continuously irradiated medium scales inversely as the dose rate to the power ~1/2, meaning that at the high dose rates common to electron microscopy (10^6–10^{10} Gy/s), the homogeneous system reaches steady state rapidly. The time to steady state for typical TEM dose rates is of the order of 1 ms –

much shorter than typical experiments. Second, note that the concentrations of almost all radiolysis products increase sub-linearly ($\beta_i < 1$) with dose rate. In other words, doubling the dose rate leads to less than a twofold increase in concentration of most radiolysis products.

7.6 Finite Beam Irradiation with Diffusion

In most liquid cell TEM experiments, only a fraction of the liquid in the cell is irradiated (Figure 7.1b). Radiolysis products generated within the irradiated region therefore diffuse outwards and reactions occur outside the irradiated volume. To examine the effects of this diffusion, we use a model system similar to that used in the thermal analysis of Section 7.3. The cylindrical electron beam of radius a is located in the center of a one-dimensional radial domain of chamber radius W. The calculations presented in this section use $a = 1$ μm and $W = 50$ μm. It is important to note that W may be as large as a few millimeters in practice. We focus here on circumstances relevant to TEM, where the electron beam is maintained at a fixed position for a prescribed time interval.

We implement the 79 kinetic reactions in COMSOL™ and account for mass transfer by diffusion [2]. In the irradiated domain $0 \leq r \leq a$, the equations include the relevant source terms R_i. In the region outside the beam $a < r \leq W$, $\psi = 0$ and therefore $R_i = 0$. Figure 7.7 shows concentrations of several key species in neat water as functions of radial position r at various times, and at the beam center ($r = 0$), edge ($r = a$), and far field ($r = W$) as functions of time.

The first notable aspect of the simulation is that species reach steady state concentrations within seconds – much longer than the equilibration time in the homogeneous case. This longer equilibration time results from the slower diffusion process that takes place over time scales on the order of the diffusion time, $t_{D,i} \approx W^2/D_i$. Typical diffusion coefficients D_i of solutes in water are on the order of 10^{-9} m²/s, giving a time to steady state of ~2 s for $W = 50$ μm. Due to the interactions between species, the actual steady state concentration is a function of position within the liquid cell.

It is very important to make two comments about the time it takes to establish a steady state. First, although it may take seconds or minutes to reach steady state throughout the entire volume, especially in a large device, the species concentrations within the irradiated volume still reach their steady state rapidly, over milliseconds, similarly to the homogeneous case. Consequently, flowing the solution through the imaging chamber is unlikely to mitigate beam-induced phenomena, as the flow rate required to sweep away primary products would need to be unrealistically large. We can estimate when flow will be important via the Damköhler number Da, or ratio of the production rate to the advective mass transfer rate. When

$$Da = \frac{R_i \, 2a}{C_i \, v} \tag{7.30}$$

is less than unity, then flow will have a significant role in the solution chemistry. For typical values of $2a$, the diameter of the irradiated medium ~1 μm, C~10^{-6} (M), ψ~10^7 (Gy/s), and G~1 (molecule/100eV), we require v, the flow velocity, to be ~1 m/s

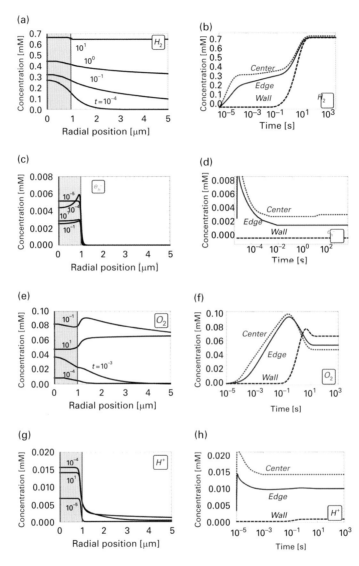

Figure 7.7. Heterogeneous model predictions for the spatial and temporal evolution of H_2 (a, b), e_h^- (c, d), O_2 (e, f), and H_3O^+ (or H^+) (g, h). The left column depicts the concentrations of the selected radiolysis products of neat water as functions of the radial distance from the center of the irradiated region at various times. The right column depicts the concentrations of the same products at the center and edge of the irradiated region and at the perimeter of the liquid cell. The beam and liquid cell radii are, respectively, 1 μm and 50 μm. The dose rate is 7.5×10^7 Gy/s. Reproduced with permission from Ref. 2. Copyright 2014 American Chemical Society.

for $Da = 1$. This means that unless the fluid velocity is above ~1 m/s in the irradiated volume, advection will not be dominant. In the absence of such a large flow velocity, radiolysis will play a significant role in determining solution chemistry.

Figure 7.7 illustrates the spatial and temporal behavior of radiolysis species in the finite beam simulation. H_2 is the most well-behaved (i.e., least reactive) species, and is continuously produced within the irradiated region and diffuses away (Figure 7.7a). At early times ($t < 10^{-4}$ s), production exceeds outdiffusion and H_2 increases rapidly at the center; at later times, production is balanced by diffusion, giving a plateau in $C_{H2}(0,t)$ (Figure 7.7b). As the H_2 concentration outside the beam builds up and the diffusive flux decreases, the H_2 concentration in the irradiated region resumes its growth to eventually approach its equilibrium value.

e_h^-, $H^\bullet$, $OH^\bullet$, H_3O^+, and O_2 are more reactive and exhibit more complex behaviors than H_2. The highly reactive e_h^-, $H^\bullet$, and $OH^\bullet$ persist mainly in the beam region and their concentrations drop quickly outside as they are consumed through chemical reactions (Figure 7.7c). The somewhat complex behavior of e_h^- (Figures 7.7c, d) can be understood by considering its interactions with O_2 (Figures 7.7e, f), the dominant scavenger of e_h^- in neat water. Oxygen production is delayed somewhat, and initially e_h^- in the irradiated region increases rapidly. But once O_2 production ramps up ($t > 10^{-5}$ s), it scavenges e_h^- and reduces its concentration. This leads to the peak at $t \sim 5$ μs. Later, the e_h^- concentration declines to its equilibrium value. Note the transient peak in e_h^- near the beam edge: it is attributable to O_2 diffusion away from the irradiated region. The temporal and spatial distributions of the oxygen concentration exhibit opposing trends to that of e_h^-.

H_3O^+ (or H^+, Figures 7.7g, h) behaves similarly to the other radical products, but with somewhat higher concentration outside the irradiated region than e_h^-, where H_3O^+ grows slowly by diffusion. Eventually the entire liquid cell will have an increased H_3O^+ concentration and hence reduced pH. Neat water initially at pH 7 exposed to a 1 μm radius beam of 1 nA at 300 kV (a dose rate of 7.5×10^7 Gy/s) will eventually drop to pH ~4.9 within the irradiated region and ~6.1 outside.

Perhaps surprisingly, given this complex behavior, it is possible to fit power laws to the concentrations at $r = 0$ as a function of dose rate. The prefactor α_i and exponent β_i ($0.3 < \beta_i < 1.5$) are listed in Table 7.1 and are applicable for systems with sufficiently large W (>10μm).

7.7 Practical Effects of Radiolysis

The results in Sections 7.5 and 7.6 have shown how radiolysis product concentrations can be calculated under various realistic circumstances. We now show experimental data in which radiolysis effects are visible. These experiments provide information on the formation of the radiolysis species and the changes that these species cause in the solution under study, and hopefully will help with interpreting beam-induced effects in other observations.

7.7.1 Hydrogen and Bubble Formation

Bubble nucleation and growth is frequently observed during liquid cell experiments [1, 26–29], occasionally with adverse effects. An example is shown in Figure 7.8. As discussed in Section 7.3, such bubbles are not caused by boiling, but instead are created by radiolytically produced H_2 that has exceeded its solubility limit [3]. The ability to predict the conditions under which bubbles will form can improve our capacity

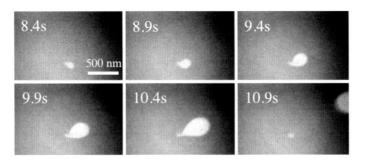

Figure 7.8. Nucleation, growth, and migration of radiolytic bubbles at a serendipitous nucleation site. Reproduced with permission from Ref. 1. Copyright 2013 American Chemical Society.

to design experiments that avoid bubble formation, or create bubbles under controlled conditions. And, as we show below, bubble observations also support the central conclusions of the radiolysis modeling in Section 7.5: the rapid rise of radiolysis products to a steady state concentration and the dependence of this concentration on dose rate.

We first use the observation in Figure 7.8 to illustrate some key points about radiolytic bubble formation. Note that the images are part of a longer sequence of heterogeneous, periodic nucleation events at relatively high dose rates [1]. The periodic behavior suggests the existence of a steady background concentration field, consistent with steady state conditions. Furthermore, prolonged irradiation under moderate dose rates is possible without the formation of bubbles, suggesting that gaseous radiolysis products do not grow without bound, hence also supporting the existence of a steady state [1, 2]. Next, consider nucleation. Here, fortuitously, a defect in the Si_xN_y membrane serves as an easy nucleation site. If no such site is available, the radiolysis model predicts that H_2 supersaturation rises everywhere in the liquid. When a bubble eventually nucleates outside the field of view it grows rapidly as it acts as a sink for a large amount of supersaturated H_2. This is consistent with the experiences of many microscopists, where imaging is suddenly interrupted by rapid appearance of a bubble from outside the field of view.

The existence of a rapidly attained steady state allows us to explore the conditions under which bubbles do or do not form. Bubble nucleation requires exceeding the solubility limit of any gaseous radiolysis product, here H_2 or O_2, by many times [30]. Using Henry's law, and assuming realistic values of the pressure in the liquid cell, one can estimate the solubility limit of the gaseous species. H_2 is less soluble than O_2. H_2 is also more plentifully produced than O_2 by radiolysis. Thus, bubble nucleation is due to hydrogen. (Though, once a bubble forms, its composition includes all gases in diffusion communication with the bubble.) Figure 7.9 depicts the concentration of radiolytic H_2 at the center of the irradiated region as a function of time in liquid cells with effective diameters of 50 µm and 1000 µm. The solubility limit is also shown at various pressures. The actual pressure in a liquid cell depends on the cell design, loading

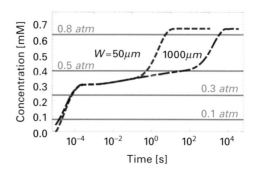

Figure 7.9. The concentration of H_2 at the center of a beam in the heterogeneous simulation. The liquid cell has dimensions of 50 μm (red line) and 1000 μm (blue line). The saturation concentration of H_2 in water at various pressures is shown as horizontal lines. Neat, de-aerated water is irradiated continuously at a dose rate of 7.5×10^7 Gy/s. The beam radius is 1 μm. Reproduced with permission from Ref. 2. Copyright 2014 American Chemical Society.

conditions, and window bowing, and is typically in the range of 0.1 to a few atmospheres. (Window deflection can reduce the pressure below atmospheric as the liquid cell is inserted into the vacuum.) Figure 7.9, or a similar graph calculated for one's own microscope parameters, can therefore be used to identify the maximum dose rate that allows for indefinite bubble-free imaging – or, alternatively, the imaging time available for bubble-free operation at a given dose rate. For the moderate dose rate chosen for Figure 7.9, prolonged, bubble-free imaging is attainable, as the supersaturation needed for bubble formation is likely much greater than the predicted steady state concentration.

Once a bubble forms, it plays a role in the local radiolysis. The concentration of species at the interface and the mass flux associated with a growing or dissolving bubble will alter solution chemistry and hence the steady state concentrations of all species. Radiolysis also occurs in gases and water vapor, but with lower production rates as the dose rate scales linearly with density. A bubble could be included in a numerical simulation to directly study these effects as long as the device pressure, the size of the bubble, and surface concentrations can be measured or estimated.

7.7.2 Radiolysis of Aerated Water

Solutes play a strong role in the steady state concentrations of radiolysis products [4, 5]. Oxygen is one solute of particular importance. Since it is not practical to de-aerate solutions in most liquid cell experiments, one will often encounter solutions with dissolved oxygen. Including the effect of this dissolved oxygen on the concentrations of the radiolysis products is achieved simply by changing the initial conditions in the numerical implementation. Aerated water is defined here as neat water saturated with oxygen at atmospheric conditions without any dissolved CO_2. The saturation concentration of oxygen is 0.255 mM, as discussed above [23]. Figure 7.10 depicts the effect of this oxygen, by comparing the steady state concentrations of other radiolysis species with and without de-aeration.

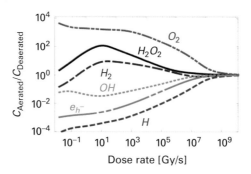

Figure 7.10. Ratio of aerated to de-aerated steady state concentrations of e_h^-, $H^\bullet$, H_2, H_2O_2, $OH^\bullet$, and O_2 as functions of dose rate. Reproduced with permission from Ref. 2. Copyright 2014 American Chemical Society.

For high dose rates ($>10^8$ Gy/s), the steady state oxygen concentration always exceeds 0.255 mM and the results are nearly independent of the initial oxygen concentration. But at lower dose rates, the presence of initial oxygen in solution has a strong effect on the steady state concentrations of all the radiolytic species. It is worth bearing this in mind when comparing the results of different experiments where the protocols for sample preparation may differ.

7.7.3 Changes in pH and the Effect of pH on Radiolysis Yields

One extremely important consequence of irradiation is the strong change in the local pH of an irradiated volume [2]. Figure 7.11 shows the steady state pH as a function of dose rate and initial pH for de-aerated water. At very low dose rates ($<10^3$ Gy/s), the pH of the solution is nearly unaffected by irradiation and is independent of the dose rate. Higher dose rates drive the system towards being more acidic. The effects are particularly notable for alkaline solutions, which may even become acidic in the beam area.

These changes in pH are expected to play an important role in many process including electrochemistry and colloidal stability. Several liquid cell studies ([26, 31, 32] among others), have shown that otherwise stable colloidal suspensions aggregate during imaging. It has been suggested [2] that aggregation occurs because the beam-induced pH change moves the colloids' surface charge closer to their isoelectric point, causing their destabilization. Simulations [26] show an expected drop in pH from 7 to 3.25 within the beam under the conditions used.

It is clear from Figure 7.11 that, at a given dose rate, the steady state pH (or H_3O^+ concentration) depends on the initial pH of the solution. Since all radiolysis products are interrelated, the steady state concentrations of the other radiolysis products also depend on the initial pH. These effects are worth considering since strong acids and bases are frequently employed in liquid cell experiments to improve salt solubility. Figure 7.12a shows the equilibrium concentrations of several key species in de-aerated water as functions of the initial solution pH while Figure 7.12b compares aerated and de-aerated

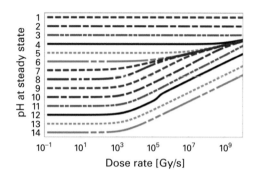

Figure 7.11. Steady state pH as a function of dose rate and initial pH prior to irradiation, for de-aerated water. Reproduced from Ref. 2. Copyright 2014 American Chemical Society.

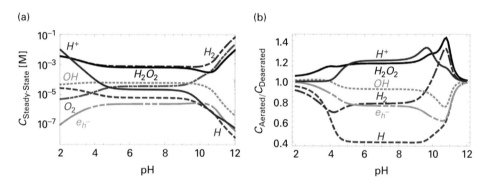

Figure 7.12. (a) Steady state concentrations of e_h^-, $H^\bullet$, H_2, H_2O_2, $OH^\bullet$, and O_2 as functions of the pH prior to irradiation, for de-aerated water under uniform irradiation (homogeneous case). (b) The ratio of radiolytic products concentrations in initially oxygen-saturated water ($C_{O_2}(0) = 0.255$ mM) and in de-aerated water as functions of initial pH. The dose rate is 7.5×10^7 Gy/s. Reproduced with permission from Ref. 2. Copyright 2014 American Chemical Society.

water. These simulations assume that the conjugate pairs to the protons/hydroxides of the added acid/base (e.g. SO_4^{2-} in sulfuric acid and K^+ in potassium hydroxide) do not significantly participate in the various reactions. At moderate pH (5 < pH < 9), the steady state concentration of all radiolysis products is nearly independent of the initial pH (plateau in Figure 7.12a), consistent with prior reports on radiolysis at low dose rates [6, 8]. But for strong acids and strong bases, steady state concentrations are greatly altered. De-aeration becomes less important in strong acids or strong bases.

7.7.4 Hydrated Electrons: Particle Nucleation, Beam Writing, Etching vs. Growth

We finally discuss a frequently observed beam-induced phenomenon, the nucleation and growth of nanocrystals in metal salt solutions [2, 33–37]. Crystal growth occurs when beam-induced species, such as hydrated electrons, reduce cations and induce precipitation under certain conditions [25, 38, 39]. A detailed discussion of the resultant structures and their growth kinetics is provided in Chapter 9. The rate of mass transfer

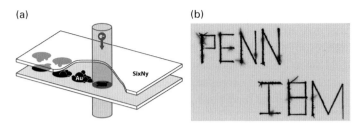

Figure 7.13. (a) Illustration of beam-induced deposition. (Figure provided by Dr. Michael Norton.) (b) Since highly reducing species do not persist outside the beam, maskless patterning of Au is possible. 30 keV electrons ($I = 0.1$ nA) irradiate 0.1 M $AuCl_3$ solution with dose rate controlled by magnification. Reproduced with permission from Ref. 1. Copyright 2013 American Chemical Society.

determines the crystal growth habit [40], so by setting the dose rate one can obtain different types of crystals and control whether materials form by single atom addition or by coalescence of clusters [16, 33, 35, 41, 42].

Since hydrated electrons are extremely reactive, they are present at significant concentrations only within the irradiated region. Precipitation will take place, therefore, mainly in the illuminated region, allowing patterning of deposits using the electron beam as a "pencil". An example is shown in Figure 7.13 [1]. An additional level of control can be achieved by applying electric fields [37].

In addition to reducing agents, radiolysis also produces strong oxidizers. These are likely to etch any metal particles that are in the irradiated area. A competition between etching and growth may be set up, with the net reaction depending on dose rate [2]. To determine whether growth or etching will dominate, one must consider the relative concentrations of the reducing and oxidizing agents and the rates of the corresponding reactions. Components other than water must also be considered. For example, Br^- may be present if the particles were originally synthesized in a process that makes use of the surfactant CTAB. Radiolysis products derived from Br^- such as $Br_2^{\bullet-}$, Br_3^-, and Br_2 can react with metals such as Au [39], so the model should include these species, as discussed in the following section.

7.8   Radiolysis beyond Neat Water

7.8.1   Other Aqueous Solutions

Additional chemistries can be included in a water radiolysis simulation to provide more accurate quantitative results in the presence of additional solutes. If the reaction pathways and associated rate constants are all known, the implementation of an augmented water radiolysis kinetic model is merely a matter of extending the set of equations to include new species and reactions according to Equation (7.23). Practically speaking, one can find the relevant reactions utilizing resources such as the NIST Solution Kinetics database [43] and published works, identify the dominant set of reactions, and implement the new reaction scheme in a numerical solver. This procedure was used in Ref. 44 to include the role of chloride and its radiolytic constituents in interpreting the formation kinetics of gold nanocrystals. The model could not include gold since the required reaction rates are unknown, but the inclusion of the chloride reactions and the appropriate pH were found essential in obtaining quantitative agreement between the model and experimental observations.

7.8.2 Non-Aqueous Solvents

Radiation chemistry in solvents other than liquid water can be treated by means analogous to the treatment of water described in this chapter. If the degradation pathways are known for the system, kinetic models can be designed to account for the radiation-induced changes in chemistry. Extensive knowledge of the effect of radiation on matter has been generated for gases, aprotic solvents, ionic liquids, and some other substances [4, 5, 45, 46]. Ionic liquids are of particular interest due to their promising material properties, especially in Li-ion batteries [47]. As discussed in Chapter 3, ionic liquids are molten salts made of a combination of organic and/or inorganic cations and anions that are liquid at low temperatures, have low vapor pressures, are combustion resistant, and have other properties making them amenable to engineered materials processing. The radiolytic behavior of many of these materials has recently been reviewed by Mincher and Wishart [46]. Similar to water, the primary form of energy deposition is in the form of ionization that cascades into the creation of radicals, excess electrons, holes, and excited states. The resulting chemistry will be derived from the constituents in the ionic liquid, and each individual material will therefore have its own radiation-induced behavior. The reaction kinetics and chemical effects of the species formed must be included for proper understanding of *in situ* observations in liquid cell electron microscopy.

7.9 Conclusions and Outlook

Understanding the fundamentals of beam–sample interactions during liquid cell electron microscopy is necessary to interpret, suppress, or exploit beam-mediated phenomena. Employing a kinetic model is a powerful tool for quantifying how irradiation alters the underlying solution chemistry in the region of liquid that is under observation. Models developed for the study of water radiation chemistry in medical imaging and nuclear reactor safety can be leveraged and extended to the extremely high dose rate regime found in liquid cell imaging. This chapter focused primarily on water, due to its prevalence in the field. As other material systems are investigated, such as ionic liquids, similar accounting must be made for the ubiquitous effects of irradiation.

In most TEM experiments, any beam-induced temperature rise is small because of conduction through the liquid layer. But for a focused intense beam and/or isolated liquid volumes (such as in the graphene liquid cells in Chapter 19), the temperature change may be significant.

However, the imaging electron beam does change the solution chemistry. The effects are *always* present, even if not visible. Of particular importance to colloidal and electrochemical aqueous systems is the change in pH in the imaged region; but for general experiments, especially with complex solutions, the effects may be difficult to predict due to the interconnected nature of the species concentrations. A key result is that theory predicts a steady state that is rapidly approached. In the usual case where only a small fraction of the cell volume is irradiated, the chemistry within the beam approaches the final value within milliseconds, although the steady state outside the beam requires longer time scales associated with diffusion.

The kinetic model described here is useful for designing experiments that minimize unwanted artifacts, as well as those that make use of the effects of irradiation. Most importantly, the model, which has been released as open source code, assists in interpreting liquid cell electron microscopy observations, and may help to establish liquid cell electron microscopy as a new tool to study fundamental radiolysis and the behavior of materials under extreme conditions.

References

1. J. M. Grogan, N. M. Schneider, F. M. Ross and H. H. Bau, Bubble and pattern formation in liquid induced by an electron beam. *Nano Lett.*, **14** (2014), 359–364.
2. N. M. Schneider, M. M. Norton, B. J. Mendel *et al.*, Electron–water interactions and implications for liquid cell electron microscopy. *J. Phys. Chem. C*, **118** (2014), 22373–22382.
3. J. W. T. Spinks and R. J. Woods, *An Introduction to Radiation Chemistry* (New York: Wiley-Interscience, 1990).
4. A. O. Allen, *The Radiation Chemistry of Water and Aqueous Solutions* (Princeton, NJ: Van Nostrand, 1961).
5. I. Draganic, *The Radiation Chemistry of Water* (New York: Elsevier, 2012).
6. B. Pastina and J. A. LaVerne, Effect of molecular hydrogen on hydrogen peroxide in water radiolysis. *J. Phys. Chem. A*, **105** (2001), 9316–9322.
7. A. J. Elliot and D. R. McCracken, Computer modeling of the radiolysis in an aqueous lithium salt blanket: suppression of radiolysis by addition of hydrogen. *Fusion Eng. Des.*, **13** (1990), 21–27.
8. J. M. Joseph, B. S. Choi, P. Yakabuskie and J. C. Wren, A combined experimental and model analysis on the effect of pH and $O_2(aq)$ on γ-radiolytically produced H_2 and H_2O_2. *Radiation Phys. Chem.*, **77** (2008), 1009–1020.
9. N. J. Carron, *An Introduction to the Passage of Energetic Particles through Matter* (Boca Raton, FL: CRC Press, 2006).
10. H. A. Bethe and J. Ashkin, Bethe: Passage of radiations through matter. In E. Segre, ed., *Experimental Nuclear Physics Vol. 1*, (New York: Wiley, 1953).
11. M. J. Berger, J. S. Coursey, M. A. Zucker and J. Chang, NIST Stopping-Power and Range Tables: Electrons, Protons, Helium Ions. Available at http://physics.nist.gov/PhysRefData/Star/Text/ESTAR.html [Accessed: 3 November 2014].
12. J. A. LaVerne and S. M. Pimblott, Electron energy-loss distributions in solid, dry DNA. *Rad. Res.*, **141** (1995), 208–215.

13. T. Tabata, A simple calculation for mean projected range of fast electrons. *J. Appl. Phys.*, **39** (1968), 5342–5343.
14. M. E. Rose, Electron path lengths in multiple scattering. *Phys. Rev.*, **58** (1940), 90.
15. D. Drouin, A. R. Couture, D. Joly et al., CASINO V2.42: a fast and easy-to-use modeling tool for scanning electron microscopy and microanalysis users. *Scanning*, **29** (2007), 92–101.
16. H. Zheng, S. A. Claridge, A. M. Minor, A. P. Alivisatos and U. Dahmen, Nanocrystal diffusion in a liquid thin film observed by in situ transmission electron microscopy. *Nano Lett.*, **9** (2009), 2460–2465.
17. H. A. Schwarz, Applications of the spur diffusion model to the radiation chemistry of aqueous solutions. *J. Phys. Chem.*, **73** (1969), 1928–1937.
18. G. V. Buxton, C. L. Greenstock, W. P. Helman and A. B. Ross, Critical-review of rate constants for reactions of hydrated electrons, hydrogen-atoms and hydroxyl radicals in aqueous solution. *J. Phys. Chem. Ref. Data*, **17** (1988), 513–886.
19. M. A. Hill and F. A. Smith, Calculation of initial and primary yields in the radiolysis of water. *Rad. Phys. Chem.*, **43** (1994), 265–280.
20. S. M. Pimblott and J. A. LaVerne, Molecular product formation in the electron radiolysis of water. *Rad. Res.*, **129** (1992), 265–271.
21. H. Christensen, Remodeling of the oxidant species during radiolysis of high-temperature water in a pressurized water reactor. *Nucl. Technol.*, **109** (1995), 373–382.
22. M. Burton, Radiation chemistry. *J. Phys. Chem.*, **51** (1947), 611–625.
23. J. Speight, *Lange's Handbook of Chemistry* (New York: McGraw-Hill Professional, 2004).
24. N. M. Schneider/Radiolysis, *github.com*. Available at https://github.com/NMSchneider/Radiolysis [Accessed: 30 June 2014].
25. E. J. Hart, The hydrated electron: properties and reactions of this most reactive and elementary of aqueous negative ions are discussed. *Science*, **146** (1964), 19–25.
26. J. M. Grogan, L. Rotkina and H. H. Bau, In situ liquid-cell electron microscopy of colloid aggregation and growth dynamics. *Phys. Rev. E*, **83** (2011), 061405.
27. U. Mirsaidov, C.-D. Ohl and P. Matsudaira, A direct observation of nanometer-size void dynamics in an ultra-thin water film. *Soft Matter*, **8** (2012), 7108–7111.
28. T.-W. Huang, S.-Y. Liu, Y.-J. Chuang et al., Dynamics of hydrogen nanobubbles in KLH protein solution studied with in situ wet-TEM. *Soft Matter*, **9** (2013), 8856–8861.
29. K. L. Klein, I. M. Anderson and N. de Jonge, Transmission electron microscopy with a liquid flow cell. *J. Microsc.*, **242** (2011), 117–123.
30. S. Jones, Bubble nucleation from gas cavities: a review. *Adv. Coll. Interf. Sci.*, **80** (1999), 27–50.
31. D. Li, M. H. Nielsen, J. R. I. Lee et al., Direction-specific interactions control crystal growth by oriented attachment. *Science*, **336** (2012), 1014–1018.
32. T. J. Woehl, J. E. Evans, I. Arslan, W. D. Ristenpart and N. D. Browning, Direct in situ determination of the mechanisms controlling nanoparticle nucleation and growth. *ACS Nano*, **6** (2012), 8599–8610.
33. K. W. Noh, Y. Liu, L. Sun and S. J. Dillon, Challenges associated with in-situ TEM in environmental systems: the case of silver in aqueous solutions. *Ultramicroscopy*, **116** (2012), 34–38.
34. J. Lee, A. Urban, X. Li et al., Unlocking the potential of cation-disordered oxides for rechargeable lithium batteries. *Science*, **343** (2014), 519–522.
35. M. Bresin, B. R. Nadimpally, N. Nehru, V. P. Singh and J. T. Hastings, Site-specific growth of CdS nanostructures. *Nanotechnology*, **24** (2013), 505305.

36. J. Park, S. Kodambaka, F. M. Ross, J. M. Grogan and H. H. Bau, In situ liquid cell transmission electron microscopic observation of electron beam induced Au crystal growth in a solution. *Microsc. Microanal.*, **18** (2012), 1098–1099.
37. M. den Heijer, I. Shao, A. Radisic, M. C. Reuter and F. M. Ross, Patterned electrochemical deposition of copper using an electron beam. *APL Materials*, **2** (2014), 022101.
38. H. Remita, I. Lampre, M. Mostafavi, E. Balanzat and S. Bouffard, Comparative study of metal clusters induced in aqueous solutions by γ-rays, electron or C^{6+} ion beam irradiation. *Rad. Phys. Chem.*, **72** (2005), 575–586.
39. W. Abidi and H. Remita, Gold based nanoparticles generated by radiolytic and photolytic methods. *Recent Patents in Eng.*, **4** (2010), 170–188.
40. W. W. Mullins and R. F. Sekerka, Stability of a planar interface during solidification of a dilute binary alloy. *J. Appl. Phys.*, **35** (1964), 444–451.
41. J. E. Evans, K. L. Jungjohann, N. D. Browning and I. Arslan, Controlled growth of nanoparticles from solution with in situ liquid transmission electron microscopy. *Nano Lett.*, **11** (2011), 2809–2813.
42. H. Zheng, R. K. Smith, Y. W. Jun *et al.*, Observation of single colloidal platinum nanocrystal growth trajectories. *Science*, **324** (2009), 1309–1312.
43. W. G. Mallard, A. B. Ross and W. P. Helman, NDRL/NIST Solution Kinetics Database on the Web: A complication of kinetics data on solution-phase reactions. Available at http://kinetics.nist.gov/solution/ [Accessed: 6 April 2015].
44. J. H. Park, N. M. Schneider, J. M. Grogan *et al.*, Control of electron beam-induced Au nanocrystal growth kinetics through solution chemistry. *Nano Lett.*, **15** (2015), 5314–5320.
45. A. Mozumder, *Fundamentals of Radiation Chemistry* (London: Elsevier Science, 1999).
46. B. J. Mincher and J. F. Wishart, The radiation chemistry of ionic liquids: a review. *Solvent Extraction and Ion Exchange*, **32** (2014), 563–583.
47. J. Y. Huang, L. Zhong, C. M. Wang *et al.*, In situ observation of the electrochemical lithiation of a single SnO_2 nanowire electrode. *Science*, **330** (2010), 1515–1520.

8 Resolution in Liquid Cell Experiments

Niels de Jonge, Nigel D. Browning, James E. Evans, See Wee Chee, and Frances M. Ross

8.1 Introduction

The key driving force behind the development of liquid cell electron microscopy has been the high spatial resolution in liquids that the technique promises compared to other microscopies that work in liquids, such as light microscopy. Depending on the sample and experiment, good temporal resolution can be a second key benefit of liquid cell electron microscopy, again compared to other microscopies that work in liquids, such as scanning tunneling microscopy. The combination of good spatial and temporal resolution is what provides liquid cell microscopy with its applicability to carry out unique experiments in a range of disciplines.

In this chapter we explore the limits of resolution in liquid cell electron microscopy. As the technique has developed, it has become clear that there is no one value of resolution for liquid cell microscopy. The resolution achievable in any experiment is a balance between competing factors. The intrinsic microscope resolution is of course important, but is by no means the only factor; the parameters of the detector, the dose sensitivity of the sample, Brownian motion within the liquid, the sample geometry, and the type of information required all affect the achievable temporal and spatial resolution. This makes even the definition of resolution somewhat complicated. Different groups report resolution according to different criteria, and all are valid depending on the context.

Here we survey the competing factors that determine spatial and temporal resolution. The type of microscopy we consider here is standard, uncorrected TEM and STEM. Liquid cell resolution in SEM is discussed separately in Chapter 4. Aberration-corrected TEM and STEM is also discussed separately in Chapter 21; aberration correction promises exciting improvements in liquid cell microscopy, with implications that extend beyond simply the increase in resolution.

We first discuss the possibly counterintuitive result that good spatial resolution is possible in even relatively thick liquid layers. We show how the spatial resolution depends on the microscope parameters, the acceptable dose to the sample, and the dynamic nature of liquid samples, particularly with regard to Brownian motion. This leads to two important trade-offs: resolution versus liquid thickness, and temporal resolution versus electron dose and beam effects. In the final sections of this chapter we discuss some closely related aspects of liquid cell electron microscopy. First we describe image simulation for liquid cell samples, essential for quantifying high resolution information to draw conclusions about the sample. Second, having explored the issues associated with resolution, we develop some practical guidelines for choosing parameters to avoid pitfalls and to optimize the information obtained from liquid cell TEM and STEM.

8.2 Spatial Resolution in Liquid Cell TEM

In TEM and STEM, the spatial resolution may be limited by spherical aberration, chromatic aberration, signal-to-noise ratio (SNR), and stability of the sample. This last parameter can dominate resolution, for example in circumstances where immersed particles undergo Brownian motion and appear blurred. We will discuss this situation in more detail in Section 8.4. Considering aberration and noise, it would seem intuitively obvious that achieving the highest resolution should be easiest in a sample (membrane plus liquid) that is as electron transparent as possible. Graphene liquid cells are no thicker than conventional solid samples, and it is not surprising that excellent spatial resolution is possible in these cells, as shown in Chapter 19. However, liquid cells with microfabricated windows are more frequently used, especially if liquid flow or electrical connections are required. A typical liquid cell sample with say 50 nm silicon nitride membranes surrounding 100 nm water can still yield lattice resolution, as shown in examples throughout this book. For microfabricated window cells, the need for the highest resolution can conflict with practical cell design. Optimizing resolution implies a membrane that is thin and made of low atomic number material, yet must withstand the vacuum environment without deflecting too much, or breaking. The liquid layer should be thin, yet thick enough to support the material or process under study.

In the following sections, we consider first spatial resolution and then temporal resolution. We base our discussion around a "typical" liquid cell sample consisting of solid particles dispersed within water. For TEM, which is covered in the following paragraphs of Section 8.2, we find that while resolution in thin liquid layers is limited by aberrations, resolution in thicker liquid layers is commonly limited by the signal-to-noise ratio. For STEM, signal to noise is the critical limiting factor under most conditions. STEM is discussed separately in Section 8.3, because of the practical differences in image formation and microscope parameters.

8.2.1 Spatial Resolution Due to Lens Aberrations and Diffraction

In light microscopy, diffraction limits the spatial resolution to approximately the wavelength λ of the light, a few hundred nanometers. In particular, the diffraction-limited point resolution includes both λ and the objective semi-angle, α:

$$d_D = 0.6\lambda/\alpha. \tag{8.1}$$

Various optical techniques can improve the resolution beyond the diffraction limit, achieving tens of nanometers for suitable samples [1]. But for high energy electrons, where energies of 100–300 keV correspond to $\lambda = 3.7$–2.0 pm, the resolution obtained in a TEM is nowhere near the diffraction-limited value. Instead, the resolution achieved is typically only around 0.2 nm. As described in Chapter 21, this value is limited by lens aberrations, or imperfections in the focusing of electrons by the lenses. Lens aberrations are larger for electrons that travel through the lens with a large angle between their path and the optic axis. In order to minimize the effects of any lens aberration, the maximum angle α allowed (beam convergence angle or objective angle) should therefore be kept to a minimum. But once the maximum angle becomes small enough, diffraction effects start to play a role via Equation (8.1) and limit the resolution. Thus, in practice, an optimum angle is chosen to balance diffraction effects and lens aberrations.

8.2.2 Spherical Aberration-Limited Resolution

For ultra-thin samples, the dominant aberration for TEM and STEM is spherical aberration (in the case of high voltage imaging and an electron source of low energy spread and high brightness). This causes electrons that travel through the lens at a large angle to the optic axis to be focused more strongly than those travelling close to the axis. Spherical aberration in the objective lens typically determines the resolution for thin samples. In liquid cell microscopy this thin sample limit only applies in a few cases, such as the graphene liquid cell (Chapter 19). For such samples, optimizing the angle α as described above leads to a point resolution that is given by:

$$d_S = C_S \alpha^3, \tag{8.2}$$

where C_S is the coefficient of spherical aberration. Typical values of C_S are in the 1–2 mm range. If the microscope is corrected for spherical aberration, the resolution can be improved, with values below 0.08 nm possible on a routine basis. The reason the resolution is not even better is because higher order aberrations or other factors then come into play.

8.2.3 Chromatic Aberration-Limited Resolution in TEM

Chromatic aberration is another key factor that can limit the image resolution in TEM. It arises because electrons with different wavelengths are focused at different distances from the lens. Chromatic aberration becomes important when there is a range of energies in the electrons that form the image (high ΔE in Equation (8.3) below), or when a lower accelerating voltage is used (small E in Equation (8.3)). Instabilities in the lenses, intrinsic properties of the source, and inelastic scattering within the sample all contribute to a spread in energy of the electrons that form the image.

If the chromatic aberration dominates, the image resolution is given by a value known as the chromatic blur, d_C:

$$d_C = \alpha C_C \Delta E / E, \tag{8.3}$$

where C_C is the chromatic aberration coefficient, ΔE is the spread in energy of the beam, and E is the beam energy.

Chromatic aberration can be dominant in situations where the spherical aberration has been corrected, or for TEM when the sample is not ultra-thin. The latter situation is very common for liquid cell imaging, particularly with microfabricated window cells. The typical thickness of a liquid cell sample (liquid plus membranes) is generally large enough to result in inelastic scattering events removing energy from the electrons so that they leave the sample with a broadened range of energies compared to their spread at the electron source. As the sample thickness increases, so does ΔE, until the chromatic blur becomes the factor that determines the image resolution.

To calculate the resolution for a given liquid cell sample, we need to relate ΔE to the inelastic scattering properties of the sample. The full-width at half-maximum of the energy distribution ΔE of transmitted electrons follows from calculations of the inelastic electron scattering cross section and is given by the following equation [2]:

$$\Delta E = \frac{N_A e^4 Z \rho T}{2\pi \varepsilon_0^2 W m_0^2 v^2}, \tag{8.4}$$

where N_A is Avogadro's number, e is the elementary charge, W the atomic weight, Z the atomic number, ρ the density and T the thickness of the liquid layer, ε_0 the permittivity of space, m_0 the mass and v the velocity of the electrons. For water, where $Z = 4.7$, we can combine the above equations [3] into the following expression for the resolution:

$$d_{C,TEM} = 6 \times 10^{12} \alpha C_C T / E^2, \tag{8.5}$$

with E in eV. For pure water, neglecting the enclosing window(s), and using typical experimental parameters, say $\alpha = 10$ mrad, $C_C = 2$ mm, $E = 200$ keV, and $T = 1$ μm, we obtain $d_{C,\,TEM} = 4$ nm. This number applies for imaging objects at the beam exit side of the liquid layer. Figure 8.1 shows this theoretical resolution limit as a function

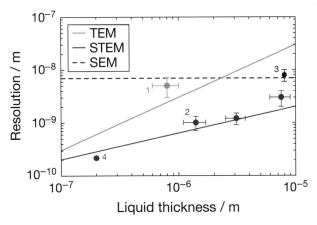

Figure 8.1. Theoretical maximum resolution versus thickness of water for TEM and STEM. The resolution was calculated for typical instrument parameters at 200 keV beam energy, for imaging Au nanoparticles at the bottom of a layer of water for TEM, and at the top of the layer for STEM. SEM resolution is shown for comparison, assuming a 30 nm Si_xN_y window [51]. The points show reported resolution obtained for Au nanoparticles in water in TEM [5] and STEM [24] and SEM with 30 nm window [51], and PbS particles in water in STEM [52]. The error bars represent experimental errors. Reprinted from Ref. 3.

of liquid thickness for pure water. Also shown are resolution values quoted in several experimental investigations.

In Equation (8.5), the thickness T corresponds to the total of the liquid plus membranes. The liquid may not be uniform in thickness, and gas bubbles may be present. These contributions deserve additional discussion.

(1) *Liquid thickness*: In microfabricated liquid cell designs, the separation between the chips is fixed but the pressure difference leads to an outward deflection or bowing of the membranes. The liquid thickness can therefore increase substantially from the edges to the center of the window. Chapter 2 showed an example calculation of membrane deflection as a function of membrane properties. Both the dimensions of the window and the thickness and mechanical properties of the membrane material affect the degree of bowing. The smallest deflection, and hence the lowest liquid thickness and highest resolution, are obtained if the window has at least one dimension that is small, and is made of a stiff and thick material. However, these criteria must be balanced against the need for a large field of view (for example in experiments involving reactions at multiple electrodes) and the need for a particular liquid thickness (for example, in modeling a physical process such as electrochemical deposition, where thin liquids do not support "realistic" diffusion fields, as discussed in Chapter 10).

(2) *The presence of bubbles*: If a bubble forms in the liquid cell [4–6], due to radiolysis or precipitation of dissolved gases, the liquid may be present only as a thin wetting layer on the two membranes. The result may be a set of images that look excellent, with high resolution, but may not reflect the physical surroundings originally anticipated for the experiment, i.e. a full liquid layer. Some publications not explicitly stating a measurement of the liquid thickness may show surprisingly high resolution as a result of the presence of a gas bubble. Interestingly, the thin liquid layer at the location of a bubble may still be thick enough to show physical phenomena suggestive of a thicker layer: for example, nanoparticles in a thin liquid layer may show movement [7]. This point will be discussed in more detail below.

(3) *Membrane thickness*: It is straightforward to calculate the relative contributions of the membranes and liquid to the chromatic blur, using materials properties such as those given in Chapter 2. The contribution for 50 nm silicon nitride membranes is less than 10% if the total liquid thickness is $T = 1$ μm, but the contribution increases rapidly for thinner liquid layers. In a sense it is only worth optimizing the window material if the thickness of the liquid to be examined is below a few hundred nanometers.

8.2.4 Effect of the Sample Depth within the Liquid in TEM

For a liquid cell sample with a reasonably thick liquid layer, the appearance of the image may be quite different depending on whether the object of interest (for example a

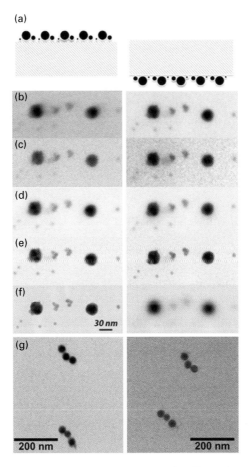

Figure 8.2. Comparison of resolution in different imaging modes. (a–f) Different imaging modes applied to 5–30 nm diameter Au nanoparticles on the electron entrance (top) and exit (bottom) surface of 500 nm of amorphous silicon nitride recorded at 200 kV: (a) schematic of sample; (b) uncorrected TEM; (c) EFTEM; (d) C_S corrected; (e) C/C_C corrected; (f) uncorrected STEM. (g) Bright field TEM images of Au nanoparticles in water imaged (left) without energy filtering and (right) in EFTEM mode using a 20 eV energy filter. (a–f) Reprinted with permission from Ref. 8. (g) Reprinted from Ref. (22).

crystal growing in the liquid) is on the entrance or exit surface of the liquid cell, considered with respect to the incident beam. This so-called top–bottom effect is well known for TEM and STEM of thick samples [2]. It is illustrated in Figure 8.2a, b for TEM of a test sample, where Au nanoparticles are imaged on either the top or bottom of a solid, amorphous layer [8]. For TEM, the highest resolution for an object on a thick sample is achieved when the object is at the electron beam exit side of the sample.

The resolution in a thick sample given by Equation (8.5) is therefore the best possible case. Objects further from the exit side of the liquid layer are imaged with poorer resolution [3] because elastic scattering of electrons in the liquid between the object and the exit window leads to additional blurring. For a TEM image of a particle at the top of a 1 μm thick water layer, Equation (8.5) predicts a resolution of 4 nm, but the additional scattering reduces the resolution to only ~12 nm. It is worth noting that diffraction techniques can be used in TEM when the sample in liquid contains repetitive units [9], and this may allow higher resolution information to be obtained. But in any case, it is worth making sure that the region of interest is close to the lower window of the liquid cell where possible. For example, electrodes used for electrochemical experiments should be patterned on the exit window; nanoparticle growth on the exit window will be imaged with the best resolution.

8.2.5 Dose-Limited Resolution in TEM

It is generally a good principle to minimize the dose to the sample, as described in Chapter 7, since the electron beam has strong effects on water and other liquids. However, lowering the dose creates a limitation on the resolution achievable in liquid cell TEM. Suppose that the sample is so beam sensitive that only a low dose can be used. The resulting image will then be noisy because of statistical variations in the low numbers of electrons at each pixel of the image. At low enough dose D (electrons per area), an image may be too noisy to be able to identify a region of the sample that contains a feature of interest.

To calculate these effects, assume that a given region has size d and on average scatters $\langle N \rangle$ electrons into the detector. And further, suppose that if this region contains a feature of interest, such as a particle, it would scatter $N + \Delta N$ electrons instead. The shot noise associated with the $\langle N \rangle$ background electrons is $\langle N \rangle^{1/2}$. Since the signal is $|\Delta N|$ and the noise $\langle N \rangle^{1/2}$, the SNR is $|\Delta N|/\langle N \rangle^{1/2}$. The recorded SNR can be defined as [10]

$$\mathrm{SNR} = (\mathrm{DQE})^{1/2}|\Delta N|/\langle N \rangle^{1/2} = (\mathrm{DQE})^{1/2} C \langle N \rangle^{1/2}. \tag{8.6}$$

Here DQE is the detector quantum efficiency, which gives a measure of the ability of the detector to measure every count from the sample. DQE = 1 for a perfect detector, a value that is actually approached for modern direct detection systems (see below). In the second part of Equation (8.6), we define SNR in terms of the contrast, or signal-to-background ratio of the sample, $C = |\Delta N|/\langle N \rangle$. To reliably detect an object above the background counts produced by the liquid layer, the SNR must be greater than some value, often taken as 5 (the Rose criterion [11]).

For samples that can only withstand a certain critical dose D_c (electrons per area) before unacceptable damage, the question then becomes, what is the resolution achievable without exceeding this dose? This situation is familiar from cryo-electron microscopy of biological samples, where samples are highly beam sensitive and in addition show low contrast, both of which imply that images are likely to be limited by noise. Only a fraction, say F, of the incident dose actually contributes to the image. Then the number of electrons detected from an area of size d is $\langle N \rangle = F D_c d^2$. For a feature of interest, of contrast C, detection is possible if its size exceeds d_D, the dose-limited resolution. This is given [10] by substituting for $\langle N \rangle$ in Equation (8.6):

$$d_D = (\text{SNR})(\text{DQE})^{-1/2} C^{-1} (FD_e)^{-1/2}. \qquad (8.7)$$

In this equation, F depends on the imaging mode, the aperture size used and whether energy filtering is used. The contrast C depends on the imaging mode and the materials composing the sample. C is low for biological materials, but can be high for samples containing heavy particles in water, for example. In the bright field conditions often used for liquid cell TEM, scattering to high angle makes denser regions of the sample appear dark and can result in good contrast. For low contrast samples, defocusing can increase contrast, at the expense of losing some localized information. C may also be increased, in principle, by using a phase plate. This will be an interesting possibility for future liquid cell experiments.

Figure 8.3 shows the dose required to achieve a given spatial resolution under reasonable assumptions about the sample and microscopes. This is compared with an

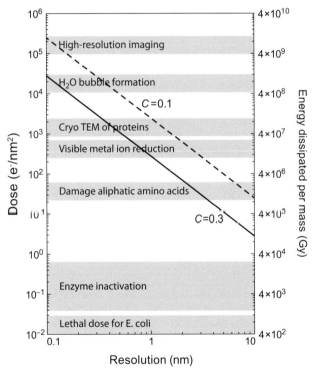

Figure 8.3. Dose required to reach a desired spatial resolution at 200 keV for typical materials contrast values of 10% and 30%. From Equation (8.7) with $C = 0.1$, 0.3, i.e. reasonable values for biological materials, DQE = 1 and $F = 1$; STEM and TEM are similar on the log scale. Tolerable dose ranges and some beam-induced phenomena are shown as horizontal bands. Also shown is the lethal dose for *Escherichia coli* (0.01 electrons/nm^2, Ref. 2, p. 468), the dose for enzyme inactivation (1 electron/nm^2) and the dose for reduction of ions in solution to form nanoparticles (Au^{3+} reduction in AuCl$_3$, Ref. 42). The dose is also given in grays, the energy deposited per unit of mass, assuming 1 electron/nm$^2 \approx 4 \times 10^4$ Gy, a typical value for water at 200 or 300 keV (Chapter 7). Adapted from D. A. Muller, private communication.

approximate "acceptable" dose D_c for water, which will be discussed below. It is clear that there is a wide range of imaging conditions where radiation damage will be the resolution-limiting factor in aqueous samples, especially if the sample contrast is low. This has some important implications. Imaging in water with atomic level resolution but "acceptable" damage may be possible for a single image. But recording a dynamic process over a series of, say, ten images, while keeping below the same total acceptable dose, would limit the spatial resolution to 1 nm or more. Spatial resolution in undamaged organic electrolytes, say, would more typically be from a few nm to tens of nm, depending on the number of images needed.

8.2.6 Acceptable Doses for Liquid Samples in TEM

Since the critical dose D_c is a key factor in determining resolution for liquid cell imaging, it is worth considering what dose is acceptable for typical liquid samples. D_c is a complicated parameter because of the different types of interaction that the sample and liquid cell membranes can experience. Radiolysis and knock-on damage are the two main damage mechanisms (see Chapter 7), but the ratio of these is material dependent, and the two mechanisms have opposite dependences on accelerating voltage. Thus there may be no optimal voltage for all the components of a complex liquid sample.

Fortunately, cryo-electron microscopists have carried out extensive studies to establish critical doses for biological materials, and low dose methods have been used for decades in structural biology. Consideration of beam-induced damage led to the establishment of a dose limit of ~10^3 electrons/nm^2 for sub-nanometer studies of protein structures via single particle tomography [12] or of ~10^4 electrons/nm^2 for nanoscale imaging of cellular ultrastructure [13]. Both electron and X-ray methods appear to require similar thresholds [14]. For electrons, the widely accepted threshold of 10^3 electrons/nm^2 was originally based on spot fading cryo-EM experiments at liquid nitrogen temperature. It was chosen for maintaining information out to a spatial resolution of ~0.4 nm at 80–300 keV. However, those experiments showed that, as the dose increases, the highest resolution spots disappear first. Thus, the acceptable dose should be considered to be even lower for atomic resolution imaging. It should be noted here that these doses are defined to avoid observable physical structural damage, but of course, the absence of observable changes in atomic positions does not guarantee that the sample is unaffected by the beam. Cryo-EM does not readily resolve the 3D electronic structure of proteins, so this type of damage would be hard to measure. We finally note that cooling all the way to liquid helium temperature may afford higher dose tolerances than standard cryo-EM at liquid nitrogen temperature, but there appear to be conflicting conclusions on its effects [15, 16].

Liquid cell electron microscopy differs significantly from cryo-EM. Observations are of course made at a higher temperature, usually at the ambient temperature within the column of 25–30 °C; the thermal conductivity of a liquid cell sample is greater than that of a typical cryo-EM sample; differences are expected in both formation rate and diffusion of radiolysis species; and liquid flow provides new opportunities for removing radiolysis species and refreshing the sample environment. Any of these differences may be important in determining how accurately the knowledge of critical doses for cryo-EM applies in liquid cell experiments. This question is an active area of research, and intriguing results are starting to appear. An original paper from the 1970s suggested a maximal dose of 10^3 electrons/nm^2 for sub-nanometer structural preservation of biomolecules in a thin liquid layer [17]. One unexpected result has been that proteins appear to be damaged more slowly during liquid cell observations, as compared to the damage rate during cryo-EM (Chapter 17). Since dry samples are damaged even faster than cryo-EM samples in ice, this may be related to the benefits of the liquid environment in allowing diffusion of reactive species away from the region of interest. The picture is even more complex, because the electron beam initiates a range of reactions in the liquid [18], and depending on the exact conditions, samples may be stable or experience large morphological changes, as was demonstrated recently for the imaging of gold nanoparticles in liquid [19]. A second active area of research involves the critical dose for whole cells: at what dose are cellular functions disrupted, even if there are physical changes in structure? The imaging of whole cells and macromolecular assemblies is discussed in more detail in Chapters 16 and 24. To measure cell function, correlative fluorescence microscopy is used to determine whether a cellular structure is still intact via, for example, a dye that would leak out of a ruptured plasma membrane [20]. Lethal doses for cells may be reached after even a single image is obtained ([2], p. 468; [21]). Thus, control and optimization of the dose is essential in liquid cell imaging of biological materials. The dose *rate* may play a role as well as the critical dose – another important area of future research.

8.2.7 Improving Resolution in Liquid Cell TEM

Exciting modern developments in electron microscopy have resulted in the ability to set spherical and chromatic aberration coefficients to zero or to a chosen value, and to energy-filter the beam so that only electrons within a narrow range of energies are used to form the image. Each of these developments provides an opportunity to improve the information obtained from liquid cell TEM under the right circumstances.

Energy filtering (Figure 8.2c, g) is routinely available in many high performance TEMs. It is particularly useful for thick liquid samples, where multiple scattering is common and the majority of electrons undergo an inelastic scattering event. By selecting only electrons that have experienced energy loss within an energy window a few tens of eV wide, the ΔE contributing to the image, and hence the chromatic blur, is reduced. Setting the energy window center on the "most probable loss" rather than zero loss is useful, since it uses the greatest fraction of the incident electrons to form the image. We anticipate much greater use of energy filtering in future liquid cell TEM experiments. The disadvantage of using energy filtering is that many electrons are not used to form the image, even at the most probable loss. This results in a higher dose to the sample, so that one may run into the considerations of dose-limited resolution.

Correction of C_S can improve image resolution in thin samples. As described in Chapter 21, delocalization effects are avoided, improving the discrimination of features such as the boundaries of particles in liquid. Correction of C_S also allows light elements to be identified, and can also produce three-dimensional information. Thin samples such as graphene liquid cells already benefit from C_S correction, with examples provided in Chapter 19, but the benefits should extend to thicker, microfabricated liquid cells also.

Correction of C_C is for now an uncommon capability, with only a few instruments worldwide capable of C_S/C_C correction. However, its role in liquid cell imaging is highly promising. We discussed previously how C_C correction returns to focus the electrons that have different energies because of source spread, instabilities, or inelastic scattering within the sample. As discussed in Chapter 21, this correction improves spatial resolution compared to C_S correction alone. But C_C correction also provides another benefit for thick liquid samples. With C_C correction, most of the electrons that were inelastically scattered in a thick sample can be used for imaging (if the sample is not so thick that too many multiple scattering events occur), rather than only low loss (or, most probable loss) electrons. Thus C_C correction allows more efficient use of the dose to the sample and improves energy-filtered imaging. Finally, C_C correction enables imaging at reasonable resolution even if the microscope is relatively low performance. This applies particularly to microscopes that have a large polepiece gap. A reasonable resolution combined with a large gap can in turn provide opportunities for additional *in situ* capabilities. These benefits are discussed in more detail in Chapter 21. Liquid cell data are not yet available from a C_S/C_C corrected TEM, but an idea of the results that will be possible can be obtained from studies carried out in model samples with thick amorphous layers. Figure 8.2d, e shows improvement in image resolution in a thick sample via C_C correction. As C_C correction becomes more commonly available, it will be exciting to see how the benefits discussed above play out in practical terms.

8.3 Spatial Resolution in Liquid Cell STEM

For STEM imaging, we start with the balance between diffraction and aberration established in Sections 8.2.1 and 8.2.2, but adapt the discussion to account for the different optics in STEM. Similarly to TEM, thin samples are imaged with a resolution determined by electron optics. This means a resolution is given by the STEM probe size, typically ~0.2 nm, or even ~0.1 nm with C_S correction. Chapters 9 and 19 show examples where crystal lattices or surface facets are resolved by STEM in liquid cells. However, for thicker samples the STEM resolution is limited either by signal to noise or by broadening of the imaging beam. These two effects are linked, and both contribute to the resolution in practical situations.

8.3.1 Probe Broadening in STEM

Probe broadening occurs because elastic scattering of the beam by the sample increases the probe diameter as the beam passes through the sample [2]. This effect can determine the image resolution in liquid samples [23–25].

Since the probe broadens progressively through the sample, there is a top–bottom effect in STEM, such that the highest resolution is achieved when the object of interest

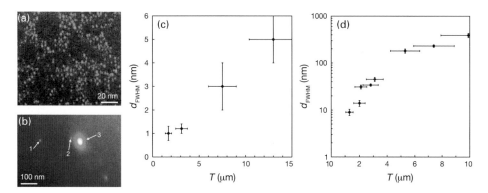

Figure 8.4. STEM imaging of nanoparticles on top of or below a water layer. (a) Dark field STEM image of Au nanoparticles with an average diameter of 1.4 nm on a liquid with thickness 3.3 μm. The signal intensity is color-coded. (b) STEM image of Au nanoparticles below a water layer of a thickness of 1.3 μm. The sample contained Au nanoparticles of different diameters. Arrows #1, #2, and #3 point to nanoparticles with respective diameters of 10, 5, and 30 nm. (c) Resolution of STEM in the top layer of a liquid measured from the full-width-at-half-maximum diameter (d_{FWHM}) of the smallest nanoparticle visible with a signal-to-noise ratio larger than 5 as function of the liquid thickness (T). (d) Resolution measured on nanoparticles below a water layer as function of T. Reprinted from Ref. 24 with permission from Elsevier.

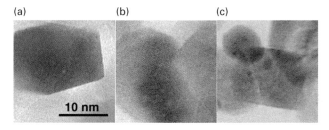

Figure 8.5. Atomic resolution imaging of particles in (a) vacuum (only windows), (b) 1 atmosphere of N_2, and (c) liquid. J. Evans, unpublished.

is at the electron beam entrance side of the sample. This effect is shown experimentally in Figure 8.2a, b in a model sample with Au particles on a thick amorphous film [8]. Figure 8.4 shows the effect in an actual liquid cell sample with Au particles above or below a thick water layer [24]. The images show large changes in spatial resolution (at fixed electron dose) as a function of the liquid thickness. In practical terms, the region of interest of the sample should be located at the entrance window of the liquid cell if possible.

The degree of broadening can be calculated using Monte Carlo simulations [26]. Such calculations show that the profile maintains a sharp central peak, consisting of unscattered electrons, but broad tails develop that consist of elastically scattered electrons. The central peak provides high resolution information even when the sample is thick, but the tails contribute an increased background level compared to images of thin samples. The effect of this background is shown experimentally in Figure 8.5.

Here, nanocrystals in a liquid cell filled with vacuum, gas at atmospheric pressure, or liquid all show lattice fringes, suggesting the same resolution. The difference between the images is the signal-to-noise ratio.

8.3.2 Noise-Limited Resolution in STEM

If probe broadening can be neglected, the resolution may instead be limited by the achievable SNR, related to the tolerable dose for the sample and the contrast difference between liquid and object. There will be some range of sample thickness, materials contrast, or electron dose at which the contrast of, say, a nanoparticle of diameter d will not be measurable above the background. To satisfy the Rose criterion, the nanoparticle needs to be large enough to be detectable at a certain dose within a liquid layer of a certain thickness and scattering properties. Analogously to Section 8.2.5 for TEM, we now determine the circumstances when resolution is limited by signal to noise due to any of these factors.

We can establish an equation for resolution for the case of standard annular dark field (ADF) imaging conditions [23, 24]. In ADF imaging, contrast is generated as the detector collects electrons that are elastically scattered out of the primary beam above a given angle β. If there are N_0 electrons in the primary beam in each pixel (a value depending on the probe current and the dwell time), the fraction N/N_0 of electrons that are scattered into the detector is given by [23]

$$\frac{N}{N_0} = 1 - \exp\left(-\frac{T}{l}\right), \quad l = \frac{W}{\sigma(\beta)\rho N_A}. \tag{8.8}$$

Here, l is the mean free path for elastic scattering and T the liquid thickness; l is a function of the elastic cross section $\sigma(\beta)$, as well as the projected atomic number $W/\rho N_A$. For water, with β = 70 mrad and E = 200 keV, l_{water} = 10.5 μm. With the same parameters, l_{Au} = 73nm.

We can calculate the thickness of Au that would be resolvable above the background noise by a linear approximation of Equation (8.8) [23]:

$$d_{STEM} = 5l_{particle}(T/l_{water}N_0)^{1/2}. \tag{8.9}$$

As an example, this suggests that a 1.4 nm diameter Au nanoparticle can be resolved on top of a water layer of T = 5 μm, using a probe current of 0.5 nA and a pixel dwell time of 10 μs. This is a remarkable result, and is consistent with the experimental data shown in Figure 8.4 [24]. It arises from the excellent sensitivity of STEM to atomic number via σ. STEM is thus very effective for imaging heavy nanoparticles in a light liquid. In general, a resolution better than 10 nm can be expected for particles positioned in the top 1 μm layer of the liquid. Figure 8.1 includes calculated STEM resolution as a function of T for typical microscope parameters. Experimental data are also shown, and Equation (8.9) provides a reasonable match for the data. Calculating the resolution under the particular conditions of the experiment clearly requires knowledge of microscope parameters, discussed in Section 8.6 below.

Equation (8.9) does not include probe broadening due to elastic scattering. For nanoparticles deeper in or below the liquid [24, 25], probe broadening must be considered. At some point (for the conditions of Figure 8.4, this is below 1 μm), beam broadening becomes the limit to the resolution. Figure 8.4 also shows the resolution to be expected for the imaging of Au nanoparticles under typical microscope settings as a function of the liquid thickness. The electron dose, radiation stability of the sample, thickness, and materials contrast all contribute to the resolution of the image.

8.4 Temporal Resolution in TEM and STEM

The liquid cell microscopy data shown throughout this book were recorded as single images, as video recordings (30 images per second, each 480 × 640 pixels), on CCD cameras operating at a few to 10 images per second with each, say, 2k × 2k pixels, or, in a few cases in Chapters 9 and 13, at higher rates of several hundred images per second. For biological samples, such as shown in Chapters 16 and 17, single images are usually obtained, since the acceptable dose precludes imaging multiple times. Sequences of images are clearly very powerful in that they permit kinetics measurements in experiments probing electrochemical deposition, phase transformations, solution-phase particle growth, and bubble dynamics, as shown in Chapters 7, 9, 11, and 13. But recording image sequences has other benefits that apply particularly to liquid samples. These are in reducing the effects of random motion, and in dose fractionation. We consider these below.

8.4.1 Brownian Motion and High Speed Image Recording

Good temporal resolution is important in every liquid cell observation, because the random movements that occur in liquids during imaging cause blurring in the image and degrade the information (Figure 8.6). In order to obtain atomic level detail in spite of random motion, the area of interest must move less than, say, 0.1 nm during imaging.

Objects a few nanometers across will move this far in microseconds due to Brownian motion, and hence, intuitively, Brownian motion should prevent any atomic level

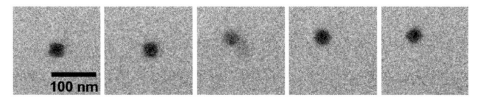

Figure 8.6. Sequence over 50 ms of a mobile Au nanocube in ~200 nm thick water acquired at 100 frames per second and employing zero loss filtering. The ability to resolve the shape is clearly limited by the 10 ms temporal resolution, but the position can be tracked reasonably well. The extracted images also suggest that the particle was spinning. Electron dose rate was $1.2–3 \times 10^4$ electrons per nm^2 per second. S. W. Chee, unpublished.

imaging of small particles for acquisition rates of even a few hundred images per second. Image acquisition times in the microsecond range, as discussed in Chapter 22, should be used to image such particles. For larger particles, for structures attached to the liquid cell walls, or for situations where lower resolution is acceptable, motion blur should be reduced to a usable level even if the image acquisition is slower [22].

However, as the first reports of nanoscale resolution appeared [23, 27], it became clear that good resolution was being obtained even without ultrafast image acquisition rates. Nanoparticles appear to move much more slowly than expected from Brownian motion when they are in close proximity to the membrane [7, 28–31]. It has been suggested that ordering of liquid near the membrane may increase the viscosity by many orders of magnitude [32]. The physics behind this is discussed further in Chapter 10, but the practical outcome is that exciting new observations appear possible in liquids in spite of Brownian motion.

8.4.2 Dose Fractionation

For beam-sensitive samples, high image acquisition rates have an additional benefit known as dose fractionation. Suppose we record a series of shorter sub-frames instead of a single exposure. Each of these sub-frames is noisy, but can be analyzed separately. The sub-frames can be aligned to reduce the effects of drift. Contrast can be tracked to determine whether the sample has started to damage under the beam, then the data obtained before this time can be combined into the final composite image.

8.4.3 Temporal Resolution Versus Noise-Limited Resolution

Assuming the camera is capable of high image acquisition rates, recording at high time resolution brings us into conflict with noise-limited spatial resolution requirements. At a fixed dose rate supplied by the electron source, reducing the exposure time will reduce the number of counts in the image until eventually the spatial resolution is limited according to Equation (8.7). Increasing the dose rate means that the sample will reach its maximum acceptable dose more rapidly.

Although this is a fundamental limitation, there is room for improvement. Improving the efficiency of the detector is a straightforward way to collect more images with the same dose to the sample. In the most commonly used detectors in electron microscopy, CCDs linked via a fiber optic, the interfaces between components cause scattering and contribute to a background. Recent developments in direct detection cameras using thinned sensors have shown greatly improved performance [33]. These cameras have lower noise due to the elimination of the fiber-optic element, and the DQE is improved over a wide frequency range. Both the fill factor and duty cycle are 100%, so virtually all electrons are counted. These cameras can provide frame rates of say 400 2k × 2k frames per second, a significant improvement over the 30 640 × 480 frames per second acquired in a standard video signal. As well as the increase in detection efficiency and low noise, a great benefit of these types of detectors is the ability to carry out dose fractionation as discussed above. One challenge with such cameras is the difficulty of handling the data. 400 2k × 2k frames per second is around 4.6 Gb/s – comparable with the data rate from CERN! Automatic image processing becomes necessary to handle information from the experiment. The opportunities and challenges of high speed imaging are discussed in more detail in Chapter 22.

8.5 Image Simulations in Liquid Cell TEM and STEM

We discussed in Section 8.3.1 calculations of the contrast in STEM images of nanoparticles in water. Such calculations were carried out using Monte Carlo methods. Electron scattering physics was implemented using the CASINO software [34] adapted for STEM accelerating voltages. By including the 3D geometry of the sample, the shape of the electron source, detector parameters, and Poisson noise, it is possible to calculate contrast levels in different STEM imaging modes and compare quantitatively with experimental data [24]. This provides key insights into the expected resolution under different experimental conditions, and hence guides experimental design and interpretation.

When the liquid cell images contain atomic-level details, such as lattice planes and nanoscale particle facets, it is also useful to adopt the approach used for interpreting conventional atomic-level STEM and TEM micrographs [35–37]. These calculations must include not only the material of interest within the liquid but also the amorphous windows and the liquid path length [38]. While not altogether straightforward, reasonable atomistic models of the windows and the liquid have been obtained from DFT/MD simulations [38]. These parts of the simulations essentially provide the diffuse background scattering on which the scattering from, say, nanocrystals in the liquid is superimposed. An example of the use of these methods for interpreting high resolution detail in liquid cell images is shown in Figure 8.7. Using such simulations we can accurately interpret contrast that comes from particles at different positions in the cell, as well as interpret contrast changes caused by orientation and composition effects.

Both types of simulations also provide another way of assessing whether our understanding of the sample is correct on a basic level. For example, is liquid even present around a nanoparticle of interest, or is the particle surrounded instead by a gas bubble? Often, the most attractive and exciting images look good only because the liquid has moved away. Quantitative comparison of image contrast can help to answer this fundamental question.

8.6 Some Practicalities and Pitfalls of Liquid Cell TEM and STEM

It is clear from the above discussions that optimizing a liquid cell experiment requires a balance between competing requirements: spatial and temporal resolution, liquid layer thickness, dose, and dose rate. But even before decisions on dose and image recording parameters are made, we also have to choose between STEM and TEM modes,

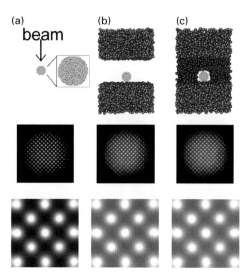

Figure 8.7. Calculated HAADF-STEM images for a galena (PbS) nanoparticle (a) in vacuum, (b) deposited at the bottom window of a gas-stage holder, and (c) deposited at the bottom window of a fluid-stage holder filled with water. A model of each system, the calculated nanoparticle image, and the calculated central 1×1 nm^2 area of the image are shown. Atomic resolution is achieved in all cases here with a window thickness and fluid/gas path length of 50 nm. In this example, noise is not included. Reprinted from Ref. 38.

and decide on the accelerating voltage. In this section we therefore discuss some practicalities of liquid cell microscopy. How can liquid thickness be measured easily, and how should we design the experiment in terms of imaging mode and voltage? Furthermore, how can we calibrate the imaging parameters, essential for modeling the experimental contrast and interpreting results? We also mention some unexpected effects that can cause difficulties: Dosing one location affects nearby areas, and prior dosing can also affect results, in other words there can be history dependence. Many other pitfalls of the liquid cell technique are specific to the material under study. For example, the electron beam can cause suspended nanoparticles to move or change morphology. Effects such as these are discussed in many of the following chapters. Overall, it is clear that the experimental approach must be designed carefully to account for the type of sample and the type of information that is required [39]. The following paragraphs may help in setting out on the path to robust and quantitative experimentation.

8.6.1 Choice of Voltage and Imaging Mode

While the high resolution TEM community pushes for lower voltages to avoid knock-on damage, higher beam energies are required for seeing through thicker liquids or imaging more quickly (see Chapter 22). Within the standard 80–300 kV energy range for many S/TEMs, it is worth varying the voltage to get a sense of the optimum value for the material or process of interest. Figure 8.8a, b [40] shows an example in which

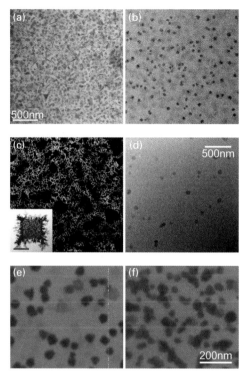

Figure 8.8. (a, b) BF STEM images of Ag nanocrystals grown from solution at, respectively, 300 kV and 80 kV. Magnification 40k×, pixel dwell time 3 ms, using a calibrated beam current of 7.1 pA; image size 1024 × 1024, corresponding to a dose of 39 e$^-$/nm^2 per frame. (c, d) Silver growth evolution for, respectively, STEM at 300 kV with a dose of 1260 e$^-$/nm^2 per frame and TEM at 300 kV with a dose of 26.4 e$^-$/nm^2 per frame. Depletion occurred in the TEM experiment. (e, f) Beam-induced nucleation and growth of Au nanoparticles from HAuCl$_4$ imaged with, respectively, STEM at 30 kV (in SEM with STEM detector) and TEM at 300 kV. Beam-induced growth creates similar morphologies in both instruments. (a–d) Reprinted from Ref. 40; (e, f) Reprinted from Ref. 41.

the morphology of beam-induced growth of Ag nanoparticles varies with voltage within this range. This is presumably related to the differences in the cross sections for the radiolysis and knock-on processes that take place when the beam interacts with the liquid. A second comparison is shown in Figure 8.8e, f, between imaging in a 300 kV TEM and in a 30 kV SEM with a STEM detector. In this case the morphology of the beam-induced Au crystals does not differ much although the density is somewhat different [41].

The choice between STEM and TEM modes is an important one and depends on the nature of the experiment. Controlling the beam is easier in STEM, and the standard high angle annular dark field (HAADF) image mode provides strong contrast between heavy inorganic materials and the liquid. However, STEM will always have slower acquisition than TEM as it is a serial technique whereas TEM is parallel, unless very small frames are recorded with STEM. Figure 8.2 compares images in the two modes directly, while Figure 8.8 illustrates differences between the two modes in observing the process of beam-induced nanoparticle growth. Another interesting effect occurs in Figures 8.8c, d, showing quite different morphologies of Ag nanoparticles in STEM and TEM imaging modes [40], both recorded at the same electron dose and within a time span not differing by more than a factor of 3. The explanation is that in the TEM experiment, the electron beam irradiated a much larger area than that viewed, and the precursor solution was rapidly depleted, while Ag nanoparticles grew during the STEM experiment in the viewing area. Although spreading the beam helps in establishing low dose conditions in the image, a large number of electrons may nevertheless be deposited in a wide region in the liquid. This may result in the depletion of reactants and a much slower apparent growth in the images, which does not scale to the increased dose in the STEM images. The total number of electrons is thus important in interpreting the results of these experiments [40], and simulations of radiolysis species concentrations need to consider the total irradiated volume (Chapter 7). In order to reduce the number of "wasted" electrons deposited in the liquid, the TEM spot should be focused into a disc only slightly larger than the recorded frame, and the intensity should be minimized by adjusting the condenser lens settings (spot size).

8.6.2 Measuring the Liquid Thickness

The thickness of the liquid plays a central role in the physics of the reaction under study, for example in controlling diffusion, bubble dynamics, and crystal nucleation sites. Thickness is a key parameter in modeling electron–sample interactions, and it also controls the image resolution. It is obviously important to know its value accurately. For example, because of bubble formation and motion, it may not even be clear whether there is actually a full liquid layer present in the cell when we perform observations. While it may appear that determining the presence of liquid should be easy, this can be a tricky part of the experiment as the formation of gas bubbles (particularly in aqueous solutions) does not always impede a reaction [42], but does improve the resolution of the images. The experiments look so much better when the liquid starts to evaporate.

An example of this type of effect is shown in Figure 8.9a [43]. Here, in the process of liquid cell drying, the material in solution self-concentrates and deposits on the windows to form a feedstock from which nanoparticles grow over time, creating local regions of material depletion (bright regions) surrounding the particles. In areas where the depletion takes place, the lack of material around the sample creates higher resolution images as the particles grow. While such an experiment does observe growth, it is not a true "*in situ* liquid" growth experiment towards the later frames of the movie: quantitative information extracted from the movie cannot be related directly to any processes occurring outside the microscope in a fully hydrated bulk specimen.

To identify and quantify the presence of liquid in the cell during the experiment, several different techniques can be used. We first discuss the use of electron energy loss spectroscopy (EELS) [44]. EELS for thickness measurement is distinct from its use for

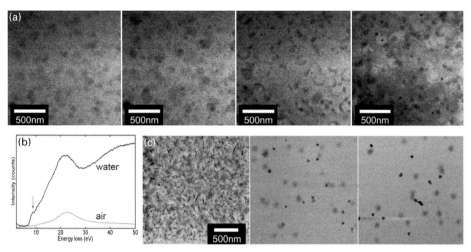

Figure 8.9. (a) Image sequence showing partial drying of nanoparticles growing in a liquid. Dried areas free from liquid give better contrast but the removal of the liquid does not stop the growth. (b) EELS low loss region spectrum acquired from an air-filled cell and a water-filled cell at the same window location. The presence of water is displayed by the increased intensity of the plasmon peak and the edge of the water valence band shown by the arrow at 9.0 eV [45]. (c) Effect of precursor solution depletion and cumulative charge during a series of consecutive Ag particle growth experiments. After a number of experiments, the initial faceted and rounded nanocrystals covering the scanned area could not be reproduced for the same microscope conditions. BF STEM images from data sets were taken at 40k×, 3 ms pixel dwell time (frame time accounting for flyback was 3.78 s), and beam currents of 6.2 pA (left and right) and 3.9 pA (middle), corresponding to 34.2 e^-/nm^2 per frame (left and right) and 21.6 e^-/nm^2 per frame (middle). (a) Reprinted from Ref. 43; (b) reprinted from Ref. 44; (c) Reprinted from Ref. 40.

chemical analysis as discussed in Chapter 20. In the energy loss spectrum, the integrated intensity ratio of unscattered to inelastically scattered electrons provides a thickness estimate in terms of the inelastic mean free path (IMFP) [46]. In Figure 8.9b, a liquid cell was assembled with a nominal thickness of 150 nm. When the cell was filled with air, its thickness was measured by EELS to be 0.71 IMFP lengths. Since a single 50 nm silicon nitride window should only account for 0.33 IMFP lengths, the increase of 7% can be attributed to the gas within the cell. When the cell was filled with water, calculating the IMFP [47] from EELS data yielded estimated thickness values of 171 nm using the material density formula [47], or 151 nm from the atomic number formula for a 7.6 mrad collection angle [48]. In the center of the cell, the liquid layer was thicker. A measured thickness of 3.69 IMFP lengths indicated the presence of a 430 nm thick water layer caused by window bulging. Thus, details from a single low loss spectrum can authenticate the existence and thickness of water within the cell during dark field imaging and can be used to quantify window bulging.

Suppose that EELS is unavailable or cannot be used because the liquid is too thick. Another method to measure the liquid thickness in STEM is by determining the total current into the ADF detector. This current is measured at the phosphor screen below the detector [24]. The fraction of the current density measured on the screen with and without a sample inserted, I_{screen}/I_0, is given by [32]

$$\frac{I_{screen}}{I_0} \cong \exp\left\{-\left(\frac{T_{SiN}}{l_{SiN}} + \frac{T_{liquid}}{l_{liquid}}\right)\right\}, \qquad (8.10)$$

with T_{SiN} and T_{liquid} the thicknesses of the two silicon nitride membranes and the liquid, respectively. In practical terms, the window contribution can be neglected for water thicknesses between 1 and 15 μm. The parameter l is defined as in Equation (8.8) above. For TEM, an analogous measurement compares the currents at the viewing screen with and without the objective aperture, removing elastically scattered electrons.

Once it can be reasonably assumed that the entire liquid cell is filled, the thickness measurement can be verified via parallax. The sample is tilted, and the relative movement of particles at the top and bottom surfaces is measured. For tilt angle α and movement Δl [24], the thickness follows from the parallax equation,

$$T = \frac{\Delta l}{\sin(\alpha)}. \qquad (8.11)$$

These two methods should provide values for thickness consistent within 30% [24].

8.6.3 Calibrating and Controlling Dose

The preceding sections have shown the importance of controlling the dose to the sample. For experiments that involve imaging of beam-sensitive materials, as well as those that probe beam-induced effects, it is essential to quantify both dose and dose rate.

As discussed in Section 8.2, understanding how the dose is applied to the sample involves a straightforward relationship between magnification, beam current, and, in the case of STEM, the dwell time for the probe [6, 49, 50]. However, for liquid samples, when a reaction is performed in one area it can affect another through diffusion. Figure 8.8c shows an example where a reactive species, generated in the irradiated area, causes reactions outside the region under observation. Chapter 7 showed that such effects are especially important for long-lived species, such as hydrogen, that can diffuse for great distances out of the beam. It is also important to note that the history of dose delivery controls the overall result; Figure 8.9c shows an example [40]. Knowing the total dose during the whole exposure is key to quantifiable and reproducible experiments.

However, calibration is not necessarily straightforward. The typical readout of beam current in a S/TEM is not calibrated and can be a factor of 2–5 different from the true current going into the sample. Calibration using a Faraday cup is essential for experiments in which dose is used to develop quantitative models for a physical phenomenon such as the reaction or diffusion-limited mechanisms controlling particle growth. Calibration is a key component of every *in situ* experiment, and liquid cell microscopy is no exception.

8.7 Summary and Outlook

Liquid cell experiments are capable of producing high resolution information: we can visualize atomic structure in nanoparticles and discriminate subtle dynamics. Although thin liquid layers provide the best looking images, thicker samples provide a better model of bulk processes, and can still be imaged with good temporal and spatial resolution. To understand the limits of spatial and temporal resolution in liquid cell TEM and STEM, we need to combine the ideas of aberration-limited and noise-limited spatial resolution with the maximum dose that a liquid sample can accept without damage, and apply these concepts to thick samples that may have low contrast. The optimum imaging conditions depend on the liquid thickness, the contrast of the materials under study, their beam sensitivity and dynamic behavior. Image simulations can provide a guide to optimum imaging conditions before starting experiments, and can help in interpretation after the fact. We have discussed the choice of imaging modes and recording strategies and have considered some of the key calibrations that need to be carried out for quantitative matching of images with models.

As liquid cell microscopy techniques develop in the future, we can expect to see improvements in the quality and quantity of information and in the matching of data with simulations and models. We anticipate wider application of microscopes with C_S/C_C correction and energy filters to liquid cell experiments. We also expect that gradual adoption of high performance direct detection cameras (which are still expensive) will make an immense difference to the practice of liquid cell electron microscopy, improving signal to noise and enabling dose fractionation. The large data rates will need corresponding advances in video processing and automatic data extraction. As with all *in situ* microscopy techniques, the early years have provided exciting and often qualitative information, while further development clarifies the possibilities for quantitative analysis, modeling and prediction.

Obtaining the best information from a liquid cell experiment requires an investment in terms of background work, but the payoff is excellent, in terms of the quantitative information achievable through this unique window into liquid phase structures and processes.

Acknowledgements

NB and JE acknowledge PNNL, a multi-program national laboratory operated by Battelle for the U.S. Department of Energy (DOE) under Contract DE-AC05-76RL01830; the Environmental Molecular Sciences Laboratory (EMSL), a national scientific user facility sponsored by the DOE's Office of Biological and Environmental Research; the Joint Center for Energy Storage Research (JCESR), an Energy Innovation Hub funded by the DOE's Office of Science, Basic Energy Sciences; the National Institutes of Health under grant numbers S10RR025032 and 5RC1GM091755 and the DOE Office of Science, Basic Energy Sciences under grant number DE-FG02-03-ER46057. NDJ acknowledges E. Arzt for his support through INM. SWC acknowledges the Singapore National Research Foundation's competitive research program (NRF-CRP9-2011-04).

References

1. S. W. Hell, Far-field optical nanoscopy. *Science*, **316** (2007), 1153–1158.
2. L. Reimer and H. Kohl, *Transmission Electron Microscopy: Physics of Image Formation* (New York: Springer, 2008).
3. N. de Jonge and F. M. Ross, Electron microscopy of specimens in liquid. *Nat. Nanotechnol.*, **6** (2011), 695–704.
4. D. B. Peckys, G. M. Veith, D. C. Joy and N. de Jonge, Nanoscale imaging of whole cells using a liquid enclosure and a scanning transmission electron microscope. *PLoS One*, **4** (2009), e8214.
5. K. L. Klein, I. M. Anderson and N. de Jonge, Transmission electron microscopy with a liquid flow cell. *J. Microsc.*, **242** (2011), 117–123.
6. T. J. Woehl, K. L. Jungjohann, J. E. Evans *et al.*, Experimental procedures to mitigate electron beam induced artifacts during in situ fluid imaging of nanomaterials. *Ultramicroscopy*, **127** (2013), 53–63.
7. E. A. Ring and N. de Jonge, Video-frequency scanning transmission electron microscopy of moving gold nanoparticles in liquid. *Micron*, **43** (2012), 1078–1084.
8. N. J. Zaluzek, The influence of Cs/Cc correction in analytical imaging and spectroscopy in scanning and transmission electron microscopy. *Ultramicroscopy*, **151** (2015), 240–249.
9. D. F. Parsons, V. R. Matricardi, R. C. Moretz and J. N. Turner, Electron microscopy and diffraction of wet unstained and unfixed biological objects. *Adv. Biol. Med. Phys.*, **15** (1974), 161–270.
10. R. F. Egerton, Control of radiation damage in the TEM, *Ultramicroscopy*, **127** (2012), 100–108.
11. A. Rose, The sensitivity performance of the human eye on an absolute scale. *J. Opt Soc Am.*, **38** (1948), 196–208.
12. J. Pierson, M. Sani, C. Tomova, S. Godsave and P. J. Peters, Toward visualization of nanomachines in their native cellular environment. *Histochem. Cell Biol.*, **132** (2009), 253–262.
13. A. Hoenger and J. R. McIntosh, Probing the macromolecular organization of cells by electron tomography. *Curr. Opin. Cell Biol.*, **21** (2009), 89–96.
14. M. R. Howells, T. Beetz, H. N. Chapman *et al.*, An assessment of the resolution limitation due to radiation-damage in X-ray diffraction microscopy. *J Electron Spectrosc. Relat. Phenomena*, **170** (2009), 4–12.
15. B. E. Bammes, J. Jakana, M. F. Schmid and W. Chiu, Radiation damage effects at four specimen temperatures from 4 to 100 K. *J. Struct. Biol.*, **169** (2010), 331–341.
16. H. Stahlberg and T. Walz, Molecular electron microscopy: state of the art and current challenges. *ACS Chem. Biol.*, **3** (2008), 268–281.
17. V. R. Matricardi, R. C. Moretz and D. F. Parsons, Electron diffraction of wet proteins: catalase. *Science*, **177** (1972), 268–270.
18. N. M. Schneider, M. M. Norton, B. J. Mendel *et al.*, Electron–water interactions and implications for liquid cell electron microscopy. *J. Phys. Chem. C*, **118** (2014), 22373–22382.
19. J. Hermannsdörfer, N. de Jonge and A. Verch, Electron beam induced chemistry of gold nanoparticles in saline solution. *Chem. Commun.*, **51** (2015), 16393–16396.
20. D. B. Peckys and N. de Jonge, Liquid scanning transmission electron microscopy: imaging protein complexes in their native environment in whole eukaryotic cells. *Microsc. Microanal.*, **20** (2014), 346–365.

21. D. B. Peckys, P. Mazur, K. L. Gould and N. de Jonge, Fully hydrated yeast cells imaged with electron microscopy. *Biophys. J.*, **100** (2011), 2522–2529.
22. S. W. Chee, D. Loh, U. Mirsaidov and P. Matsudaira, Probing nanoparticle dynamics in 200 nm thick liquid layers at millisecond time resolution. *Microsc. Microanal.*, **21** (Suppl 3) (2015), 267–268.
23. N. de Jonge, D. B. Peckys, G. J. Kremers and D. W. Piston, Electron microscopy of whole cells in liquid with nanometer resolution. *Proc. Natl. Acad. Sci. USA*, **106** (2009), 2159–2164.
24. N. de Jonge, N. Poirier-Demers, H. Demers, D. B. Peckys and D. Drouin, Nanometer-resolution electron microscopy through micrometers-thick water layers. *Ultramicroscopy*, **110** (2010), 1114–1119.
25. T. Schuh and N. de Jonge, Liquid scanning transmission electron microscopy: nanoscale imaging in micrometers-thick liquids. *C. R. Phys.*, **15** (2014), 214–223.
26. H. Demers, N. Poirier-Demers, D. Drouin and N. de Jonge, Simulating STEM imaging of nanoparticles in micrometers-thick substrates. *Microsc. Microanal.*, **16** (2010), 795–804.
27. H. Zheng, S. A. Claridge, A. M. Minor, A. P. Alivisatos and U. Dahmen, Nanocrystal diffusion in a liquid thin film observed by in situ transmission electron microscopy. *Nano Lett.*, **9** (2009), 2460–2465.
28. H. Zheng, R. K. Smith, Y. W. Jun *et al.*, Observation of single colloidal platinum nanocrystal growth trajectories. *Science*, **324**, (2009), 1309–1312.
29. X. Chen and J. Wen, In situ wet-cell TEM observation of gold nanoparticle motion in an aqueous solution. *Nano. Res. Lett.*, **7** (2012), 598.
30. E. R. White, M. Mecklenburg, B. Shevitski, S. B. Singer and B. C. Regan, Charged nanoparticle dynamics in water induced by scanning transmission electron microscopy. *Langmuir,* **28** (2012), 3695–3698.
31. Y. Liu, X.-M. Lin, Y. Sun and T. Rajh, In situ visualization of self-assembly of charged gold nanoparticles. *J. Am. Chem. Soc.*, **135** (2013), 3764–3767.
32. A. Verch, M. Pfaff and N. De Jonge, Exceptionally slow movement of gold nanoparticles at a solid:liquid interface investigated by scanning transmission electron microscopy. Langmuir, **31** (2015), 6956–6964.
33. D. Contarato, P. Denes, D. Doering, J. Joseph and B. Krieger, High speed, radiation hard CMOS pixel sensors for transmission electron microscopy. *Phys. Procedia*, **37** (2013), 1504–1510.
34. D. Drouin, A. R. Couture, R. Gauvin *et al.*, CASINO V2.42: a fast and easy-to-use modeling tool for scanning electron microscopy and microanalysis users. *Scanning*, **29** (2007), 92–101.
35. J. M. LeBeau, A. J. D'Alfonso, S. D. Findlay, S. Stemmer and L. J. Allen, Quantitative comparisons of contrast in experimental and simulated bright-field scanning transmission electron microscopy images. *Phys. Rev. B*, **80** (2009), 174106.
36. E. J. Kirkland, R. F. Loane and J. Silcox, Simulation of annular dark field STEM images using a modified multislice method. *Ultramicroscopy*, **23** (1987), 77–96.
37. K. Ishizuka, A practical approach for STEM image simulation based on the FFT multi-slice method. *Ultramicroscopy*, **90** (2002), 71–83.
38. D. A. Welch, R. Faller, J. E. Evans and N. D. Browning, Simulating realistic imaging conditions for in-situ liquid microscopy. *Ultramicroscopy*, **135** (2013), 36–42.
39. N. de Jonge, M. Pfaff and D. B. Peckys, Practical aspects of transmission electron microscopy in liquid. *Adv. Imag. Electron Phys.*, **186** (2014), 1–37.
40. P. Abellan, T. J. Woehl, L. R. Parent *et al.*, Factors controlling quantitative liquid (scanning) transmission electron microscopy. *Chem. Commun.*, **50** (2014), 4873–4880.

41. J. M. Grogan, J. H. Park, X. Ye *et al.*, Liquid cell in-situ electron microscopy: interfacial phenomena and electrochemical deposition. *Microsc. Microanal.*, **18** (2012), 1160–1161.
42. J. H. Park, N. M. Schneider, J. M. Grogan *et al.*, Control of electron beam-induced Au nanocrystal growth kinetics through solution chemistry. *Nano Lett.*, **15** (2015), 5314–5320.
43. J. P. Patterson, P. Abellan, M. S. Denny, Jr. *et al.*, Observing the growth of metal–organic frameworks by in situ liquid cell transmission electron microscopy, *J. Am. Chem. Soc.*, **137** (2015), 7322–7328.
44. K. L. Jungjohann, J. E. Evans, J. A. Aguiar, I. Arslan and N. D. Browning, Atomic scale imaging and spectroscopy for in situ liquid scanning transmission electron microscopy. *Microsc. Microanal.*, **18** (2012), 621–627.
45. D. Grand, A. Bernas and E. Amouyal, Photo-ionization of aqueous indole – conduction band edge and energy gap in liquid water. *Chem. Phys.*, **44** (1979), 73–79.
46. R. F. Egerton, *Electron Energy Loss Spectroscopy* (New York: Plenum, 1996).
47. T. Malis, S. C. Cheng and R. F. Egerton, EELS log-ratio technique for specimen thickness measurement in the TEM. *J. Electron Microsc. Tech.*, **8** (1988), 193–200.
48. K. Iakoubovskii, K. Mitsubishi, Y. Nakayama and K. Furuya, Mean free path of inelastic electron scattering in elemental solids and oxides using transmission electron microscopy: atomic number dependent oscillatory behavior. *Phys. Rev. B*, **77** (2008), 104102.
49. M. Hahn, J. Seredynski and W. Baumeister, Inactivation of catalase monolayers by irradiation with 100kV electrons. *Proc. Natl. Acad. Sci. USA*, **73** (1976), 823–827.
50. T. J. Woehl, J. E. Evans, I. Arslan, W. D. Ristenpart and N. D. Browning, Direct in-situ determination of the mechanisms controlling nanoparticle nucleation and growth, *ACS Nano*, **6** (2012), 8599–8610.
51. H. Nishiyama, M. Suga, T. Ogura *et al.*, Atmospheric scanning electron microscope observes cells and tissues in open medium through silicon nitride film. *J. Struct. Biol.*, **169** (2010), 438–449.
52. J. E. Evans, K. L. Jungjohann, N. D. Browning and I. Arslan, Controlled growth of nanoparticles from solution with *in situ* liquid transmission electron microscopy. *Nano Lett.*, **11** (2011), 2809–2813.

Part II
Applications

9 Nanostructure Growth, Interactions, and Assembly in the Liquid Phase

Hong-Gang Liao, Kai-Yang Niu, and Haimei Zheng

9.1 Introduction

The growth of nanoparticles in solution phases can be traced back to the 1850s, when Michael Faraday prepared his ruby gold by reducing gold chloride with phosphorus in water [1]. But in the past two decades, solution based synthesis of nanoparticles with tailored properties has been developing rapidly. Nanostructures have been created in many materials and with a variety of shapes – sphere, cube, cuboctahedron, octahedron, tetrahedron, decahedron, icosahedron, thin plate, rod, and wire [2–4]. However, the nucleation and growth mechanisms of many nanoparticles are not well understood. This means that (in the absence of a hard template) solution-based methods require precise tuning of growth conditions to achieve controlled synthesis. The poor predictability of size, morphology, and structure as a function of synthesis conditions limits practical applications of nanoparticles with optimized performance. Important growth parameters include the reduction potential, temperature, precursor concentration, and presence of surfactants. Our understanding of the nucleation and complex growth steps involved in achieving hierarchical functional structures is especially limited.

The primary barrier for obtaining a sufficiently detailed knowledge of nanomaterials formation arises from the difficulty in "seeing through" the liquid to probe chemical and physical events. This also leads to limitations in understanding the relationship between structure and functionality for applications. With the recent technical advancements in electron microscopy and nanofabrication, liquid cell transmission electron microscopy has made it possible to image through liquids with nanometer resolution and beyond. Liquid cell TEM has been applied to the study of many different nanoparticle systems, and by tracking single-particle trajectories, novel growth mechanisms have been identified. Liquid cell TEM forms a unique platform for the study of nanoparticle formation, assembly, and sintering in the liquid phase [5].

In this chapter, we discuss nanoparticle growth as seen by liquid cell TEM. Most of the reported instances of nanoparticle growth in liquid cells are beam-induced, i.e. the imaging beam triggers the formation of particles. However, we also introduce examples where heating or other stimuli are used to drive growth. Factors controlling the morphology of the growing nanoparticles, such as electron dose, surfactants, neighboring particles, solvent, and precursor concentration, will also be discussed. We show two growth modes, monomer attachment and coalescence, then describe the growth of heterogeneous nanostructures and arrays, and briefly touch on measurements of nanoparticle diffusion and interactions and the etching of solid materials. At the end, an outlook on the future development and application of liquid cell TEM for the study of nanoparticles in liquids is provided. We focus here on experiments carried out using liquid cell TEM. Particle nucleation and growth occur equally in liquid cell SEM, and electron beam-induced deposition in SEM can be applied to a variety of materials to create complex patterns. Such experiments are discussed in Chapter 14.

9.2 Formation of Nanoparticles in TEM

9.2.1 Radiolytic and Thermal Formation of Particles

Nucleation and growth of nanoparticles starts from a supersaturated solution. Different stimuli can initiate the formation of nanoparticles, including heat, chemical reduction, and irradiation by the electron beam, other energetic particles, or a laser (Figure 9.1). Several of these processes have been examined using liquid cell TEM. The majority of

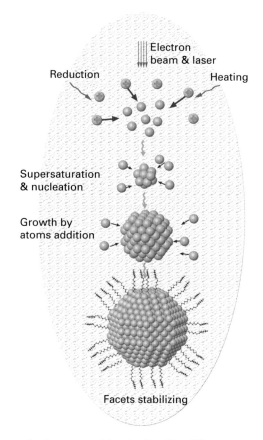

Figure 9.1. The nucleation and growth of a nanoparticle stimulated by different means.

studies have involved nanoparticle formation by electron beam irradiation, although growth has also been observed under thermal heating in liquid cell TEM, using a commercial heating sample stage and compatible liquid cell [6, 7]. During liquid cell TEM experiments, nanoparticles can nucleate heterogeneously on the walls of a liquid cell, and are then fixed in position, or they can nucleate homogeneously in liquid and are then free to move around. In this chapter, we focus on the growth of nanoparticles and assemblies. The information on particle motion that is also provided by these experiments will be discussed in Chapter 14. It has been noted in several studies, discussed below, that nanoparticles grown under an electron beam during observation in liquid cell TEM share many similarities to those grown *ex situ* by flask synthesis. This means that the *in situ* information obtained on nanoparticle growth mechanisms is relevant to nanoparticle synthesis under other circumstances.

9.2.2 Information Available from Liquid Cell EM

All liquid cells for electron microscopy, whether fabricated using Si wafers with Si_xN_y membranes or using graphene, enclose a thin liquid film between the membranes. As the nanoparticles nucleate, grow, and move in the liquid film, the process can be imaged in real time (Figure 9.2). Several types of information can be extracted, especially if the experiments are assisted by analytical studies using EELS, XEDS, or electron diffraction. By tracking single particle growth trajectories, we can measure size

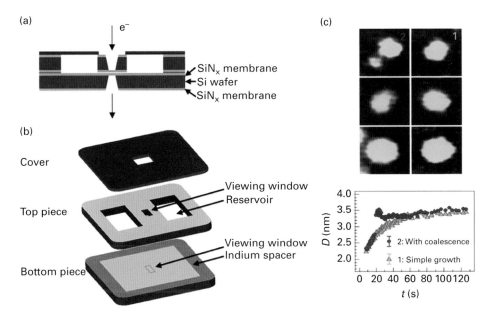

Figure 9.2. (a) A schematic of a self-contained liquid cell. (b) The liquid cell components: the top and bottom chips and a cover to seal the liquid cell. (c) Sequential images and trajectories of the growth of Pt nanoparticles. Two growth modes are visible, monomer attachment and nanoparticle coalescence, which can lead to identical particle sizes. From Ref. 8. Reprinted with permission from AAAS.

versus time of each particle (Figure 9.2c); particle shape and facet evolution; the dynamics of motion, coalescence, and rotation of nanoparticles; and the compositional or structural changes of each particle during growth.

9.2.3 Radiolysis and Other Effects of the Beam

When using liquid cell TEM to study any dynamic materials processes, the high energy electron beam induces changes in the sample, as discussed in Chapter 7. The most important, for the studies described below, is radiolysis: Nanoparticle growth occurs when the electron beam creates hydrated electrons that reduce metal ions in solution, forming metal atoms that nucleate metal clusters. Nanoparticle growth by radiolytic reduction of metal ions often requires a certain dose threshold, as all the radiolysis species are interlinked by reaction pathways and have complex dependences on dose and dose rate.

However, it is also worth remembering that the electron beam triggers many different reactions in any given liquid sample. For example [9], for water, typical imaging conditions lead to radiolysis products such as hydrogen as well as hydrated electrons. At a sufficiently high dose rate, the gaseous products can reach high enough concentrations to nucleate bubbles. The presence of bubbles alters diffusion pathways and may therefore change particle growth kinetics. Furthermore, the electron beam can in principle heat the sample, although, as discussed in Chapter 7, heating effects are expected to be insignificant.

The electron beam also causes other types of effect on the sample that can alter the process of nanoparticle growth, often by modifying the interface between the liquid and the enclosing membranes. These effects can be complex and difficult to predict. For example, beam-induced copper clusters grow from copper sulfate solution containing chloride ions, but only on gold electrodes and only under a range of applied potentials [10]. In this case, the potential is thought to drive electrochemical deposition of copper, but only on regions where the beam has changed the surface chemistry of the substrate by reducing an adsorbed copper chloride surface layer to copper. Other beam effects can be caused by charging. The Si_xN_y membranes (or the liquid if it has low conductivity) can become charged via emission of secondary electrons, resulting in a perpendicular electric field along the beam direction that allows nanoparticle growth at the membrane. The electron beam can change the interactions between nanoparticles, resulting in aggregation [11], repulsion, or changes in the crystal lattice [12]. These side effects of the beam should always be considered in analysis of nanoparticle growth phenomena.

9.3 Single Particle Growth Trajectories

Classical theory considers that monomer attachment from solution is the main mechanism for nanoparticle growth. It was also assumed that growth by coalescence (i.e. attachment of one nanoparticle to another) should be avoided if we want to achieve

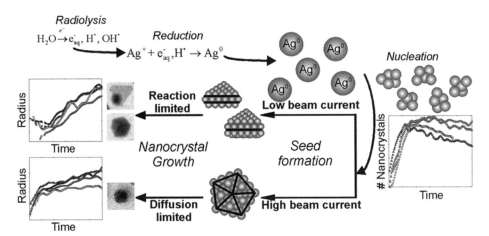

Figure 9.3. Silver nanoparticle growth under different electron beam dose rates. Reprinted with permission from Ref. 15. Copyright 2012 American Chemical Society.

a very sharp distribution of particle sizes. However, liquid cell TEM experiments have shown that coalescence is a significant growth pathway [8], implying that individual nanoparticles can act as building blocks to create complex structures. The experiments involved tracking the growth of individual Pt nanoparticles in an organic solvent (Figure 9.2c) and showed that the coalesced nanoparticles can eventually reshape into spherical nanoparticles with sizes similar to those of particles that had grown by monomer attachment. Both growth mechanisms can play a role simultaneously. For example, during ZnO nanoparticle precipitation [11], Ostwald ripening and coalescence both occur at early stages of growth, while coalescence is dominant at later times. The growth of Ag nanoparticles [12] similarly involves multiple processes, where individual nanoparticles grow by monomer attachment but the ensemble large-scale growth is dominated by aggregation. Direct measurements of particle size distributions revealed that growth follows a Smoluchowski model. By capturing both the mean growth rates and particle size distribution, this illustrates the power of liquid cell observations in untangling complex growth mechanisms.

During growth by monomer attachment, the rate-limiting steps of the attachment process itself can be probed via the relationship between nanoparticle morphologies, growth rates, and electron beam parameters [15]. Low beam currents facilitate reaction-limited growth, which yields faceted nanocrystals, while higher beam currents promote diffusion-limited growth, which yields spherical nanocrystals (Figure 9.3). This type of information leads to more reproducible control over growth mechanisms and hence the morphology of the structures produced [16].

During growth by coalescence, a key question is how the particles reshape, by surface diffusion or by grain boundary diffusion. These processes can be untangled by imaging coalescence events between similarly sized and unequally sized nanoparticles [17]. Both of these mass transport mechanisms are active during the relaxation period after coalescence, with a duration that increases with particle size. There is competition between the time required for complete coalescence and the collision time,

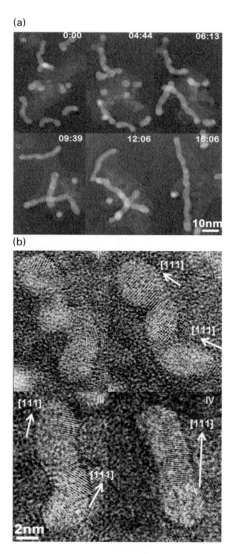

Figure 9.4. (a) Sequential TEM images showing the formation of a Pt_3Fe nanowire in a liquid cell. (b) Sequential HRTEM images (I to IV) show both crystal orientation and shape changes during the straightening of a twisted nanoparticle chain. Reprinted with permission from Ref. 13. Copyright 2013 American Chemical Society.

where collision time is determined by factors such as nanoparticle concentration and the speed of nanoparticle movement, while the electron beam influences coalescence. Coalescence can also produce interesting non-spherical geometries, as we show in Figure 9.4. Here, Pt_3Fe grows in nanowire form [13] by nanoparticle attachment to the ends of previously grown chains of nanoparticles, a process that can be attributed to the dipolar interaction between nanoparticles. Growth can take forms such as dendrites, as seen [14] after nucleation of Pd clusters with a focused probe.

The interaction between nanoparticles is clearly important during nanoparticle growth by coalescence. It may involve van der Waals forces, hydrophobic interactions, magnetic force, charge interaction, and other phenomena. As particles approach, liquid cell microscopy shows the details of the interaction. Examples include anisotropic attractive interactions and size-dependent segregation, reported during CTA-Au nanoparticle self-assembly into nanoparticle chains [11], and electrostatic dipolar interaction, invoked for explaining the growth of the Pt$_3$Fe chains described above [13]. During iron oxide formation, growth occurs by oriented attachment [18]. The nanoparticles undergo continuous rotation and interaction before they find the perfect lattice match with the same orientation. Measured translational and rotational accelerations show that strong, highly direction-specific interactions drive crystal growth via oriented attachment. These phenomena appear to be generally important in biomineralization processes, and are discussed in more detail in Chapter 15. In small faceted Pt nanoparticles, observed at high resolution using graphene liquid cells [19], coalescence is site selective, with lowest energy or lowest ligand-coverage facets being preferred during nanoparticle attachment. The details of the atomic rearrangement after coalescence can be resolved as the coalesced nanoparticles form a single crystal or a twinned nanoparticle.

It is clear from the above examples that nanoparticle growth mechanisms display a rich complexity, and that liquid cell TEM can contribute uniquely to understanding the details behind the mechanisms of coalescence and attachment. However, it is important to note that the nature of the liquid cell walls should be considered in interpreting growth kinetics. Nanoparticles of Pt–Pd alloy grown in a graphene oxide liquid cell show no visible coalescence process, and kinetics consistent with a surface-limited reaction [32]. In comparison with the systems described above showing strong coalescence effects, it is possible that the graphene oxide substrate may pin clusters and limit particle coalescence, or it may be that Pt–Pd behaves differently from pure Pt. To resolve such questions, it can be useful to compare measured reaction kinetics in liquid cells with different membrane materials.

9.4 Important Factors in Nanoparticle Growth

The shape of nanocrystals strongly influences their performance in catalysis, sensing, and many other surface-enhanced applications [2, 3]. Although complex shapes can be created, crystallization is a complicated process involving the arrangement of thousands of atoms or molecules near the surface region, further complicated by the interactions between the atoms and the surrounding liquid medium. Modifying the environment via temperature, precursor, surfactant concentration, or solvent allows greater control over growth and hence the ability to generate different shapes and structures. Liquid cell TEM provides an effective method to examine shape-control mechanisms during colloidal synthesis.

The Wulff construction [20, 21] is generally used to predict the equilibrium shape of nanocrystals, where the length of a normal vector drawn from the crystal center to an external surface is proportional to the surface free energy. It is widely accepted that in the growth of nanocrystals the high energy facets grow at a higher rate than the low energy facets; therefore, the fast-growing facets will eventually disappear, resulting in

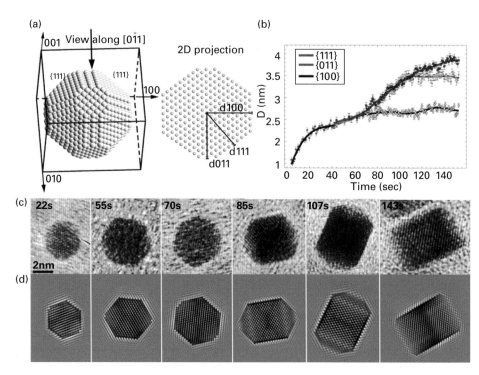

Figure 9.5. Facet development in a Pt nanocube. (a) Atomic model of a truncated Pt nanocube and its projection along the [011] zone axis. The distance from the crystal center to each of the {100}, {011}, and {111} facets is highlighted. (b) The measured average distance from the crystal center to each facet as a function of time. (c) Sequential images extracted from the movie show the growth of the Pt nanocube. (d) Simulated TEM images of the particle in (c). From Ref. 25. Reprinted with permission from AAAS.

a nanocrystal terminated with low energy facets [4, 22]. The commonly used surfactants are assumed to modify the energy of specific facets through preferential adsorption, influencing the relative growth rates of facets and hence the shape of a nanocrystal [23, 24]. Exploring the evolution of nanoparticle morphology provides direct information on these assumptions and the shape control mechanisms of nanocrystals. Recent liquid cell observations of Pt nanocube growth with high spatial and temporal resolution [25] show, in fact, that surface energy minimization rules break down at the nanoscale (Figure 9.5). The growth rates of all the low index facets, i.e. {100}, {110}, and {111}, are similar until the {100} facets stop growing. The {110} facets continue to grow until they reach a limit at the edge of the cube. The {111} facets then grow until they fill the corners of the cube. Rather than following surface energy minimization rules, the shape appears to arise through selective facet arrested growth.

Surfactants can modify certain facets and control the final shape of nanoparticles. But they can also inhibit nanoparticle aggregation and prevent coalescence. Liquid cell experiments allow surfactant concentration effects to be explored, for example during Pt nanoparticle growth [8], where dendritic crystals form at lower surfactant

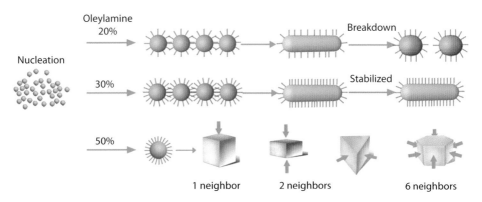

Figure 9.6. A summary of oleylamine surfactant effects and stereo-hindrance effects during PtFe nanoparticle growth. In the precursor solution, 20% oleylamine led to nanoparticles forming a chain, but eventually breaking into individual nanoparticles; 30% oleylamine resulted in stable nanowires; 50% oleylamine prevents nanoparticles from merging together and leads to shapes influenced more by neighboring particles than by surfactants. Reprinted with permission from Ref. 26. Copyright 2013 American Chemical Society.

concentrations rather than compact particles, or PtFe growth [26], where the surfactant concentration determines whether chains, particles, or separate nanowires form (Figure 9.6).

Temperature is a fundamental parameter during crystal growth. During the growth of Bi nanoparticles at an elevated temperature of 180 °C, achieved using a commercial heating sample stage and 3 mm liquid cell [7], growth took place with oscillatory dynamics. Individual and pairwise growth oscillations and collective fluctuations at global length scales were observed, and a depletion zone was identified around particles. These results suggest the interesting possibility of counteracting the energetically favorable Ostwald ripening process to prevent nanoparticle coarsening. This could be beneficial to industrial reactions such as heterogeneous nanoparticle catalysis.

Finally, the nature of the solvent is a critical factor for morphology control in nanoparticle growth. The three commonly used solvents are water, organic liquids, and ionic liquids. Aqueous solutions have been the most commonly studied in liquid cell TEM. However, many results have been obtained using organic solvents, such as pentadecane, dichlorobenzene, or benzene ether, for the growth of Pt spheres and cubes, Pt_3Fe nanowires, Bi spheres, and other nanoparticles [5, 6, 8, 13, 17, 25, 26], or an o-dichlorobenzene/oleylamine mixture, used for Pt nanoparticle growth [19]. Ionic liquids have benefits because they avoid evaporation and charging, and can be directly used in an open cell in the vacuum environment, as described in Chapter 3. An example is the use of 1,3-diallylimidazolium bromide ($C_9H_{13}BrN_2$) in the form of a thin film supported across a holey substrate by surface tension [27]. In this material, the growth of $NaClO_3$ crystals could be recorded with simultaneous formation and dissolution of pre-nucleation clusters and the development of different phases. The roles of solvent molecules in the growth of nanoparticles needs to be further explored using liquid cell TEM.

9.5 Growth of Materials Architectures

9.5.1 Formation of Heterogeneous Particles

Liquid cell TEM is well suited for characterizing the formation of complex nanostructures composed of different materials. The ability to measure the evolution of individual nanostructures allows growth mechanisms to be understood and controlled, for example as the various materials are added sequentially. Core-shell structured nanoparticles provide an excellent illustration of this concept. Gold particles can act as seeds on which Pd can be added heterogeneously by radiolysis or galvanic replacement [28, 29]. Liquid cell imaging shows that the size and shape of the Au seeds determine the morphology of the Pd shells (Figure 9.7a). The results can be explained by consideration of the numbers of nucleation sites for Pd on large and small Au particles, and the number of Pd ions available via diffusion [29]. Furthermore, an analysis of reaction kinetics demonstrates that the redox reaction that deposits the Pt is first order with respect to the concentration of hydrated electrons. Galvanic replacement reactions between Ag nanoparticles and aqueous Pd salt solution create a different type of nanostructure (Figure 9.7b) in which the Ag particles transform into hollow Ag–Pd nanostructures [28]. A comparison with *ex situ* growth kinetics shows that the electron beam strongly affects the process, even though there is a galvanic driving force. Such *in situ/ex situ* comparisons are important in interpreting liquid cell data.

If the growth can be observed as a function of position on highly symmetric particles, the details of the shell growth mechanism itself become accessible. For example, during Au shell formation on icosahedral Pt particles, Au first deposits at corner sites, then diffuses to the terraces followed by layer by layer growth to complete the shell [31]. The real-time data even allow diffusion parameters to be estimated for some of these

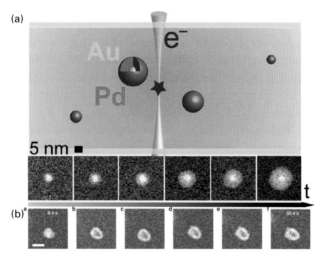

Figure 9.7. (a) Schematic and images of heterogeneous growth of Pd on Au seeds, forming Au–Pd core–shell structures. Reprinted with permission from Ref. 29. Copyright 2013 American Chemical Society. (b) Evolution of a Ag particle in aqueous PdCl$_2$ solution. Reprinted with permission from Ref. 28. Copyright 2012 MacMillan Publishers Ltd.

processes. When multi-element deposition experiments, such as those described above or the alloy growth mentioned in Section 9.4 [32], are combined with simultaneous measurement of composition, as has been demonstrated for Cu deposition on Ag nanowires or Pd nanotubes [30], the results are likely to be even more informative.

9.5.2 Growth of Extended Structures

As well as forming compact morphologies, it is possible to create extended nanostructures by tuning growth conditions. Under certain conditions, Au dendrites will form from Au nanoparticle seeds [33], with dendrite tip velocities consistent with diffusion-limited growth. Similarly, 2D Pd dendrites grow under the beam with fractal dimensions consistent with diffusion-limited aggregation and monomer addition [14]. Gold can grow as highly faceted platelike crystals, and STEM observations [35] suggest that to achieve this the supply rate of the Au monomers must be below a critical rate, otherwise growth is dominated by kinetic effects and is unfaceted. Gold can also form fivefold twinned crystals; for these, the growth pathways were identified from liquid cell observations and an important role of sintering proposed [34]. Experiments such as these can delineate the regimes under which particular types of complex morphology can be expected to form.

9.5.3 Synthesis in a Template

Template synthesis has been a widely used method to produce materials with specific shapes and sizes. However, new materials are usually developed by a trial and error process, with the growth mechanism inferred from the final synthetic structure. *Postmortem* analysis can provide only limited information, especially for the complex rheological fluids used in preparing nanoporous materials. A fundamental understanding of nanostructure evolution is needed at the length scales of the confined volumes in which the particles are growing.

Direct observation of synthesis in a highly viscous lyotropic liquid crystal template has indeed been demonstrated using liquid cell TEM [36]. Nanoparticles can be imaged as they nucleate and grow to a few nanometers, after which connections are made with other nanoparticles and clusters form. The clusters in turn grow and become mobile in the template, eventually resulting in porous spherical nanoparticles. Mesoporous particles also form from palladium in a solvated block copolymer template [37], where the electron beam induces nucleation in the aqueous phase around the micelles, eventually forming an ordered mesoporous structure. In multi-step processes like these, it is important to compare the liquid cell grown materials with those synthesized *ex situ*. The ability to observe synthesis at the length scales relevant to rheological fluids provides new opportunities for materials synthesis via an understanding of the processes at work.

9.6 Nanoparticle Diffusion and Assembly

We now move from discussing nanoparticle growth mechanisms to consider the motion of particles in liquid. We discuss the information that can be obtained by quantification of particle motion and the interactions between particles when they assemble into larger arrays.

9.6.1 Tracking Single Particle Diffusion

Nanoparticles are often observed to move around vigorously in the liquid cell. Several phenomena may contribute to this: Brownian motion, chemical reactions inducing local environment changes, liquid flow, or electron beam effects. Understanding the physics and origin of nanoparticle motion is important for interpreting liquid cell TEM data. But it also provides useful information on properties of the liquid, such as its viscosity. In particular, since liquid cell TEM is carried out in thin liquid layers, particle motion may provide information on any changes in liquid properties as the thickness decreases. Approaching the nanometer scale in thickness, several factors can become important. These include solvent surface fluctuations, the presence of the liquid–substrate interface or the vapor–liquid interface (if a bubble is present), and any intrinsic differences in the relaxation and transport properties in an ultra-thin liquid film compared to its bulk. In an ultra-thin liquid film, the viscosity can be higher than in the bulk, and interactions between the solvent molecules and the substrate can affect nanoparticle diffusion. The diffusion of nanoparticles in such thin liquid films is largely beyond the predictive capabilities of current theoretical computation.

Many experimental observations of nanoparticle motion by liquid cell TEM have shown intriguing clues to nanoscale liquid behavior. These include experiments in self-contained liquid cells [8] and flow cells [38, 39, 44], cells in which the liquid is micrometers in thickness [42] and cells in which liquid evaporation occurs during observation leaving ultra-thin layers [40]. In this last case, tracking real-time diffusion of Au nanocrystals in a thin film of a water–15% glycerol mix reveals complex movements such as rolling motions coupled to large-step movements and macroscopic violations of the Stokes–Einstein relation for diffusion, as well as motion dominated by the retraction of the liquid front on drying. Most particle motion data relate to individual nanoparticles, but the relative motion of pairs of Au nanoparticles attached by DNA [41] reveals information on the DNA properties. When interpreting particle dynamics, it is important to include electron beam effects, such as electrophoretic charging, that can drive particles towards the edge of the irradiated area [43]. Interpretation therefore requires care, but these observations are leading to a deeper understanding of liquids at the nanoscale. This will be discussed in Chapter 13.

9.6.2 Observation of Nanoparticle Self-Assembly

The self-assembly of nanoparticles into arrays has the potential to fabricate hybrid systems from different types of materials and produce new collective properties. Three-dimensional self-assembled superstructures can be formed with different packing densities and arrangements of nanoparticles. Two-dimensional arrays on large substrates have shown potential as an alternative to top-down fabricated devices. Colloidal nanoparticles often provide assemblies that are homogeneous and intrinsically tunable

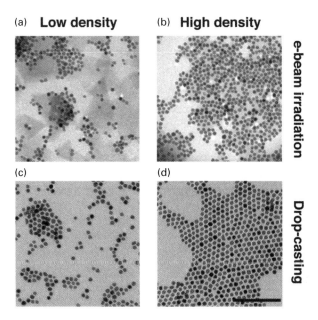

Figure 9.8. TEM images of nanoparticle assemblies formed (a, b) under electron beam irradiation, and (c, d) by drop casting on a Si_xN_y TEM grid. The first image of each pair has a lower particle density. The scale bar is 100 nm. Reprinted with permission from Ref. 47. Copyright 2012 American Chemical Society.

with respect to their composition, size, and shape [45, 46]. However, most nanoparticle assemblies are formed under empirically optimized, perhaps uncontrolled, drying conditions. Understanding the fundamental mechanisms active during assembly may help guide the creation of large-scale arrays suitable for effective device architectures.

Significant efforts have therefore been devoted to both experimental and theoretical approaches for understanding self-assembly mechanisms. The experimental tools include *in situ* optical spectroscopy and small-angle X-ray scattering (SAXS) techniques. These allow monitoring of nanoparticle self-assembly, but information on individual nanoparticles is difficult to obtain due to the spatial resolution. In addition, SAXS provides reciprocal space data and requires complex analysis to extract real space structural information. Liquid cell TEM provides a direct imaging platform to visualize how nanoscale objects assemble into arrays. Although the diffusion in a thin liquid film near the surface may differ from the real bulk liquid, as discussed in Section 9.6.2 above, the processes observed *in situ* provide a useful probe of interactions between, and assembly of, nanoparticles.

Figure 9.8 shows the formation of an ordered nanoparticle superlattice from randomly distributed nanoparticles [47]. Here, both capillary forces and local fluctuations in solvent geometry due to electron beam-induced local evaporation appear to play a role during assembly. Particles can also assemble into chains and rings [26]. More irregular, extended aggregates can also form [48] and, for these, the kinetics and fractal dimension are consistent with three-dimensional cluster–cluster diffusion-limited aggregation. Interestingly, while individual particles may move too quickly to be visible, the aggregates move more slowly and can be characterized in detail.

The electron beam appears to play a fundamental role in these assembly processes. As well as causing particle motion, as described in Section 9.6.1 above, it can change the assembly dynamics via interactions with any surfactant present. For example, Au nanoparticles coated with positively charged CTA^+ or negatively charged citrate ions behave differently during illumination [11], presumably because of the difference in surface charge: positively charged Au nanoparticles form chains, while negatively charged particles remained stationary regardless of the electron beam intensity. As with any *in situ* experiment, it is worth comparing results with and without illumination.

9.6.3 Manipulation of Nanoparticles with the Electron Beam

The highly focused electron beam can be used to manipulate nanoparticles, acting as "electron tweezers", and to probe the interaction forces between nanoparticles while their dynamic motion is observed. Early work on electron tweezers in liquids [49] involved trapping and steering solid, 20–300 nm Al particles inside a molten Al–Si eutectic alloy. Around the same time, Batson *et al.* proposed another method to manipulate nanoparticles with the electron beam [50, 51], controlling the position of Au nanoparticles on an amorphous carbon film with the STEM probe.

Figure 9.9 shows nanoparticle manipulation in a microfabricated liquid cell [52, 53]. The degree of nanoparticle manipulation is very strong: single or multiple nanoparticles can be trapped inside the beam and their global movements then follow the movement of the beam. A trapping force of 1 pN was measured via the probability distribution of the nanoparticles with respect to the beam center. Nanoparticles can also be attracted outside the electron beam [54] via an electrostatic driving force that depends on the particle-to-beam distance.

The nature of interaction between nanoparticles and a highly focused electron beam needs to be further studied. While electron beam tweezing of nanoparticles may bring useful applications, the study of nanoparticle diffusion and assembly using liquid cell

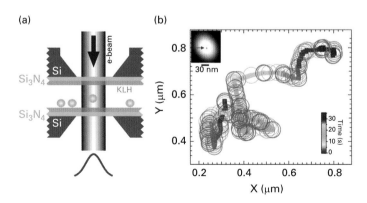

Figure 9.9. Manipulation and imaging of Au nanoparticle movements in a liquid cell using an electron beam. (a) In the experimental setup, an electron beam passes through the Si_xN_y window and traps Au nanoparticles inside the beam. (b) Trajectories of the electron beam movement and the position of the Au nanoparticle. Reprinted with permission from Refs. 52 and 53. Copyright 2015 American Chemical Society.

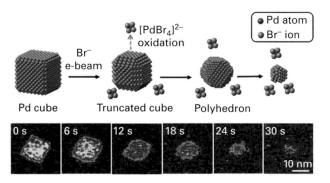

Figure 9.10. Schematic and *in situ* observation of cubic Pd nanoparticle shrinking due to oxidative etching. Reprinted with permission from Ref. 55. Copyright 2014 American Chemical Society.

TEM requires careful control of electron beam perturbation. In addition to reducing the electron beam dose, it is necessary to be aware of the critical electron beam current value for each system, above which nanoparticle motion is dominated by electron beam effects.

9.7 Etching and Corrosion

The study of etching processes is important in nanoparticle synthesis since many nanoparticles can shrink or dissolve by slightly varying the growth conditions. Oxidative etching prevails in the synthesis of nanocrystals. Figure 9.10 shows an example of the oxidative etching of cubic Pd nanocrystals [55] in the liquid cell. This is a beam effect, since etching was realized with oxidative radiation reactants produced by radiolysis in the presence of Br^- ions. Such experiments allow dissolution dynamics of monodispersed and aggregated nanocrystals to be investigated and compared to provide quantitative information on reaction pathways. Experiments on the corrosion of larger scale structures, such as thin metal films with or without electrical bias, are discussed in Chapter 12. The importance of compositional heterogeneity to corrosion is well known, and it would be interesting to examine the dissolution of heterogeneous nanoparticles such as those discussed in Section 9.5.1 above.

9.8 Conclusions and Outlook

The recent development of liquid cell TEM technologies has provided unique capabilities for the study of nanoparticle growth, diffusion, and assembly in liquids with high spatial and temporal resolution. It has opened tremendous opportunities for the study of materials and dynamic processes in liquid environments that were previously not reachable. Publications on liquid cell TEM studies of nanoparticle growth, dynamics, and assembly in liquids have increased rapidly over the past several years. However, there is still a lot of room for improvement of liquid cell TEM. Liquids with high vapor pressure, including aqueous solutions, are hard to handle since they can easily dry out during sample loading or under an electron beam. This limits the applications of liquid cell TEM techniques. Colloidal synthesis in a chemistry lab is often achieved by hot injection, mixing precursor solutions together at a specific temperature, but reactions by solvent mixing have not yet been achieved by liquid cell TEM. Additionally, although atomic-scale resolution has been achieved in some liquid cell experiments by incorporating ultra-thin silicon nitride or graphene membranes, improving the spatial resolution is still necessary in many experiments. Probing nanoparticle growth and tracking the interactions and assembly of nanoparticles with liquid cell TEM is expected to experience revolutionary growth in the near future. Liquid cell TEM will allow the imaging of atomic or molecular processes of materials at solid–liquid interfaces under an applied external stimulus, as reliable control of the liquid environment and electron beam effects is achieved.

Acknowledgements

We thank group members Yilong Zhu and Tao Xu for their help in collecting useful information for this chapter. HZ thanks the DOE Early Career Research Program for support.

References

1. M. Faraday, The Bakerian lecture: experimental relations of gold (and other metals) to light. *Phil. Trans. R. Soc. Lond.*, **147** (1857), 145–181.
2. M. C. Daniel and D. Astruc, Gold nanoparticles: assembly, supramolecular chemistry, quantum-size-related properties, and applications toward biology, catalysis, and nanotechnology. *Chem. Rev.*, **104** (2004), 293–346.
3. C. J. Murphy T. P. Sau, A. M. Gole, *et al.*, Anisotropic metal nanoparticles: synthesis, assembly, and optical applications. *J. Phys. Chem. B*, **109** (2005), 13857–13870.
4. Y. Xia, Y. Xiong, B. Lim and S. E. Skrabalak, Shape-controlled synthesis of metal nanocrystals: simple chemistry meets complex physics? *Angew. Chem. Int. Ed.*, **48** (2009), 60–103.
5. H.-G. Liao, K. Niu and H. Zheng, Observation of growth of metal nanoparticles. *Chem. Commun.*, **49** (2013), 11720–11727.
6. K.-Y. Niu, J. Park, H. Zheng and A. P. Alivisatos, Revealing bismuth oxide hollow nanoparticle formation by the Kirkendall effect. *Nano Lett.*, **13** (2013), 5715–5719.
7. H. L. Xin and H. Zheng, In situ observation of oscillatory growth of bismuth nanoparticles. *Nano Lett.*, **12** (2012), 1470–1474.
8. H. Zheng, R. K. Smith, Y.-W. Jun, *et al.*, Observation of single colloidal platinum nanocrystal growth trajectories. *Science*, **324** (2009), 1309–1312.
9. J. M. Grogan, N. M. Schneider, F. M. Ross and H. H. Bau, Bubble and pattern formation in liquid induced by an electron beam. *Nano Lett.*, **14** (2013), 359–364.

10. M. den Heijer, I. Shao, A. Radisic, M. C. Reuter and F. M. Ross, Patterned electrochemical deposition of copper using an electron beam. *APL Materials*, **2** (2014), 022101.
11. Y. Liu, X.-M. Lin, Y. Sun and T. Rajh, In situ visualization of self-assembly of charged gold nanoparticles. *J. Am. Chem. Soc.*, **135** (2013), 3764–3767.
12. T. J. Woehl, C. Park, J. E. Evans, *et al.*, Direct observation of aggregative nanoparticle growth: kinetic modeling of the size distribution and growth rate. *Nano Lett.*, **14** (2013), 373–378.
13. H.-G. Liao, L. Cui, S. Whitelam and H. Zheng, Real-time imaging of Pt_3Fe nanorod growth in solution. *Science*, **336** (2012), 1011–1014.
14. G. Zhu, Y. Jiang, F. Lin, *et al.*, In situ study of the growth of two-dimensional palladium dendritic nanostructures using liquid-cell electron microscopy. *Chem. Commun.*, **50** (2014), 9447–9450.
15. T. J. Woehl, J. E. Evans, I. Arslan, W. D. Ristenpart and N. D. Browning, Direct in situ determination of the mechanisms controlling nanoparticle nucleation and growth. *ACS Nano*, **6** (2012), 8599–8610.
16. J. E. Evans, K. L. Jungjohann, N. D. Browning and I. Arslan, Controlled growth of nanoparticles from solution with in situ liquid transmission electron microscopy. *Nano Lett.*, **11** (2011), 2809–2813.
17. K.-Y. Niu, H.-G. Liao and H. Zheng, Visualization of the coalescence of bismuth nanoparticles. *Microsc. Microanal.*, **20** (2014), 416–424.
18. D. Li, M. H. Nelson, J. R. Lee, *et al.*, Direction-specific interactions control crystal growth by oriented attachment. *Science*, **336** (2012), 1014–1018.
19. J. M. Yuk, J. Park, P. Ercius, *et al.*, High-resolution EM of colloidal nanocrystal growth using graphene liquid cells. *Science*, **336** (2012), 61–64.
20. G. Wulff, On the question of speed of growth and dissolution of crystal surfaces. *Z. Krystallogr. Mineral.*, **34** (1901), 449–530.
21. J. W. Gibbs, H. A. Bumstead, R. G. Van Name and W. R. Longley, *The Collected Works of J. Willard Gibbs* (London: Longmans, Green and Co., 1902).
22. N. Tian, Z.-Y. Zhou, S.-G. Sun, Y. Ding and Z. L. Wang, Synthesis of tetrahexahedral platinum nanocrystals with high-index facets and high electro-oxidation activity. *Science*, **316** (2007), 732–735.
23. E. Ringe, R. P. Van Duyne and L. D. Marks, Wulff construction for alloy nanoparticles. *Nano Lett.*, **11** (2011), 3399–3403.
24. C. R. Bealing, W. J. Baumgardner, J. J. Choi, T. Hanrath and R. G. Hennig, Predicting nanocrystal shape through consideration of surface-ligand interactions. *ACS Nano*, **6** (2012), 2118–2127.
25. H.-G. Liao, D. Zherebetskyy, H. Xin, *et al.*, Facet development during platinum nanocube growth. *Science*, **345** (2014), 916–919.
26. H.-G. Liao and H. Zheng, Liquid cell transmission electron microscopy study of platinum iron nanocrystal growth and shape evolution. *J. Am. Chem. Soc.*, **135** (2013), 5038–5043.
27. Y. Kimura, H. Niinomi, K. Tsukamoto and J. M. García-Ruiz, In situ live observation of nucleation and dissolution of sodium chlorate nanoparticles by transmission electron microscopy. *J. Am. Chem. Soc.*, **136** (2014), 1762–1765.
28. E. Sutter, K. Jungjohann, S. Bliznakov *et al.*, *In situ* liquid-cell electron microscopy of silver-palladium galvanic replacement reactions on silver nanoparticles. *Nat. Commun.*, **5** (2014), 4946.

29. K. Jungjohann, S. Bliznakov, P. Sutter, E. A. Stach and E. Sutter, In situ liquid cell electron microscopy of the solution growth of Au–Pd core–shell nanostructures. *Nano Lett.*, **13** (2013), 2964–2970.
30. E. A. Lewis, S. J. Haigh, T. J. A. Slater, *et al.*, Real-time imaging and local elemental analysis of nanostructures in liquids. *Chem. Commun.*, **50** (2014), 10019–10022.
31. J. Wu, W. Gao, J. Wen *et al.*, Growth of Au on Pt icosahedral nanoparticles revealed by low-dose in situ TEM. *Nano Lett.*, **15** (2015), 2711–2715.
32. A. De Clercq, W. Dachraoui, O. Margeat, *et al.*, Growth of Pt–Pd nanoparticles studied in situ by HRTEM in a liquid cell. *J. Phys. Chem. Lett.*, **5** (2014), 2126–2130.
33. T. Kraus and N. de Jonge, Dendritic gold nanowire growth observed in liquid with transmission electron microscopy. *Langmuir*, **29** (2013), 8427–8432.
34. H.-G. Liao, Y. Shao, C. M. Wang, *et al.*, TEM study of fivefold twinned gold nanocrystal formation mechanism. *Mater. Lett.*, **116** (2014), 299–303.
35. D. Alloyeau, W. Dachraoui, Y. Javed, *et al.*, Unravelling kinetic and thermodynamic effects on the growth of gold nanoplates by liquid transmission electron microscopy. *Nano Lett.*, **15** (2015), 2574–2581.
36. L. R. Parent, D. B. Robinson, T. J. Woehl, *et al.*, Direct in situ observation of nanoparticle synthesis in a liquid crystal surfactant template. *ACS Nano*, **6** (2012), 3589–3596.
37. L. R. Parent, D. B. Robinson, P. J. Cappillino, *et al.*, In situ observation of directed nanoparticle aggregation during the synthesis of ordered nanoporous metal in soft templates. *Chem. Mater.*, **26** (2014), 1426–1433.
38. X. Chen and J. Wen, In situ wet-cell TEM observation of gold nanoparticle motion in an aqueous solution. *Nanoscale Res. Lett.*, **7** (2012), 1–6.
39. E. A. Ring and N. de Jonge, Microfluidic system for transmission electron microscopy. *Microsc. Microanal.*, **16** (2010), 622–629.
40. H. Zheng, S. A. Claridge, A. M. Minor, A. P. Alivisatos and U. Dahmen, Nanocrystal diffusion in a liquid thin film observed by in situ transmission electron microscopy. *Nano Lett.*, **9** (2009), 2460–2465.
41. Q. Chen, J. M. Smith, J. Park, *et al.*, 3D motion of DNA-Au nanoconjugates in graphene liquid cell electron microscopy. *Nano Lett.*, **13** (2013), 4556–4561.
42. N. de Jonge, N. Poirier-Demers, H. Demers, D. B. Peckys and D. Drouin, Nanometer-resolution electron microscopy through micrometers-thick water layers. *Ultramicroscopy*, **110** (2010), 1114–1119.
43. E. R. White, M. Mecklenburg, B. Shevitski, S. B. Singer and B. C. Regan, Charged nanoparticle dynamics in water induced by scanning transmission electron microscopy. *Langmuir*, **28** (2012), 3695–3698.
44. C. Mueller, M. Harb, J. R. Dwyer and R. D. Miller, Nanofluidic cells with controlled pathlength and liquid flow for rapid, high-resolution in situ imaging with electrons. *J. Phys. Chem. Lett.*, **4** (2013), 2339–2347.
45. F. Li, D. P. Josephson and A. Stein, Colloidal assembly: the road from particles to colloidal molecules and crystals. *Angew. Chem. Int. Ed.*, **50** (2011), 360–388.
46. J. L. Baker, A. Widmer-Cooper, M. F. Toney, P. L. Geissler and A. P. Alivisatos, Device-scale perpendicular alignment of colloidal nanorods. *Nano Lett.*, **10** (2009), 195–201.
47. J. Park, H. Zheng, W. C. Lee, *et al.*, Direct observation of nanoparticle superlattice formation by using liquid cell transmission electron microscopy. *ACS Nano*, **6** (2012), 2078–2085.
48. J. M. Grogan, L. Rotkina and H. H. Bau, In situ liquid-cell electron microscopy of colloid aggregation and growth dynamics. *Phys. Rev. E*, **83** (2011), 061405.

49. V. P. Oleshko and J. M. Howe, Are electron tweezers possible? *Ultramicroscopy*, **111** (2011), 1599–1606.
50. P. E. Batson, A. Reyes-Coronado, R. G. Barrera, *et al.*, Nanoparticle movement: plasmonic forces and physical constraints. *Ultramicroscopy*, **123** (2012), 50–58.
51. P. E. Batson, A. Reyes-Coronado, R. G. Barrera, *et al.*, Plasmonic nanobilliards: controlling nanoparticle movement using forces induced by swift electrons. *Nano Lett.*, **11** (2011), 3388–3393.
52. H. Zheng, U. M. Mirsaidov, L.-W. Wang and P. Matsudaira, Electron beam manipulation of nanoparticles. *Nano Lett.*, **12** (2012), 5644–5648.
53. H. Zheng, Using molecular tweezers to move and image nanoparticles. *Nanoscale*, **5** (2013), 4070–4078.
54. Y.-T. Chen, C.-Y. Wang, Y.-J. Hong, *et al.*, Electron beam manipulation of gold nanoparticles external to the beam. *RSC Adv.*, **4** (2014), 31652–31656.
55. Y. Jiang, G. Zhu, F. Lin, *et al.*, In situ study of oxidative etching of palladium nanocrystals by liquid cell electron microscopy. *Nano Lett.*, **14** (2014), 3761–3765.

10 Quantifying Electrochemical Processes Using Liquid Cell TEM

Frances M. Ross

10.1 Introduction

Electrochemical processes play an essential role in energy storage and materials synthesis. In the microelectronics industry, electrochemical deposition is used to fabricate the copper interconnects that enable high density packing of transistors; packaging and read-write heads for disk drives are other examples where electrochemical processes are used to create complex structures. Electrochemical deposition can produce hard, dense coatings or porous materials suitable for sensors or filters. Subtractive processes such as anodic etching can increase the range of structures achievable. The low cost and simplicity of electrochemical processing, the absence of complex vacuum requirements, the wide range of materials accessible, and the ability to conformally coat a substrate or deposit within a deep channel all make electrochemistry a key process for a wide range of applications. If we include the ubiquitous effects of corrosion and the importance of battery operation, which relies on electrochemical reactions at interfaces between carefully designed materials, it is clear that electrochemical processes have an enormous impact in everyday life.

The ability to control the location and kinetics of electrochemical reactions and the morphology and composition of the resulting material is critical to all applications involving electrochemical processes. Yet an understanding of even some very basic features of electrochemical deposition remains elusive, so that development of new processes can require a lengthy trial and error effort. Electrochemical reactions are generally studied on the benchtop, measuring parameters such as voltage and total current, with only the end results imaged using microscopy.

Direct observation is so exciting for electrochemistry because it can capture the kinetics of the transient phenomena, following composition and structure as they develop over a few seconds and at nanometer length scales. Such detailed information can be used to test and extend electrochemical growth models, which in real life apply to individual structures (nuclei, pits, dendrites) yet had to be developed based on ensemble measurements such as cluster density, average film thickness, and current per unit area. Liquid cell TEM has a key role to play. Other techniques [1–6] have better spatial or temporal resolution (Figure 10.1), but liquid cell TEM occupies a unique position, with reasonable resolution in both space and time coupled with the simultaneous measurement of electrochemical parameters. This niche drives interest in the use of liquid cell TEM for electrochemical processes. Indeed, the modern silicon nitride liquid

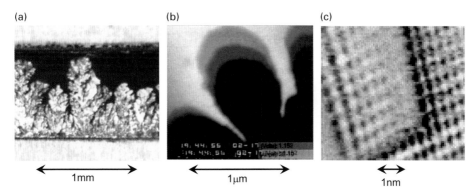

Figure 10.1. Comparison of imaging techniques for electrochemistry illustrating the length scales addressed by each. (a) Light microscopy: Deposition of Zn in alkaline electrolyte, from a movie recorded at 30 images per second. Reproduced with permission from Ref. 6, Copyright 2003, The Electrochemical Society. (b) Liquid cell TEM: Progression of an irregular growth front during potentiostatic Cu deposition from $H_2SO_4/CuSO_4$. Three images are superimposed showing the advancing growth front. Adapted from Ref. 21. (c) STM: Cu(100) ledge flow in 0.01 M HCl, −0.23 V with respect to a saturated calomel electrode, from a sequence recorded at 10 images per second. Reprinted from Ref. 3, Copyright 2001, with permission from Elsevier.

cell design was initially developed for electrochemical experiments [7] and the earliest experiments using this cell involved nucleation and growth of Cu [8].

In this chapter we discuss the key design considerations for electrochemical measurements using closed liquid cell TEM. We illustrate, mainly through the example of Cu electrodeposition, the range of measurements that can be made. Electrochemical liquid cell TEM has become readily available and we describe some of the innovative commercial and home-made liquid cells now available. But it is important to use these capabilities in a quantitative way that considers artifacts. We therefore address how well liquid cell microscopy can work, given the small electrolyte volumes, small electrode areas, and complexity of beam effects. We illustrate this through several examples of the use of liquid cell TEM for basic electrochemical studies: nucleation and growth mechanisms, growth front stability, and diffusion of ions in solution. Two application areas that have received particular attention are described in the next two chapters, battery and fuel cells (Chapter 11) and corrosion (Chapter 12). We conclude with an outlook on future areas in which we expect liquid cell TEM to make a strong impact.

10.2 Design of Liquid Cells for Quantitative Electrochemical Experiments

10.2.1 Historical Context

Figure 10.2 shows an early design of a liquid cell for studying an electrochemical reaction: the formation of porous silicon [9]. Porous silicon, created by anodic etching

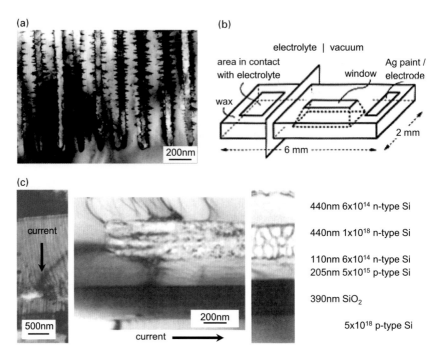

Figure 10.2. Liquid cell TEM of porous silicon formation. (a) Bright-field TEM image showing pores in 0.1 Ω cm n-type Si formed at 50 mA cm^{-2} in 20 wt% HF. Unpublished. (b) Sample geometry for *in situ* pore formation using a window formed in a silicon-on-insulator chip. Adapted from Ref. 9. (c) Porous silicon that had been formed in the sample holder, imaged post growth in plan view and after making cross sections in the pore direction and parallel to the pore. Adapted from Ref. 9.

of crystalline silicon using HF electrolyte, is a nanostructured material of interest for optoelectronics, sensors, and biomedical applications. The structures produced by the electrochemical dissolution process have morphology and dimension that depend on the electrochemical parameters as well as the doping type and level of the silicon. With the idea of observing etching as it happens, the reaction was confined in a highly doped Si layer sandwiched between two low-doped Si layers that did not etch. The sample and holder were not well enough sealed from the microscope vacuum to produce quantitative results, but the experiment proved useful in developing a more usable liquid cell based on closely spaced silicon nitride windows. One reason for showing this early design is that modern closed liquid cells for electrochemistry are mostly based on aligned pairs of Si_xN_y window chips, as described in Chapter 2. But the use of other materials, as in the Si multilayer shown in Figure 10.2c, could provide different opportunities for delivering liquid and current/voltage in well-defined geometries. We now discuss conventional closed liquid cells, in particular the features that must be optimized for electrochemical measurements: the electrodes, electrolyte, and external electrochemical control.

10.2.2 The Working and Counter Electrodes

Electrochemical liquid cells have multiple electrodes controlled by an external potentiostat (Figures 10.3 and 10.4). Since the electrodes are the key feature that enables electrochemical studies, they must be designed carefully to enable quantitative measurements that can be interpreted to provide information relevant to "real life" reactions over larger electrode areas and in larger volumes of electrolyte. Three electrodes are required for most measurements, although, as we discuss below, useful functionality is also possible with two [10]. Typically, the reaction of interest occurs at one electrode, which we refer to as the working electrode (WE). The WE should be patterned to overlap the electron-transparent window. The second electrode, or counter electrode (CE), generally has a larger area than the WE. It is often not necessary to observe

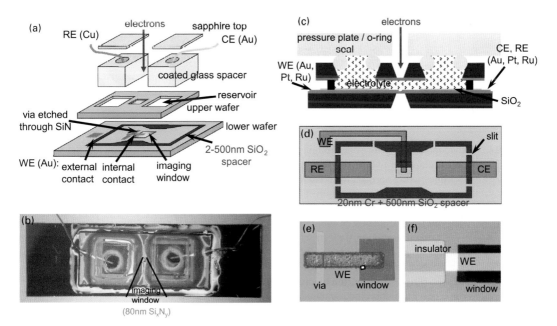

Figure 10.3. Early electrochemical liquid cell designs: (a) Schematic view of a complete Si_xN_y window cell with patterned working electrode (WE) and wire counter (CE) and reference electrodes (RE). Adapted from Ref. 8. (b) Photograph of completed liquid cell showing reservoirs, wire electrodes, and viewing window. (c) Schematic showing later generation design of liquid cell with multiple electrodes on the lower window chip and o-ring rather than glue sealing. (d) Second generation design in which all electrodes are patterned on the lower chip. Note the use of the spacer layer to define the active area of the WE. The slits are included to avoid short-circuits between electrodes, since the spacer layer involves a conductive adhesion layer. (e) Photograph of WE in the first generation design showing the via used to allow the WE electrode to make electrical contact with the Si wafer. Image shown after deposition of Cu, illustrating the active WE area. The rectangle indicates the typical field of view in TEM. (f) Photograph of WE in the second generation design, showing alignment issues and invisible area of WE. (b, d–f) reprinted with permission from Ref. 34.

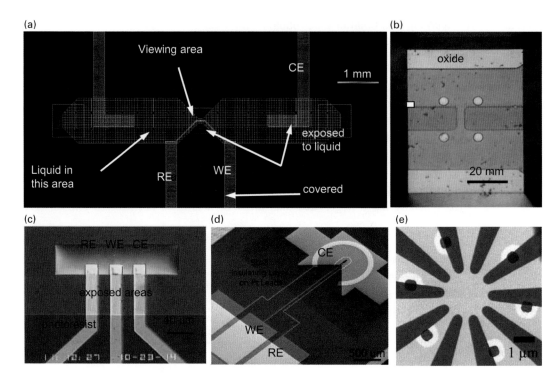

Figure 10.4. Developments in electrode design. (a) The Nanoaquarium: Overview showing the wafer bonded chip and two of the four external pads that are used to connect the holder to the electrodes using spring clips. The electrodes are 20 nm polycrystalline Pt. Reprinted from Ref. 21. (b) The Nanoaquarium: Enlarged view looking down on the window and showing the two visible electrodes. The small circles are posts that maintain membrane separation. The typical area imaged in the TEM is indicated by a small white square. Reprinted from Ref. 15. (c) Hummingbird electrochemical cell: Light microscope image of the central part of a chip in which three 150 nm polycrystalline Au electrodes extend into the viewing window. Chip imaged after electrochemical deposition to illustrate the opportunities for post-growth analysis. Below the line, the electrodes are masked by photoresist. Unpublished. (d) Protochips Poseidon electrochemical cell: Light microscope image of a chip showing electrodes in a Hele–Shaw geometry with a glassy carbon WE, a Pt RE nearby (within 100 μm) and larger area CE placed 500 μm away from the WE. The thin Pt tracks that connect the three electrodes to the microchip contact pads are covered. Reprinted from Ref. 36. (e) Sandia CINT electrochemical cell: Bright field STEM image showing ten electrodes composed of W tracks. These are masked with Al_2O_3 with small regions exposed (bright halos) and Au or 35 nm Ti patterned to create electrodes each with area ~1 μm². © 2015 IEEE. Reprinted, with permission, from Ref. 24.

reactions on the CE, so it can be located away from the window. The third electrode or reference electrode (RE) is discussed in Section 10.2.3. The RE should be close to the WE, but in principle no reactions take place at the RE so it also does not need to be observed in the window. Additional electrodes are often included, either for redundancy (in case one connection does not work) or for more complex electrochemical analysis.

We now consider the fabrication, materials, and dimensions of the WE and the CE. Both WE and CE are often formed from metals such as Au, Pt, or Ni, which can be deposited in the form of a polycrystalline layer by evaporation or sputtering on the silicon nitride interior surface of the liquid cell chip (Chapter 2). Typical metal thicknesses are between 10 nm and 500 nm, and the electrode dimensions are of the order of tens of micrometers. In recent experiments, glassy carbon has been used as an amorphous electrode [11, 12]. Many metals require an adhesion layer to avoid peeling off the Si_xN_y surface. It is worth choosing the adhesion layer material carefully: Ti is usually acceptable [13], but Cr can be electrochemically active and distort the results of the experiment. The electrodes are defined lithographically using photoresist via a technique such as lift-off. Because each electrode needs to interface with a contact somewhere on the liquid cell holder, it is generally patterned in the form of a long track extending from (or near) the viewing area towards the edge of the chip. Often this track is covered with an insulating layer, leaving an exposed area at the tip of the electrode that is well defined and relatively small (Figure 10.3f).

For the WE there are two design strategies, depending on the nature of the experiment. The WE itself can be made of an electrochemically active material, if a reaction on its surface or along its edge is the object of the investigation. These "active" WEs can be electron transparent (e.g. 10–20 nm thick if made of Au or Pt), allowing imaging of reactions in plan view on the electrode surface as well as in projection around its edges [7, 12–25]. Alternatively, the WE can be thicker so that, rather than being imaged in plan view, the reaction of interest is always imaged off its edges [26–30]. Thicker electrodes are perhaps easier to fabricate reliably using conventional deposition and lithographic techniques, and may incur a smaller voltage drop along their length due to a lower resistance.

The second strategy is to use a WE material that is inert under the conditions of the experiment to deliver the current or voltage to an active material that is attached to the WE. The active material may be formed by printing [11] or depositing particles or nanostructures, say by drop casting [31–33], so that the attached structure extends sideways for imaging. The active material can also be cut and manipulated using a focused ion beam, as for the corrosion and battery experiments described in Chapters 11 and 12.

In both of these WE strategies, the electrode typically does not block the entire area of the viewing window. This makes it possible to image through the electrolyte alone, to see any phenomena that may be occurring there. Furthermore, part of the WE is hidden from view as it extends away from the viewing area. These hidden parts of the electrode are often coated with an insulating layer to minimize reactions on the area that cannot be imaged (as in Figures 10.3f and 10.4). Most or all of the reaction is then visible within the window, but the very small electrode area can be difficult to work with because the total current that flows is small and can be hard to measure.

The CE is usually designed with a larger area than the WE [8, 23, 30], and can be distant from the viewing window since the reaction that occurs there does not generally need to be observed. In some electrode designs, however, the CE can be similar to the

WE and may even be placed within the viewing window [24, 25, 28] to allow reactions on the neighboring electrodes to be imaged.

10.2.3 The Reference Electrode

Many electrochemical experiments in the liquid cell, especially in some of the first publications, used two electrodes [7, 20, 26–29, 31]. A two-electrode design is easier to implement, and allows for experiments in which a controlled current flows through the cell to trigger, say, deposition of metals under galvanostatic conditions. However, including a third, reference electrode [12, 16–19, 22, 23, 30, 34], or even greater numbers of electrodes [21, 24, 25], allows a wider range of measurements. It is possible to carry out deposition under controlled voltage (i.e. potentiostatic conditions) and quantify the effect of solution resistance, the double layer capacitance, ohmic potential drops, or cell impedance [10]. This makes it easier to compare the results of the liquid cell experiment with data from standard electrochemical experiments. In particular, ohmic drops can be significant in liquid cell TEM experiments because of the long tracks leading to the electrodes and the relatively large distances between electrodes compared to the small thickness of the electrolyte layer.

For the most interpretable electrochemical results, the RE should be close to the WE so that it is in the same electrochemical environment and ohmic potential drops are minimized [12, 23, 30]. Since conventional electrochemical REs (saturated calomel electrodes, hydrogen electrodes) are physically too bulky for the liquid cell, experiments rely on pseudo-reference electrodes made from a material such as Cu [16–19], Ti [24, 25], Au [22, 23, 30, 34], or Pt [12, 21]. The pseudo-reference potential has a fixed value that can be calibrated as a function of conditions. It is also possible to use a RE that is physically distant but connected by an electrolyte bridge [33]. A wider range of REs is then feasible, but the ohmic drop between RE and WE has to be accounted for in the analysis.

10.2.4 Putting Together the Complete Cell

To illustrate how the ideas above are implemented into a complete electrochemical liquid cell, we consider an early design [8], Figure 10.3. The WE was electron transparent, made from 20 nm Au sputtered over a 5 nm Ti adhesion layer. The CE and RE were not patterned onto the lower wafer, but instead were made from wires glued into the cell that extend into the electrolyte. These wires required an extra assembly step but had several advantages. The WE, CE, and RE could be made from different materials; for Cu deposition a good choice is Au for the WE and Cu for the CE and RE. The wires provided large surface areas for the CE (and RE). And since only the WE was patterned onto the lower Si window chip, it was possible to connect the WE electrically, using a via in the Si_xN_y, through the Si of the lower window chip and out to an external contact pad (Figure 10.3e). The size of the WE was then easily defined and a long connecting track and insulating layer were not required.

To assemble this cell, the top and bottom window chips were glued using a patterned spacer layer of SiO_2 (see Chapter 2) and glass reservoirs were glued above the top wafer to increase the volume of liquid, Figure 10.3a, b [7, 8, 13]. A curable epoxy was used for this step [13]. Then the electrolyte was introduced with a small syringe before sealing the cell with a UV-cure glue to avoid any further heating. The wires were attached to contacts on the sample holder using silver paint. The entire, hermetically sealed cell was loaded into the TEM vacuum. The multiple gluing steps and complexity of the WE were disadvantages, but the cells were able to provide quantitative data on nucleation and growth processes of Cu [8, 13, 14, 16–18]. Later versions had three or four electrodes patterned on the lower wafer, Figure 10.3d, f, with tracks covered with an insulating layer of SiO_2 to define their active areas [22, 34]; o-ring sealing proved more reliable than glue (Figure 10.3c).

With the development of commercial systems and other home-built liquid cells, a wide variety of electrode designs, materials, and assembly steps have since been explored [35]. Several electrode designs are shown in Figure 10.4, all patterned onto one of the paired window chips. Many designs are symmetrical to allow for redundancy while others include design features of benchtop electrochemical cells. Different materials can be used for each electrode with extra lithography steps. The electrode areas are defined by an insulating layer, and the window dimensions are optimized to control liquid thickness, including the use of ultra-small windows [24]. The loading and sealing of the liquid has also been greatly simplified through innovative holder designs that avoid the use of glue (Chapter 2). Liquid flow has not been widely used in electrochemical experiments to date [36], but the chip and holder designs that enable flow can be used to provide additional control over the reaction under study.

10.2.5 External Electrochemical Control

Once the liquid cell is loaded into the sample holder, the external connections are made using contact pads, usually on the periphery of one of the liquid cell chips. Spring-loaded contacts are convenient, but wire bonds or connections made using conductive epoxy can also be a good solution. These contacts are connected to electrical leads that feed through the sample holder rod to a potentiostat. The potentiostat must be capable of measuring the small currents that flow in liquid cells. A typical current may be below the 10^{-7} A range (for example, 5 mA cm^{-2} is a typical current density for Cu deposition and the electrode area is 2000 μm^2 in Figure 10.3e). Low impedance electrical wires and connections are required for minimizing noise in the measurements at these low currents [36], and avoidance of ground loops is important for reproducible results.

To verify the performance of an electrochemical system, electrochemists perform standard measurements. Figure 10.5 illustrates such tests for the case of Cu deposition from acidified sulfuric acid. A key first step is cyclic voltammetry (CV), where the potential is ramped from one value to another and back, with various scan rates. CV is used to identify the reactions present, determine kinetics, reversibility, diffusion parameters, and reduction potentials. It shows qualitatively the regimes in which diffusion and surface reaction are the rate-limiting processes. Other techniques include

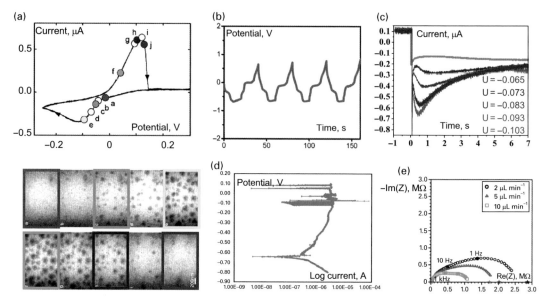

Figure 10.5. Examples of electrochemical measurements. (a) Cyclic voltammetry: Frames from a bright field movie (below) correlated with current vs. voltage as the voltage is ramped during deposition and stripping of Cu in 0.1 M $CuSO_4$ + 0.18 M H_2SO_4. Each image shows a small area (1.3 μm × 1.8 μm) of the 100 μm × 20 μm WE. Note the characteristic nucleation and diffusion-limited growth peak at −0.095 V and stripping peak at 0.11 V on the reverse scan, consistent with values expected from larger area electrodes. The relatively high contact resistance to the WE in this cell design causes some distortion in the scan. Subsequent scans showed good reproducibility. Adapted with permission from Ref. 17. Copyright 2006 American Chemical Society. (b) Controlled current: Voltage vs. time measured in a liquid cell during application of a square wave in current using Pt electrodes and the same electrolyte as in (a). Adapted from Ref. 50. (c) Controlled voltage: Current vs. time measured at a set of different (constant) voltages (U), using Au WE and CE and Cu RE and the same electrolyte as in (a). Reprinted with permission from Ref. 16. Copyright 2006 American Chemical Society. (d) Polarization curve (Tafel plot): Potential vs. log current during a slow voltage ramp obtained for Al in 0.1 M NaCl in a liquid cell. Unpublished.
(e) Electrochemical impedance spectroscopy Nyquist plot of [$Fe(CN)_6$] obtained in a liquid cell with a glassy carbon WE, recorded with different flow rates and constant potential of 0 V. Reprinted from Ref. 36.

chronopotentiometry, chronoamperometry, and electrical impedance spectroscopy (EIS), all of which provide information on reaction kinetics. For example, chronoamperometry can be used to measure diffusion coefficients of ions to compare with bulk values [18, 36] and to test for any effects of liquid flow [36].

After verifying the electrochemical operation of the liquid cell, and checking that the electrodes and electrolyte are visible through the window (a common problem, for example, is a bubble in the field of view), we can now start to probe electrochemical processes in action. The following examples are based on two types of experiment, plan view nucleation studies and lateral growth front propagation experiments. These illustrate the broad range of physics that can be accessed and suggest new opportunities in other materials systems.

10.3 Electrochemical Nucleation and Growth in Plan View

Figure 10.6 shows a selection of plan view nucleation and growth experiments. These data were obtained during deposition of Cu from acidified $CuSO_4$ onto a polycrystalline electrode [8, 16–18]. The use of electrochemically deposited Cu for interconnects in microelectronics [37] requires good control over nucleation in order to deposit thin but continuous layers. In Figure 10.6a, in a bright field image series acquired during potentiostatic deposition, Cu nuclei can be imaged as they form, appearing as dark regions superimposed on the background from the electrode. The nuclei can be distinguished via mass contrast over the background from the windows, electrode, and electrolyte even when relatively small, a few nm in diameter. Note the simultaneous acquisition of images and electrochemical parameters in Figure 10.6b. Here the current was measured as a function of time at fixed potential using a Cu pseudo-reference electrode.

Quantitative analysis of this type of data requires reproducibility in the experiments. A reversible system is helpful since we can perform multiple deposition experiments under the same conditions. Applying a positive potential redissolves (strips) Cu off the electrode and back into solution, so deposition can be repeated. Reproducibility is improved if the electrode is cleaned immediately prior to deposition by applying a positive potential for 2 s (Figure 10.6b). Current vs. time is then similar for repeated depositions at one potential, and deposition can be carried out at a series of potentials. We now discuss several different types of information that can be extracted from the combination of video images and time-varying electrochemical parameters.

10.3.1 Tests of Nucleation and Growth Models

The experiment in Figure 10.6a can provide a stringent test of commonly used models for island nucleation and growth during electrochemical deposition [38]. These models were developed to explain experimentally accessible parameters, such as $I(t)$ (current versus time) during potentiostatic deposition, by making assumptions about the nucleation kinetics and rate-limiting steps for growth. If we assume nucleation is followed by diffusion-limited growth, we can predict a distinctive form for the $I(t)$ curve. When the voltage is first applied there is a fast induction time where the charging of the double layer capacitance creates a current spike. Islands then nucleate and begin to grow. The islands are assumed to be hemispheres that grow by addition of ions that have diffused from the solution to the island surface. Nucleation can be either instantaneous, i.e. completed in a short interval, or progressive, taking place over a finite time. As more islands appear and the surface area of each increases, the current increases, with details dependent on the type of nucleation. In this early stage growth, ions are in plentiful supply. But in later stage growth the electrolyte around each island becomes locally depleted of ions. Diffusion fields develop and the growth rate of each island slows down; the current starts to decrease. At longer times the diffusion fields from neighboring islands overlap. The final situation is a planar diffusion profile with a simple $I(t)$ dependence that is independent of the details of nucleation.

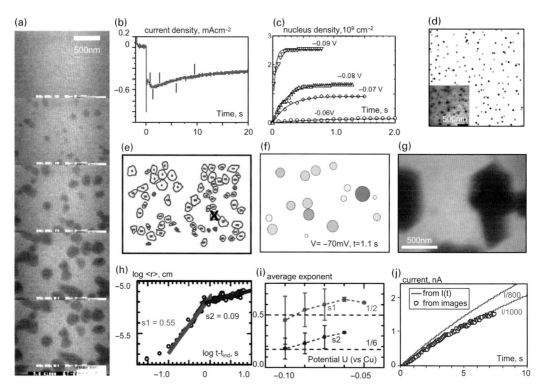

Figure 10.6. Plan view nucleation and growth. (a) Series of bright field images obtained during potentiostatic deposition of Cu islands on Au at −70 mV with respect to Cu RE, in 0.1 M $CuSO_4$ + 0.18 M H_2SO_4. Reprinted with permission from Ref. 16. Copyright 2006 American Chemical Society. (b) Current vs. time for the movie in (a) with the time of each image indicated. Note that this graph is one of the set shown in Figure 10.5c. Reprinted with permission from Ref. 16. Copyright 2006 American Chemical Society. (c) The number density of nuclei measured from the images in (a) as well as other image series recorded at different potentials (indicated). The fits show Avrami kinetics. Reprinted with permission from Ref. 16. Copyright 2006 American Chemical Society. (d) Nucleation positions for multiple cycles. The center-of-mass positions of Cu clusters are indicated during four deposition experiments (different colors) on the same area of electrode during galvanostatic deposition at 50 mA cm^{-2} in 0.3 M $CuSO_4$. Inset is the micrograph that corresponds to the black circles. Reprinted from Ref. 8. (e) Single movie frame segmented to show nuclei. From deposition at −90 mV after 1 s. Reprinted from Ref. 14. (f) Single frame coded to show island growth rate and size. Darker shading indicates faster rate of area increase. From deposition at −70 mV after 1.1 s. Reprinted from Ref. 14. (g) Shape change due to additives. Recorded during growth of Cu on Ru in the presence of Cl^- ions added as 50 ppm (1.4×10^{-3} M) HCl. Unpublished. (h) Radius vs. time for the individual island marked as X in (e), shown as a log-log plot. Two growth exponents are visible, indicated as s1 and s2. Adapted with permission from Ref. 17. Copyright 2006 American Chemical Society. (i) Average values ⟨s⟩ of the two gradients s1 and s2, obtained by averaging measurements made from the ensemble of islands visible during deposition at each potential (2.6 μm^2 field of view). Clusters grown at −0.05 V do not show two clear growth regimes. The expected values of s1 and s2 in diffusion-limited growth are indicated by dashed horizontal lines. Reprinted with permission from Ref. 16. Copyright 2006 American Chemical Society. (j) Current required to deposit the volume of Cu that was observed during deposition in 0.1 M $CuSO_4$ + 0.18 M H_2SO_4 at −0.07 V in a 2.6 μm^2 field of view. The volume was calculated from the cluster size by assuming hemispherical islands (blue circles). The current density through the cell is also shown. The electrode area is 2000 mm^2 or 800× larger than the imaged area. The best fit is at ratio 1000×. Unpublished.

For this model of nucleation and diffusion-limited growth, analytical solutions for $I(t)$ have been derived for a few limiting cases [38]. These equations involve two unknown parameters, the nucleation rate constant, k_n (for progressive nucleation), and the island density at large times, N_0, and take different forms depending on whether nucleation is instantaneous or progressive. The equations also involve known parameters such as the bulk concentration, diffusion coefficient, and valence of the metal ion.

In a typical electrochemical analysis we assume that the model applies, then fit the experimental $I(t)$ with the model equation to obtain the unknown parameters. But the liquid cell experiments allow us to approach the problem in a different way. We can first verify whether the model assumptions are in fact valid. Examples include the type of nucleation kinetics and the diffusion-limited growth assumed for individual islands. We can then measure directly the key parameters N_0 and k_n and compare with values obtained from fitting $I(t)$. In this example imaging allowed direct verification of several assumptions of the model, but the island density N_0 calculated from fitting $I(t)$ turned out to be substantially different from the observed value. We describe these results below and show that this discrepancy implies the need to modify a fundamental assumption on which nucleation and diffusion-limited growth models are based.

10.3.2 The Diffusion Coefficient

The behavior at long times expected from diffusion-limited growth models is a useful test of liquid cell data. $I(t)$ takes a simple form, known as the Cottrell equation, which includes the diffusion coefficient and a $t^{1/2}$ dependence. Plots of I vs. $t^{1/2}$ yield the diffusion coefficient of the ions [18, 36]. For the data of Figure 10.6b, we find that at long times the current does indeed show a $t^{1/2}$ dependence, and the diffusion coefficient is in excellent agreement with bulk literature values. We are not using the imaging ability of the liquid cell directly here, but rather performing a test of the validity of the liquid cell technique that helps to provide confidence in other measurements from these data. However, it is important to note that apparent values for parameters such as diffusion coefficients depend on the system fulfilling assumptions of the model. For example [18] at less negative potentials, the diffusion coefficient deviates from the bulk value, suggesting that deposition is not purely diffusion limited, but may be limited by the rate of the reaction at the island surfaces (i.e. "kinetically limited").

10.3.3 Nucleation Sites: Preferred Locations and Avrami Kinetics

On a single crystal substrate, nucleation is known, in many materials systems, to take place at heterogeneous sites such as step edges (e.g. [1]). Scanning probe microscopy provides an excellent tool to study the atomic level details of such nucleation events. But nucleation on a polycrystalline surface is perhaps more relevant for industrial processes such as the formation of interconnects or coatings. Conveniently, polycrystalline films form spontaneously when many metals are deposited onto Si_xN_y. Although atomic resolution has not been achieved on polycrystalline electrodes, in principle one can distinguish, for example, nucleation at grain boundaries or triple points versus on the surfaces of grains.

Given an inhomogeneous substrate, an important question is whether certain features of the surface act as preferred nucleation sites. We can therefore cycle between depositing and stripping of nuclei under conditions that do not alter the electrode surface [8]. For Cu deposition on polycrystalline Au, the nuclei appear in different positions each time ([8]; Figure 10.6d). Since the grain size in the electrode (15 nm) is much smaller than a typical separation between nuclei (400 nm), the density of surface defects potentially available (steps, grain boundaries, and triple points) is much higher than the actual nucleus density. Thus, the uncorrelated arrangement shows that nucleation is not dominated by a small population of highly preferred sites. Instead, we infer that many sites have similar nucleation probability, and each random nucleation event blocks neighboring sites via its diffusion field.

As well as the spatial location of each nucleation event, we can also measure its location in time [14, 16, 18]. The results in Figure 10.6c directly verify a key model assumption, that nucleation follows first order Avrami kinetics with island density $N(t) = N_0(1 - \exp(-k_n t))$. The measured $N(t)$ can be fed directly into models to improve on the simpler forms often used.

Experimentally, nucleation occurs more rapidly as the voltage becomes more negative: the nucleation rate k_n increases approximately exponentially with potential. The variation of k_n with V provides rich information on the processes that control growth, including, from classical electrocrystallization theory [39], the critical nucleus size. The full data set from which Figure 10.6c was taken yields a critical nucleus size of 3 [18], in agreement with findings from STM [1].

10.3.4 Correlations between Neighboring Islands

An example of some image processing steps used to analyze correlations between neighboring islands is shown in Figure 10.6e, f [14]. In terms of the spatial correlations, statistical analysis shows that second and third near-neighbor distance distributions are consistent with complete spatial randomness (Poisson distribution), indicating no long-range interactions between nuclei [14]. However, the nearest neighbor distances are larger than expected from the Poisson distribution, revealing some short-range correlation. This is partly due to the non-zero size of previously nucleated islands, and also reflects interactions between nuclei via diffusion fields. Similarly, one can in principle correlate the nucleation time and growth rate of each island with its local neighborhood. This can provide a powerful tool for investigating the physical mechanisms that control growth. A practical issue is the need to obtain a large field of view yet sufficient spatial resolution of each island. Advances in detectors will improve the opportunities from these types of studies.

10.3.5 Shapes of Crystals

So far we have treated each island as a point that grows into a hemisphere. However, liquid cell data can show clearly whether an island is faceted – or, for that matter, whether it is polycrystalline or a single crystal. Facets are expected when the rate at which metal ions are reduced to metal at the surface of the island is slow compared to surface diffusion. Both surface diffusion and the surface reaction rate depend on potential, and we generally expect faceting at small potentials. The faceting observed in liquid cell experiments on Cu deposition [17, 18] is consistent with conditions where cyclic voltammetry indicates that the surface reaction rate is limiting. Imaging at various potentials allows direct visualization of island shapes in the diffusion-limited and kinetically limited growth regimes. It also allows the effects of additives on island shapes (Figure 10.6g; see Section 10.3.8) and details of the growth rates of different faces to be measured.

10.3.6 Growth Rates of Individual Islands

Electrochemical models are based on average behavior: the measured current density is obtained from the average number density and the average growth rate. By measuring individual objects we can go beyond these average parameters. In Figure 10.6h we show the radius of an individual island as a function of time. On the log-log plot, the island appears to show two growth exponents. The diffusion equation for a hemispherical island predicts growth proportional to $t^{1/2}$. Later, when the diffusion fields have overlapped, the exponent ½ should decrease to ⅙. Thus, for this island, the kinetics confirm in detail the behavior expected for diffusion-limited growth. Furthermore, we can compare the exponents at different voltages. One approach is to average the early and late time exponents, Figure 10.6i. The values deviate from the expectations from diffusion-limited growth at small V, where the deposition process is under mixed kinetic/diffusion control.

Diffusion-limited growth thus appears to be a good model for the detailed behavior of individual islands. But not all aspects of growth are consistent with the model [16]. In the example in Figure 10.6a–c, there is a discrepancy between the parameters (nucleus density, growth rates) obtained from fitting $I(t)$ and the parameters measured from the movies. The diffusion-limited growth rate (the prefactor, as well as the exponent ½) is determined exactly by the diffusion coefficient and concentration. But islands grow ~10 times more slowly than expected, and at densities ~10^3 higher than expected. This discrepancy suggests that processes are contributing to growth that are not included in the model. It appears that two processes occur in parallel: growth via direct attachment from the electrolyte (included in the models), plus growth via Cu that adsorbs onto the electrode surface and diffuses to growing islands along the surface. This contribution is more important at early times. These two parallel processes can explain the shape of the $I(t)$ curve, as well as the high nucleation density and slow growth rate. It is possible that surface diffusion processes can explain $I(t)$ data in other materials as well.

10.3.7 Additive and Substrate Chemistry

In many commercial electrodeposition processes, additives control the morphology of the deposited film. In the case of Cu, a cocktail of additives is used to modify growth at the individual nucleus level, changing crystal shape and nucleation density by orders of magnitude and affecting coalescence [40]. Liquid cell TEM can show the effects of additives [34], providing information that is complementary to that from scanning probe microscopies, which allow additive effects to be measured on individual crystal facets [4]. The example in Figure 10.6g shows Cu deposited on a Ru substrate forming highly faceted islands in the presence of chloride ions. In the case of Cu on Ru, liquid cell data obtained with different sets of additives directly confirms a "differential inhibition" mechanism that was initially derived from current measurements [41] and is important in creating uniform thin deposits in trenches. Organic additives can be sensitive to radiolysis and may for example crosslink under the beam. However, there is great scope for direct observation of the effects of many additives on morphology and kinetics during deposition.

10.4 Growth Front Propagation via Lateral Measurements

The above sections illustrate the rich variety of information possible from observations made in "plan view" geometry. But as nuclei coalesce and the film grows thicker, further measurements become difficult. The deposited material may grow too thick to see through, or even fill the space between the electrode and the top window of the liquid cell. Instead, we can redesign the experiment as shown in Figure 10.7a to provide information in a lateral geometry and explore other growth regimes. Such experiments can produce quantitative data addressing questions of growth front stability and the formation of dendrites, the production of gaseous species and the development of diffusion fields in the electrolyte. Lateral growth measurements interrogate length and time scales that are difficult to access by other means. In particular, highly non-planar growth is not amenable to scanning probe techniques, and light microscopy cannot resolve initial stages or key nanoscale phenomena such as tip splitting.

Lateral growth begins once deposition on the electrode surface has depleted ions from the thin electrolyte layer above, so that the electrode edge becomes more favored. The growth front propagates smoothly during slow growth, as seen for Ni [26], Cu [42], or Pb [20] deposition, and the crystallinity and growth kinetics can be obtained. Quantification of the measured lateral growth rates under different conditions is useful in understanding where current flows, given the complex geometry of the liquid cell.

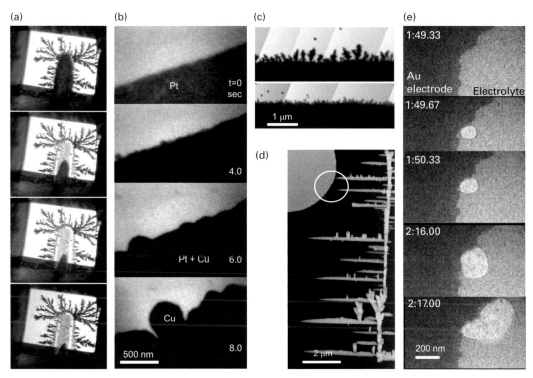

Figure 10.7. Lateral growth. (a) Light microscope images recorded during deposition of Cu onto polycrystalline TiN from an electrolyte containing 0.1 M $CuSO_4$ + 0.18 M H_2SO_4 + 50 ppm Cl^- + 0.3g/l PEG + 10 ppm MPS at −0.3 V relative to the open circuit potential, followed by stripping to remove Cu from above the electrode, disconnecting the dendrites from the circuit. The whole of the electron-transparent window is shown and the dendrites formed are tens of micrometers in length. (b) Potentiostatic deposition of Cu onto polycrystalline Pt from the same electrolyte as in Figure 10.6a, at −1.2 V relative to the open circuit potential. t = 0 when the potential is applied. Rapid coverage of the electrode is followed by unstable lateral growth. The field of view moves to track the growth front. From Ref. 21. (c) Galvanostatic deposition of Zn onto polycrystalline Au. 40 nA was applied in a pulsed mode (20 cycles of 10 s on, 10 s off). The electrolyte is 0.5 M $ZnSO_4$ + 0.1 M H_2SO_4 with additive in the lower image, 0.25 mM Bi_2O_3. Bi does not suppress dendrites, but forces Zn growth in the gaps between dendrites. The model is different from levelling in other systems (ohmic blocking), since Zn grows on a pristine electrode rather than on previously grown Zn that now has a Bi layer. Unpublished. (d) STEM image showing a Pb dendrite that grows from 1.5 M $PbNO_3$ to short-circuit two electrodes. The polycrystalline Au WE is the large circle. A similar CE is outside the field of view. The electrodes are shorted when contact is made at the point circled. Reprinted with permission from Ref. 20. Copyright 2012 American Chemical Society. (e) TEM image series showing nucleation and growth of electrolytic H_2 gas in 0.1 M H_2SO_4 at an Au WE electrode. Gas nucleates at −2 V, compared to −1.6 V in bulk. The earliest visible bubble is away from the electrode, suggesting that it nucleates homogeneously in the liquid and subsequently wets the electrode. Reproduced from Ref. 28 with permission from The Royal Society of Chemistry.

10.4.1 Dendritic or Ramified Growth

Electrochemical models predict that increasing the deposition rate will change this smooth, compact growth to a ramified or dendritic morphology. This transition is indeed observed [20, 25, 42]. The driving force is diffusion. Any asperities that form will capture more of the incoming flux of ions, destabilizing the growth front. The formation of ramified, irregular, or dendritic structures has been observed in materials including Cu [8, 19, 21, 42, 43], Pb [20, 27], and Li [12, 23, 25, 30], and some examples are given in Figure 10.7b–e. Direct observation provides details that are difficult to obtain otherwise, helping to understand the mechanisms that contribute to growth instabilities. Growth rates can be measured at the tips of asperities, tip splitting can be resolved directly, and the crystal orientation or polycrystalline nature of the growing material can be determined. The nature of the nucleation sites for dendrites can be examined through repeated cycling [20]. Dendritic features can also be dissolved by reversing the applied potential. A common observation, shown for example in Figure 10.7a, is that dendrites dissolve at their bases, becoming electrically disconnected and even floating away [8, 19, 20, 23, 25]. Dendritic growth is discussed in more detail in Chapter 11 because of the importance of growth front morphology to performance in secondary batteries. For example, dendrite formation has been linked to battery shorting and failure, while disconnected dendrites result in lost capacity when cycling secondary batteries.

Quantitative studies of dendritic structures and growth kinetics are in their infancy, with many opportunities available given the ability to measure individual features. The onset of instability, the multiple length scales involved, and the power laws governing the development of roughness are accessible from liquid cell experiments and can be correlated with electrochemical conditions. Initial measurements [42, 44] show distinct regimes in growth front behavior, consistent with models for kinetic roughening and diffusional instabilities, and can perhaps provide stringent tests of models similar to that achieved for nucleation and growth studies.

10.4.2 Controlling Growth Front Stability

In practical terms, it is important to develop strategies to control growth front morphology. Ramified growth is useful in high surface area applications such as sensors, while compact and smooth growth is preferred when fabricating coatings or charging batteries. Several strategies are known to prevent or slow the onset of ramified growth. Liquid cell TEM is starting to address these strategies to develop an understanding of basic physics as well as to form useful structures.

Since unstable growth is driven by diffusion fields, one strategy for suppressing it, known as pulse plating, is to periodically pause growth and allow diffusion to catch up so that subsequent deposition is smooth. A variant is pulse-reverse plating, in which some material is removed. Liquid cell TEM can visualize these processes at work and promises quantitative analysis of the development of individual asperities [42, 44]. A second strategy is to use additives, as was discussed in Section

10.3.8 for control of nucleation. Additive effects have not yet been widely examined, but Figure 10.7c shows an example where Bi slows but does not suppress the formation of instabilities during Zn growth. Pb additives smooth the growth front for Cu [42]. Many industrially relevant additives could be examined using this approach.

10.4.3 Gaseous Products of Electrochemical Reactions

Lateral deposition experiments provide a good view of gaseous reaction products, as shown in Figure 10.7e [28]. Hydrogen and oxygen are common electrochemical products during aqueous processes. The images suggest that the gas may dissolve into solution and then nucleates near, but not on, the electrodes, subsequently moving to wet the electrodes [28]. Quantitative analysis of gaseous products in electrochemical liquid cell TEM cannot neglect radiolytically produced gas (Chapter 7), which presumably can diffuse and enter electrochemically generated bubbles, and must also consider the high pressures expected within such nanoscale bubbles.

10.4.4 Detecting Ion Concentration Gradients

Figure 10.8 shows a remarkable result [20]. The intensity of the electron beam transited *through the electrolyte* varies during an electrochemical deposition and stripping cycle.

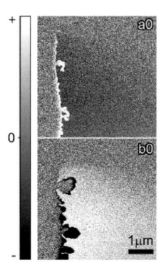

Figure 10.8. Time evolution of the Pb^{2+} ionic concentration. Images a0 and b0 are constructed by subtracting two consecutive 1.14 s frames recorded during Pb deposition and stripping, giving lighter or darker regions where the intensity increases or decreases, respectively. The STEM beam is scanning from top to bottom. During the acquisition of image a0 (b0) the electrode potential switches polarity and begins plating (stripping) lead, which creates the white (black) regions on the electrode boundary. Reprinted with permission from Ref. 20. Copyright 2012 American Chemical Society.

The brightness depends on position, because in these STEM images different parts of the image are recorded at different times. In Figure 10.8, during deposition of Pb, the image becomes brighter (in bright field conditions) because Pb^{2+} ions are removed from the solution. The brightness is correlated with the rate of Pb deposition and the current passed. Even for lighter elements, simple scattering calculations for TEM suggest that the transmitted intensity may be sensitive to ion concentration. For example, 0.1M Cu^{2+} in a 1 μm thick liquid layer contains 60 Cu ions per nm^2 in projection, or 2.5 monolayers of Cu. This would reduce the intensity of transmitted electrons by ~3% under typical conditions. We expect mapping ion concentration during deposition and dissolution to provide unique information on the liquid phase diffusion processes that control growth locally.

10.5 Experimental Challenges

Sections 10.3 and 10.4 have highlighted applications of liquid cell TEM for electrochemical processes. In analyzing such unique data, it is especially important to design the experiment to avoid or minimize artifacts. We need to think not just about the pitfalls of conventional liquid cell TEM, discussed in Chapter 8, but also about challenges that are specific to electrochemical control. Addressing these issues can contribute to robust experimental design and more interpretable results.

10.5.1 Choice of the Electrolyte

Liquid cell TEM has made use of a relatively limited range of electrolytes to date. When working with an unfamiliar one, there are several factors to consider.

Reactivity: Clearly liquids that react with interior components of the cell should be avoided. Silicon nitride is relatively inert, but the liquid may also come into contact with silicon, o-rings, and the metal and tubing of the holder. It is a good idea to test all the cell components in the electrolyte first, although reactivity may in principle change under the electron beam.

Electron beam sensitivity: Many of the complex electrolytes used for example in Li-ion batteries are beam sensitive and degrade under irradiation [25, 45, 46]. Simple aqueous solutions can show strong effects, depending on the exact species present. For example, metal ions in solution can be reduced into metal nanoparticles by the beam (Chapter 9), while alkaline solutions are expected to experience large pH changes under standard illumination conditions (Chapter 7). We expect relatively less pH change for acidic solutions. Beam effects are discussed in more detail in Section 10.5.6 below.

Conductivity: The liquid cell volume is small, with relatively long and narrow path lengths through the electrolyte. This can result in large ohmic drops that will distort the results. Higher conductivity will minimize such effects.

Ion concentration: If deposition takes place on a large electrode but the ion concentration is low, the electrolyte will become depleted of ions locally. It has been suggested that liquid flow cells can provide advantages in such experiments [36].

Atmospheric sensitivity: Liquid cell TEM electrochemistry is in general less well controlled than electrochemical experiments in dedicated systems. For example, it is awkward to de-aerate the electrolyte prior to loading in a static cell or in the pump for a flow cell, leading to poor control of dissolved oxygen. This can affect the reactions taking place during copper plating, for example. Other electrolytes such as those used in Li-ion batteries may be sensitive to air or water, requiring a glove box for loading the liquid into the cell. All of the components within the cell can potentially contaminate the electrolyte, so the sensitivity of the results to such contamination should be assessed.

10.5.2 The Electrolyte Thickness: Balancing Resolution with Diffusion

Diffusion often dominates electrochemical behavior. When we try to set up a "realistic" electrochemical process, a thicker liquid layer will support diffusion fields that are more representative of the bulk process. However, a thick layer conflicts with the requirements for good image resolution (Chapter 8). Choosing the right balance of resolution and diffusion is very important when designing the electrochemical experiment. For example, an electrolyte 1 μm thick reproduces a bulk process, say growth on an electrode surface, better than a 100 nm electrolyte layer, but the process will be imaged with lower resolution. The balance point depends on what information is needed. Lower quality images, as in Figure 10.6, can work well to establish growth rates, nucleation positions, crystal shapes and similar parameters. One way to quantify the effect of liquid thickness is to carry out simulations of diffusion fields under the experimental conditions, such as is shown in Figure 10.9 [16]. We can then determine the conditions under which liquid thickness dominates the results.

10.5.3 The Hidden Electrode

In most electrochemical liquid cell designs, part of the working electrode is hidden from view beyond the edge of the window. Does this part undergo the same reactions as the observable region? For example, consider the experiment in Figure 10.6a. The movie shows an area 1.85 μm × 1.3 μm, or 2.64 μm^2, about 760 times smaller than the total electrode area (20 μm × 100 μm). On each movie frame, we calculate the volume of Cu

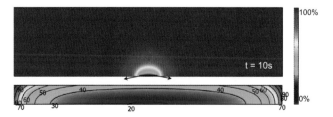

Figure 10.9. Finite element calculation of the diffusion field above a 20-μm wide electrode after different times of growth at −0.03 V overpotential from 0.1 M CuSO$_4$ + 0.18 M H$_2$SO$_4$. Scale is 0 to 100% of the bulk concentration. A diffusion layer extends from the electrode but it does not reach the boundaries of the cell before 10 s of deposition. Reprinted from Ref. 34.

deposited by assuming hemispherical islands (verified by post-growth AFM), then scale to the volume on the whole electrode area. Since two electrons flow per deposited Cu atom, this provides the charge that must have passed to achieve this growth. If we compare with the integrated current, the agreement is surprisingly good, within 20% (Figure 10.6i). This suggests that the entire electrode behaves relatively uniformly, consistent with results suggested by post-growth AFM. A consistency check like this is important for any experiment.

10.5.4 Length and Volume Scales

All electrodes in the liquid cell, and the electrolyte itself, have dimensions (area, volume) that are small compared to conventional benchtop electrochemical systems. In several of the examples in Figures 10.3 and 10.4, the WE is of the order of 10^3–10^4 μm^2 in area, but electrodes below 1 μm^2 have been fabricated by careful etching of a passivation layer (Figure 10.4d). Typical electrolyte volumes in static cells are of the order of microliters. These small dimensions can present a challenge when we try to make detailed comparisons between liquid cell TEM results and benchtop experiments [8, 24, 36, 44]. Electrochemical theory dealing with small electrodes is an interesting topic in itself, and liquid cell TEM is well positioned to examine size effects.

In a liquid cell, a small WE area is helpful for several reasons. Any reaction is more likely to be spatially uniform, helpful in making the types of reality checks described in Section 10.5.3. Also, given the small electrolyte volume available, a small WE is also less likely to create large changes in the electrolyte composition (for example generating gaseous species that form bubbles). As the WE area is reduced, however, the currents become smaller and more difficult to measure, so a compromise has to be reached.

10.5.5 Liquid Flow

We have discussed the use of a flow cell to minimize problems associated with depletion of the small electrolyte volume available in a static liquid cell. However, liquid flow has a separate benefit, the ability to change the electrolyte chemistry during the experiment. This does not appear to have been explored much to date, but galvanic replacement reactions (e.g. [47]) would be clear beneficiaries of this capability. After setting up the imaging conditions while flowing water, the reaction could be initiated by changes in solution chemistry rather than a voltage or electron beam stimulation. The path lengths involved with flow cells are long so the chemical change is not especially sharp and the solution is likely to take several minutes to change composition. However, experimental design continues to improve, and we can imagine using this approach to image a wide range of important electrochemical processes.

10.5.6 Electron Beam Effects

In electrochemical experiments, the electron beam is not the primary stimulus driving the reaction, as it is in the beam-induced growth discussed in Chapter 9. However,

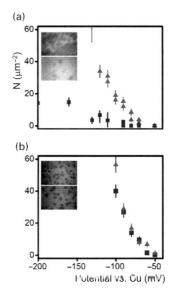

Figure 10.10. Beam effects in chloride but not plain solutions. (a) The nucleus density measured after potentiostatic deposition within (red) and outside (blue) the irradiated area using an electrolyte containing 0.1 M CuSO$_4$ 0 + 0.18 M H$_2$SO$_4$ +1.4 mM HCl. Three separate sets of depositions were measured on the same electrode at different times. Between measurements the deposit was stripped by applying +400 mV for several seconds. Insets are images obtained at −110 mV during deposition (upper image) and in an unirradiated area after deposition (lower image). (b) The nucleus density measured during potentiostatic deposition in an electrolyte without chloride. Two separate deposition experiments were performed for each potential, one with the electron beam focused to irradiate only the field of view (2 μm diameter) (red) and the other with weak illumination, with the beam spread to over 10 μm diameter (blue). Inscts are images obtained after 2 s at −80 mV with focused (upper image) and spread (lower image) beam. Reprinted with permission from Ref. 22. Copyright 2014, AIP Publishing LLC.

irradiation usually affects the reaction in some way, subtly changing kinetics or dramatically changing solution chemistry and reaction pathways. It is hard to give general rules for what to expect because the effects are so dependent on the materials involved. Figure 10.10 gives an example of this [22]. Here, deposition of Cu from acidified CuSO$_4$ solution is not modified much by the beam, but when chloride ions are added, irradiation changes the voltage range over which deposition takes place. The beam is thought to disrupt a CuCl surface layer that forms on and passivates the Au electrode in the presence of chloride. A practical consequence is a voltage range in which beam-induced patterning of Cu nanostructures is possible [22]. In another example, beam-accelerated surface reactions have been seen during Li deposition along with changes in deposition potentials and morphology [25].

The beam may affect an electrochemical experiment via radiolytic changes to electrolyte chemistry, changes to the electrode (as in the example above), heating, or deposition of charge into the sample.

In terms of *radiolysis*, reduction of metal ions by the electron beam is a likely scenario in electrolytes used for typical deposition experiments. Beam-induced deposition can be especially confusing because it takes place at the same time as the intended, electrochemically induced deposition. Islands growing off the electrode, where there is no electrical circuit, are probably beam-induced. Other radiolysis effects include changes in pH (especially in alkaline solutions, as mentioned in Section 10.5.1) and crosslinking of organic additives under the beam [34]. Electrolyte beam-induced degradation is well documented [25, 45, 46] – indeed, it has been suggested that this may indicate the stability of the electrolyte in a battery [45].

Electron beam-induced *heating* is generally slight, but electrochemical reactions can be very sensitive to temperature since temperature changes the effective potential.

Electrostatic charging by the electron beam due to the production of secondary electrons could change the electrochemical potential or the charge distribution in the capacitive double layer. As the charge flows away, it can also add to the current measured by the potentiostat. It is always worth recording electrochemical parameters such as CV scans with the beam off as well as on to assess whether such effects are relevant. This is especially important where the currents are low due to small electrode areas.

It is impossible to eliminate beam effects, but important to characterize them. Remembering that the beam only covers a small area of the electrode; we have many opportunities after each reaction to compare the result in irradiated and distant, unilluminated areas, and to quantify the effects of different dose rates and electrolyte chemistries. A rigorous comparison with *ex situ* and benchtop experiments increases confidence in the results from liquid cell TEM.

10.6 Outlook

Real-time, spatially resolved observations of electrochemical growth provide unique measurements of morphology and kinetics. Electrochemical liquid cell TEM can help to develop nucleation and growth models and perhaps provide insights into growth front instabilities and other processes of importance in energy storage. Acquiring quantitative data is the key – so the liquid geometry must be controlled, results must be compared with *ex situ* experiments, and beam effects must be understood. Given these caveats, there are many areas where we anticipate electrochemical liquid cell TEM will produce new insights.

Developments in technique will offer new opportunities. A key advance has been the use of more complex samples. Attaching structures to simple electrodes allows "real" materials such as stainless steels to be examined (Chapter 12). Complex battery electrode materials can be deposited directly onto the window [48] to increase the range of materials available. Analytical techniques (Chapter 21) allow tracking of chemical changes. Improved detectors reduce the dose and sample damage. Correlative experiments can access a wider range of length scales, important in developing electrochemical models. An intriguing possibility is the correlation of electrochemical reactions with stress state. Electrochemical reactions on a cantilevered electrode are known to produce stress and hence bending [49]. We have observed flexing of electrodes when a voltage is applied, suggesting that stress may be accessible in a way that is complementary in scale to wafer bending experiments. Finally, heating has not yet been integrated with electrochemical capabilities, but temperature-dependent electrochemical processes provide rich opportunities, for example in battery operation or reactions involving molten salts under extreme environments.

Addressing fundamental questions in electrochemistry has provided the framework for the exciting progress in two key application areas: batteries (Chapter 11) and corrosion (Chapter 12). Future developments in these areas are discussed in the respective chapters. In terms of fundamentals, we foresee progress in several directions. One is the use of electrochemical liquid cell TEM on a wider range of materials and processes. New materials may include electrochemically deposited polymers or alloy or multilayer films deposited from electrolytes with multiple species. Alloy deposition, for example, shows fascinating anomalous behavior that would benefit from a direct view. Electroless deposition, in which a metal is deposited in the presence of a reducing agent but without an applied voltage, has applications in coating technology and in the formation of magnetic layers such as NiFe that are used in magnetic read heads. The demonstration of core-shell structures formed by galvanic replacement [47] suggests that liquid cell TEM can yield interesting results. For magnetic materials, it may be particularly illuminating to combine liquid cell microscopy with holography to visualize the development of magnetic fields as the material is deposited. Bipolar deposition is another interesting process in which a disconnected electrode (a patch or particle of metal) placed between two electrodes becomes sufficiently polarized to drive deposition on one end and dissolution on the other end. This is useful for making Janus particles and other nanostructures, and presents fascinating questions for real-time imaging. In the longer term, a fundamental challenge for liquid cell TEM is to image the electrochemical double layer, especially as it changes during electrochemical processes. Achieving this goal will require a combination of developments, such as an atomically flat electrode parallel to the beam and adaptation of holographic techniques to allow electric field measurement over the relevant length scale of a few nanometers. As we improve our experience in using liquid cell microscopy for electrochemical processes, it becomes possible to imagine experiments of even greater complexity and excitement.

References

1. M. H. Hölzle, V. Zwing and D. M. Kolb, The influence of steps on the deposition of Cu onto Au(111). *Electrochim. Acta*, **40** (1995), 1237–1247.
2. M. H. Hölzle, C. W. Apsel, T. Will and D. M. Kolb, Copper deposition onto Au(111) in the presence of thiourea. *J. Electrochem. Soc.*, **142** (1995), 3741–3749.
3. O. M. Magnussen, L. Zitzler, B. Gleich, M. R. Vogt and R. J. Behm, In-situ atomic-scale studies of the mechanisms and dynamics of metal dissolution by high-speed STM. *Electrochim. Acta*, **46** (2001), 3725–3733.

4. O. M. Magnussen, W. Polewska, L. Zitzler and R. J. Behm, In situ video-STM studies of dynamic processes at electrochemical interfaces. *Faraday Discuss.*, **121** (2002), 43–52.
5. M. Azhagurajan, R. Wen, A. Lahiri et al., Direct evidence of homoepitaxial growth in the electrodeposition of Au observed by ultra-high resolution differential optical microscopy. *J. Electrochem. Soc.*, **160** (2013), D361–D365.
6. J. W. Gallaway, D. Desai, A. Gaikwad et al., A lateral microfluidic cell for imaging electrodeposited zinc near the shorting condition. *J. Electrochem. Soci.*, **157** (2010), A1279–A1286.
7. F. M. Ross, Growth processes and phase transformations studied by in situ transmission electron microscopy. *IBM J. Res.*, **44** (2000), 489–501.
8. M. J. Williamson, R. M. Tromp, P. M. Vereecken, R. Hull and F. M. Ross, Dynamic electron microscopy in liquid environments. *Nat. Mater.*, **2** (2003), 532–536.
9. F. M. Ross and P. C. Searson, In situ microscopy of the anodic etching of silicon. In G. W. Bailey, M. H. Ellisman, R. A. Hennigar and N. J. Zaluzec, eds., *Proceedings of the 53rd Annual MSA Meeting, Kansas City, August 1995, 232–233* (New York: Jones and Begell Publishing, 1995).
10. A. J. Bard and L. R. Faulkner, *Electrochemical Methods: Fundamentals and Applications*, 2nd edn. (Hoboken, NJ: Wiley, 2001).
11. M. E. Holtz, Y. Tu, D. Gunceler et al., Nanoscale imaging of lithium ion distribution during in situ operation of battery electrode and electrolyte. *Nano Lett.*, **14** (2014), 1453–1459.
12. R. L. Sacci, J. M. Black, N. Balke et al., Nanoscale imaging of fundamental Li battery chemistry: solid-electrolyte interphase formation and preferential growth of lithium metal nanoclusters. *Nano Lett.*, **15** (2015), 2011–2018.
13. M. J. Williamson, Investigations of materials issues in advanced interconnect structures. Ph.D. Thesis, University of Virginia (2002).
14. A. Radisic, Electrochemical nucleation and growth of copper, Ph.D. Thesis, The Johns Hopkins University (2005).
15. J. M. Grogan, Ph.D. Thesis, University of Pennsylvania (2012).
16. A. Radisic, P. M. Vereecken, J. B. Hannon, P. C. Searson and F. M. Ross, Quantifying electrochemical nucleation and growth mechanisms from real-time kinetic data. *Nano Lett.*, **6** (2006), 238–242.
17. A. Radisic, F. M. Ross and P. C. Searson, In situ study of the growth kinetics of individual islands during electrodeposition of copper. *J. Phys. Chem. B*, **110** (2006), 7862–7868.
18. A. Radisic, P. M. Vereecken, P. C. Searson and F. M. Ross, The morphology and nucleation kinetics of copper islands during electrodeposition. *Surf. Sci.*, **600** (2006), 1817–1826.
19. F. M. Ross, Electrochemical nucleation, growth and dendrite formation in liquid cell TEM. *Microsc. Microanal.*, **16** (2010), 326–327.
20. E. R. White, S. B. Singer, V. Augustyn et al., In situ transmission electron microscopy of lead dendrites and lead ions in aqueous solution. *ACS Nano*, **6** (2012), 6308–6317.
21. J. M. Grogan, N. M. Schneider, F. M. Ross and H. H. Bau, The Nanoaquarium: a new paradigm in electron microscopy. *J. Indian Inst. Sci.*, **92** (2012), 295–308.
22. M. den Heijer, X. Shao, A. Radisic, M. C. Reuter and F. M. Ross, Patterned electrochemical deposition of copper using an electron beam. *APL Mater.*, **2** (2014), 022101.
23. B. L. Mehdi, J. Qian, E. Nasybulin et al., Observation and quantification of nanoscale processes in lithium batteries by operando electrochemical (S)TEM. *Nano Lett.*, **15** (2015), 2168–2173.

24. A. J. Leenheer, J. P. Sullivan, M. J. Shaw and C. T. Harris, A sealed liquid cell for in situ transmission electron microscopy of controlled electrochemical processes. *J. Microelectromech. Syst.*, **24** (2015), 1061–1068.
25. A. J. Leenheer, K. L. Jungjohann, K. R. Zavadil, J. P. Sullivan and C. T. Harris, Lithium electrodeposition dynamics in aprotic electrolyte observed in situ via transmission electron microscopy. *ACS Nano*, **9** (2015), 4379–4389.
26. X. Chen, K. W. Noh, J. G. Wen and S. J. Dillon, In situ electrochemical wet cell transmission electron microscopy characterization of solid–liquid interactions between Ni and aqueous $NiCl_2$. *Acta Mater.*, **60** (2012), 192–198.
27. M. Sun, H.-G. Liao, K. Niu and H. Zheng, Structural and morphological evolution of lead dendrites during electrochemical migration. *Sci. Rep.*, **3** (2013), 2227.
28. Y. Liu and S. J. Dillon, In situ observation of electrolytic H_2 evolution adjacent to gold cathodes. *Chem. Commun.*, **50** (2014), 1761–1763.
29. Z. Zeng, W.-I. Liang, H.-G. Liao et al., Visualization of electrode-electrolyte interfaces in $LiPF_6$/EC/DEC electrolyte for lithium ion batteries via in-situ TEM. *Nano Lett.*, **14** (2014), 1745–1750.
30. R. L. Sacci, N. J. Dudney, K. L. More et al., Direct visualization of initial SEI morphology and growth kinetics during lithium deposition by in situ electrochemical transmission electron microscopy. *Chem. Commun.*, **50** (2014), 2104–2107.
31. M. Gu, L. R. Parent, L. Mehdi et al., Demonstration of an electrochemical liquid cell for operando transmission electron microscopy observation of the lithiation/delithiation behavior of Si nanowire battery anodes. *Nano Lett.*, **13** (2013), 6106–6112.
32. B. L. Mehdi, M. Gu, L. R. Parent et al., In situ electrochemical transmission electron microscopy for battery research. *Microsc. Microanal.*, **20** (2014), 484–492.
33. R. R. Unocic, X. G. Sun, R. L. Sacci et al., Direct visualization of solid electrolyte interphase formation in lithium-ion batteries with in situ electrochemical transmission electron microscopy. *Microsc. Microanal.*, **20** (2014), 1029–1037.
34. M. den Heijer, In situ transmission electron microscopy of electrodeposition: technical development, beam effects and lithography. M.Sc. Thesis, University of Leiden (2008).
35. N. de Jonge and F. M. Ross, Electron microscopy of specimens in liquid. *Nat. Nanotechnol.*, **6** (2011), 695–704.
36. R. R. Unocic, R. L. Sacci, G. M. Brown et al., Quantitative electrochemical measurements using in situ ec-S/TEM devices. *Microsc. Microanal.*, **20** (2014), 452–461.
37. P. C. Andricacos, C. Uzoh, J. O. Dukovic, J. Horkans and H. Deligianni, Damascene copper electroplating for chip interconnections. *IBM J. Res. Devel.*, **42** (1998), 567–574.
38. B. Scharifker and G. Hills, Theoretical and experimental studies of multiple nucleation. *Electrochimica Acta.*, **28** (1983), 879–889.
39. A. Milchev, *Electrocrystallization, Fundamentals of Nucleation and Growth* (New York: Springer US, 2002).
40. P. M. Vereecken, R. A. Binstead, H. Deligianni and P. C. Andricacos, The chemistry of additives in damascene copper plating. *IBM J. Res. Devel.*, **49** (2005), 3–18.
41. L. Yang, A. Radisic, M. Nagara et al., Multi-scale modeling of direct copper plating on resistive non-copper substrates. *Electrochimica Acta*, **78** (2012), 524–531.
42. N. M. Schneider, J. H. Park, J. M. Grogan et al., Visualization of active and passive control of morphology during electrodeposition. *Microsc. Microanal.*, **20** (S3) (2014), 1530–1531.

43. F. M. Ross, M. den Heijer, M. J. Williamson and D. Steingart, Correlating light microscopy and electron microscopy for measuring microstructural evolution during electrochemical deposition. *Adv. Imag. Electron Phys.,* **179** (2013), 180–182.
44. N. M. Schneider, Liquid cell electron microscopy with the nanoAquarium: radiation and electrochemistry (January 1, 2015). Dissertations available from ProQuest, Paper AAI3721631.
45. P. Abellan Baeza, B. L. Mehdi, L. R. Parent *et al.*, Probing the degradation mechanisms in electrolyte solutions for Li-ion batteries by in-situ transmission electron microscopy. *Nano Lett.*, **14** (2014), 1293–1299.
46. Z. Zeng, W.-I. Liang, Y.-H. Chub and H. M. Zheng, In situ TEM study of the Li–Au reaction in an electrochemical liquid cell. *Faraday Discuss.*, **176** (2014), 95–107.
47. E. Sutter, K. Jungjohann, S. Bliznakov *et al.*, In situ liquid-cell electron microscopy of silver-palladium galvanic replacement reactions on silver nanoparticles. *Nat. Commun.*, **5** (2014), 4946.
48. R. R. Unocic, L. Baggetto, G. M. Veith *et al.*, Probing battery chemistry with liquid cell electron energy loss spectroscopy. *Chem. Commun.*, **51** (2015), 16377–16380.
49. Y. Nagai, J. D. Carbajal, J. H. White *et al.*, An electrochemically controlled microcantilever biosensor. *Langmuir*, **29** (2013), 9951–9957.
50. N. M. Schneider, J. H. Park, J. M. Grogan *et al.*, In situ electrochemical measurements in the Nanoaquarium. *Microsc. Microanal.*, **19** (S2) (2013), 433–434.

11 Application of Electrochemical Liquid Cells for Electrical Energy Storage and Conversion Studies

Raymond R. Unocic and Karren L. More

11.1 Introduction

The ability to probe nanoscale electrochemical processes under realistic operating conditions can provide unprecedented insight into the fundamental mechanisms that determine the functionality of electrochemical energy storage and conversion devices. This idea has been the cornerstone for the development of novel *in situ* liquid cell microscopy-based characterization techniques that have enabled quantitative electrochemistry while permitting the simultaneous visualization of microstructural evolution via scanning and transmission electron microscopy (S/TEM) at high spatial and temporal resolution. Significant advances have been made in the development of liquid cell TEM systems specialized for electrochemical measurements, as described in Chapter 10. These systems are based on closed cell silicon microchip devices that incorporate integrated microelectrodes immersed in the liquid that interface with an external potentiostat for conducting quantitative electrochemical measurements. The development and application of these *in situ* S/TEM electrochemical platforms provides the means to understand complex electrochemical reaction mechanisms and to elucidate reaction kinetics that control energy storage and transport in electrical energy storage systems. Several experimental limitations and challenges remain in this growing field of research that must be understood and overcome for data acquisition and interpretation. In this chapter, we summarize the current status regarding development and application of *in situ* electrochemical S/TEM (ec-S/TEM) for research into electrical energy storage and energy conversion.

11.2 Electrical Energy Storage and Conversion Systems: Challenges and Opportunities

A grand challenge in battery research has been the development of *operando* methods to understand the nanoscale structural and chemical changes that occur during operation of energy conversion and storage systems and devices, for example at battery electrodes or at the electrode–electrolyte interfaces and other interfaces within electrodes. Electrochemical energy storage and conversion systems such as rechargeable batteries or fuel cells convert electrochemical energy into a useful form of electrical energy that can be harnessed to power a wide range of systems, from portable electronic devices to electric vehicles [1, 2]. For example, in rechargeable Li-ion batteries, Li ions shuttle back and forth between the anode and cathode through a liquid or solid electrolyte via an intercalation, conversion, or alloying mechanism [3]. Complex electrochemical reactions occur during repeated electrochemical charge/discharge cycling (as for electrochemical processes in general), causing structural and chemical changes within the electrode and the solid electrolyte interphase (SEI) [4]. In fuel cells, electrocatalysts are used to convert H_2 gas into electrical energy by a series of electrochemical oxidation and reduction reactions. During fuel cell operation, several issues affect the overall fuel cell performance that are often linked to the stability of the electrocatalyst and support materials [5]. Understanding the fundamental processes that control degradation is key in the development of materials with increased energy densities and prolonged lifetimes for energy storage and conversion technologies.

Significant developments in the application of *in situ* S/TEM techniques towards investigating the structural and chemical changes in battery electrode materials during ion intercalation have been enabled by the use of TEM vacuum-compatible ionic liquid electrolytes or solid electrolytes as a medium for ionic transport under an applied potential [6]. This particular *in situ* TEM experimental setup, often referred to as an open cell configuration, is discussed in greater detail in Chapter 3. Experiments have demonstrated the lithiation behavior for a range of electrode materials (e.g. Si nanowires, SnO_2) and provided key insight regarding the lithiation mechanism and accompanying phase transformations at or near atomic resolution. It is also possible to couple analytical electron microscopy methods such as EELS and XEDS with *in situ* imaging to investigate chemical changes during lithiation/delithiation processes. However, there are limitations on performing quantitative electrochemical measurements using the open cell approach. This is because the electrodes are extremely small, in either single particle or nanowire form with typical dimensions on the nanometer scale. As a result, there is difficulty in accurately measuring ultra-low currents (typically pA–fA range) for potential cycling experiments, or applying small currents for galvanostatic cycling experiments. Such information is necessary to correlate ec-S/TEM observations with factors such as storage capacity and the amount of charge passed during each electrochemical cycle. Characterization of the structural and chemical evolution of electrode materials and electrode–electrolyte interfaces under working or *operando* conditions requires an alternative strategy for sealing high vapor pressure electrolytes and for performing quantitative electrochemical measurements. These requirements formed the primary basis for implementation of the "closed cell" ec-S/TEM system for energy storage and conversion studies, which is the focus of this chapter.

11.3 Closed Cell Electrochemical-S/TEM for Energy Storage and Conversion Studies

Closed cell liquid cell S/TEM has been applied to characterize a broad range of liquid phase reaction processes at high spatial and temporal resolution [7]. A general description of the technique and the fabrication of closed liquid cells was presented in Chapters 1 and 2. There have been major advances in transforming the basic liquid cell concept towards electrochemical studies, most notably by integration of microelectrodes within the microchip platforms to enable investigation of interfacial electrochemical reactions. Quantitative electrochemical measurements can be made while dynamic electrochemical processes are imaged at the electrodes through the electron-transparent region of the cell. The earliest electrochemical liquid cells, described in Chapter 10, were created by epoxy bonding two microfabricated silicon wafers containing electron-transparent silicon nitride (Si_xN_y) membranes and integrated gold working electrodes [8]. The dynamics of copper nanocrystal nucleation and growth on an Au working electrode, visualized during galvanostatic electrodeposition using this system, paved the way for further advancements in electrochemical cell design, microfabrication methods for silicon-based electrochemical microchips, and development of vacuum-tight *in situ* TEM holders with electrical biasing contacts for interfacing the cell to an external potentiostat/galvanostat for electroanalytical measurements [9–15].

Various types of ec-S/TEM characterization systems have already been described in Chapter 10. For electrochemical measurements on energy storage and conversion systems, it is convenient to use an electrochemical cell that is sealed within the tip of a vacuum-tight *in situ* TEM holder, and has an integrated microfluidic delivery system and electrical contacts. The complex interfacial processes occurring at the electrodes or within the liquid electrolyte layer are simultaneously captured during electrochemical measurement via imaging (TEM and/or STEM mode), electron diffraction, and/or spectroscopic methods such as EELS, XEDS, and energy-filtered transmission electron microscopy (EFTEM). The ability to flow liquids into the cell enables the transfer of highly volatile electrolytes. The interface with the external potentiostat/galvanostat is critical for quantitative electrochemical measurements. Low impedance electrical biasing wires and connectors ensure that low currents can be measured, since typical current measurements are within the nanoampere regime [16].

11.3.1 Electrochemical Microchip Platforms

Major advances in microfabrication techniques have contributed to the design of current silicon microchip platforms used for ec-S/TEM. Si_xN_y is used as the membrane material in electrochemical studies because it is chemically inert over a wide range of electrolytes commonly used in battery applications, and has sufficient strength to withstand the pressure differential between the liquid cell interior and the TEM vacuum. The electrochemical microchip portion of the electrochemical cell is fabricated using additional masking and metal deposition processes that pattern microelectrodes directly onto the microchip surface. Electrodes are typically composed of gold, platinum, titanium, silver, or glassy carbon and are patterned in a variety of optimized geometrical configurations. A three-electrode design is typically used for battery or related experiments, wherein a working, reference, and counter electrode are microfabricated directly onto the chip. Further details of two- or three-electrode configurations and the associated electrochemical measurements are discussed in Chapter 10.

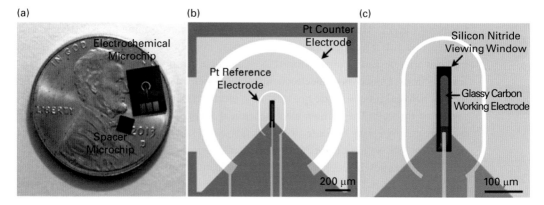

Figure 11.1. Microfabricated electrochemical microchip devices. (a) Image of commercially available microchip devices (positioned on a U.S. penny for scale). (b) Scanning electron microscopy image showing the detailed microelectrode layout design of the electrochemical microchip platform with circular platinum counter and reference microelectrodes surrounding a glassy carbon working electrode. (c) Higher magnification SEM image showing the details of the glassy carbon working electrode on the silicon nitride viewing window.

Figure 11.1 shows a typical electrochemical microchip device with a three-electrode configuration, used in many energy conversion and storage studies. In the SEM images, Figure 11.1b, c, the circular reference and counter electrodes are composed of platinum microfabricated directly onto the Si_xN_y/Si support. The working electrode is composed of glassy carbon, which is microfabricated onto the Si_xN_y viewing window portion of the device (Figure 11.1c). A large area of the electrochemical microchip and electrical contact leads is covered with a thin layer of SU-8, which is used to mask a portion of the tracks such that the electrochemical measurements are confined to areas of the electrodes that are exposed to the electrolyte and are observable through the Si_xN_y membranes. Electrochemical reactions are then directly imaged on the surface and edges of the working electrode. Different electrode layout configurations are commercially available, such as parallel electrodes, for certain energy conversion and storage experiments.

11.3.2 Microelectrode Materials Selection

The choice of electrode material is an important consideration for energy conversion and storage experiments. The working electrode can be involved in the electrochemical reaction (i.e. through ion intercalation or alloying) or it can be electrochemically inert. Noble metals such as Au or Pt are readily microfabricated onto the electrochemical microchip and are stable and non-reactive for a wide range of electrochemical reactions. However, noble metals are typically not suitable as electrode materials in Li-ion battery research due to the formation of alloys with Li at high reducing potentials [17]. For example, Au–Li alloying has been reported when a microfabricated Au electrode reacts with a $LiPF_6$ EC:DEC electrolyte [18, 19]. During this process, an interfacial reaction front develops as a result of electrolyte decomposition when the cell potential is reduced, which is followed by Au–Li alloying and the dissolution of the Au electrode with continued alloying. A similar behavior is expected for a Pt working electrode since Pt will form a Pt–Li alloy [20]. Metals such as Ti, Cu, or Al can be better choices for non-reactive electrode materials, depending on the electrochemical potentials used. A glassy (vitreous) carbon working electrode is an even better electrode material since it is electrochemically inert, in other words, it will not intercalate Li ions or form an alloy, and this material has been used for a number of fundamental battery research studies [21, 22]. More details of ec-S/TEM studies using the glassy carbon working electrode microchip are presented in Section 11.5.

11.3.3 Methods for Interfacing Battery Electrodes to Microfabricated Electrodes

There are several methods for testing battery electrode materials using ec-S/TEM. The battery electrode can be directly integrated on the electrochemical microchip, employing the same microfabrication techniques used to fabricate the silicon microchip devices. But other methods are required for interfacing general anode materials (such as graphitic carbons, silicon, silicon dioxide, etc.) or cathode materials ($LiFePO_4$, $LiMn_2O_4$, etc.) onto the microfabricated electrode contacts. Focused ion beam microfabrication/manipulation, sputtering, drop casting, ink printing, electrodeposition, or a combination of techniques, can be used. A FIB instrument can be used to extract, mill, and weld micrometer-sized electrode materials, such as highly oriented pyrolitic graphite (HOPG), directly onto the electrode contact of the electrochemical microchip [23]. Electrical contacts can also be fabricated using electron beam-induced deposition or ion beam-induced deposition from gaseous precursors of C, Pt, or W. For example, FIB patterning has been used to join Si nanowire electrodes onto a Pt microelectrode contact [14]. Sputtering methods, such as RF magnetron sputtering or pulsed laser deposition, require a precision-machined shadow mask for the site-specific deposition of the battery electrode materials onto the working electrode contact. A slurry printing approach has been used to deposit $LiFePO_4$ cathode particles onto a glassy carbon working electrode and to deposit activated carbon onto the Pt counter electrode [13].

The overall thickness of the electrode materials deposited on or attached to the electrode biasing contacts should be kept sufficiently thin so that the electrochemical processes within the electrode materials or at the electrode/electrolyte interfaces can be imaged through the liquid layer without further compromising image resolution. It is worth pointing out that the techniques described above to interface materials onto liquid cell electrodes have other applications as well. Examples of interfaced materials in corrosion studies are provided in Chapter 12.

11.3.4 Microfluidic Transfer of Air/Moisture-Sensitive Electrolytes into the Electrochemical Cell

Some liquid electrolytes have special handling requirements for transport into the electrochemical cell. Commercial Li-ion battery electrolytes generally consist of Li salts (e.g. $LiPF_6$ or $LiClO_4$) dissolved within a combination of organic, aprotic polar solvents such as ethylene carbonate (EC), dimethyl carbonate (DMC), diethyl carbonate (DEC), or propylene carbonate (PC) [24]. Unlike the ionic liquid-based electrolytes used in the open cell S/TEM experiments described in Chapter 3, commercial organic electrolytes have a high vapor pressure and will rapidly evaporate in the high vacuum column of the TEM. For this specific reason, the closed cell approach must be used to seal volatile Li-ion battery electrolytes between the electrochemical microchips. Further complications can arise when using commercial battery electrolytes since they are chemically reactive with air and moisture. For example, a solid $LiPF_6$ electrolyte will decompose to form HF in the presence of a trace amount of water. The chemistry of the electrolyte can radically change when HF is produced, and HF can react with the microelectrodes and other cell components. The decomposition of $LiPF_6$ can also produce PF_6 gas, which can form nanobubbles and become a concern for dewetting or membrane rupturing.

It is thus typically necessary to prepare the electrochemical cells for ec-S/TEM experiments in an inert atmosphere, i.e. in a glove box. Several techniques are effective for transporting the electrolyte into the liquid cell. In the most basic technique, a droplet of electrolyte is dispensed on the bottom microchip using a micropipette. The top microchip is then placed atop the bottom microchip to encapsulate the electrolyte between the membranes. While this task may seem trivial, there is inherent difficultly in assembling the extremely small-scale microchip devices in the liquid cell holder while working in a glovebox. An alternative method exploits the liquid flow capabilities of the liquid cell [25], using the microfluidic delivery system to transport the air/moisture sensitive electrolyte into the assembled electrochemical microchips in the holder through microfluidic tubing, via a syringe and microfluidic syringe pump. The electrolyte is delivered to the cell through an inlet port and excess electrolyte is collected in a container on the outlet port. This is particularly advantageous since the microchip assembly in the liquid cell holder can be prepared outside the glovebox, then filled inside the glovebox. The liquid electrolyte delivered into the inlet port is drawn between the viewing windows by capillary forces [19, 26]. The inlet and outlet ports can be subsequently sealed using vacuum epoxy to create a closed system.

11.4 Electroanalytical Measurement Techniques

Numerous electroanalytical techniques can be used to evaluate the performance, durability, and cycle-life of electrical energy storage and conversion materials. Chronopotentiometry, cyclic voltammetry (CV), and electrical impedance spectroscopy (EIS) are well-established techniques that are often employed to determine the kinetics of interfacial reactions, rate-limiting steps associated with oxygen reduction and oxygen evolution reactions, and electron/charge/mass transfer mechanisms. However, these techniques alone do not allow for the simultaneous observation or chemical analysis of time-resolved electrochemical reactions that occur locally at critical solid–solid, gas–solid, or liquid–solid interfaces. The ec-S/TEM method enables simultaneous

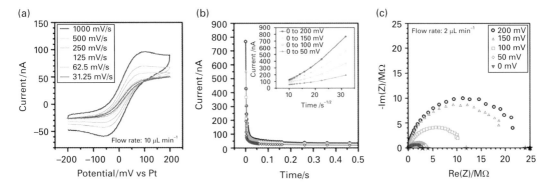

Figure 11.2. Electrochemical characterization methods often used to evaluate the performance characteristics and electrochemical processes in batteries and fuel cell materials. (a) Cyclic voltammetry, (b) chronoamperometry, and (c) electrochemical impedance spectroscopy acquired for a $[Fe(CN)_6]^{3-/4-}$ redox system. Reprinted with permission from Ref. 16.

quantitative electrochemical measurements and direct visualization of dynamic, site-specific electrochemical reactions at high spatial and temporal resolution.

Electroanalytical measurement techniques are often used in tandem to evaluate the performance characteristics of batteries and fuel cells. Here we focus on the quantitative electrochemical measurements that have been used for ec-S/TEM experiments on these devices. For detailed descriptions of electrochemical measurements in general, the reader is directed to the textbook by Bard and Faulkner [27]. Figure 11.2 shows examples of CV, chronoamperometry, and EIS measurements for the $[Fe(CN)_6]^{3-/4-}$ redox couple [16]. For CV measurements, the cell potential is linearly scanned at a given rate between two set potentials while monitoring current. Typical information obtained from CV measurements includes the redox processes associated with electrolyte reduction potentials, ion intercalation potentials, the double layer capacitance, the electrochemical window of the electrolyte, and oxidation state changes. The dependence of scan rate on mass transport mechanisms and diffusion-limited processes can also be derived from CV. For the fast CV scan rates shown in Figure 11.2a the two distinct features correspond to the cathodic peak at −60 mV and the anodic peak at 100 mV. The cathodic peak correlates to the reduction of $[Fe(CN)_6]^{3-}$ to $[Fe(CN)_6]^{4-}$ while the anodic peak correlates to the oxidation of $[Fe(CN)_6]^{4-}$ to $[Fe(CN)_6]^{3-}$. Chronoamperometry is often used to determine the diffusion coefficient of electroactive species, in addition to interpreting diffusion-controlled electrochemical processes. Figure 11.2b shows typical chronoamperometric measurements for $[Fe(CN)_6]^{4-}$ oxidation. Since the analyte concentration and electroactive surface area are known for this particular electrochemical cell geometry, the diffusion coefficient can be calculated from the current and time measurements using the Cottrell equation, as described in Chapter 10. EIS is used to evaluate and separate the effects of electron/charge/mass transport in batteries and fuel cells. Typical EIS data for $[Fe(CN)_6]^{3-/4-}$ in the electrochemical cell are presented in Figure 11.2c. Although not shown in Figure 11.2, chronopotentiometry is a constant current measurement technique used in galvanostatic charge/discharge cycle experiments of battery insertion materials to extract coulombic efficiencies (charge ratio) and to determine irreversible capacity loss as a function of charging rate. The charging current is determined by the theoretical capacity of the electrode material and the rate at which Li ions intercalate into the host intercalation compound. The voltage profiles (voltage vs. capacity) provide information on the capacity and capacity retention of the electrode material when Li ions intercalate and de-intercalate in the structure. Furthermore, the influence of rate capability (C-rate, i.e. the charge or discharge rate) on capacity retention can also be extracted from chronopotentiometry data. Any decrease in capacity as a function of the number of charge/discharge cycles indicates how well the electrode material performs over time.

11.5 Application of Electrochemical-S/TEM for Battery Research

The ongoing interest in using ec-S/TEM to gain a fundamental understanding of interfacial electrochemical reactions and in energy storage systems has had two main focus areas to date. These are directly identifying structural and chemical changes within the battery electrode materials, and on the surface of the patterned electrode when electrolyte breakdown occurs during electrochemical charge and discharge cycling.

11.5.1 The Solid Electrolyte Interphase in Li-ion Batteries

The solid electrolyte interphase is one of the most studied interfacial reactions in battery research [4, 28]. The SEI is an important interfacial film that forms and evolves during electrochemical cycling and is associated with irreversible capacity loss and increased cell impedance during cycling [29]. The SEI is composed of inorganic and organic electrolyte species that solidify on the electrode as a result of electrolyte decomposition [24, 28]. Cyclic voltammetry in ec-STEM (0 to -4V at 20 mV/s) has been used to investigate SEI formation on a Au electrode from a commercial 1.2 M $LiPF_6$ EC:DMC electrolyte [30]. The onset potential (-1.25 V vs. Au) for EC solvent reduction was correlated with dynamic observations of SEI nucleation and growth at the Au electrode–electrolyte interface. Figure 11.3 shows the result of scanning the potential towards more negative potentials: a dendritic SEI formed, which was attributed to the electric field distribution that developed around the Au electrode during the CV measurements. Similar experiments investigating interfacial reactions in commercial battery electrolytes have been reported using a two-electrode cell configuration [18, 19]. Cyclic voltammetry was used to induce electrolyte reduction while imaging the Au electrode–electrolyte interface and the local structural changes during Li alloying with the Au electrode under high reducing potentials. Initial formation of a compact SEI layer at the Au electrode surface was followed by gas evolution (presumably a PF_5 gaseous byproduct of $LiPF_6$ reduction). Since the two-electrode configuration does not include a reference electrode, it is difficult to correlate standard electrochemical reduction potentials to the Au–Li alloying potentials reported.

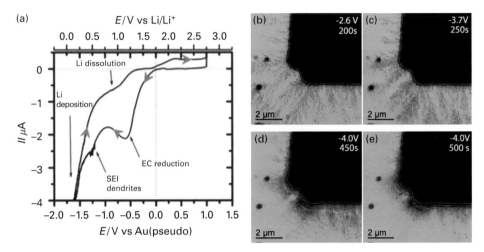

Figure 11.3. Direct observations of SEI formation and dendritic growth on a Au working electrode. (a) CV acquired at a sweep rate of 20 mV/s from a 1.2 M LiPF$_6$ EC:DMC electrolyte. (b–e) False-colored bright field TEM images showing SEI evolution with potential change. Reproduced from Ref. 12 with permission from The Royal Society of Chemistry.

Even though gold (or other noble metals) can readily be microfabricated onto the electrochemical microchip platform, it is not used as an electrode material or current collector in Li-ion batteries because of the formation of Au–Li alloys discussed in Section 11.3.2 [31]. Graphitic carbons, on the other hand, are intercalation compounds and are the most commonly used anode materials in rechargeable Li-ion batteries [32]. Carbon has a theoretical energy density of 372 mA h/g and Li ions can intercalate between the graphitic basal planes. Graphitic carbons are thermodynamically unstable in liquid electrolytes, and a kinetically passivating SEI forms during the first charging cycle. By interfacing a graphite anode to microfabricated Au contacts in the electrochemical cell, as described in Section 11.3.3 above, the formation and evolution of the SEI can be observed with reductions in the cell potential [33]. A three-electrode cell configuration was used for these experiments, with the graphite/gold serving as the working electrode inside the cell. Lithium metal as the external counter and reference electrodes was connected via an electrolyte bridge with microfluidic PEEK tubing. A large ohmic potential drop was expected due to the location of the Li counter and reference electrodes relative to the working electrode and the limited electrolyte volume. However, the SEI growth could still be imaged. Figure 11.4 shows the SEI with a particle-like morphology extending from the graphite electrode, its outer layer breaking apart during a reduction in the cell potential, then quickly restructuring through a nanoparticle attachment mechanism. In a more recent study, SEI formation was imaged during CV at a 100 mV/s sweep rate in a three-electrode cell configuration with a glassy carbon working electrode and Pt counter and reference electrodes [11]. The SEI formed during negative potentials at the glassy carbon electrode–electrolyte interface and the SEI thickness increased with continued sweep-holds at more negative

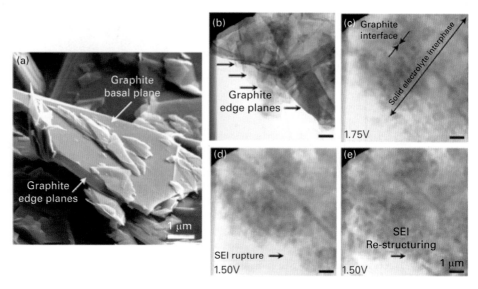

Figure 11.4. Characterization using ec-S/TEM of SEI formation on natural graphite. (a) SEM image of natural graphite flakes. BF-TEM images showing (b) graphite working electrode imaged in the cell without liquid electrolyte and (c–e) evolution of the SEI on the working electrode with reduction in cell potential within a 1 M LiPF$_6$ EC:DEC electrolyte. Reprinted with permission from Ref. 15.

potentials. The significance of these findings is the nanometer-scale, real-time imaging of the evolving nature of the SEI during repeated *in situ* electrochemical charge and discharge cycling.

As discussed in Chapter 7, electron beam-induced radiation damage readily occurs when a highly energetic electron beam interacts with a liquid [34–37]. Complex radical species and hydrated (or solvated) electrons are generated, with concentrations that are electron-dose and dose-rate sensitive [34]. It has been shown that the electron beam can be used to probe the breakdown mechanisms of organic battery electrolytes [38]. The hydrated electrons are highly reducing in nature and will decompose battery electrolytes, effectively separating the electrolyte salts (LiAsF$_6$, LiPF$_6$, and LiTf) from the organic solvents (DOS, DMSO, EC, and DMC) in which they are dissolved. For example, in 1 M LiAsF$_6$ in DMC electrolyte, a critical electron-dose threshold (3.4 e$^-$/nm^2 per frame) was determined to initiate electrolyte breakdown. As shown in the BF-STEM images in Figure 11.5a, under this constant electron dose the electrolyte salts dissociate from the solvent, and then nucleate and grow on the Si$_x$N$_y$ membrane. Figures 11.5b–d show measurements of particle growth rate as a function of cumulative electron dose. From the chemically sensitive ADF STEM image (Figure 11.5b) and line intensity profile measurements (Figure 11.5c), it was shown that the smaller primary particles have a higher intensity (higher atomic number) while the larger secondary particles have a lower intensity (lower atomic number). This was interpreted as the primary particles being composed of inorganic materials such as LiF and the secondary particles, polymerized organic compounds. Such electron beam irradiation studies of

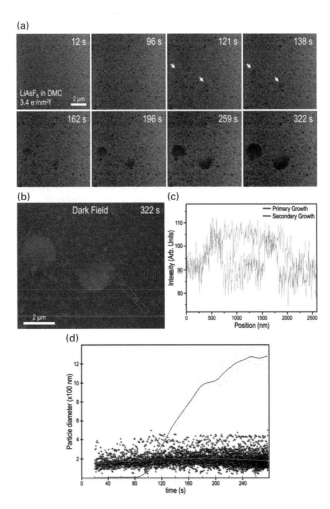

Figure 11.5. Influence of electron dose on battery electrolyte breakdown mechanisms. (a) Series of BF-STEM images showing the influence of electron dose on the breakdown of 1M $LiAsF_6$ in DMC. (b, c) ADF-STEM image and line intensity profiles for analysis of particle size measurements. (d) Particle diameter as a function of cumulative electron dose. Adapted with permission from Ref. 38. Copyright (2014) American Chemical Society.

electrolyte degradation are an integral part of understanding and interpreting ec-S/TEM experiments, and are needed to distinguish electron beam damage from electrochemical reactions induced by electrical stimuli; however, the exact mechanism(s) of SEI formation from electron beam-induced radiation vs. electrochemical SEI formation in a Li-ion battery may be completely different.

11.5.2 Lithiation and Delithiation Mechanisms in Battery Electrodes

The lithiation and delithiation of active battery electrode materials have been studied via ec-S/TEM, providing unique information on the volumetric and phase changes

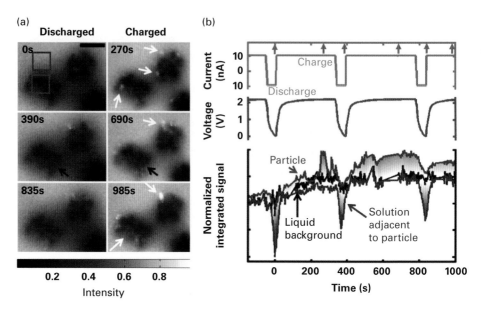

Figure 11.6. *In situ* electrochemical charge–discharge of LiFePO$_4$ cathode material. (a) EFTEM images (5 eV energy window) showing the electrochemical lithiation and delithiation behavior of nm-scaled LiFePO$_4$ in a 0.5 M Li$_2$SO$_4$ electrolyte. (b) Current and voltage cycling profiles and integrated intensity from different regions of the field of view. Scale bar is 400 nm. Adapted with permission from Ref. 13. Copyright (2014) American Chemical Society.

associated with the reactions [14, 39, 40] that is complementary to the open cell results shown in Chapter 3. To track the lithiation-induced volumetric expansion of a Si nanowire [14], FIB was used to create an ohmic contact between a single Si nanowire electrode (Cu-coated to increase conductivity) and a Pt contact on the microchip. Lithium metal was positioned in the tip of the TEM holder and interfaced with a holder-integrated Pt biasing contact to serve as a Li counter electrode. The Si nanowire was positioned over the electron-transparent Si$_x$N$_y$ viewing window, sealed within a LiClO$_4$ EC:DMC liquid electrolyte layer, and subjected to electrochemical lithiation. As Li intercalates into the Si nanowire, the nanowire undergoes a significant volumetric expansion; the Cu layer, which does not lithiate, acts as an effective fiducial marker to track the lithiation kinetics of the Si nanowire. Delithiation kinetics were also measured on Si nanowire electrodes prepared with EBID Pt fiducial markers.

An alternative method to study lithiation/delithiation of a cathode material [39] is shown in Figure 11.6. LiFePO$_4$ nanoparticles were inkjet slurry-printed directly onto the microfabricated glassy carbon working electrode and activated carbon was inkjet slurry-printed onto the circular shaped Pt counter electrode, producing a configuration that mimicked a full battery cell. A 0.5 M Li$_2$SO$_4$ aqueous-based electrolyte was used as the medium to transport Li into and out of the LiFePO$_4$ cathode under constant-current charge–discharge cycling at a 10C-rate. Valence EELS and EFTEM were used to track the chemical changes in LiFePO$_4$ during lithiation and delithiation events. Enhanced contrast is obtained from individual LiFePO$_4$ nanoparticles as Li is extracted from the structure of $FePO_4$. Chemical imaging through spectroscopic methods such as EELS and EFTEM can be combined with imaging to elucidate the nanoscale structural and chemical changes during repeated electrochemical lithiation and delithiation.

11.5.3 Lithium Dendrite Nucleation and Growth Mechanisms

Dendrite formation is a major concern for the implementation of Li-ion batteries from a safety and failure point of view. Cell short-circuiting occurs when a dendrite nucleates at the anode and grows across the electrolyte layer to make contact with the cathode. A number of *in situ* techniques have been used to study Li dendrite growth [41–44]; however, most of these techniques lack the high spatial resolution imaging capability of ec-S/TEM, which allows visualization of dendrites, such as those formed from Pb in aqueous $PbSO_4$ electrolyte solutions [45, 46], and correlation of their morphology with changes in current density. The *in situ*-grown Pb dendrites exhibit the typical branched structure, with primary dendrites growing at a faster rate in the expected $\langle 100 \rangle$ crystallographic growth direction than the secondary branched dendrite arms [45, 46].

Lithium dendrite formation has also been studied using ec-S/TEM, but with an aprotic organic liquid battery electrolyte commonly used in commercial Li-ion batteries [20, 47]. Figure 11.7 shows the nucleation and growth of Li dendrites on the surface and edge of a glassy carbon working electrode imaged in a 1.2 M $LiPF_6$ EC:DMC organic battery electrolyte [11]. The nucleation and growth process seen during a -4.5 V potential hold for 60 s confirms the preferential nucleation of Li metal at the glassy carbon working electrode and electrolyte interface. The arrow in Figure 11.7b marks the glassy carbon working electrode–electrolyte interface where the Li dendrites nucleated. By measuring the current and charge passed during Li electrodeposition, it was concluded that Li electrodeposition is mass transport limited: at highly negative overpotentials, Li ion transport is limited by the rate at which it can diffuse through the electrolyte and SEI.

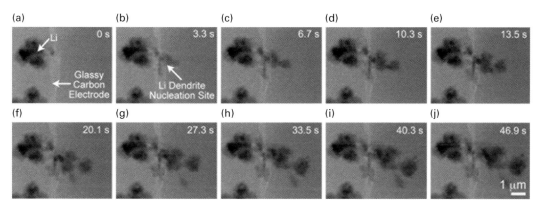

Figure 11.7. Time-lapse ADF STEM images of the preferential nucleation and growth of Li dendrites at the edge of a glassy carbon working electrode within a 1.2 M $LiPF_6$ EC:DMC electrolyte during a linear sweep voltammetry hold at -4.5 V. Scale bar is 1 μm. Adapted with permission from Ref. 11. Copyright (2015) American Chemical Society.

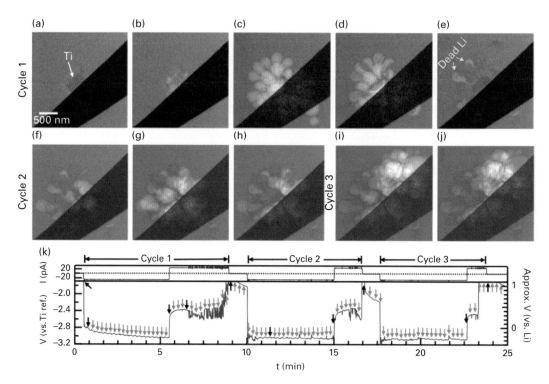

Figure 11.8. Series of BF-STEM images showing the evolution of Li dendrites on a Ti electrode pad, stimulated by cyclic chronopotentiometry from 1.0 M LiPF$_6$ EC:DMC electrolyte. In BF-STEM, the Li dendrites have a lighter contrast than the Ti pad and masked W electrode. Adapted with permission from Ref. 47. Copyright (2015) American Chemical Society.

Similar ec-S/TEM experiments have been carried out using Pt [20] and Ti [47] working electrodes and either a 1 M LiPF$_6$ PC or 1 M LiPF$_6$ EC:DMC electrolyte. Platinum is not typically used as an electrode material in Li-ion batteries due to the tendency to form a Pt–Li alloy, but Li dendrite growth and dissolution on Pt could be observed during repeated CV with preferential nucleation at the tip of the electrode, in localized regions of high electric field [20]. For dendrite formation on Ti, shown in Figure 11.8, chronopotentiometry provided information on the interplay between current density, electron dose, and the resultant morphology of the electrodeposited Li [47]. It is clear from these examples that ec-S/TEM has a powerful role to play in understanding dendrite formation and perhaps contributing to its control.

11.5.4 Characterizing Chemical Changes via Analytical Electron Microscopy

The atomic number dependence of the contrast in ADF-STEM images can be used to determine the chemical constituents of a sample in a liquid cell. For example, in recent ec-S/TEM experiments on Li electrodeposition [20, 47], the contrast in STEM images was used to qualitatively distinguish metallic Li and Li compounds (such as LiF and

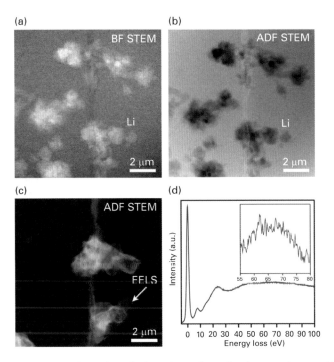

Figure 11.9. STEM imaging of Li metal electrodeposited on the surface of a glassy carbon working electrode. (a) BF-STEM, (b) ADF-STEM, (c) ADF-STEM with LiPF$_6$ EC:DMC electrolyte removed and (d) EEL spectra identifying metallic Li and LiF from the region marked EELS in (c). Note contrast difference between the two imaging modes. Adapted with permission from Ref. 11. Copyright (2015) American Chemical Society.

Li$_2$CO$_3$, part of the SEI phase) from the electrolyte. When metallic Li or Li compounds are imaged within the battery electrolyte they appear brighter in BF-STEM images and darker in ADF-STEM images. Figure 11.9 shows these clear contrast differences [11]. The image intensity in ADF-STEM images follows an inverse proportional relationship to the scattering mean free path; Li has a larger mean free path than the surrounding electrolyte; thus, Li is expected to appear darker than the liquid in ADF-STEM. When the liquid electrolyte is displaced from the cell, the contrast reverses, as expected, and the Li exhibits brighter contrast in ADF-STEM imaging [11]. Removing the liquid electrolyte also allows the structures to be confirmed as Li and LiF using EELS measurements from the Li K-edge.

As discussed in Chapter 21, XEDS and EELS are important analytical electron microscopy (AEM) methods that can be performed in conjunction with *in situ* liquid cell studies. For energy storage materials, the primary use of AEM is to correlate localized changes in chemistry with electrochemical measurements. In Li-ion battery research, XEDS cannot be used to identify Li or Li compounds within the SEI because of the inherent light-element detectability limits. EELS is an alternative; however, thickness-dependent effects through the whole cell obscure core-loss ionization edges as a result of multiple inelastic scattering from the electron beam. If the total cell is

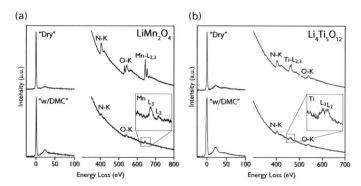

Figure 11.10. Liquid cell EELS measurements of battery electrodes. Simultaneously acquired low loss and core-loss spectra of (a) LiMn$_2$O$_4$ and (b) Li$_4$Ti$_5$O$_{12}$ acquired with and without dimethyl carbonate. Low loss EELS is used for measuring cell thickness and for removing plural scattering effects from core-loss EELS with Fourier ratio deconvolution. The Mn and Ti L$_{2,3}$ white lines are used for oxidation state determination. Reproduced from Ref. 50 with permission from The Royal Society of Chemistry.

sufficiently thin, EELS can be used for the chemical identification of Li metal and SEI chemistry using the Li K-edge at 55 eV. Furthermore, EELS can be used to directly determine the oxidation state of electrode materials. For transition metals, this is done by analyzing the characteristic sharp spectral features (white-lines) of the L$_{2,3}$ ionization edge, which originate from the electronic transition from 2p$_{3/2}$ and 2p$_{1/2}$ states to the unoccupied 3d states. The ratio between the L$_3$ and L$_2$ intensities determines the oxidation state [48, 49]. Figure 11.10 shows how this works for two materials commonly used in Li-ion batteries, LiMn$_2$O$_4$ (cathode) and Li$_4$Ti$_5$O$_{12}$ (anode). Low loss and core-loss spectra were simultaneously acquired from thin films deposited onto the Si$_x$N$_y$ membrane using radio-frequency magnetron sputtering. The spectra were acquired from dry films and with films in dimethyl carbonate, a commonly used battery solvent. Sufficient signal to noise is achieved (provided that t/λ is small) to distinguish the white lines from the background. Using the Mn L$_{2,3}$ and Ti L$_{2,3}$ white lines, the oxidation state of the electrode material was determined to be mixed Mn$^{3+/4+}$ for LiMn$_2$O$_4$ and Ti^{4+} for Li$_4$Ti$_5$O$_{12}$ following a Fourier-ratio deconvolution and background subtraction to account for plural scattering effects. This work shows the feasibility of EELS for chemical analysis; however, the influence of the electron dose on potentially altering the chemistry of the materials within the liquid cell must be understood to differentiate beam effects from changes induced by chemical or electrochemical reactions.

11.6 Application of ec-S/TEM for Fuel Cell Research

Fuel cells are an important energy conversion technology. In polymer electrolyte membrane (PEM) fuel cells, Pt-based nanoparticle electrocatalysts are typically supported on carbon and used to oxidize H$_2$ on the anode side and reduce O$_2$ on the cathode side to create electrical energy. Due to the high cost of Pt, non-precious metal catalysts and Pt alloys are being developed as alternative materials for fuel cell catalyst applications [5]. It is important to understand how structure and chemistry evolve in these catalyst systems and to correlate observed changes with electrocatalytic activity and performance under working conditions. Several degradation mechanisms of the catalyst and support occur, such as dissolution, coarsening, and carbon corrosion, and can significantly reduce electrocatalyst performance. Identical location TEM (IL-TEM) is a method used to characterize the same catalyst particles on a TEM grid before and after electrochemical testing [51]. While IL-TEM may be useful for inferring a degradation

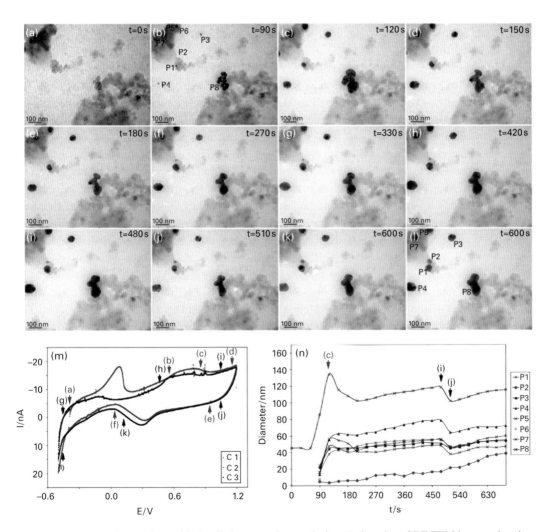

Figure 11.11. *In situ* observations of fuel cell electrocatalyst evolution. (a–l) series of BF-TEM images showing the coarsening behavior of Pt–Fe electrocatalysts that are correlated with (m) CV measurements performed at 10 mV/s in 0.1 M $HClO_4$. (n) Quantitative image analysis measurements showing the coarsening behavior of individual Pt–Fe catalyst nanoparticles labeled P1–P8. Adapted with permission from Ref. 52. Copyright (2015) American Chemical Society.

mechanism, the technique cannot be used to image the dynamics of reaction and/or degradation processes. The ec-S/TEM technique has therefore been used to study such degradation mechanisms.

Direct observation of the evolution of Pt–Fe electrocatalysts during CV measurements is shown in Figure 11.11 [52]. Coarsening behavior is visible in the Pt catalysts during electrochemical cycling. By correlating TEM images (Figure 11.11a–l) with cell potential, the conditions at which the Pt–Fe electrocatalysts coarsen can be determined (Figure 11.11m, n). In the future we anticipate a wider role of ec-S/TEM in understanding materials evolution in fuel cells.

11.7 Summary

Electrochemical experiments in closed liquid cells in S/TEM have developed to the point where they can probe nanoscale structural and chemical changes in materials for electrical energy storage and conversion systems. The ec-S/TEM approach enables new *operando* investigations of electrochemical reactions and kinetics for a wide range of materials systems, the outcomes of which can be used to interpret how material properties are linked to microstructural evolution. The primary advantage of this technique is the ability to contain a thin layer of electrolyte and to image through highly volatile liquids while simultaneously conducting quantitative electrochemical measurements. In this way, the dynamics of electrochemical processes can be visualized and correlated with testing variables in real time and at high spatial resolution. Advances in the design of electrochemical cells have allowed for quantitative electrochemical measurements to be obtained in a three-electrode electrochemical cell. By coupling spectroscopic techniques such as EELS and XEDS, chemical evolution can be tracked and correlated with electrochemical measurements and structural changes. The combination of imaging, spectroscopy, and electrochemical measurements in a liquid opens a new pathway for materials characterization. As ec-S/TEM continues to evolve through further improvements in cell design and greater understanding of beam-induced effects, new insight into the fundamental processes that control electrochemical reactions for a wide range of materials systems will inevitably lead to new discoveries in the materials used to power electrical energy storage systems.

Acknowledgements

Support provided by Oak Ridge National Laboratory's Center for Nanophase Materials Sciences, which is a U.S. Department of Energy, Office of Science User Facility. The authors would like to acknowledge the following researchers who contributed to *in situ* ec-STEM research at Oak Ridge National Laboratory: Dr. Robert L. Sacci, Dr. Loïc Baggetto, Dr. Gabriel M. Veith, Dr. Nancy J. Dudney, Dr. Jennifer M. Black, Dr. Xiao-Guang Sun, and Dr. Sheng Dai.

References

1. J.-M. Tarascon and M. Armand, Issues and challenges facing rechargeable lithium batteries. *Nature*, **414** (2001), 359–367.
2. M. Armand and J.-M. Tarascon, Building better batteries. *Nature*, **457** (2008), 652–657.
3. A. S. Arico, P. Bruce, B. Scrotasi, J.-M. Tarascon and W. Van Schalkwijk, Nanostructured materials for advanced energy conversion and storage devices. *Nat. Materials*, **4** (2005), 366–377.
4. J. B. Goodenough and Y. Kim, Challenges for rechargeable Li batteries. *Chem Mater.*, **22** (2010), 587–603.
5. M. K. Debe, Electrocatalyst approaches and challenges for automotive fuel cells. *Nature*, **486** (2013), 43–51.
6. C. M. Wang, In situ transmission electron microscopy and spectroscopy studies of rechargeable batteries under dynamic operating conditions: a retrospective and perspective view. *J. Mater. Res.*, **30** (2014), 326–339.
7. N. de Jonge and F. M. Ross, Electron microscopy of specimens in liquid. *Nat. Nanotechnol.*, **6** (2011), 695–704.
8. M. J. Williamson, R. M. Tromp, P. M. Vereecken, R. Hull and F. M. Ross, Dynamic microscopy of nanoscale cluster growth at the solid–liquid interface. *Nat. Mater.*, **2** (2003), 532–536.
9. R. R. Unocic, R. L. Sacci, G. M. Brown et al., Quantitative electrochemical measurements using in situ ec-S/TEM devices. *Microsc. Microanal.*, **20** (2014), 452–461.
10. Z. Zeng, W.-I. Liang, H.-G. Liao et al., Visualization of electrode–electrolyte interfaces in $LiPF_6$/EC/DEC electrolyte for lithium ion batteries via in situ TEM. *Nano Lett.*, **14** (2014), 1745–1750.
11. R. L. Sacci, J. M. Black, N. Balke et al., Nanoscale imaging of fundamental Li battery chemistry: solid-electrolyte interphase formation and preferential growth of lithium metal nanoclusters. *Nano Lett.*, **15** (2015), 2011–2018.
12. R. L. Sacci, N. J. Dudney, K. L. More et al., Direct visualization of initial SEI morphology and growth kinetics during lithium deposition by in situ electrochemical transmission electron microscopy. *Chem. Commun.*, **50** (2014), 2104–2107.
13. M. E. Holtz, Y. Yu, D. Gunceler et al., Nanoscale imaging of lithium ion distribution during in situ operation of battery electrode and electrolyte. *Nano Lett.*, **14** (2014), 1453–1459.
14. M. Gu, L. R. Parent, B. L. Mehdi et al., Demonstration of an electrochemical liquid cell for operando transmission electron microscopy observation of the lithiation/delithiation behavior of Si nanowire battery anodes. *Nano Lett.*, **13** (2013), 6106–6112.
15. R. R. Unocic, X.-G. Sun, R. L. Sacci et al., Direct visualization of solid electrolyte interphase formation in lithium-ion batteries with in situ electrochemical transmission electron microscopy. *Microsc. Microanal.*, **20** (2014), 1029–1037.
16. R. R. Unocic, R. L. Sacci, G. M. Brown et al., Quantitative electrochemical measurements using in situ ec-S/TEM devices. *Microsc. Microanal.*, **20** (2014), 452–461.
17. M. Moshkovich, M. Cojocaru, H. E. Gottlieb and D. Aurbach, The study of the anodic stability of alkyl carbonate solutions by in situ FTIR spectroscopy, EQCM, NMR and MS. *J. Electroanal. Chem.*, **497** (2001), 84–96.
18. Z. Zeng, W.-I. Liang, Y.-H. Chu and H. Zheng, In situ TEM study of the Li–Au reaction in an electrochemical liquid cell. *Faraday Discuss.*, **176** (2014), 95–107.

19. Z. Zeng, W.-I. Liang, H.-G. Liao et al., Visualization of electrode–electrolyte interfaces in LiPF$_6$/EC/DEC electrolyte for lithium ion batteries via in situ TEM. *Nano Lett.*, **14** (2014), 1745–1750.
20. B. L. Mehdi, J. Qian, E. Nasybulin et al., Observation and quantification of nanoscale processes in lithium batteries by operando electrochemical (S)TEM. *Nano Lett.*, **15** (2015), 2168–2173.
21. M. Tang, S. Lu and J. Newman, Experimental and theoretical investigation of solid-electrolyte-interphase formation mechanisms on glassy carbon. *J. Electrochemi. Soc.*, **159** (2012), A1775–A1785.
22. M. Tang and J. Newman, Transient characterization of solid-electrolyte-interphase using ferrocene. *J. Electrochem. Soc.*, **159** (2012), A281–A289.
23. R. Unocic, L. Adamczyk, N. Dudney et al., In-situ TEM characterization of electrochemical processes in energy storage systems. *Microsc. Microanal.*, **17** (2011), 1564–1565.
24. K. Xu, Nonaqueous liquid electrolytes for lithium-based rechargeable batteries. *Chem Rev.*, **104** (2004), 4303–4418.
25. E. A. Ring and N. de Jonge, Microfluidic system for transmission electron microscopy. *Microsc. Microanal.*, **16** (2010), 622–629.
26. J. M. Grogan and H. H. Bau, The Nanoaquarium: a platform for in situ transmission electron microscopy in liquid media. *J. Microelectromech. Syst.*, **19** (2010), 885–894.
27. A. J. Bard and L. R. Faulkner, *Electrochemical Methods: Fundamentals and Applications*, 2nd edn. (New York: John Wiley & Sons, 2001).
28. P. Verma, P. Maire and P. Novák, A review of the features and analyses of the solid electrolyte interphase in Li-ion batteries. *Electrochimica Acta*, **55** (2010), 6332–6341.
29. M. Winter, The solid electrolyte interphase: the most important least understood solid electrolyte in rechargeable Li batteries. *Z. Phys. Chem.*, **223** (2009), 1395–1406.
30. R. L. Sacci, N. J. Dudney, K. L. More et al., Direct visualization of initial SEI morphology and growth kinetics during lithium deposition by in situ electrochemical transmission electron microscopy. *Chem. Commun.*, **50** (2014), 2104–2107.
31. M. Moshkovich, Y. Gofer and D. Aurbach, Investigation of the electrochemical windows of aprotic alkali metal (Li, Na, K) salt solutions. *J. Electrochem. Soc.*, **148** (2001), E155–E167.
32. D. Aurbach, M. D. Levi, E. Levi et al., Common electroanalytical behavior of Li intercalation processes into graphite and transition metal oxides. *J. Electrochem. Soc.*, **145** (1998), 3024–3034.
33. R. R. Unocic, X.-G. Sun, R. L. Sacci et al., Direct visualization of solid electrolyte interphase formation in lithium-ion batteries with in situ electrochemical transmission electron microscopy. *Microsc. Microanal.*, **20** (2014), 1029–1037.
34. N. M. Schneider, M. M. Norton, B. J. Mendel et al., Electron–water interactions and implications for liquid cell electron microscopy. *J. Phys. Chem. C.*, **118** (2014), 22373–22382.
35. P. Abellan, T. J. Woehl, L. R. Parent et al., Factors influencing quantitative liquid (scanning) transmission electron microscopy. *Chem. Commun.*, **50** (2014), 4873–4880.
36. T. J. Woehl, J. E. Evans, I. Arslan, W. D. Ristenpart and N. D. Browning, Direct in situ determination of the mechanisms controlling nanoparticle nucleation and growth. *ACS Nano*, **6** (2012), 8599–8610.
37. T. J. Woehl, C. Park, J. E. Evans et al., Direct observation of aggregative nanoparticle growth: kinetic modeling of the size distribution and growth rate. *Nano Lett.*, **14** (2014), 373–378.

38. P. Abellan, B. L. Mehdi, L. R. Parent et al., Probing the degradation mechanisms in electrolyte solutions for Li-ion batteries by in situ transmission electron microscopy. *Nano Lett.*, **14** (2014), 1293–1299.
39. M. E. Holtz, Y. Yu, D. Gunceler et al., Nanoscale imaging of lithium ion distribution during in situ operation of battery electrode and electrolyte. *Nano Lett.*, **14** (2014), 1453–1459.
40. K. W. Noh and S. J. Dillon, Morphological changes in and around Sn electrodes during Li ion cycling characterized by in situ environmental TEM. *Scripta Materialia*, **69** (2013), 658–661.
41. R. Bhattacharyya, B. Key, H. Chen et al., In situ NMR observation of the formation of metallic lithium microstructures in lithium batteries. *Nat. Mater.*, **9** (2010), 504–510.
42. K. Nishikawa, T. Mori, T. Nishida, Y Fukunaka and M. Rosso, Li dendrite growth and Li+ ionic mass transfer phenomenon. *J. Electroanal. Chem.*, **661** (2011), 84–89.
43. D. R. Ely and R. E. Garcia, Heterogeneous nucleation and growth of lithium electrodeposits on negative electrodes. *J. Electrochem. Soc.*, **160** (2013), A662–A668.
44. T. Nishida, K. Nishikawa, M. Rosso and Y. Fukunaka, Optical observation of Li dendrite growth in ionic liquid. *Electrochimica Acta*, **100** (2013), 333–341.
45. E. R. White, S. D. Singer, V. Augustyn et al., In situ transmission electron microscopy of lead dendrites and lead ions in aqueous solution. *ACS Nano*, **6** (2012), 6308–6317.
46. M. Sun, H.-G. Liao, K. Niu, H. Zheng, Structural and morphological evolution of lead dendrites during electrochemical migration. *Sci. Rep.*, **3** (2013), 3227.
47. A. J. Leenheer, K. L. Jungjohann, K. R. Zavadil, J. P. Sullivan and C. T. Harris, Lithium electrodeposition dynamics in aprotic electrolyte observed in situ via transmission electron microscopy. *ACS Nano*, **9** (2015), 4379–4389.
48. T. Riedl, T. Gemming and K. Wetzig, Extraction of EELS white-line intensities of manganese compounds: methods, accuracy, and valence sensitivity. *Ultramicroscopy*, **106** (2006), 284–291.
49. M. Varela, M. Oxley, W. Luo et al., Atomic-resolution imaging of oxidation states in manganites. *Phys. Rev. B.*, **79** (2009), 085117.
50. R. R. Unocic, L. Baggetto, G. M. Veith et al., Probing battery chemistry with liquid cell electron energy loss spectroscopy. *Chem. Commun.*, **51** (2015), 16377–16380.
51. J. C. Meier, C. Galeano, I. Katsounaros et al., Degradation mechanisms of Pt/C fuel cell catalysts under simulated start–stop conditions. *ACS Catal.*, (2012), 832–843.
52. G.-Z. Zhu, S. Prabhudev, J. Yang et al., In situ liquid cell TEM study of morphological evolution and degradation of Pt–Fe nanocatalysts during potential cycling. *J. Phys. Chem. C.*, **118** (2014), 22111–22119.

12 Applications of Liquid Cell TEM in Corrosion Science

See Wee Chee and M. Grace Burke

12.1 Introduction

Corrosion is broadly defined as the degradation of materials (primarily metals) due to interactions with the environment [1]. The detrimental impact of corrosion can range from failure of structural materials in large-scale infrastructures to breakdown of miniature components in microelectronics. These hazards are mitigated through strategies that either control corrosion rates or predict the service life of a material. Yet our efforts are hampered by the lack of fundamental insight into how a material behaves in various corrosive environments [2, 3]. The challenge here is that corrosion is caused by a complex interplay of material and environmental factors, where many relationships remain largely unresolved.

Although most materials are susceptible to corrosion, research efforts have focused on commercial metals and alloys due to their widespread use and propensity to corrode. Metallic corrosion is an electrochemical process that involves metal oxidation and reduction of other species within an electrolyte with the corresponding charge transfer. In general, the corrosion resistance of a metal or alloy is determined by its microstructure and composition. However, techniques that can probe chemical and structural changes during corrosion are limited because these reactions take place at the interface between a metal surface and its aqueous environment – or, for the case of atmospheric corrosion, within a thin liquid film [4]. Materials characterization techniques with the desired resolution or analytical capabilities often require samples to be placed in a vacuum environment, but a hydrated surface will desiccate in vacuum and artifacts may be introduced. Existing *in situ* techniques used to study corrosion include synchrotron X-ray radiation and variants of scanning probe microscopy [2]. Hence, the development of new techniques, such as *in situ* electron microscopy, to probe the dynamics of corrosion at high resolution while maintaining the aqueous environment will be advantageous [2, 3]. Electron microscopy has traditionally provided *post-mortem* analyses of sites affected by corrosion and their surrounding structures, for example identifying corrosion products or secondary phases in the microstructure that promote localized corrosion. In contrast, liquid cell transmission electron microscopy can benefit corrosion science by providing direct visualization of the critical changes that occur during corrosion events.

With liquid cell EM, nanometer-scale events in liquids can be captured with exceptional spatial and temporal resolution [5]. Recent studies have also demonstrated the ability to perform chemical analyses in liquids using EELS [6–8] and XEDS [9, 10].

Furthermore, coupling liquid cell EM with electrochemical methods has seen significant advances in battery research and related areas, as described in Chapters 10 and 11. Simultaneous imaging, chemical analysis, and electrochemical measurement can provide valuable information about the underlying mechanism(s) of corrosion processes. However, translating the methodology is not straightforward and the application of liquid cell EM to corrosion science is still in its infancy. This chapter aims to provide a framework for future development in this challenging and important field by presenting an overview of current research and a discussion of the experimental considerations, compiled from experiences of research groups working in the area.

12.2 Studying Corrosion in Aqueous Environments

Corrosion behavior can be classified according to the nature of the environment, the characteristics of the corrosion phenomena, or the material system of interest. The reader is referred to one of many excellent textbooks on corrosion, such as Ref. 11. Aqueous corrosion of metals is commonly divided into two categories, uniform and localized. In uniform corrosion, the attack proceeds evenly over the exposed area and is generally less damaging because the corrosion rate can be measured and predicted relatively easily. Localized corrosion is the non-uniform attack of metals and alloys that are protected by a passive film, typically the metal oxide. Most corrosion-resistant alloys in use today, such as stainless steels and Al and Ti alloys, are protected this way. Localized corrosion is dangerous because it is difficult to detect and can lead to sudden failure. Other frequently encountered forms of corrosion include galvanic corrosion, or the preferential dissolution of one metal when two dissimilar metals are in electrical contact within an electrolyte, environment enhanced corrosion such as erosion corrosion, microbial corrosion, and stress corrosion cracking, where tensile stress interacts with the chemical environment.

Two key concepts will be elaborated in this section. First is the understanding of localized corrosion, an area where we believe that liquid cell TEM will have the most impact, and second is the integration of electrochemical techniques, which is a necessary development if we are to quantify the processes that lead to corrosion.

12.2.1 Localized Corrosion

Localized corrosion is a major concern in structural materials. The accelerated dissolution of underlying metal is often insidious because the microscopic pits can be obscured by the corrosion product, hindering early detection. Different forms of localized corrosion [12] include:

(i) Pitting corrosion: corrosion at specific sites due to breakdown of the metal oxide film caused by aggressive species in the environment, such as the ubiquitous chloride ion;
(ii) Crevice corrosion: corrosion aggravated by occluded areas; and
(iii) Intergranular corrosion: increase in corrosion susceptibility due to the presence of secondary phases or compositional depletion or segregation.

Despite considerable research efforts, we are still plagued by questions about how localized corrosion develops [12, 13]. For instance, it will be a significant advancement to elucidate the mechanism(s) of pit initiation. It has been postulated that the selective attack is due to defects in the oxide film, but the causes of passive film failure and early stages of pit formation remain unclear [13, 14]. This is partly due to the limited analytical tools that can characterize the structural and chemical nature of nanometer thick passive films in the aqueous environment. Direct correlation of pit locations with intrinsic microstructural features in the metal or the oxide film is also difficult for other reasons [12]. One is that the initiation sites are nanometers in size, requiring high resolution analysis under wet conditions. Another reason is that initiation is rare and stochastic, making it impossible to predict where and when a pit will form in real-time experiments, impeding direct high resolution observations. And finally, the rate of pit growth after initiation is very high, with the chemical species in solution and local potentials varying at the same time. It is not easy to capture these rapid changes in chemistry and morphology. Liquid cell TEM has tantalizing potential to improve our understanding of localized corrosion with its unique combination of spatial resolution, temporal resolution, and analytical capability.

12.2.2 Electrochemical Techniques

Open circuit (free corrosion) experiments in the liquid cell can allow study of corrosion structures, but they are time consuming, taking at least a few hours. Exposing a sample to the electron beam for these extended durations is not ideal, as discussed in Chapter 7, with respect to minimizing beam-induced effects [15]. Moreover, pit initiation events are difficult to capture when neither the location of a pit nor the time needed to initiate pitting can be controlled. Therefore, control of the electrochemical parameters of the experiment is crucial for *in situ* studies of corrosion behavior.

The voltage–current characteristic of an electrolyte–electrode interface is probed by imposing either a potential (the more common choice) or a current. Chapters 10 and 11 showed how such techniques are applied in liquid cell TEM and that they work: for example, several electrochemical measurement techniques have been validated *in situ* [16] for a standard one-electron transfer redox couple $[Fe(CN)_6]^{3-/4-}$. For conventional corrosion studies, several electrochemical techniques are useful, as described in sources such as Refs. 4 and 17–20. When the potential of a sample is biased away from the corrosion potential, it is referred to as being polarized. Anodic polarization occurs when the potential is driven in the positive direction, causing oxidation/corrosion reactions to dominate at the working electrode. Cathodic polarization occurs if the potential is forced in the negative direction and reduction reactions on the sample surface, such as oxygen reduction, dominate instead. Three techniques are commonly applied for corrosion measurements:

(i) Potentiodynamic polarization: The potential is stepped over a wide range (a few volts) so that different regions of corrosive behavior (so-called active, passive, and transpassive) can be qualitatively identified. Useful information can be obtained about passive metals, such as the corrosion current in the passive regime. During anodic polarization, current transients below the pitting potential are associated with the formation of metastable pits.

(ii) Linear polarization and Tafel extrapolation: These measurements scan the potential over tens to a few hundreds of millivolts around the corrosion potential so that the polarization resistance (linear polarization) and the Tafel constants (Tafel extrapolation) can be extracted from the anodic or cathodic polarization curves. Using these parameters, corrosion rates are estimated by using the Stern–Geary equation to calculate the corrosion current.

(iii) Cyclic polarization: Cyclic polarization measures the tendency of a sample to pit. Here, the sample is scanned in the anodic direction until there is a large increase in current, giving the pitting potential. Then, the scan is reversed. The potential at which the hysteresis loop closes is the re-passivating potential. The likelihood of pitting is given by the difference between the pitting and re-passivating potentials.

These techniques can all be integrated with liquid cell TEM, as will be discussed in Section 12.5.

12.3 Studies of Corrosion using Liquid Cell TEM

In this section we discuss some corrosion phenomena that have been examined in the liquid cell. In the laboratory setting, corrosion has been induced chemically by exposure to a corrosive environment, or electrochemically by applying a potential or current within an electrochemical cell. Both provide insights into corrosion processes.

12.3.1 Corrosion of Thin Metal Films

Thin films are a natural match with liquid cell TEM because the films can be directly deposited over the electron-transparent windows and patterned using standard masking techniques. The corrosion of thin metal films is also relatively well studied in conventional corrosion science [21–26], providing an extensive literature of known behaviors for comparison with liquid cell results.

Figure 12.1 shows the corrosion structures induced in Al thin films by exposure to NaCl solutions [27–29]. The morphologies observed depend on NaCl concentrations and exposure times. In 0.01 M NaCl solution blisters are observed in the Al film, and in 1.0 M NaCl solution meandering corrosion tracks are seen, both corresponding to structures observed previously using other methods [21, 22]. In 0.1 M NaCl solution, an intermediate morphology of anomalous enlarged grains was observed. While these results show how corrosion structures develop, the evolution of morphology was only captured as snapshots in time. Continuous imaging was not used [15] because these structures required hours to days to form. In Figure 12.2, we show for comparison the pits formed in a Cu film after 2 hours. Copper films in concentrated brine [29] and Fe films in acetic acid solutions (Khalid Hattar, personal communication) have been

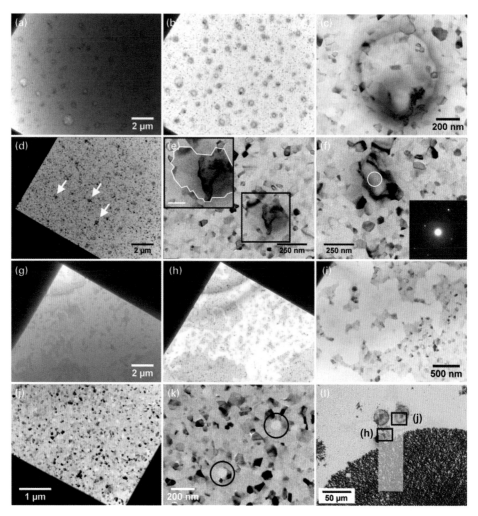

Figure 12.1. Corrosion structures observed in Al thin film immersed in NaCl solutions. (a)–(c) TEM bright field images of the film after 15 hour exposure to 0.01 M NaCl solution, (d)–(f) after 5 hour exposure to 0.1 M NaCl solution, (g)–(k) after 4 hour exposure to 1.0 M NaCl solution. (a) and (g) were taken with liquid still in the cell. For the other images, the liquid was dewetted from the Al surface using a water–alcohol mixture under a fully spread beam to improve image resolution [30]. (c) High magnification image of a blister. (e) and (f) High magnification images of the anomalous enlarged grains. Insert in (e) outlines the grain in white and insert in (f) is a selected area diffraction pattern of another grain, showing that it is single crystalline. These features were formed over the entire film (three of them are highlighted in (d) with white arrows). (i) Higher magnification image taken from the area shown in (h) where the film had corroded away in the form of corrosion tracks. (k) Higher magnification image taken from the opposite corner of the windowed area, which is shown in (j), showing pit formation (highlighted with black circles). (l) Light microscope image taken after the liquid cell is separated, showing that (h) is part of a larger fractal pattern whereas (j) belongs to an area of more intact film. Adapted from Refs. 28 and 29 with permission from the Royal Society of Chemistry.

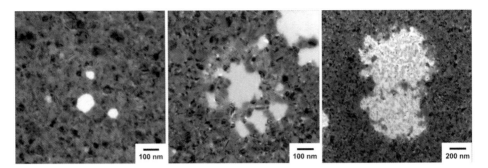

Figure 12.2. Corrosion of Cu thin film. Pits of different sizes are observed in 50 nm Cu films in flowing 6 M NaCl solution for 2 hours. Excerpted from Ref. 29 with permission from the Royal Society of Chemistry.

studied in real time. It is clear that broad opportunities exist to examine the details of how corrosion processes initiate and propagate in thin film microstructures.

12.3.2 Corrosion of Nanoparticles

Nanoparticle nucleation and growth has been investigated extensively in liquid cell TEM [31], as described in Chapter 9. In contrast, the dissolution of nanoparticles in corrosive environments has received less attention. Materials examined have included the dissolution of Pd nanoparticles in potassium bromide [32] and Pt in chloroauric acid [33]. Galvanic replacement reactions have also been visualized in the liquid cell. An example is the replacement of Ag by Pd, which transforms the initial nanoparticles into hollow structures [34]. Comparison with *ex situ* results suggests that the replacement reactions are accelerated under the electron beam. More significantly, the authors were able to modify the observed reactions by using isopropanol as a scavenger of the hydroxyl radical, demonstrating that electron beam effects can be mitigated (or at least altered) with scavenging strategies.

12.3.3 Polarization of Thin Films within a Microfluidic Corrosion Cell

The ability to apply a voltage to electrodes integrated into the liquid cell brings an additional level of control to corrosion studies. As a proof-of-concept, Figure 12.3 shows the polarization curve of an Al film mounted in a liquid cell holder but measured on the laboratory bench [29]. In this rudimentary corrosion cell design, the Al film was deposited over the surface of a chip except near the Ti leads which extend outside the holder. Selective masking ensured that one lead made contact with the film, and this was designated as the working electrode. The other two leads were designated as reference and counter electrodes. All three leads were exposed to the solution flowing through the cell while biased using a potentiostat. The polarization curve obtained in 0.01 M NaCl is shown in Figure 12.3d. As compared to a measurement in a chip without Al, the curve for the Al-covered chip shows a higher current density as well as features associated with the breakdown of passivity. Figure 12.4 shows the results of carrying out this

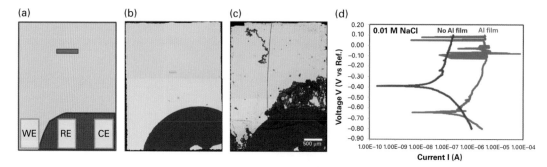

Figure 12.3. (a) Schematic of a basic corrosion cell for TEM. The microfluidic chip is selectively masked during deposition such that only one Ti lead (yellow) is in contact with the film (designated as the working electrode). The other two leads are used as the reference and counter electrodes. Light microscope images of the chip are shown (b) before and (c) after one cycle of polarization. (d) Polarization curves obtained from the chip in (b) compared against a blank chip with no film, both exposed to 0.01 M NaCl solution. The fluctuations at the positive potentials are likely due to gas bubbles adhering to the reference electrode. From Ref. 29 with permission from the Royal Society of Chemistry.

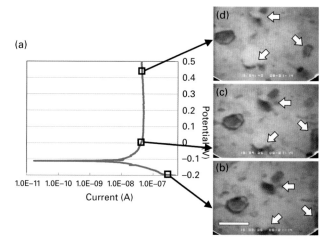

Figure 12.4. (a) Polarization curve of an Al thin film in 0.1 M Na_2SO_4 and 0.001 M NaCl solution. (b–d) The film microstructure at different points of the polarization (from −200 mV to 500 mV) showing the dissolution of single grains (indicated with white arrows). Scale bar is 50 nm. Unpublished data.

measurement within the TEM, allowing the dissolution of individual grains to be tracked during anodic polarization [35].

Since liquid cell TEM can only view a small region of the chip, it is important to relate observations such as those in Figure 12.4 to the larger film area that is exposed to the electrolyte. After recording the polarization curve, the two chips of the liquid cell can be separated and examined using light microscopy. Doing this for the sample in Figure 12.3 shows the presence of meandering corrosion tracks similar to those in Figure 12.1. However, the area where the Al and Ti made contact (lower left corner of

Figure 12.3c) was severely corroded. Such corrosion prevents any extended polarization experiments using this simple design, but could clearly be addressed with a more sophisticated design. Other metals have already been examined in thin film form under applied bias, such as Ni and Au [36]. Exciting future opportunities for *in situ* electrochemical studies of corrosion will be driven by further improvements to liquid cell and chip designs.

12.3.4 *In Situ* Corrosion Studies of Alloys

Alloys provide rich opportunities for spatially resolved corrosion studies, since selective leaching of minor constituents or de-alloying reactions can lead to structural effects such as weakening. Sputtered AgAu and CuAu thin films have been studied in sulfuric acid under applied potential [37] to understand the mechanisms of crack formation during de-alloying. The results illustrate the power of combining *in situ* TEM with other analytical techniques. After measuring the as-deposited grain boundary chemistry using XEDS, the authors employed a combination of *ex situ* SEM and *in situ* TEM to observe microstructural evolution as a function of alloy composition and applied potential. Gold segregates at the grain boundary in Cu–Au alloys. *Ex situ* SEM showed, for Cu_3Au films, de-alloying reactions propagating along grain boundaries, as well as localized suppression of grain boundary crack formation; *in situ* TEM showed these reactions in more detail. The authors were able to estimate that grain boundaries with increased resistance to chemical attack had concentrations of ~40% Au. If such enriched grain boundaries inhibit grain boundary corrosion, this experiment implies that the distribution of grain boundaries with this critical concentration strongly influences de-alloying behavior in these materials.

It is also possible to map compositional changes from corrosion attack directly from within the microfluidic liquid cell. Figure 12.5 shows the dissolution of MnS inclusions

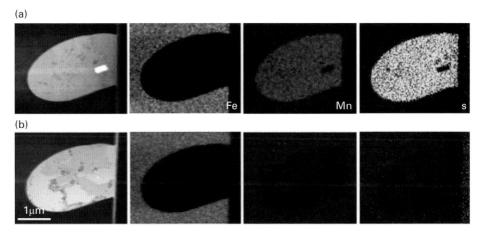

Figure 12.5. Bright field STEM and XED spectrum imaging showing dissolution of a MnS inclusion after *in situ* exposure to water: (a) in 1 bar air; (b) after 24 h in water. From Ref. 10.

in Type 304 austenitic stainless steel studied using XED spectrum imaging [10]. In this case, the chemical changes were tracked against time of exposure to water. It is clear that the combination of analytical information, especially *in situ*, with dynamic behavior during corrosion is a powerful tool in understanding corrosion mechanisms.

12.4 Considerations Pertaining to Studying Corrosion with Liquid Cell TEM

Several experimental considerations are important when setting up a chemical or electrochemical corrosion experiment in the liquid cell. Here, we discuss some experience gained in materials choice and sample preparation as well as artifacts caused by electron beam-induced effects. Other factors, such as flow rate and temperature, are expected to be important but so far have not been investigated in much detail.

12.4.1 Sample Selection

There are two primary limitations in sample selection: the thickness required to maintain electron transparency (the liquid layer and Si_xN_y membranes add to the effective thickness through which electrons must be transmitted) and the issue of accommodating the sample within the microfluidic cell. Depending on the density of the material of interest, the sample typically cannot be more than a few hundred nanometers thick [5] in the imaged region.

Studying aqueous corrosion in model systems such as thin films and nanoscale flakes, powders, or particles is the most straightforward. As highlighted in Sections 12.3.1 and 12.3.3, the merits of vapor-deposited thin films include direct deposition over the windows and possibilities for patterning. The sample thickness can also be adjusted via deposition conditions, and there is some control over surface morphology with different deposition methods. To expand the range of possible experiments, thin films can be patterned to include junctions between dissimilar metals, or co-deposited metal films can be heated to induce phase separation. Nanomaterials that can be dispersed in solution are usually introduced into liquid cells by pipetting into reservoirs fabricated in the cell [31] or by capillary flow [32]. The mixture may also be drop cast or directly slurry-printed onto one of the two chips [7]. A wide range of materials can be studied in this way with minimal processing.

On the other hand, the observation of conventional "bulk" metals and alloys has been limited by these sample size requirements. The optimum technique for preparing electron-transparent metallic samples is twin-jet electropolishing, but this method normally uses 3 mm diameter discs with initial thicknesses ~100 μm. Such a disc sample could not be placed between liquid cell chips and still maintain an acceptably small spacing between the windows. Alternatively, focused ion beam (FIB) sample preparation techniques can be utilized to extract a thin lamella from a bulk material [38]. Recent experiments on battery materials, described in Chapter 11, have shown that thin sections can be cut with a gallium beam and welded on the pre-patterned electrodes in the liquid cell using ion beam-induced deposition of Pt [39]. However, corrosion processes are extremely sensitive

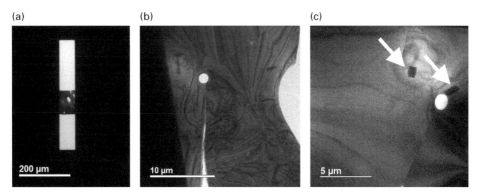

Figure 12.6. (a) Low magnification TEM image of "hybrid" sample on liquid cell window (400 μm × 50 μm), (b) sample imaged without water, and (c) sample in water with crystalline deposits formed during exposure under the electron beam. From Ref. 40.

to surface chemistry. Thus, potential artifacts can arise from Ga^+-ion implantation into the sample, as well as ion beam damage [38]. FIB-induced damage of TEM samples will not only modify the surface composition but also create dislocation loops and "black spot" damage that may affect the electrochemical response of the metal.

In order to facilitate the study of conventional metals and alloys, a hybrid preparation method has been developed [40] that can extract bulk metal samples while minimizing contamination from the FIB. The procedure involves using conventional twin-jet electropolishing to prepare electron-transparent thin-foil samples, which are analyzed to document suitable areas for *in situ* study. Then, the selected electron-transparent regions are extracted by cutting large site-specific areas (~100 μm by 100 μm) using the FIB. To attach these sections to liquid cell chips, either Pt "welding" or a vacuum compatible epoxy can be used. Figure 12.6 shows results from this procedure as applied to sections of cold-rolled Type 304 austenitic stainless steel where oxide crystal formation is observed and identified in water using analytical TEM. This is an exciting development in extending the application of liquid cell TEM to "real" materials.

12.4.2 Surface Preparation

Corrosion behavior can be influenced strongly by the nature of the surface of the sample. In TEM samples, surface conditions are primarily determined by the preparation technique. Metal thin films deposited using vapor deposition can be considered pristine in terms of surface contamination and sample preparation artifacts, but the morphology depends on the deposition conditions. For materials prepared using FIB, the quality of the sample surface/near-surface region and the amount of Ga^+-ions implanted are determined by details of the procedure employed. Plasma cleaning or glow discharge treatments are commonly used to remove surface hydrocarbons [41], but may damage or change certain materials.

The sample can also be altered by exposure to the ambient environment in the liquid cell before observations begin. For instance, residual moisture in the fluid tubing can lead to unwanted corrosion [30]. In conventional corrosion measurements, the test sample is sometimes left in the electrolyte for extended durations to reach steady open circuit potential, but this may not be practical for the thin samples used in TEM. If the liquid cell has electrochemical biasing capabilities, it may be possible to obtain a fresh metal surface or initiate pitting by brief application of a higher potential to the sample. The starting potential and the scan direction in electrochemical measurements can also modify the surface characteristics if they induce reactions such as oxide formation or reduction.

12.4.3 Choice of Electrolyte

As a practical consideration, compatibility between the electrolyte and liquid cell sample holder must be taken into account. This is extremely important due to the aggressive nature of many liquids used in corrosion studies. All the materials that make up the microfluidic cell and the sample holder have to be evaluated – the holder body, the microchips, and any o-rings or bonding agents used to make the vacuum seal. Solvents that can swell the o-rings or dissolve the bonding agents should be avoided or used with great care. Solutions that corrode the holder material or silicon nitride membranes at a significant rate will be detrimental to vacuum integrity when the holder is in the microscope. Even slow reactions with holder materials can lead to artifacts, as the dissolved species may change the solution chemistry and affect the phenomena under study.

Another practical concern is how the electrolyte wets the interior surfaces of the cell. Silicon nitride, the most common membrane material, is usually hydrophobic as-manufactured [41]. Water in a cell with hydrophobic surfaces is more likely to dewet during imaging in the TEM. Plasma cleaning or glow discharge treatments have been used to render the Si_xN_y surface hydrophilic [41], but the wetting characteristics of aqueous solutions can change with solute concentration and for different surfaces. Concentrated NaCl solutions have been observed to dewet rather easily on Al thin film surfaces [29].

To mitigate this issue, the cell can be pre-wetted with a liquid that wets the surface or a chemical treatment can be used to render the surface hydrophilic. The materials used in these procedures of course should not react with the sample or affect the corrosion process. Poly-L-lysine [42] and alcian blue [43] have been shown to be effective wetting pre-treatments for Si_xN_y surfaces. Alcohols, such as ethanol, are also useful for holders with flow channels because the liquid cell can be assembled dry and loaded into the TEM for preliminary inspection of the sample [30]. Ethanol can then be pumped through the fluid tubing and into the viewing area. It will fill a dry cell much more easily than water (presumably due to its lower surface tension) and act as a "wick" for the introduction of aqueous solutions. However, carbon contamination can be a potential problem with this approach.

The key consideration for corrosion studies is that the solutions should be of appropriate aggressiveness to corrode the material under study within reasonable time frames. While long waiting times are perhaps undesirable, overly aggressive solutions can corrode the sample faster than the time required to assemble the cell and load the holder into the TEM. Weaker organic acids can replace mineral acids to slow down the corrosion rate of the material of interest. Acetic acid, as an example, attacks Fe (Khalid Hattar, personal communication). For electrochemical studies, the solution also has to be sufficiently conductive to avoid ohmic potential drops during testing. Otherwise, the uncompensated ohmic resistance can introduce considerable errors in the potential between reference and working electrode during potentiostatic measurements [17]. To increase the conductivity, ions that do not take part in the corrosion reactions can be added to the electrolyte by dissolving other salts. In addition, it is necessary to decide whether to de-aerate the electrolyte by bubbling Ar or N_2 gas through it prior to flowing into the liquid cell. De-aeration alters reactions of the electron beam with the liquid (Chapter 7) but, importantly for corrosion experiments, allows more of the anodic region of a polarization curve to be measured by reducing the contribution of the oxygen reduction reaction.

12.4.4 Imaging Conditions and Electron Beam-Induced Effects

For any *in situ* TEM experiment it is important to determine the optimal conditions of accelerating voltage, beam current, illumination conditions, and imaging method [44], which together also determine the electron dose rate to the sample. Both resolution and image contrast are degraded by diffuse scattering of the electrons when liquid is present in the cell. Conventional TEM with state-of-the-art digital imaging systems offers higher frame rates for imaging dynamic reactions, whereas STEM has been shown to allow higher resolution imaging of certain types of sample through thicker liquid layers [45]. Since most materials of interest for corrosion experiments are polycrystalline, diffraction contrast in the TEM can enhance the imaging of individual grains. Other methods to improve the quality of the acquired image include careful choice of objective aperture or the use of energy-filtered imaging to reduce the impact of diffuse scattering. It is also possible to displace the liquid by forming a gas bubble deliberately with a focused electron beam, providing higher quality images of the sample [45].

The liquid geometry under which the experiment is to be performed – the liquid thickness, as set by the thickness of the spacers between the liquid cell chips and window bowing, and whether the liquid is static or flowing – also require careful deliberation. The desire to use thinner spacers and liquid layers so that fine-scale pit initiation events can be resolved must be balanced against the need to maintain a liquid layer thick enough for the chemical processes to be representative of the bulk phenomena under study [46]. Narrow spacers can also constrain flow through the cell. Continuous flow should, in principle, sweep reaction products from the electrodes, disperse radiolytic species generated by the electron beam, and replenish the solution. It is good practice to have the counter electrode downstream of the working electrode to reduce any impact of species produced at the counter electrode on the experiment [47].

A careful consideration of electron beam-induced effects is required for correct interpretation of *in situ* results. The cascade of species, such as hydrogen radicals, oxygen radicals, and solvated electrons [48], described in Chapter 7, can clearly affect corrosion processes. The beam can also enhance the rate of surface contamination [41]. In most cases, beam heating is not significant for typical imaging conditions [49]. Low dose protocols for imaging biological materials are currently being explored in the liquid cell community, but even this level of illumination will still generate a perceptible concentration of radiolytic species that may modify corrosion behavior in the irradiated area. Besides, imaging in aqueous solutions may also create new radiolytic species. For example, radiolysis of NaCl solutions by ionizing radiation can form chlorite and hypochlorite [50] as byproducts; the latter turns alkaline solutions into strongly oxidizing media. The corrosion products themselves, for example metal ions released into solution, can in principle also take part in radiolytic reaction sequences. Given the sensitivity of corrosion processes to the electrolyte chemistry and potential, behavior observed *in situ* should be correlated with *ex situ* experiments to identify possible artifacts. Non-irradiated areas of the sample should be compared with areas that received the highest irradiation doses (bearing in mind the long diffusion distances of some of the radiolytic species). These comparisons will ensure that liquid cell studies provide the best model for the corrosion that takes place under real-world conditions.

The electron beam can have additional effects in experiments involving electrochemical measurements. It has been shown [51] that the incident electron beam (under conventional bright field TEM imaging conditions) can induce a significant potential shift from the free corrosion potential. We have measured a reproducible shift of approximately +500 mV when a Type 304 austenitic stainless steel sample in water was imaged in the TEM, with rapid return to the open circuit potential when the beam was turned off. This shows the importance of comparisons with and without illumination in interpretation of electrochemical data.

12.5 Microfluidic Cell Design for Electrochemical Corrosion Experiments

The challenge with integrating electrochemical techniques into liquid cell experiments is how to translate the bench-top experiments used in corrosion science to the microfluidic platform available for liquid cell TEM. Currently, electrochemical liquid cell design is somewhat restricted. The existing electrochemical cells provided by commercial manufacturers of liquid cell holders are not designed with corrosion experiments in mind.

For conventional electrochemical cells, guidelines for the placement of electrodes include having a counter electrode with a surface area larger than the working electrode and placing the reference electrode close to the working electrode to minimize ohmic potential drops. Microfluidic electrochemical cells attempt to fulfill these requirements, as described in Chapter 10. However, it is debatable whether designs used in electrochemical liquid cell experiments (such as [52]) are optimal for corrosion research. The main concern here is that the working electrode for corrosion studies is the material of interest itself, not the inert electrodes used in most other electrochemical studies. Furthermore, having an active metal, such as Al, on top of the commonly used electrode materials, Au and Pt, can lead to galvanic corrosion if the interface is exposed to liquid [53]. Figure 12.7 shows this effect using Al thin film squares immersed in a beaker of

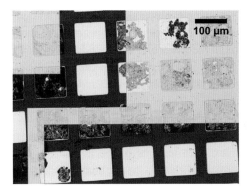

Figure 12.7. Selective corrosion of Al thin film squares in touch with pre-fabricated Au electrodes after exposure to 0.1 M NaCl solution for 4 hours. Unpublished data.

0.01 M NaCl solution for 4 hours. It is clear that squares touching the Au electrode have corroded, whereas isolated squares on bare Si_xN_y remain largely intact.

To mitigate galvanic corrosion, the sample needs to be selectively masked. It is also highly desirable to have only the areas around the window exposed to the electrolyte so that the physical phenomena observed *in situ* can be correlated more directly with the electrochemical data recorded from the liquid cell during the experiment. In conventional benchtop experiments, the sample may be physically masked to define the area exposed to electrolyte, and junctions of connecting wires and other portions of the sample can be covered with epoxy or lacquer for isolation. This is tricky to accomplish in existing microfluidic cells because adding layers of insulating material to mask areas on the pre-fabricated chips can increase the eventual window separation and is perilous for the fragile Si_xN_y membranes. The best solution to overcome these issues is to build the corrosion cell from the bottom up, but this will incur significant costs and time investment.

Finally, the choice of materials for the reference and counter electrodes can also be further explored. The application of potential can induce the dissolution of a gold thin film counter electrode [54] so the materials must be chosen for stability during the measurements. The development of a reference electrode that is analogous to the widely used Ag/AgCl or saturated calomel electrode in corrosion studies will be essential for obtaining the highest quality electrochemical data from liquid cell experiments.

12.6 Outlook

The study of corrosion reactions using liquid cell electron microscopy is an emerging field and there are numerous opportunities for groundbreaking and important work.

So far, promising results have been obtained from metal films and foils that are exposed to liquids under controlled electrochemical conditions during observation *in situ*. To capture and study corrosion events in real time, future developments are likely to be in the direction of improved liquid cell designs that are optimized for corrosion, in sample preparation methods, and in understanding the chemistry of electrolytes under practical electron beam conditions.

Corrosion in the real world is also induced by combinations of environmental inputs. It will be fascinating if experiments can be replicated in the liquid cell with multiple *in situ* stimuli, such as temperature and stress, in addition to solution chemistry and electrical potential. The ability to perform corrosion experiments at elevated temperatures can be invaluable because most passive metals have a critical pitting temperature below which pitting will not occur. On the other hand, there are also possibilities for more complex environments in the liquid cell, especially in the area of microbial-induced corrosion where a better understanding of biofilms and their influence on corrosion is needed [55].

Finally, advances in our ability to probe the chemistry of a solid–liquid interface using analytical techniques in the liquid cell can also benefit corrosion science. Remembering that corrosion is the sum of coupled oxidation and reduction reactions, worthy avenues of investigation include measuring the local solution composition in a growing pit and understanding how the metal surface, bare metal, or oxide film catalyzes the reduction reactions.

Acknowledgements

The authors gratefully acknowledge Professor David Duquette (Rensselaer Polytechnic Institute) for assistance with the manuscript, and Dr. Khalid Hatter (Sandia National Laboratory) for helpful discussions. SWC also acknowledges support from Professors Robert Hull and David Duquette and from the National Science Foundation (#DMR-1309509), MGB acknowledges support from the UK Engineering and Physical Sciences Research Council (EP/I003290/1).

References and Notes

1. B. Shaw and R. Kelly, What is corrosion? *Interface, Electrochem. Soc.*, Spring (2006), 24–26.
2. D. Duquette and R. Schafrik, *Research Opportunities in Corrosion Science and Engineering* (Washington, D.C.: The National Academies Press, 2011).
3. G.-L. Song, The grand challenges in electrochemical corrosion research. *Front. Mater,* **1** (2014), 2.
4. G. Frankel, Electrochemical techniques in corrosion: status, limitations, and needs. *J. ASTM Int.*, **5** (2008), 1–27.
5. N. De Jonge and F. M. Ross, Electron microscopy of specimens in liquid. *Nat. Nanotechnol.*, **6** (2011), 695–704.
6. K. L. Jungjohann, J. E. Evans, J. A. Aguiar, I. Arslan and N. D. Browning, Atomic-scale imaging and spectroscopy for in situ liquid scanning transmission electron microscopy. *Microsc. Microanal.*, **18** (2012), 621–627.

7. M. E. Holtz, Y. Yu, J. Gao, H. D. Abruna and D. A. Muller, In situ electron energy loss spectroscopy in liquids. *Microsc. Microanal.*, **19** (2013), 1027–1035.
8. R. R. Unocic, L. Baggetto, K. Unocic et al., Coupling EELS/EFTEM imaging with environmental fluid cell microscopy. *Microsc. Microanal.*, **18** (2012), 1104–1105.
9. N. J. Zaluzec, M. G. Burke, S. J. Haigh and M. A. Kulzick, X-ray energy-dispersive spectrometry during in situ liquid cell studies using an analytical electron microscope. *Microsc. Microanal.*, **20** (2014), 323–329.
10. S. Schilling, A. Janssen, Z. L. Zhong, N. J. Zaluzec and M. J. Burke, Liquid in situ analytical electron microscopy: examining SCC precursor events for Type 304 stainless steel in H2O. *Microsc. Microanal.*, **21** (2015), 1291–1292.
11. E. McCafferty, *Introduction to Corrosion Science* (New York: Springer, 2010).
12. G. Frankel and N. Sridhar, Understanding localized corrosion. *Mater. Today*, **11** (2008), 38–44.
13. G. Frankel, Pitting corrosion of metals. *J. Electrochem. Soc.*, **145** (1998), 2186–2198.
14. J. Soltis, Passivity breakdown, pit initiation and propagation of pits in metallic materials: review. *Corros. Sci.*, **90** (2015), 5–22.
15. Accelerated dissolution of thin metal films had been observed during continuous imaging under the electron beam.
16. R. R. Unocic, R. L. Sacci, G. M. Brown et al., Quantitative electrochemical measurements using in situ ec-S/TEM devices. *Microsc. Microanal.*, **20** (2014), 452–461.
17. G. Frankel and M. Rohwerder, Electrochemical techniques for corrosion. In *Encyclopedia of Electrochemistry* (Weinheim, Germany: Wiley-VCH, 2007).
18. R. G. Kelly, J. R. Scully, D. Shoesmith and R. Buchheit, *Electrochemical Techniques in Corrosion Science and Engineering* (New York: Marcel Dekker, 2013).
19. G. Frankel, *Techniques for Corrosion Quantification in the Characterization of Materials*, 2nd edn. (Hoboken, NJ: John Wiley & Sons, 2012), pp. 850–864.
20. M. Keddam, Application of advanced electrochemical techniques and concepts to corrosion phenomena. *Corrosion*, **62** (2006), 1056–1066.
21. G. Frankel, The growth of 2-D pits in thin film aluminum. *Corros. Sci.*, **30** (1990), 1203.
22. L. Balazs and J. Gouyet, Two-dimensional pitting corrosion of aluminium thin layers. *Phys. A Stat. Mech. Appl.*, **217** (1995), 319–338.
23. G. Frankel, Pit growth in thin metallic films. *Mater. Sci. Forum*, **247** (1997), 1–8.
24. J. Proost, M. Baklanov and R. Verbeeck, Morphology of corrosion pits in aluminum thin film metallizations. *J. Solid State Electrochem.*, **2** (1998), 150–155.
25. S. Hernandez, A. Griffin Jr., F. Brotzen and C. Dunn, The effect of thickness on the corrosion susceptibility of Al thin film metallizations. *J. Electrochem. Soc.*, **142** (1995), 1215–1220.
26. Y.-P. Zhao, C.-F. Cheng, G.-C. Wang and T.-M. Lu, Characterization of pitting corrosion in aluminum films by light scattering. *Appl. Phys. Lett.*, **73** (1998), 2432–2434.
27. S. W. Chee, F. M. Ross, D. Duquette and R. Hull, Studies of corrosion of Al thin films using liquid cell transmission electron microscopy. *MRS Proc.*, **1525** (2013), mrsf12-1525-ss11-03.
28. S. W. Chee, D. J. Duquette, F. M. Ross and R. Hull, Metastable structures in Al thin films before the onset of corrosion pitting as observed using liquid cell transmission electron microscopy. *Microsc. Microanal.*, **20** (2014), 462–468.
29. S. W. Chee, S. H. Pratt, K. Hattar et al., Studying localized corrosion using liquid cell transmission electron microscopy. *Chem. Commun.*, **51** (2015), 168–171.
30. S. W. Chee, R. Hull and F. M. Ross, Liquid cell TEM of the corrosion of metal films in aqueous solutions. *Microsc. Microanal.*, **18** (2012), 1110–1111.

31. H.-G. Liao, K. Niu and H. Zheng, Observation of growth of metal nanoparticles. *Chem. Commun.*, **49** (2013), 11720–11727.
32. Y. Jiang, G. Zhu, F. Lin, H. Zhang and C. Jin, In situ study of oxidative etching of palladium nanocrystals by liquid cell electron microscopy. *Nano Lett.*, **14** (2014), 3761–3765.
33. J. Wu, W. Gao, H. Yang and J.-M. Zuo, Imaging shape-dependent corrosion behavior of Pt nanoparticles over extended time using a liquid flow cell and TEM. *Microsc. Microanal.*, **20** (2014), 1508–1509.
34. E. Sutter, K. Jungjohann, S. Bliznakov *et al.*, In situ liquid-cell electron microscopy of silver-palladium galvanic replacement reactions on silver nanoparticles. *Nat. Commun.*, **5** (2014), 4946.
35. S. W. Chee, J.-H. Park, A. Pinkowitz *et al.*, Liquid cell TEM of Al thin film corrosion under potentiostatic polarization. *Microsc. Microanal.*, **21** (2015), 973–974.
36. J. H. Park, S. W. Chee, S. Kodambaka and F. M. Ross, In situ LC-TEM studies of corrosion of metal thin films in aqueous solutions. *Microsc. Microanal.*, **21** (2015), 1291–1292.
37. K. W Noh, K. Tai, S. Mao and S. J. Dillon, Grain boundary parting limit during dealloying. *Adv. Eng. Mater.*, **17** (2015), 157–161.
38. J. Mayer, L. A. Giannuzzi, T. Kamino and J. Michael, TEM sample preparation and FIB-induced damage. *MRS Bull.*, **32** (2007), 400–407.
39. R. Unocic, L. Adamczyk, N. Dudney *et al.*, In-situ TEM characterization of electrochemical processes in energy storage systems. *Microsc. Microanal.*, **17** (2011), 1564–1565.
40. X. Zhong, M. G. Burke, S. Schilling, S. J. Haigh and N. J. Zaluzec, Novel hybrid sample preparation method for in situ liquid cell TEM analysis. *Microsc. Microanal.*, **20** (2014), 1514–1515.
41. T. J. Woehl, K. L. Jungjohann, J. E. Evans *et al.*, Experimental procedures to mitigate electron beam induced artifacts during in situ fluid imaging of nanomaterials. *Ultramicroscopy*, **127** (2013), 53–63.
42. E. A. Ring and N. de Jonge, Microfluidic system for transmission electron microscopy. *Microsc. Microanal.*, **16** (2010), 622–629.
43. S. M. Hoppe, D. Y. Sasaki, A. N. Kinghorn and K. Hattar, In-situ transmission electron microscopy of liposomes in an aqueous environment. *Langmuir*, **29** (2013), 9958–9961.
44. P. Abellan, T. J. Woehl, L. R. Parent *et al.*, Factors influencing quantitative liquid (scanning) transmission electron microscopy. *Chem. Commun.*, **50** (2014), 4873–4880.
45. K. L. Klein, I. M. Anderson and N. de Jonge, Transmission electron microscopy with a liquid flow cell. *J. Microsc.*, **242** (2011), 117–123.
46. The liquid layer thicknesses quoted in atmospheric corrosion studies are normally in the tens of micrometers.
47. R. L. Sacci, N. J. Dudney, K. L. More and R. R. Unocic, In operando transmission electron microscopy imaging of SEI formation and structure in Li-ion and Li-metal batteries. *Microsc. Microanal.*, **20** (2014), 1598–1599.
48. N. M. Schneider, M. M. Norton, B. J. Mendel *et al.*, Electron–water interactions and implications for liquid cell electron microscopy. *J. Phys. Chem. C.*, **118** (2014), 22373–22382.
49. J. M. Grogan, N. M. Schneider, F. M. Ross and H. H. Bau, Bubble and pattern formation in liquid induced by an electron beam. *Nano Lett.*, **14** (2013), 359–364.
50. M. Kelm, E. Bohnert and I. Pashalidis, Products formed from alpha radiolysis of chloride brines. *Res. Chem. Intermed.*, **27** (2001), 503–507.

51. M. E. Holtz, Y. Yu, D. Gunceler *et al.*, Nanoscale imaging of lithium ion distribution during in situ operation of battery electrode and electrolyte. *Nano Lett.*, **14** (2014), 1453–1459.
52. S. Schilling, A. Janssen, M. G. Burke *et al.*, In situ analytical electron microscopy: imaging and analysis of steel in liquid water. 18th International Microscopy Congress (2014), www.microscopy.cz/proceedings/all.html#abstract-2947.
53. Bi-metallic exposure in the electrolyte frequently leads to galvanic corrosion but the effects of coupled metals are not so straightforward. Depending on the metals that are connected, it is possible that the more active metal becomes more resistant to corrosion. The reader is referred to general texts on corrosion for clarification.
54. J.-H. Park, M. C. Reuter, S. Kodambaka and F. M. Ross, Electric field induced Au nanocrystal formation in aqueous solutions. *Microsc. Microanal.*, **20** (2014), 1598–1599.
55. S. M. Hoppe, B. A. Hernandez-Sanchez, K. Hattar and D. Y. Sasaki, Progress towards in situ TEM of biofouling. *Microsc. Microanal.*, **20** (2012), 1132–1133.

13 Nanoscale Water Imaged by *In Situ* TEM

Utkur Mirsaidov and Paul Matsudaira

13.1 Introduction

Our understanding of fluid dynamics stops just above the nanometer scale. We lack a theory that properly accounts for the nanoscale fluid-mediated processes that are ubiquitous in physical, chemical, and biological systems. Unlike microscopic fluid behavior, nanoscopic fluid dynamics cannot be described by a simple, universal continuum theory [1]. At nanometer scales the free energies and configurational entropy between solvent molecules, between solvent and solute molecules, and between solvent molecules and solid interfaces become important and compete to dominate. Phenomenological descriptions of nanoscale fluid mechanics rely on experimental observations instead of being derived from first principles.

A prime example is the double layer theory, which describes how a first layer of charged fluid molecules that are immobilized on solid surfaces are then electrically screened from the bulk fluid by a second intervening diffuse layer of counter-ions [2]. The double layer theory, which explains the stability of colloidal suspensions as well as charge transport on electrodes, contains system-specific empirical constants that can only be measured experimentally [3]. Further, the double layer delineates a nanoscale boundary layer where bulk fluid transitions into a glassy, surface-immobilized fluid. Although the double layer is an accepted concept, our understanding of boundary layer mechanics has been limited by the conventional experimental methods that are available for studying nanoscale fluid behavior.

Several methods have been developed for visualizing the boundary layer. Visible light-based microscopy can measure dynamic behavior but only to a resolution of about a hundred nanometers [4], while force probe microscopy resolves surface topography of liquids with poor temporal resolution [5] or the response of fluids to external force [6]. A promising alternative is to use transmission electron microscopy based techniques to directly visualize fluid boundary dynamics at sub-nanometer resolution with chemical characterization. Electron microscopy has been applied successfully to study phenomena at the solid–liquid interface, including solidification or ordering in liquids at crystalline interfaces [7–12], with good spatial and temporal resolution. However, these experiments have involved low vapor pressure liquids such as metal–semiconductor eutectics. It is only with the advent of liquid cell TEM techniques that water has been accessible for study in the TEM.

Here, we present the capabilities of closed liquid cell TEM as an experimental imaging platform to study the structure and dynamics of volatile nanoscale fluids. In this chapter, we discuss how nanoscale water moves on a solid substrate, how nanoscale objects move inside the nanoscale interfacial water, and how nanoscale voids form in thin liquid films.

13.2 Interfacial Fluids

The flow of the first few nanometers of water at the interface is encountered in a wide range of physical, chemical, and biological phenomena. While much is known about the movement of bulk water from experiments, most of what is known about interfacial water results from modeling and computational simulations. For example, simulations that extend to nanosecond time scales, probing droplets consisting of a handful of water molecules, show that water can slip at the water–solid interface on hydrophilic surfaces depending on the surface composition [13]. On the other hand, state-of-the-art AFM measurements show that dynamic properties of water such as viscosity can drastically change under nanoscale confinement [6, 14]. Both modeling and experiments provide valuable information about the possible differences between bulk and interfacial water. However, the direct effects of an interface on water flow, or motion of nanoscale objects in nanoscale water near liquid–solid interfaces, are beyond their reach. Here, we attempt to illustrate how these things can potentially be studied with TEM.

13.2.1 Stick-Slip Flow of Nanoscale Droplets

One of the intriguing questions is how water flows near interfaces. To address this, we turn to electron microscopy to visualize the flow of nanoscale interfacial liquids on surfaces. Pure water is pipetted into a microfabricated liquid cell platform for *in situ* TEM imaging. The liquid cell is made up of two ~20 nm thick Si_xN_y hydrophilic membrane windows separated and sealed by a ~200 nm spacer. When irradiated with 120 keV electrons, the liquid water between membranes charges up and retracts, initially leaving an ultra-thin patch of water film (~5–20 nm) on both top and bottom hydrophilic membranes. Its thickness can be crudely approximated from electron scattering by comparing the electron intensity at wet and dry regions [15–17]. This thin water film subsequently dewets, leaving cap-like nanodroplets with diameters in the range of 10–80 nm on the membrane, as shown in Figure 13.1a [17].

Figure 13.1b, c shows that these nanodroplets move in discrete steps on the hydrophilic surface in response to the electron beam; they do not flow continuously on the surface. From the image contrast, it appears that water starts accumulating at the rim to form a toroidal structure of non-uniform radial thickness (Figure 13.1d). Next, the steepest side of the nanodroplet (with greater contact angle) forms a thin wetting precursor film as shown in Figure 13.1d, e. The whole mass of water then shifts from the droplet into this thin precursor film, flattening the torus shape. As this happens the rear of the nanodroplet recedes and follows the rest of the water mass in the

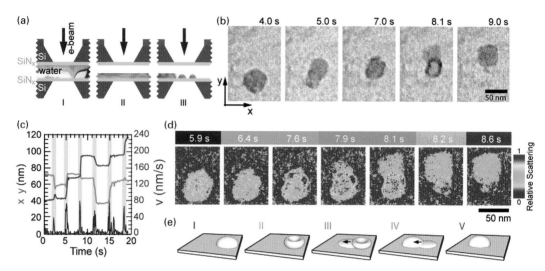

Figure 13.1. Stick-slip translocation of water nanodroplets. (a) Diagram of the sample geometry. (b) TEM image series of a moving water nanodroplet on a Si_xN_y membrane surface. The stationary droplet stretches towards the moving direction, shifts its mass to a new position, and comes to rest. (c) x (green curve), y (red curve) position and instantaneous velocity (black curve) of the droplet's center-of-mass position illustrates distinct steps taken during the translocation. The peak velocity is 20–80 nm/s. (d) TEM image of the same droplet where the color corresponds to relative thickness. The water flows and redistributes inside the droplet resulting in a shift of liquid to the rim, which is then followed by formation of a thin wetting film from the rim. The rest of the droplet water moves to cover this thin wetting film, which slowly advances forward. (e) Schematic illustration of the beam-induced translocation process. Reproduced from Ref. 17.

nanodroplet, causing it to regain its cap-like shape. This process is cyclic over six steps. The movement of nanodroplets is driven by electron beam-induced charging of the droplets, but the beam is uniformly spread over the imaged area and does not guide droplet motion in any particular direction.

This type of translocation is characteristic of a stick-slip movement [18, 19]. The nanodroplet slips when the redistribution of water overcomes interfacial forces between the nanodroplet and the substrate. Redistribution is caused by charging of the droplet that pushes water outward forming the toroid structure. Thus, charging provides a driving force much as thermal [20], chemical [21], and topographic gradients [22] do, and can drive nanodroplet movement. The non-uniformity around the rim is probably set by variations in local charge distribution of the membrane surface under the droplet, while the contact angles at which both the advancing (θ_{adv}) and receding (θ_{rec}) three-phase (liquid–gas–solid) lines slip are set by molecular interactions, surface roughness, and chemical heterogeneities that pin this three-phase contact line [23, 24]. The slippage of nanodroplets is common on hydrophobic surfaces [25, 26], but recent simulations show that such slippage can occur on hydrophilic surfaces [13]. These TEM observations clearly reveal the mechanism of such translocation, with de-pinning of the advancing contact line assisted by formation of the thin wetting film. The formation of precursor wetting films has been observed in the spreading of macroscopic droplets on surfaces, but their role in de-pinning of the contact line and in the origin of contact angle hysteresis has not been recognized previously [27].

13.2.2 Diffusion of Particles in Nanoscale Droplets

From the stick-slip dynamics exhibited by nanodroplets, we infer that interfacial interactions dominate the fluidity of water near solid surfaces. In this section we further explore how the fluidic characteristics of these interfacial droplets, as well as ultra-thin films of water, differ from bulk water.

Fluidity of a liquid is described by its viscosity, a measure of the fluid's resistance to flow. In bulk liquids, the diffusive movement of particles is generally governed by the viscosity, as it provides a resistive force for random diffusive movement driven by thermal energy. For a spherical nanoparticle of radius $R = 2$ nm in water, the classical Stokes–Einstein equation gives the bulk diffusion coefficient expected at room temperature [28]:

$$D_t = k_B T/(6\pi\eta R) \approx 1 \times 10^{-10} \text{ m}^2\text{s}^{-1}. \tag{13.1}$$

Here, k_B is Boltzmann's constant, $T = 300$ K, and the viscosity of water $\eta = 9 \times 10^{-4}$ Pa s. What this means in terms of imaging dynamics in TEM is that in the time interval of 10 ms, a particle with 2 nm radius will diffuse a distance of 1 μm. This would suggest that unless one acquires images at rates higher than gigahertz, nanoparticle images in water will be severely blurred. However, as has already been described in Chapter 9, nanoparticles can be imaged in water layers by TEM with much slower acquisition rates. Unexpectedly slow diffusion behavior has been observed in thin liquid films in several experiments [29–33]. An example is given in Figure 13.2. The particles appear to move much more slowly than expected; diffusion is substantially suppressed.

The same highly damped diffusive movement of nanoparticles occurs in interfacial nanodroplets. Figure 13.3 shows an analysis of the motion of Au nanoparticles, whose diameters are less than 10 nm, inside a nanodroplet [31]. The diffusion coefficients D_{MSD} of individual particles inside this droplet can be estimated from the slopes of their mean squared displacement (MSD) plots as a function of time t (Figure 13.3e): $D_{MSD} = \text{MSD}/(4t)$. The trajectories of these nanoparticles are random and erratic as seen from their uncorrelated instantaneous speeds. The velocity distributions fit Gaussian profiles (Figure 13.3d) centered about zero, as expected of diffusion in the absence of persistent drift currents. The computed time-dependent MSD of individual nanoparticles (Figure 13.3e) indicate an average diffusion coefficient of $D_{MSD} \approx 5 \times 10^{-20}$ m^2 s^{-1}. These diffusion coefficients are 10^9 times less than what we would expect from Equation (13.1).

The fact that the diffusion is reduced both in thin unbounded liquid films and in nanodroplets suggests that in-plane particle confinement by the lateral boundary of the nanodroplet cannot explain the deviation from bulk diffusion D. The suppression of the diffusion coefficient must reflect some change in the properties of water near solid

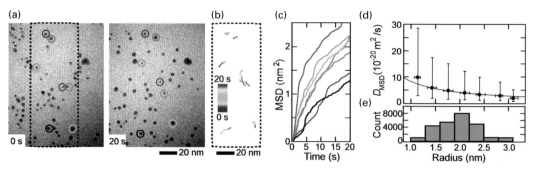

Figure 13.2. Gold nanoparticles diffusing in an ultra-thin water film. (a) TEM snapshots of Au nanoparticles enclosed in water film. (b) Path traces for seven selected particles shown in (a). (c) Mean square displacement of nanoparticles versus time. Curve colors correspond to circle colors enclosing nanoparticles as shown in (a) at $t = 0$ s. (d) Diffusion coefficients computed from individual particles' mean-square-displacement are associated with their instantaneous particle radii, $D_{MSD}(R)$. These are averaged (black discs with blue error bars) by radius categories regardless of particle label and observation time. The full set of $D_{MSD}(R)$ (not just the averaged values) are fitted (red line) to the Einstein–Stokes equation where $D_{MSD} \sim R^{-1}$. (e) Histogram of the number of observation counts in each radius category in (d). Reproduced from Ref. 31.

interfaces. The strongly damped diffusion most likely arises from interactions between particle and substrate at the liquid–solid interface. The ultra-thin layers that show damped particle motion are typically 10–20 nm thick. Since this is comparable to the nanoparticle radii, it could dominate the viscous drag of particles close to the membrane. However, another explanation is that within the ultra-thin film, the few monolayers of water molecules closest to the membrane have a viscosity that is several orders of magnitude higher than in bulk water [14, 34, 35]. Direct contact between particles and the solid substrate is unlikely to be relevant since nanoparticles in the absence of liquid do not show noticeable diffusion. The recent studies of nanoparticle dynamics in the graphene liquid cell (Chapter 19), i.e. in a thin liquid film sandwiched between two graphene layers, reveal similar damped motion. This suggests that the nature of the membrane material is irrelevant [36, 37] and supports the idea of drastic changes in fluid properties at a liquid–solid interface, which was recently proposed by Verch *et al.* [33]. In conclusion, while strongly damped diffusion in interfacial liquids is the crucial factor that enables high resolution TEM imaging of particles, its origin is still being explored.

13.3 Nanodroplet Condensation

In situ TEM has the potential to resolve the initial stages of nanodroplet nucleation and growth where vapor condenses into a liquid, a process that is very familiar to many of us macroscopically. The growth dynamics of water nanodroplets visualized by environmental scanning electron microscopy [38–42] already tell us that chemical and topographical surface heterogeneities have an impact on droplet condensation [40, 43]. However, in order to visualize the nucleation and initial phases of nanodroplet growth

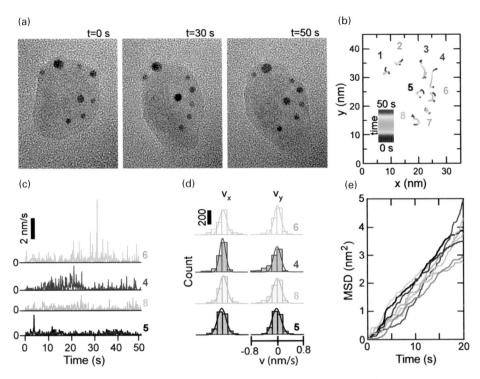

Figure 13.3. Gold nanoparticles diffusing in a nanodroplet. (a) Time series images of eight Au nanoparticles enclosed in a nanodroplet imaged with TEM. (b) The path traces of eight nanoparticles during 50 s. (c) Instantaneous velocities for nanoparticles 6 (bright blue), 4 (red), 8 (bright green), and 5 (black) as labeled in (b). The uncorrelated instantaneous speeds are $v = \sqrt{v_x^2 + v_y^2}$, with $v_x = \Delta x/\Delta t$ and $v_y = \Delta y/\Delta t$ the x and y velocity components of the particles' center of mass respectively, as determined from consecutive images separated by time interval $\Delta t = 62.5$ ms. (d) Instantaneous velocity distribution in x (left column) and y (right column) directions for the same four particles shown in (c) (solid curves are Gaussian fits). (e) Mean square displacement of eight particles as a function of time. The curve colors correspond to label colors in (b). Reproduced from Ref. 31.

we need to push the resolution beyond 50 nm and therefore we will benefit from TEM imaging.

TEM captures the nucleation of water nanodroplets at 10 nm Au nanoparticles and the subsequent growth dynamics (Figure 13.4a) [44]. A hydrophilic Au nanoparticle serves as a nucleation site for a droplet to form from water vapor enclosed in a liquid cell. A detectable amount of water is formed on the nanoparticle in about 7 s and then grows to ~40 nm in diameter in the next 9 s. The growth dynamics of the droplet area fits a power law $A(t) \sim R^2(t) \sim (t - t_0)^{2\beta}$, as shown in Figure 13.4b, where $\beta \sim 0.22$. The water in the nanodroplet most likely originates from water clusters on the membrane surface. Water clusters that are mobile on the surface form by condensation of vapor from the ambient. Once these diffusing surface clusters reach a suitable site, they attach (Figure 13.4b schematic). This process occurs in the presence of the electron

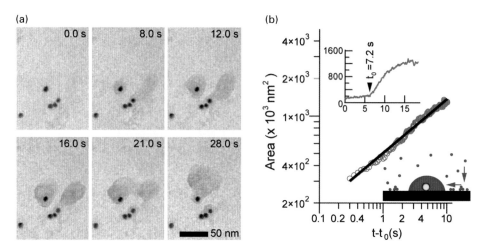

Figure 13.4. Nanodroplet condensation. (a) In the span of 16 s a water nanodroplet nucleates on Au nanoparticles and grows. Next, it undergoes coalescence with a nearby nanodroplet. (b) A power law, $A(t) \sim (t - t_0)^{2\beta}$ with $\beta \approx 0.22$ fits the time evolution of the nanodroplet area, where $t_0 = 7.2$ s as shown in the upper inset. The inset schematic shows the proposed mechanism of condensation whereby water molecules in a liquid cell condense on the surface and diffuse towards the nucleation site. Attachment to the growing nanodroplet results in a depleted region. Reproduced from Ref. 44.

beam. In these experiments charging of the surface by the beam probably assists polar water molecules to adhere to the substrate surface, allowing for nucleation and growth.

Depending on the exact model chosen for diffusion and attachment of water molecules, one will obtain different exponents to a power law describing the nanodroplet growth. For example, the assumption of attachment rate proportional to perimeter gives a power law $R(t) \sim (t - t_0)^{\beta}$ with $\beta = 1/2$ [45]. A constant growth rate leads to a different power law, $\beta = 1/3$ [46]. A time-dependent concentration of diffusing monomers can give a power law with $\beta = 1/4$ [47]. The widely used model with constant growth rate [46] is known to fit a good range of macroscopic observations [48]. The observations in Figure 13.4, with $\beta < 1/3$, suggest that the concentration of water monomers decreases as the droplet enlarges, perhaps due to formation of a depletion region as the droplet absorbs nearby clusters. TEM also shows that the growing droplet is not exactly centered on the nanoparticle. This could imply an asymmetric flux of the water, possibly due to nanoscale inhomogeneity of the surface. The nanoscale view of the growth dynamics provided by liquid cell TEM can therefore play a role in developing models that take into account heterogeneities in the surfaces across which water clusters diffuse, and in the local concentration of the water clusters that form a nanodroplet.

13.4 Fluids in Nanochannels

So far we have described how liquid cell platforms that sandwich liquid between two flat thin membranes enable us to visualize nanoscale dynamics in fluids. However,

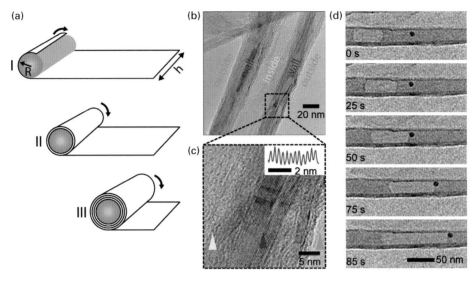

Figure 13.5. Movement of a nanoparticle in a graphene nanochannel. (a) Graphene scrolls around nanometer-size water in order to lower the total surface energy of the graphene+water complex, then subsequent rolling of the graphene to form a multiwall nanoscroll is driven to lower the graphene's surface energy. (b) A nanochannel with a 32 nm inner diameter. (c) Higher magnification image showing the graphene layers in the scroll. Inset is an intensity profile across the red line illustrating the 0.346 nm lattice spacing. Yellow, red, and green arrowheads indicate the inside, scroll wall, and outside of the channel. (d) Time series of TEM images show a bubble and accompanying water column flowing from left to right. At ~50 s, the leading edge of the bubble encounters a nanoparticle and transports it at the liquid–gas interface along the graphene nanochannel. Reproduced from Ref. 54.

the geometry of these liquid cells is such that the electron beam is perpendicular to the membrane windows. This limits us to top-down views; we only see the liquid–gas periphery of the nanodroplets and we cannot see along the liquid–solid interface. In order to overcome this limitation, an alternative technique is needed to image interfacial processes in-plane. The best approach is to use thin tubular structures from low atomic number materials so we can see processes occurring near the tube walls with high contrast and resolution. Carbon nanotubes (CNTs) meet this requirement and can also retain liquids inside the TEM vacuum. CNTs have been shown to be extremely useful in probing properties of nanoscale liquids [49–52]. For example, by examining the contact angle between a CNT wall and a confined nanoscale liquid it has been shown that the inner wall of a CNT is hydrophilic, which is quite surprising for carbon surfaces [52]. However, these nanotubes are difficult to load with the liquid. An easier approach is to form a nanochannel around the nanoscale liquid.

Water can be encapsulated in tubular nanochannel structures formed by scrolling from planar graphene layers [53]. Figure 13.5a shows that as water dries on a sheet of graphene, the interplay between water capillarity and graphene elasticity induces the planar graphene sheets to roll into tubes [53]. The walls of the nanochannels consist of multiple layers with a spacing of 3.46 Å between the layers (Figure 13.5c). This spacing is consistent with the interplanar spacing of graphene layers in graphite and confirms that the tube is formed from a graphene sheet rolling onto itself several times into a scroll [54]. These nanochannels encapsulate water and nanoscale objects within the water as shown in Figure 13.5d. The tubular graphene system is extremely stable: we have trapped water for up to three months.

The nanochannel configuration allows us to observe interactions between a liquid–gas interface and a 10 nm nanoparticle (Figure 13.5d). The movement of a bubble trapped in a channel is initiated by the electron beam. The nanoparticle remains stationary as displaced liquid flows around it at $v \sim 5$ nm/s, but is dislodged when it comes in contact with the liquid–gas interface. The drag force on the moving nanoparticle in liquid is $F = 6\pi\eta Rv \approx 4 \times 10^{-19}$ N, while the surface tension force $F \sim \gamma_w R = 4 \times 10^{-10}$ N acts at the liquid–gas interface. Here, $\gamma_w = 72$ mJ/m^2 is the surface tension, and the viscosity of water is as given in Section 13.2.2. The nine orders of magnitude larger surface tension force explains why the particle moves when it contacts the gas–liquid interface but not prior to contact. To achieve a Stokes force on a particle comparable to the surface tension force requires flow faster than 1 m/s! The strong forces exerted by three-phase contact lines are clearly effective for dislodging and removing particles, but also have potential to aid in nanoscale assembly. For example, observations in a Si$_x$N$_y$ liquid cell showed that nanoparticles assembled [37, 55] in response to the force exerted by a moving contact line. The capillary force at the liquid–solid interface can also re-orient anisotropic nanoparticles [55].

The nanoscroll liquid cell is ideal for imaging the initial stages of nanodroplet condensation, as discussed in Section 13.3, in side view (Figure 13.6). As with the data shown in Figure 13.4, again the presence of the electron beam may affect the growth

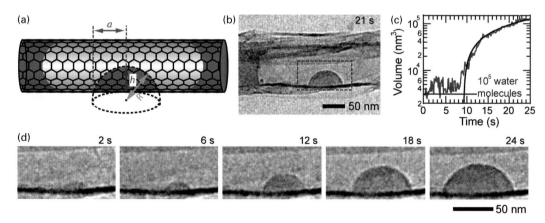

Figure 13.6. Nanodroplet condensation inside a graphene channel. (a) Schematics of graphene nanochannel scroll around the liquid with a hemispherical cap-shaped nanodroplet forming on the inner wall. (b) TEM image of a nanodroplet on a channel wall. (c) Volume of the droplet in (b) calculated assuming a hemispherical cap (a) (red curve). The fit line (black curve) for $V(t) \sim (t - 9.2 \text{ s})^{3\beta}$ gives an exponent of $\beta = 0.25$. The blue line corresponds to 100 000 water molecules. (d) Time-lapse images of the condensation of a water nanodroplet on the nanochannel wall. Reproduced from Ref. 44.

dynamics. At the earliest time at which the droplet is clearly visible, it is 10 nm in diameter and contains an estimated 10 000 water molecules. The droplet initially shows a three-phase contact angle of ~40° but this angle approaches 90° as the droplet grows. If we approximate the droplet as a spherical cap with volume $V = \pi h(3a^2 + h^2)/6$ (Figure 13.6a, b), we can fit a power law $V(t) \sim (t - 9.2 \text{ s})^{3\beta}$ (Figure 13.6c) that yields an exponent of $\beta = 0.25$, similar to the observation in Section 13.3 of growth on Si_xN_y membranes. This experiment shows the potential for the ultra-thin carbon nanochannel liquid cell to quantify the dynamics of interface-induced molecular effects in liquids with good image contrast and with cross-sectional imaging geometry.

13.5 Voids and Nanobubbles in Liquid Films

The dynamics and stability of thin liquid films have significant implications for lubrication and coating technologies and have been studied intensively for the past 20 years [56]. The stability of thin liquid films can be compromised by the formation of gas bubbles or voids. Liquid cell TEM imaging can examine the criteria that govern the dynamics of such voids and the dewetting of thin interfacial liquid films.

Bubbles can form in liquid films in response to local heating [57], with an example shown in Section 6.4.1, or as a result of radiolysis of water to produce hydrogen gas [58, 59]; an example is shown in Section 7.7.1. In both cases liquid cell TEM can provide data on the bubble geometry and the dependence of bubble volume on conditions, enabling modeling of the processes involved. For example, for nanobubbles growing due to temperature and radiolysis, the shrinking of nanobubbles as the temperature or beam current is lowered can be understood via the inward Laplace pressure that causes the gas in the nanobubble to diffuse out into the surrounding water faster than it is being replenished [57–59]. In order to sustain a nanobubble in steady state, outward gas flow and gas generation must be balanced.

Nanovoids form instead of bubbles in much thinner liquids (~10 nm), causing the equivalent of dewetting at the nanoscale [16]. Figure 13.7a shows these small nanovoids forming spontaneously under the electron beam but then shrinking and disappearing. Macroscopic and microscopic void formation in liquid films can be understood in terms of nucleation theory [60–62]. We evaluate the free energy change, ΔG, of a film when a cylindrical void is formed. ΔG includes competing terms that arise from the change of surface area and the energy required for formation of the three-phase contact line. The graph in Figure 13.7b shows the dependence of ΔG on radius. As in conventional nucleation theory, a local maximum occurs in ΔG when the critical void radius is equal to

$$r_0 \approx \frac{2\kappa + \gamma_w h}{\gamma_w}, \quad (13.2)$$

or approximately 24 nm [16] for a $h = 10$ nm thick water film, for γ_w as given in Section 13.4, and $\kappa \approx 7 \times 10^{-11}$ J/m [63]. Nanovoids of radius $r < r_0$ are expected to shrink. This is consistent with the experimental results in Figure 13.7a and d, where all voids

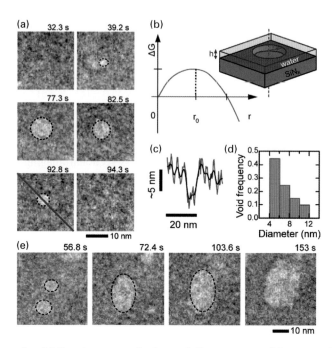

Figure 13.7. Nanovoid dynamics. (a) Spontaneous nucleation and disappearance of the nanovoid in a 10 nm thick water film. (b) Change of free energy on formation of a cylindrical void of radius r in a liquid film of thickness h. When void radius is less than the critical radius r_0, shrinking is favorable, and when $r > r_0$ growth is favored. (c) Depth profile of the void in water across the red line shown in (a) at $t = 92.8$ s. (d) Void count of the maximum diameter reached by unstable voids shows that these voids can reach 12 nm in diameter before spontaneously closing up. (e) Time-series showing the dynamics of a coalescence event and subsequent growth in a thin water film. Reproduced from Ref. 16.

below 12 nm in diameter collapse. Nanovoids with radius $r > r_0$ are expected to grow, and, indeed, Figure 13.7e shows two small voids coalescing into a nanovoid larger than the critical diameter, which then grows. Thus, small diameter voids (<12 nm) are unstable and either close up or coalesce with other voids. Such a simple phenomenological model confirms that there is an energy barrier that predicts shrinkage of small voids and unbounded growth of larger voids. This has an important implication for thin film dewetting: if a spontaneously formed void exceeds the critical radius set by the film thickness, liquid and substrate properties, then the hole in the film will grow larger and the film will dewet.

13.6 Outlook

The study of nanoscale fluid dynamics has broad impact in many important processes that occur at fluid-surface boundary layers: diffusive transport through boundary layers in heterogeneous catalysis [64], charge transport through electrolyte boundary layers in batteries [65, 66], dynamic sub-nanometer hydration layers around proteins [67], lubrication theory, and tribology. The physics of fluid boundary layers is still poorly understood and actively researched [68]. For example, the widely assumed no-slip boundary layer in fluid mechanics cannot be derived from first principles, and many exceptions have been observed [1].

In his famous 1985 review article, de Gennes pointed out that the inability to image liquid–solid interfaces with electron microscopy is one of the challenges in understanding interfacial fluids [27]. Since then, *in situ* TEM has proven capable of imaging dynamic processes in nanoscale liquids. It complements the knowledge we acquire from other long-established methods. In the next few years of liquid cell electron microscopy, we may expect to resolve how properties of interfacial liquids are tied to their nanoscale architecture – the hydration layer, ordering of water near interfaces, the double layer – potentially enabling us to develop an atomistic theory of fluids. For example, *in situ* TEM imaging can help to map out the dynamic consequences of the theory of colloid stability by directly probing liquid mediated interactions between nanoscale particles, which so far have only been validated by force probe microscopy or other indirect methods. It would also be extremely interesting to observe directly how liquids crystallize so that we can describe the atomic mechanism of phase boundary formation between two phases composed of the same material. The ongoing improvements in instrumentation, such as the capability to change temperature, already suggest that probing phase transitions such as freezing of water at the atomic scale is not far away, and that these methods can be extended to other liquid–solid phase transitions [69]. With the development of more sensitive and fast detectors we are poised to discover the mechanisms that govern the dynamics of these transitions in and from the liquid phase.

References

1. E. Lauga, M. Brenner and H. Stone, Microfluidics: the no-slip boundary condition. In C. Tropea, A. Yarin, J. Foss, eds., *Springer Handbook of Experimental Fluid Mechanics* (Berlin, Heidelberg: Springer, 2007) pp. 1219–1240.
2. J. Israelachvili, Electrostatic forces between surfaces in liquids. In *Intermolecular and Surface Forces* (New York: Academic Press, 2011) pp. 291–337.
3. R. Parsons, The electrical double layer: recent experimental and theoretical developments. *Chem. Rev.*, **90** (1990), 813–826.
4. C. U. Chan and C.-D. Ohl, Total-internal-reflection-fluorescence microscopy for the study of nanobubble dynamics. *Phys. Rev. Lett.*, **109** (2012), 174501.
5. K. Xu, P. Cao and J. R. Heath, Graphene visualizes the first water adlayers on mica at ambient conditions. *Science*, **329** (2010), 1188–1191.
6. R. C. Major, J. E. Houston, M. J. McGrath, J. I. Siepmann and X. Y. Zhu, Viscous water meniscus under nanoconfinement. *Phys. Rev. Lett.*, **96** (2006), 177803.
7. S. H. Oh, Y. Kauffmann, C. Scheu, W. D. Kaplan and M. Rühle, Ordered liquid aluminum at the interface with sapphire. *Science*, **310** (2005), 661–663.
8. S. K. Eswaramoorthy, J. M. Howe and G. Muralidharan, In situ determination of the nanoscale chemistry and behavior of solid-liquid systems. *Science*, **318** (2007), 1437–1440.

9. J. M. Howe, *Interfaces in Materials: Atomic Structure, Thermodynamics and Kinetics of Solid-Vapor, Solid-Liquid and Solid-Solid Interfaces* (New York: Wiley-Interscience, 1997).
10. W. D. Kaplan and Y. Kauffmann, Structural order in liquids induced by interfaces with crystals. *Annu. Rev. Mater. Res.*, **36** (2006), 1–48.
11. S. E. Donnelly, R. C. Birtcher, C. W. Allen *et al.*, Ordering in a fluid inert gas confined by flat surfaces. *Science*, **296** (2002), 507–510.
12. B. J. Kim, J. Tersoff, S. Kodambaka *et al.*, Kinetics of individual nucleation events observed in nanoscale vapor-liquid-solid growth. *Science*, **322** (2008), 1070–1073.
13. T. A. Ho, D. V. Papavassiliou, L. L. Lee and A. Striolo, Liquid water can slip on a hydrophilic surface. *Proc. Natl. Acad. Sci. USA*, **108** (2011), 16170–16175.
14. Y. Zhu and S. Granick, Viscosity of interfacial water. *Phys. Rev. Lett.*, **87** (2001), 096104.
15. T.-W. Huang, S.-Y. Liu, Y.-J. Chuang *et al.*, Self-aligned wet-cell for hydrated microbiology observation in TEM. *Lab Chip*, **12** (2012), 340–347.
16. U. Mirsaidov, C.-D. Ohl and P. Matsudaira, A direct observation of nanometer-size void dynamics in an ultra-thin water film. *Soft Matter*, **8** (2012), 7108–7111.
17. U. M. Mirsaidov, H. Zheng, D. Bhattacharya, Y. Casana and P. Matsudaira, Direct observation of stick-slip movements of water nanodroplets induced by an electron beam. *Proc. Natl. Acad. Sci. USA*, **109** (2012), 7187–7190.
18. P. A. Thompson and M. O. Robbins, Origin of stick-slip motion in boundary lubrication. *Science*, **250** (1990), 792–794.
19. M. Urbakh, J. Klafter, D. Gourdon and J. Israelachvili, The nonlinear nature of friction. *Nature*, **430** (2004), 525–528.
20. H. A. Zambrano, J. H. Walther, P. Koumoutsakos and I. F. Sbalzarini, Thermophoretic motion of water nanodroplets confined inside carbon nanotubes. *Nano Lett.*, **9** (2008), 66–71.
21. J. D. Halverson, C. Maldarelli, A. Couzis and J. Koplik, A molecular dynamics study of the motion of a nanodroplet of pure liquid on a wetting gradient. *J. Chem. Phys.*, **129** (2008), 164708–164712.
22. A. Moosavi, M. Rauscher and S. Dietrich, Motion of nanodroplets near edges and wedges. *Phys. Rev. Lett.*, **97** (2006), 236101.
23. E. Rio, A. Daerr, F. Lequeux and L. Limat, Moving contact lines of a colloidal suspension in the presence of drying. *Langmuir*, **22** (2006), 3186–3191.
24. P. Brunet, J. Eggers and R. D. Deegan, Vibration-induced climbing of drops. *Phys. Rev. Lett.*, **99** (2007), 144501.
25. C. Cottin-Bizonne, B. Cross, A. Steinberger and E. Charlaix, Boundary slip on smooth hydrophobic surfaces: intrinsic effects and possible artifacts. *Phys. Rev. Lett.*, **94** (2005), 056102.
26. C. Sendner, D. Horinek, L. Bocquet and R. R. Netz, Interfacial water at hydrophobic and hydrophilic surfaces: slip, viscosity, and diffusion. *Langmuir*, **25** (2009), 10768–10781.
27. P. G. de Gennes, Wetting: statics and dynamics. *Rev. Mod. Phys.*, **57** (1985), 827–863.
28. H. Berg, *Random Walks in Biology* (Princeton, NJ: Princeton University Press, 1993).
29. H. Zheng, S. A. Claridge, A. M. Minor, A. P. Alivisatos and U. Dahmen, Nanocrystal diffusion in a liquid thin film observed by in situ transmission electron microscopy. *Nano Lett.*, **9** (2009), 2460–2465.
30. E. R. White, M. Mecklenburg, B. Shevitski, S. B. Singer and B. C. Regan, Charged nanoparticle dynamics in water induced by scanning transmission electron microscopy. *Langmuir*, **28** (2012), 3695–3698.

31. J. Lu, Z. Aabdin, N. D. Loh, D. Bhattacharya and U. Mirsaidov, Nanoparticle dynamics in a nanodroplet. *Nano Lett.*, **14** (2014), 2111–2115.
32. J. M. Grogan, L. Rotkina and H. H. Bau, In situ liquid-cell electron microscopy of colloid aggregation and growth dynamics. *Phys. Rev. E*, **83** (2011), 061405.
33. A. Verch, M. Pfaff and N. de Jonge, Exceptionally slow movement of gold nanoparticles at a solid/liquid interface investigated by scanning transmission electron microscopy. *Langmuir*, **31** (2015), 6956–6964.
34. T.-D. Li, J. Gao, R. Szoszkiewicz, U. Landman and E. Riedo, Structured and viscous water in subnanometer gaps. *Phys. Rev. B*, **75** (2007), 115415.
35. K. B. Jinesh and J. W. M. Frenken, Capillary condensation in atomic scale friction: how water acts like a glue. *Phys. Rev. Lett.*, **96** (2006), 166103.
36. Q. Chen, J. M. Smith, J. Park *et al.*, 3D motion of DNA-Au nanoconjugates in graphene liquid cell electron microscopy. *Nano Lett.*, **13** (2013), 4556–4561.
37. J. Park, H. Zheng, W. C. Lee *et al.*, Direct observation of nanoparticle superlattice formation by using liquid cell transmission electron microscopy. *ACS Nano*, **6** (2012), 2078–2085.
38. Z. Barkay, Wettability study using transmitted electrons in environmental scanning electron microscope. *Appl. Phys. Lett.*, **96** (2010), 183109–183103.
39. Z. Barkay, Dynamic study of nanodroplet nucleation and growth on self-supported nanothick liquid films. *Langmuir*, **26** (2010), 18581–18584.
40. K. Rykaczewski and J. H. J. Scott, Methodology for imaging nano-to-microscale water condensation dynamics on complex nanostructures. *ACS Nano*, **5** (2011), 5962–5968.
41. K. Rykaczewski, J. H. J. Scott, S. Rajauria *et al.*, Three dimensional aspects of droplet coalescence during dropwise condensation on superhydrophobic surfaces. *Soft Matter*, **7** (2011), 8749–8752.
42. N. Miljkovic, R. Enright and E. N. Wang, Effect of droplet morphology on growth dynamics and heat transfer during condensation on superhydrophobic nanostructured surfaces. *ACS Nano*, **6** (2012), 1776–1785.
43. K. Rykaczewski, Microdroplet growth mechanism during water condensation on superhydrophobic surfaces. *Langmuir*, **28** (2012), 7720–7729.
44. D. Bhattacharya, M. Bosman, V. R. S. S. Mokkapati, F. Y. Leong and U. Mirsaidov, Nucleation dynamics of water nanodroplets. *Microsc. Microanal.*, **20** (2014), 407–415.
45. R. N. Leach, F. Stevens, S. C. Langford and J. T. Dickinson, Dropwise condensation: experiments and simulations of nucleation and growth of water drops in a cooling system. *Langmuir*, **22** (2006), 8864–8872.
46. A. Steyer, P. Guenoun, D. Beysens and C. M. Knobler, Growth of droplets on a substrate by diffusion and coalescence. *Phys. Rev. A*, **44** (1991), 8271–8277.
47. T. M. Rogers, K. R. Elder and R. C. Desai, Droplet growth and coarsening during heterogeneous vapor condensation. *Physi. Rev. A*, **38** (1988), 5303–5309.
48. I. O. Ucar and H. Y. Erbil, Use of diffusion controlled drop evaporation equations for dropwise condensation during dew formation and effect of neighboring droplets. *Coll. Surf. A: Physicochem. Eng. Aspects*, **411** (2012), 60–68.
49. M. Whitby and N. Quirke, Fluid flow in carbon nanotubes and nanopipes. *Nat. Nano*, **2** (2007), 87–94.
50. N. Naguib, H. Ye, Y. Gogotsi *et al.*, Observation of water confined in nanometer channels of closed carbon nanotubes. *Nano Lett.*, **4** (2004), 2237–2243.
51. D. Mattia and Y. Gogotsi, Review: static and dynamic behavior of liquids inside carbon nanotubes. *Microfluid Nanofluid*, **5** (2008), 289–305.

52. M. P. Rossi, H. Ye, Y. Gogotsi *et al.*, Environmental scanning electron microscopy study of water in carbon nanopipes. *Nano Lett.*, **4** (2004), 989–993.
53. N. Patra, B. Wang and P. Král, Nanodroplet activated and guided folding of graphene nanostructures. *Nano Lett.*, **9** (2009), 3766–3771.
54. U. Mirsaidov, V. R. S. S. Mokkapati, D. Bhattacharya *et al.*, Scrolling graphene into nanofluidic channels. *Lab Chip*, **13** (2013), 2874–2878.
55. M. J. Dukes, B. W. Jacobs, D. G. Morgan, H. Hegde and D. F. Kelly, Visualizing nanoparticle mobility in liquid at atomic resolution. *Chem. Commun.*, **49** (2013), 3007–3009.
56. R. V. Craster and O. K. Matar, Dynamics and stability of thin liquid films. *Rev.Mod. Phys.*, **81** (2009), 1131–1198.
57. E. R. White, M. Mecklenburg, S. B. Singer, S. Aloni and B. C. Regan, Imaging nanobubbles in water with scanning transmission electron microscopy. *Appl. Phys. Express*, **4** (2011), 055201.
58. J. M. Grogan, N. M. Schneider, F. M. Ross and H. H. Bau, Bubble and pattern formation in liquid induced by an electron beam. *Nano Lett.*, **14** (2013), 359–364.
59. T.-W. Huang, S.-Y. Liu, Y.-J. Chuang *et al.*, Dynamics of hydrogen nanobubbles in KLH protein solution studied with in situ wet-TEM. *Soft Matter*, **9** (2013), 8856–8861.
60. C. Redon, F. Brochard-Wyart and F. Rondelez, Dynamics of dewetting. *Phys. Rev. Lett.*, **66** (1991), 715–718.
61. M. Elbaum and S. G. Lipson, How does a thin wetted film dry up? *Phys. Rev. Lett.*, **72** (1994), 3562–3565.
62. U. Thiele, M. Mertig and W. Pompe, Dewetting of an evaporating thin liquid film: heterogeneous nucleation and surface instability. *Phys. Rev. Lett.*, **80** (1998), 2869–2872.
63. T. Pompe and S. Herminghaus, Three-phase contact line energetics from nanoscale liquid surface topographies. *Phys. Rev. Lett.*, **85** (2000), 1930–1933.
64. J. R. H. Ross, *Heterogeneous Catalysis: Fundamentals and Applications* (Kidlington, UK: Elsevier, 2011).
65. T. Reddy, *Linden's Handbook of Batteries*, 4th edn. (New York: McGraw-Hill Professional, 2010).
66. J. M. Tarascon and M. Armand, Issues and challenges facing rechargeable lithium batteries. *Nature*, **414** (2001), 359–367.
67. S. Ebbinghaus, S. J. Kim, M. Heyden *et al.*, An extended dynamical hydration shell around proteins. *Proc. Natl. Acad. Sci. USA*, **104** (2007), 20749–20752.
68. T. M. Squires and S. R. Quake, Microfluidics: fluid physics at the nanoliter scale. *Rev. Mod. Phys.*, **77** (2005), 977–1026.
69. K. Tai, Y. Liu and S. J. Dillon, In situ cryogenic transmission electron microscopy for characterizing the evolution of solidifying water ice in colloidal systems. *Microsc. Microanal.*, **20** (2014), 330–337.

14 Nanoscale Deposition and Etching of Materials Using Focused Electron Beams and Liquid Reactants

Eugenii U. Donev, Matthew Bresin, and J. Todd Hastings

14.1 Overview of Gas-Phase Focused Electron Beam-Induced Processing (FEBIP)

14.1.1 Background

Focused electron beam-induced deposition (FEBID) and etching (FEBIE) employ gas-phase reactants to locally add or subtract material where the beam impinges on a substrate. In this chapter, we discuss how both deposition and etching may be enhanced through the use of *liquid-phase* reactants, and the key role of liquid cell electron microscopy in these developments.

The processes of FEBID and FEBIE are shown in Figure 14.1. Deposition occurs when the electron beam causes an adsorbed precursor molecule to dissociate and leave behind the desired material (Figure 14.1a). Etching results when an electron-induced reaction between the adsorbed molecule and the substrate produces a volatile product (Figure 14.1b). Focused electron beam-induced processing (FEBIP) encompasses both deposition and etching, and is a term commonly used to describe beam-initiated or beam-assisted processes from an overall standpoint. In FEBIP, primary, secondary, and backscattered electrons can contribute to the processes, as illustrated in Figure 14.1c, although a clear consensus is yet to emerge regarding the relative roles of electrons with different energies in the precursor dissociation mechanisms [1]. FEBIP has been performed in scanning electron microscopes (SEM), electron beam lithography (EBL) systems, and scanning transmission electron microscopes (STEM). Depending on the system and electron beam energy, the beam is focused to typically 1–10 nm for SEM (1–10 keV) and EBL (20–100 keV) and 0.2 nm for STEM (200+ keV); for a given system, the beam diameter increases as the energy is lowered [2], hence decreasing the FEBIP resolution. Excellent overviews of FEBIP can be found in review articles by van Dorp and Hagen [1], Randolph *et al.* [3], Botman *et al.* [4], and others [2, 5–7], while the texts by Silvis-Cividjian and Hagen [8] and Utke *et al.* [9] provide more in-depth treatments.

FEBIP has found application in both research and manufacturing because it enables one-step nanoscale patterning of materials with *in situ* imaging and metrology. In the research domain, FEBIP is routinely used to electrically connect chemically synthesized nanostructures to external systems, to prototype nanoscale electronic, photonic, magnetic, and mechanical devices, and to modify scanning probes such as atomic force microscope and near-field optical microscope tips. Industrial application of FEBIP has

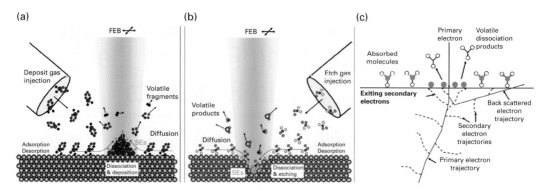

Figure 14.1. Processes active in focused electron beam processing: molecules adsorb, desorb, and diffuse at the surface and are dissociated under electron impact. (a) FEBID: non-volatile dissociation products form a deposit growing coaxially into the beam. Volatile fragments are pumped away. (b) FEBIE: surface-adsorbed molecules dissociate under electron impact into reactive species that react with the substrate material to form volatile compounds. (c) Interactions of primary electrons generating an emitted flux of secondary electrons and backscattered electrons; all electrons can dissociate surface adsorbed molecules via electronic excitation. Reprinted with permission from Ref. 2, © 2008 American Vacuum Society.

focused on integrated-circuit mask repair, with the Zeiss line of MeRiT tools as the primary example. There is also emerging interest in using FEBIP for integrated-circuit editing and debugging, a field traditionally dominated by focused ion beam-induced processing.

14.1.2 Challenges

Despite their success, electron beam-induced processes still have a number of limitations. For example, electron beam-induced deposition from metalorganic precursors typically yields metal grains embedded in an amorphous matrix. With the exception of Co deposited from dicobalt octacarbonyl, these materials rarely exceed 50 at.% purity and their properties differ markedly from their bulk counterparts [4]. Modest improvements in purity for Au, Pt, and Ni have been achieved by using fluoro- and chlorophosphine-based precursors; however, significant phosphorus incorporation remains problematic and the lack of commercial availability, along with the toxicity, instability, and expense of these precursors, limits their use. Post-deposition annealing or irradiation can also improve purity, but often at the expense of geometric alteration of the pattern.

A broad spectrum of materials has been deposited from gas-phase precursors including many transition metals, along with oxides of Si, Ti, Fe, Ni, and Pt. However, the deposition of alloys and other compounds is much more limited, with the exceptions of GaAs [10], GaN [11], FePt alloys [12], and recently, $Pb_xC_yO_\delta$ superconductors [13]. Although instrumentation for simultaneous injection of multiple gases is commonly available, it can be challenging to optimize conditions for the reliable deposition of compounds given the widely varying residence times, dissociation cross sections, and contaminating byproducts of the precursors.

For electron beam-induced etching, the primary challenge is the identification of reactants that produce volatile products. Oxygen and water vapor work well for carbonaceous materials, and xenon difluoride and halogens are appropriate for silicon-based materials, but many technologically important transition metals, along with their oxides and nitrides, cannot be etched with current electron beam-induced processes. As a result, gas-assisted focused ion beam milling is widely used, but introduces the additional challenges of material redeposition, reduced resolution, orientation dependence, ion implantation, bubble formation, and poor etch selectivity between materials.

14.1.3 Liquid-Phase Alternatives to Gas-Phase FEBIP

The challenges of gas-phase FEBIP, combined with the recent advances in electron microscopy in liquids, make it natural to consider focused electron beam-induced processing with liquid-phase reactants (LP-FEBID). Replacing traditional gas-phase precursors and etchants with liquid reactants offers a number of advantages. For deposition, a vastly enlarged range of materials, including alloys and compounds, can be deposited from liquid solutions, and the choice of solution can be guided by previous results from electrochemistry and radiation chemistry. In addition, the use of soluble metal-ion complexes instead of volatile metalorganics should greatly reduce contamination and enable deposition of higher-purity materials. Finally, when using metal salts in aqueous solutions, the precursors can be far lower in cost, less toxic, more varied, and more stable than their gas-phase counterparts. For etching, the use of liquids enables etching of materials with soluble reaction products, and liquids can greatly enhance the concentration of reactants near the beam interaction point, leading to higher-throughput processing. Finally, the liquids can also be electrically conductive and help dissipate charge from insulating substrates. Despite these promising considerations, liquid-phase reactants introduce many new scientific questions and technological challenges that must be addressed. This chapter highlights the techniques used for LP-FEBIP, the materials that have been processed to date, and the possible reaction mechanisms involved.

14.2 Methods for LP-FEBIP

The novel approach of exposing liquid-phase reactants to a focused electron beam presents an obvious technical challenge, since most liquids are simply too volatile to last long enough inside a high vacuum system for deposition or etching processes to occur. Thus far, three main methods have been employed in LP-FEBIP to overcome this difficulty: (i) isolating the liquid from the vacuum chamber of an SEM or TEM instrument in a closed cell [14–36]; (ii) thermodynamically stabilizing the liquid precursors via "cell-less" processing [37, 38] in an environmental scanning electron microscope (ESEM) equipped with differentially pumped vacuum chambers and specialized detectors; and (iii) using ionic liquids as precursors [39–42] with vapor pressures low enough to withstand vacuum pumping. These three LP-FEBIP methods also represent the approaches used for electron microscopy in liquids [43, 44], as described elsewhere in this book: for TEM, imaging in closed cells and in low pressure open cells (with ionic liquids) are described in detail in Chapters 2 and 3, respectively, while high pressure open cell (i.e. "cell-less") imaging is mentioned also in Chapter 3 and elsewhere; for SEM, closed cells are described in Chapter 4.

14.2.1 Closed Cells

A number of closed (encapsulated) liquid cell designs for electron microscopy are detailed in other portions of this text. Similar cells have been used for FEBIP. As an example, Figure 14.2a illustrates FEBID in a QuantomiX WetSEM cell. Deposition occurs on the bottom side of a polyimide membrane in contact with a liquid precursor.

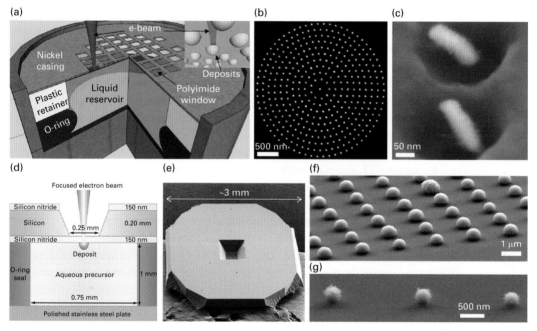

Figure 14.2. (a) Schematic of commercial liquid cell (QuantomiX WetSEM [45]; see also Chapter 4 in this book) with polyimide membrane. Liquid precursors are contained within the sealed capsule and are accessible through the membrane. Inset: electron beam penetrates the membrane, inducing deposition on the liquid side. (b) *In situ* SEM image of Pt nanoparticles deposited via LP-FEBID on polyimide membrane (20 mM H_2PtCl_6, 20 keV, 0.16 nA, 25 pC/dot). (c) *Ex situ* (liquid side of membrane) tilted-view SEM image of Pt nanorods (20 mM H_2PtCl_6, 20 keV, 0.2 nA, 110 pC/dot). (d) Cross-sectional schematic of home-built liquid cell with top Si_xN_y window and deep (1 mm) reservoir. (e) Tilted-view SEM image of Si chip (~3 mm diameter) with Si_xN_y window. (f, g) *Ex situ* tilted-view SEM images of Pt particles deposited on Si_xN_y membranes using the liquid cell in (d) (20 mM H_2PtCl_6, 20 keV, 1.3 nA, ~3–5 nC/dot). (a) Reproduced with permission from Ref. 14. © 2013 Wiley-VCH Verlag GmbH & Co. KGaA, Weinheim. (b) From Ref. 20. © 2009 IOP Publishing. Reproduced with permission. All rights reserved. (c, d, g) From Ref. 22. Reproduced by permission of the Royal Society of Chemistry.

The design and operation of this type of cell are detailed in Chapter 4. Such cells are commercially available and compatible with standard electron microscopes, making them excellent for exploring electron beam-induced processes such as the deposition of Pt nanoparticles shown in Figure 14.2b. Although quite robust, the polyimide membranes in these cells can be damaged during electron beam-induced processing, as seen in Figure 14.2c. Under these conditions the membrane was etched away as Pt nanorods were deposited. It is hypothesized that modification of the polyimide membrane occurs due to radiation damage by high energy electrons, which induce scissioning of molecular chains prior to crosslinking, followed by swelling of the membrane as the smallest and more volatile molecular species accumulate in the polymer matrix as dissolved gas [31]. Even when such large damaged regions are not observed, it is common for deposits to begin embedded in the membrane and grow outward (e.g. see Ref. 21 and Figure 14.7c); however, this is certainly not universal behavior, and Figure 14.9a contains a counterexample.

A variety of custom liquid cells has also been used to study electron beam-induced processes. One particularly intriguing design, described in Chapters 2 and 19, is the graphene liquid cell [34] in which liquid droplets are trapped between graphene sheets. Graphene is impermeable to liquid, essentially non-reactive, and minimally degrades resolution. It is currently unclear whether STEM or SEM could write deterministic patterns on graphene because the beam-induced nanocrystals that have been formed in the graphene liquid cell [34] remain mobile. Nevertheless, the nearly ideal qualities of the graphene membrane recommend this system for further investigation.

More common than graphene for liquid cell SEM and STEM is to use silicon nitride for the membrane [44]. Deposition and etching experiments have been carried out in these types of cells as well. For SEM systems, a cell with a large liquid reservoir can be easily constructed using a single low stress (Chapter 2) Si_xN_y membrane [46], an o-ring, and a polished plate, as shown in Figure 14.2d, e. As in the QuantomiX WetSEM cell, deposition occurs on the bottom side of the membrane in contact with the liquid precursor. However, deposition can proceed quite differently on the Si_xN_y, as seen by comparing the Pt deposits in Figure 14.2f, g with those of Figure 14.2b, c. On Si_xN_y the Pt initially deposits as a nearly spherical particle (Figure 14.2g), and then grows into a hemispherical particle as electron dose increases (Figure 14.2f). It is also important to note that Si_xN_y does not appear to experience the same membrane damage as polyimide, at least for Pt deposition from chloroplatinic acid.

For both STEM and SEM, Si_xN_y membranes can also form cells with thin, electron-transparent liquid layers. Deposition then takes place on the opposite surface, which can be a thick substrate such as a Si wafer (Figure 14.3a), or a second membrane to achieve a transmission configuration (Figure 14.3c). As was the case in early microscopy cell development (Chapter 2), gap control and membrane bowing make repeatable use of such cells somewhat challenging. One example of this can be seen in Figure 14.3b, in which deposition proceeded on both the top Si_xN_y membrane and the underlying Si wafer producing structures that ultimately bridged the vertical liquid gap. Similarly, Figure 14.3c depicts Si deposition from $SiCl_4$ on the top and bottom Si_xN_y membranes of a transmission cell, with an AFM image of one such deposit shown in Figure 14.3d.

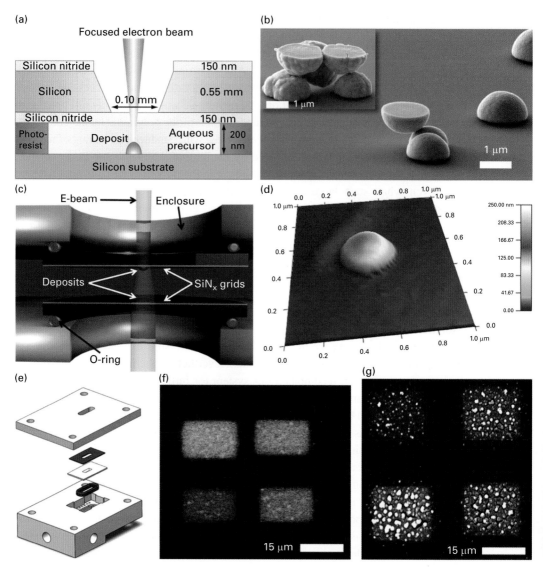

Figure 14.3. (a) Cross-sectional schematic of home-built liquid cell with top Si_xN_y window and shallow (nominally 200 nm, actually ~2 μm) reservoir; compare with Figure 14.2d. (b) *Ex situ* tilted-view SEM images of Pt particles deposited on Si_xN_y membranes using the cell in (a) (20 mM H_2PtCl_6, 20 keV, 1.7 nA, ~10 nC/dot), showing that deposits formed on the top window can detach upon cell disassembly. (c) Cross-sectional schematic of home-built liquid cell with top *and* bottom Si_xN_y windows and shallow reservoir (nominally 100 nm, actually ~10 μm). (d) AFM image of Si nanoparticle deposited on Si_xN_y membrane using the cell in (c) ($SiCl_4$, 30 keV, 0.74 nA, 7.4 nC). (e) Exploded-view schematic of electrochemical SEM cell comprising six parts (bottom to top): holder with channels, PDMS ring, chip with Pt electrodes and Si_xN_y window, another PDMS ring, clamping lid, and circuit board (not shown). (f) *In situ* and (g) *ex situ* SEM images of Ni patches deposited in the cell in (e) (10 keV, 0.86 A/m^2, 15–60 s irradiation). (c, d) From Ref. 28, © 2012 IOP Publishing. Reproduced with permission. All rights reserved. (e, f, g) Reprinted from Ref. 25, © 2013, with permission from Elsevier.

Si_xN_y membranes have also been used to construct more sophisticated cells similar to their microscopy counterparts, which contain liquid channels and biased electrodes. Figure 14.3e, f, g shows such a cell and the corresponding beam-induced deposition of Ni.

In summation, LP-FEBID has been achieved in both SEM and STEM closed cell configurations, using thick (in SEM only) and thin layers of liquid precursors, yielding deposits that float freely in the liquid precursor (see Chapter 9) or – and this is the regime we discuss in this chapter – grow with different morphologies on different substrates, which may, in turn, undergo either little to none or substantial electron beam-induced damage.

14.2.2 "Cell-Less" LP-FEBIP: *In Situ* Hydration, Liquid Injection, Ionic Liquids

LP-FEBID may be performed without the need for cells, by forming or maintaining liquid precursors within an SEM chamber. The formation of liquid precursors is facilitated by low vacuum or environmental SEM, where control of the chamber pressure and stage temperature (Peltier stage) is possible. To form a viable deposition source, a water soluble precursor is first placed on a substrate and then hydrated *in situ* by creating a 100% relative humidity condition within the SEM. This hydrated precursor may be subsequently patterned with the electron beam, similarly to a closed cell. The precursor hydration process is illustrated schematically in Figure 14.4a–d, with results for Ag and Cu patterning shown in Figure 14.4e and Figure 14.4f, respectively.

The resolution of deposits formed by cell-less LP-FEBID is partially dependent on the thickness of the formed liquid droplet, which is a function of the contact angle and the distance from the droplet edge. The contact angle of water droplets on Si in an ESEM has been reported as 40° [47] and 47° [38] and is roughly similar to the atmospheric contact angle. From Monte Carlo simulations, the penetration range of a 20 kV electron beam in water is approximately 10 μm. Taking the electron penetration range along with the droplet contact angle, the practical limitation for deposition is 10 μm from the droplet edge. However, scattering of the electron beam as it traverses the liquid droplet would conceivably result in significant lateral broadening of deposits. Indeed, early results in cell-less LP-FEBID showed deposit broadening as a function of distance from the droplet edge, similar to the deposits shown in Figure 14.4i. One method to increase the resolution of the deposits is to reduce the droplet contact angle, which is achieved with the introduction of a chemical surfactant. An additional benefit of co-hydrating a surfactant is the increase of practical patterning distance away from the droplet edge. It has been shown that thinner droplets result in significant improvement to patterning resolution for Ag and Cu deposition compared to thicker droplets, where 70 nm line resolution has been achieved [38]. While it has been demonstrated that thinner droplets (with an estimated contact angle of 3–8°) yield higher resolution deposits, process optimization for cell-less LP-FEBIP has been beyond the scope of the extant literature. Many factors would ultimately contribute to process optimization,

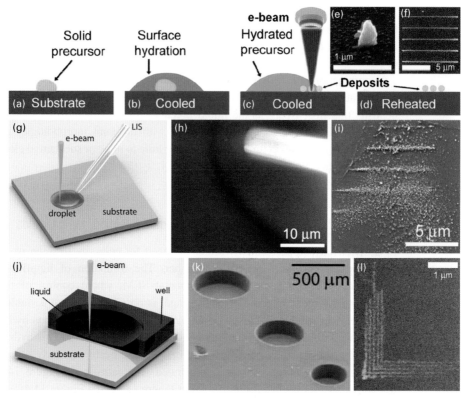

Figure 14.4. Schematics and SEM images of LP-FEBID on bulk substrate (Si) without a closed cell. (a) Solid precursor placed on substrate. (b) Substrate cooled to 3 °C and chamber H_2O vapor pressure increased to 5.7 Torr, causing surface condensation. (c) Hydrated precursor patterned with electron beam on thin edge. (d) Substrate returned to room temperature, removing hydrated precursor while leaving irradiated deposits, such as (e) 250 nm diameter Ag particle (250 mM $AgNO_3$ + 4.0 mM SDS surfactant, 20 keV, 0.15 nA, 30 pC/dot) and (f) Cu lines (250 mM $CuSO_4$ + 4.0 mM SDS surfactant, 20 keV, 0.15 nA, 0.2 µC/cm). (g) Schematic of liquid insertion and subsequent electron beam patterning using liquid injector system (LIS), and (h) image of stable droplet and capillary tube (bright area) just before patterning along droplet edge. (i) Image of LIS-deposited Cu lines (0.3 mM $CuSO_4$, 20 keV, 0.15 nA, 0.8 µC/cm). (j) Schematic and (k) image of liquid microwells in 100 µm thick SU-8 photoresist on Si substrate. (l) Image of dense Cu lines (~50 nm width, 70 nm half-pitch) deposited in microwell. (a–f, h, i) Reproduced from Ref. 38, © 2014 Microscopy Society of America.

such as the beam interaction volume, the rate of generation of intermediate species, the diffusivities of reactants and products as well as the droplet thickness.

Specialized liquid injectors may also be employed to deliver the precursor to the substrate (Figure 14.4g), similar to gas-phase injectors. A pulled glass capillary, with a micrometer-sized inner diameter, can selectively be frozen by evaporative cooling and unfrozen by contact with a substrate. Selective freezing allows the liquid injector system (LIS) to have precise control over the amount of liquid deposited as well as droplet placement. This precise control of the droplet is a major advantage when compared to the *in situ* hydration technique, since the latter can expand or contract the droplet with pressure control, but cannot reliably shift the droplet's position. Furthermore, liquid injectors are potentially compatible with high vacuum SEM [38]. A proof-of-concept example of droplet delivery with an FEI Company liquid injector is shown in Figure 14.4h, while a line pattern is shown in Figure 14.4i.

Ionic liquids may be used in place of water in high vacuum systems. Ionic liquids are typically used as solvents for chemical catalysis, where their inherent low vapor pressure makes them ideal for thermal phase separation. For LP-FEBID, the low vapor pressure makes these liquids compatible with standard SEMs, which may make ionic liquids an ideal choice for evaluating potential chemistries where an ESEM is not available. Figure 14.6e shows the deposition of dendritic Ag structures using the ionic liquid 1-butyl-3-methylimidazolium tetrafluoroborate ([BMIM][BF$_4$], p_{vapor} $< 1 \times 10^{-4}$ Torr) and AgBF$_4$ as the precursor [40]; see also chapter 3. Gold nanoparticles have also been produced using a variety of ionic liquids and the precursor [NaAuCl$_4$][2H$_2$O] while using a low energy electron flood gun [41, 42].

14.3 Survey of LP-FEBID of Transition Metals

14.3.1 Platinum, Gold, Silver, and Copper

Platinum: The earliest reports [20, 21] of controlled electron beam-induced nanopatterning from a bulk liquid precursor (an aqueous solution of chloroplatinic acid, H$_2$PtCl$_6$) demonstrated at the outset several characteristic features of LP-FEBIP: high resolution, high fidelity deposits (Figure 14.5a–d), history-dependent collateral deposition (e.g. see Figure 2a in Ref. 21), modification of the polymer membrane of the liquid cell (Figure 14.2c), and, crucially, great enhancements in the growth rate and compositional purity (Figure 14.5e) of the deposited material. For instance, Pt structures have been deposited by LP-FEBID at a rate of 1000 nm per C/cm^2 [21], whereas only 50 nm per C/cm^2 was reported [48] for a gas-phase metalorganic Pt precursor at the same electron beam energy, albeit on a Si rather than polymer substrate. Furthermore, gas-phase FEBID has been reported to yield deposits of 10–15 at.% Pt from metalorganic and 35 at.% Pt from inorganic precursors [4], compared with 80–95 at.% Pt for deposits produced via LP-FEBID from a dilute aqueous precursor [21, 22]. The high purity of Pt via LP-FEBID even exceeds that of gas-phase ion beam-induced deposits with Pt concentrations of 30–55 at.% [49–53].

As a practical matter, the composition of LP-FEBID materials is typically determined by X-ray energy-dispersive spectroscopy, in conjunction with Monte Carlo simulations [54, 55] of the interactions of keV-electrons with the sample materials, including elastic and inelastic scattering and X-ray generation. In this chapter, we show compositional measurements through comparisons of experimental and simulated XEDS spectra for LP-FEBID deposits of Pt (Figure 14.5e), Cr (Figure 14.7a), Au–Pt alloy (Figure 14.8c), Au–Ag alloy (Figure 14.8d), and CdS (Figure 14.9a).

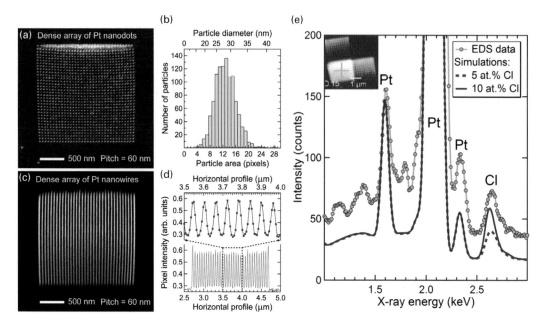

Figure 14.5. Resolution, pattern fidelity, and purity of Pt nanostructures deposited via LP-FEBID on polyimide membranes. (a) SEM image of dense array (60 nm pitch) of small Pt dots (20 mM H_2PtCl_6, 20 keV, 0.16 nA, 8 pC/dot), with (b) lateral-size distribution plotted, showing that 80% of all dots have diameters between 20 and 30 nm. (c) SEM image of dense array (60 nm pitch) of parallel Pt lines (0.34 nA, 7 pC/dot with vertically stigmated beam), and (d) horizontal intensity profile extracted from middle of image; 25 ± 5 nm full linewidth at half maximum. (e) Experimental XEDS spectrum of Pt deposits (inset: SEM image), along with Monte Carlo simulations (NIST DTSA-II software [54]) of free-standing films of $Pt_{(1-x)}Cl_x$, where $x = 0.10$ atomic fraction is the upper-bound estimate of Cl contamination. (a–d) From Ref. 20, © 2009 IOP Publishing. Reproduced with permission. All rights reserved. (e) Adapted with permission from Ref. 21, © 2009 American Chemical Society.

Gold: Gas-phase FEBID (without post-processing) of Au using metalorganic precursors typically yields 5–50 at.% Au in a carbonaceous matrix [4, 56], with the notable exception of >90 at.% Au from an unstable and commercially unavailable chloro-Au-trifluorophosphine (PF_3AuCl) precursor [57]. On the other hand, LP-FEBID attained at least 70 at.% Au from a disulfitoaurate complex ($[Au(SO_3)_2]^{3-}$). Figure 14.6a shows Au nanoparticles deposited from this precursor [32], commonly used in Au electro-deposition [58]. Similarly, 95 at.% Au was attained from a chloroauric acid ($HAuCl_4$) precursor [32]. Both acidic ($HAuCl_4$) and neutral ($NaAuCl_4$) aqueous solutions of chloroaurate ions produced LP-FEBID deposits of similar morphology. Of the three liquid precursors, the $[Au(SO_3)_2]^{3-}$ solution yielded the "cleanest" deposits with the least amount of undesirable collateral deposition in unexposed regions [32].

Silver: Ag structures deposited with silver nitrate ($AgNO_3$) via LP-FEBID have also exhibited purity levels of >85 at.% Ag [26, 31, 38]; micrographs of these deposits are shown in Figure 14.4e and Figure 14.6b–d. It is worth noting that no known precursors for gas-phase FEBID of Ag exist at present, which makes LP-FEBID the technique of choice for controlled direct-write fabrication of Ag nanostructures. Deposition of

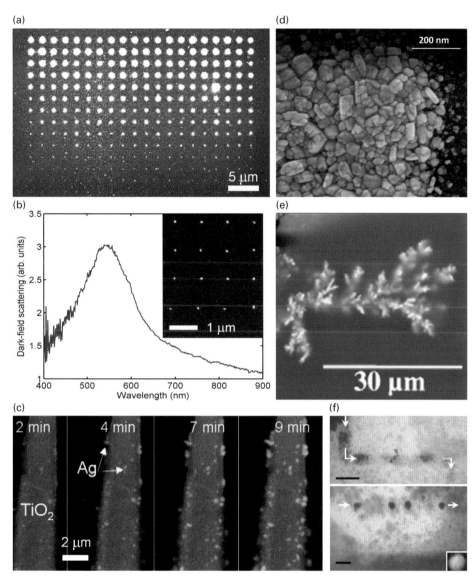

Figure 14.6. SEM/TEM images of LP-FEBID deposits on polyimide membranes in (a, b, d), TiO$_2$ nanowire in (c), TiO$_2$ substrate in (e), and Si$_x$N$_y$ membrane in (f). (a) Dose array of Au nanoparticles deposited from aqueous solution of disulfitoaurate complex (1 mM [Au(SO$_3$)$_2$]$^{3-}$, 20 keV, 0.2 nA, 30–350 pC/dot from bottom to top). (b) Dark-field light scattering from Ag deposits (inset: micrograph of Ag dot array; 0.75 mM AgNO$_3$, 20 keV, 0.16 nA, 150 pC/dot), exhibiting localized surface-plasmon resonance characteristic of good-quality Ag nanoparticles embedded in dielectric medium (polyimide membrane). (c) Sequential images of TiO$_2$ nanowire in aqueous solution of AgNO$_3$ (0.1 mM), under continuous scanning at 120 s/frame (30 keV, ~1 nA, 2–9 min irradiation), with Ag nanoparticles nucleating and growing at nanowire surface. (d) Image of dense Ag deposit (1 mM AgNO$_3$, 5 keV, 0.005 nA, 50 pC/dot). (e) Image of dendritic Ag structures deposited on anatase TiO$_2$ surface from ionic liquid ([BMIM] [BF$_4$] + AgBF$_4$, 25 keV, 20 µA, 45 min irradiation). (f) Beam-induced electrochemical deposition of

materials such as Ag and Au may find applications in the vast field of plasmonics; a scattering spectrum of the localized surface-plasmon resonance of Ag nanoparticles deposited via LP-FEBID is shown in Figure 14.6b. Furthermore, LP-FEBID of Ag illustrates one of the major advantages of the technique with respect to gas-phase FEBID, namely that the lack of gas-phase precursors limits the potential materials available, while the abundance of liquid-phase precursors allows for much more chemical exploration.

Copper: Recently, Cu structures have been successfully deposited from $CuSO_4$ via cell-less LP-FEBID in an ESEM (Figure 14.4f, i, l) [37, 38] and beam-assisted electrochemical deposition in a closed TEM cell (Figure 14.6f) [19]. The purity of the cell-less deposits was conservatively estimated at 70 at.% Cu [38], comparing favorably with Cu deposited via gas-phase FEBID, which has not yet exceeded 11 at.% [59, 60]. The primary contaminant was oxygen [38], which suggests some portion of the deposit contained an oxide; however, since elemental quantification with XEDS was not performed with a known CuO_x standard, further analysis would be required to determine the majority phase. Other studies that use an electrochemical approach [19, 37], did not explicitly discuss the possibility for oxide phases, and it remains unclear if the electrochemical vs. non-electrochemical approaches yield differing compositions. Regardless, provided the oxide content is minimal, the ability to site-specifically deposit high quality Cu has tantalizing possibilities for circuit edit, particularly in combination with the Cu-etch chemistry shown in Section 14.5.2.

One generally undesirable feature of FEBID processes, observed for both gas-phase and liquid-phase precursors, is the so-called collateral ("parasitic") deposition or proximity effect. It occurs when already-deposited structures accumulate additional material due to scattered electrons from nearby deposition or diffusion of reduced metallic clusters or particles in the bulk solution [1]. In other words, collateral deposits accrue in locations other than the area directly targeted by the beam, as can be seen in Figure 14.4i and, to a lesser extent, in Figure 14.6a. For example, Ref. 21 shows how a nominally identical dose can produce varying amounts of deposited material depending on the depositions that follow; that is, deposits earlier in time continue to grow during later depositions at different locations. The amount and extent of collateral deposition in LP-FEBID seem to depend on deposited material, electron beam dose [20], precursor [32], and substrate [22]. Notwithstanding this proximity effect, the LP-FEBID technique has successfully produced densely packed nanostructures, such as the Ni nanoparticles shown in Figure 14.7c, or the Pt nanoparticles and nanowires shown in Figure 14.5a, c, whose size distribution and grating profile are plotted in Figure 14.5b, d [20].

Caption for **Figure 14.6.** (*cont.*) lines of Cu clusters produced at negative potential (-100 mV) while moving substrate in TEM under fixed electron beam (inset: beam size during scanning) at different scan speeds (100 mM $CuSO_4$ + 180 mM H_2SO_4 + 1.4 mM HCl, 300 keV); scale bars: 1 μm. (a) From Ref. 32, © 2011 IOP Publishing. Reproduced with permission. All rights reserved. (c) Adapted from Ref. 26, © 2010 American Chemical Society. (d) Reprinted with permission from Ref. 31, © 2008 American Vacuum Society. (e) Reprinted from Ref. 40, © 2009, with permission from Elsevier. (f) Reprinted with permission from Ref. 19, © 2014 AIP Publishing LLC.

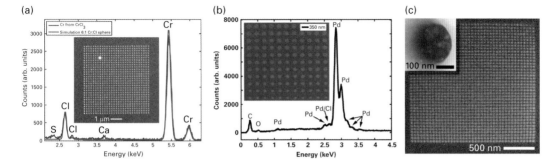

Figure 14.7. XEDS spectra and SEM images of transition metals deposited via LP-FEBID on polyimide membranes. (a) Array of Cr nanoparticles (1 M CrCl$_3$), showing good match between experimental and simulated spectra for Cr:Cl ratio of 6:1. (b) Dose array of Pd nanoparticles (20 mM Na$_2$PdCl$_4$). (c) Dense array (50 nm pitch) of high resolution Ni nanoparticles (1 mM NiSO$_4$, 20 keV, 3 pC/dot), with (inset) *ex situ* (i.e. liquid-side) tilted-view image of one Ni-filled indentation in the membrane.

14.3.2 Chromium, Palladium, and Nickel

Chromium has been a long-standing material of interest for FEBIP because of its importance in semiconductor photomasks, although mask repair now involves a large variety of materials. For gas-phase processes, Cr(CO)$_6$ is a standard precursor and produces relatively high quality deposits [61]. There has been very limited work with liquid precursors, but Cr has been deposited on polyimide membranes from aqueous solutions of CrCl$_3$. As shown in Figure 14.7a, this process results in a deposit with moderate chlorine contamination [62]. Gas-phase FEBID of Pd has been reported from a Pd bis(hexaflouroacetylacetonate) precursor, achieving structures with Pd content up to 20 at.% [63]. In LP-FEBID, Pd structures were deposited on a polyimide membrane from a 20 mM aqueous solution of Na$_2$PdCl$_4$, with a representative array of nanoparticles shown in the inset SEM image in Figure 14.7b. Figure 14.7b also presents an XEDS spectrum confirming that the deposits are largely Pd with a Pd:Cl ratio of at least 19:1. Carbon and oxygen contamination are more difficult to quantify because of their presence in the membrane. Nickel deposition on a polyimide membrane has thus far achieved the highest resolution for LP-FEBID: the Ni nanoparticles shown in Figure 14.7c are ~30 nm in diameter with a pitch of 50 nm. The inset SEM image in Figure 14.7c reveals the morphology of one isolated Ni nanoparticle. It differs from that of the Pt nanoparticles in Figure 14.2c, yet a common feature for both deposits is the indentation of the polyimide membrane.

14.4 Multi-Element LP-FEBID

Multi-element FEBID constitutes a much smaller portion of the FEBID literature to date. For gas-phase deposition, as previously noted, the complexities in tuning the molecular parameters make multi-element FEBID a challenging task. However,

LP-FEBID of multi-element materials is readily accomplished using mixtures. It has been performed with combinations of single-reactant LP-FEBID precursors or by modifying recipes from wet chemical synthesis techniques.

14.4.1 AuPt and AuAg Alloys

Noble metal alloys have been deposited using mixtures of single-reactant precursors. Investigation into the resulting atomic concentration of the deposits as a function of the initial liquid solution ratios reveals a strong correlation with the kinetics of hydrated (solvated) electrons. In the case of AuPt and AuAg mixtures, the resulting deposit concentrations are consistent with first-order kinetics of hydrated electron interactions with the individual ionic complexes. By changing the initial concentration of the individual reactants, the resulting deposit composition is predictable, as demonstrated in Figure 14.8. This result strongly implies that hydrated electrons are produced by the electron beam and react in parallel with the individual metal-ionic complexes. As demonstrated, these reactions can be exploited for compositionally tunable deposits. Furthermore, the large extant literature on hydrated electron rate kinetics [64] can be used to rapidly explore new material sets.

14.4.2 Cadmium Sulfide (CdS) and Lead Sulfide (PbS)

An exciting prospect for LP-FEBID is the deposition of materials that do not have obvious, or non-toxic, gas-phase precursors. One such class of materials relates to semiconductors and photovoltaics. These have been difficult to access with FEBIP, as many of these materials are formed in bulk through wet chemical techniques such as chemical bath deposition (CBD). Using LP-FEBID, it has been shown that a liquid precursor containing cadmium sulfate, thiourea, and ammonium sulfate results in site-specific deposition driven by the electron beam. This precursor mixture is derived from standard CBD, while hydroxide-forming reagents, such as ammonia, are intentionally left out of the mixture to prevent spontaneous deposition. The resulting deposits from LP-FEBID are compositionally and structurally similar to hexagonal CdS deposited via wet chemical techniques, as shown by the TEM and photoluminescence results in Figure 14.9 [15]. Critically, the CdS deposits are formed in the vicinity of the electron beam and do not form spontaneously outside the patterned region.

The reaction mechanism for LP-FEBID of CdS is more complex than reported for AuPt and AuAg alloys. While the reaction can be attributed to first-order kinetics of hydrated-electron reactions for alloys, the same reaction kinetics do not apply for CdS deposition. If such reaction kinetics did dominate, the Cd:S ratio would heavily favor Cd. Rather, the hydrated electrons in the case of CdS are believed to initiate secondary reactions, similar to results from pulse radiolysis studies [65], which then form stoichiometric CdS.

A similar reaction is proposed for the formation of PbS nanocrystallites from a liquid solution containing lead acetate and thioacetamide. Hydrated electron reactions have been proposed [66] to liberate S ions from the thioacetamide, while the final chemical

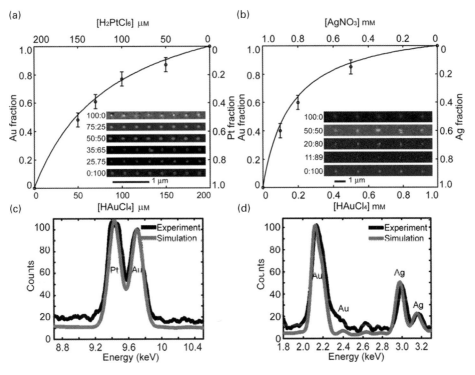

Figure 14.8. Predicted (with no fitting parameters) vs. experimental composition (determined by XEDS) of bimetallic deposits obtained via LP-FEBID on polyimide membranes (20 keV, 1.3 nA). (a) Experimental data and relative rate curve (Equations (1) and (2) in Ref. 14) for AuPt deposits; inset: SEM images of AuPt patterns deposited from solutions of $HAuCl_4$ and H_2PtCl_6 (0.2 mM total concentration, 100 pC/dot); Au:Pt solution concentration ratios listed next to each set of deposits. (b) Experimental data and relative rate curve (Equations (1) and (4) in Ref. 14) for AuAg deposits; inset: SEM images of AuAg patterns deposited from solutions of $HAuCl_4$ and $AgNO_3$ (1.0 mM total concentration, 100 pC/dot); Au:Ag solution concentration ratios listed next to each set of deposits. (c) Representative XEDS spectrum from bimetallic AuPt deposit, well matched by simulated spectrum with 50:50 Au:Pt ratio (Au:Pt ratio in solution was 25:75). (d) Representative spectrum from bimetallic AuAg deposit, well matched by simulated spectrum with 40:60 Au:Ag ratio (Au:Ag ratio in solution was 10:90). Adapted with permission from Ref. 14, © 2013 Wiley-VCH Verlag GmbH & Co. KGaA, Weinheim.

reaction is between free Pb and S ions. Figure 14.9d–i shows PbS nanocrystallites deposited from a similar liquid solution in an *in situ* TEM holder [67]. The morphology of the PbS deposits is controlled by adjusting the Pb:S ratio of compounds and the overall dilution factor. Morphology differences are attributed to selectivity of kinetic and thermal growth modes at specific solution conditions.

LP-FEBID deposition of CdS and PbS demonstrates the ability to explore reaction chemistry that would be difficult to access with traditional gas-phase FEBID. Indeed, one of the primary drawbacks of gas-phase FEBID is the lack of available precursors; hence, deposits containing Cd have not been previously reported with FEBID. In addition, the CdS and PbS results also suggest that reactive species produced by the

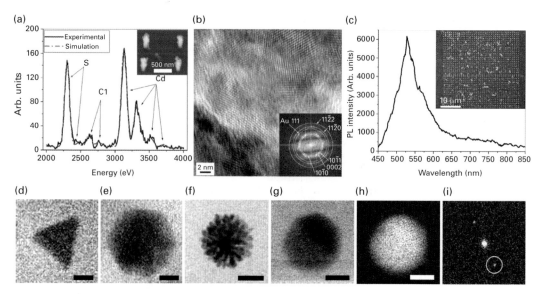

Figure 14.9. Characterization of material composition, phase, and optical properties of CdS and PbS nanostructures deposited, respectively, via LP-FEBID on polyimide membranes and *in situ* STEM. (a) XEDS analysis indicates approximately 1:1 ratio of Cd to S, with 6.4 at.% Cl contamination; inset: SEM image of rod-like CdS deposits (20 mM $CdCl_2$ + 15 mM $SC(NH_2)_2$ + 67.5 mM NH_4Cl, 20 keV, 0.25 pA, 40 pC/dot). (b) High resolution TEM image, detailing internal nanostructure of CdS deposits as predominately polycrystalline; inset: indexed Fourier transform of micrograph in (b), indicating crystallites consistent with hexagonal CdS.
(c) Photoluminescence (PL) spectrum of CdS dot array (inset: SEM image), exhibiting primary peak (528 nm) in green PL band, consistent with donor–acceptor recombination. (d–g) Various PbS nanocrystallite geometries grown with *in situ* STEM; (h) dark field and (i) corresponding FFT. Scale bars: 12.5 nm in (d, e), 25 nm in (f), and 2.5 nm in (g, h). (a–c) From Ref. 15, © 2013 IOP Publishing. Reproduced with permission. All rights reserved. (d–i) Reprinted with permission from Ref. 67, © 2011 American Chemical Society.

electron beam interaction with water may have chemical selectivity, while solution concentration tuning may affect the particle growth modes.

14.5 Liquid-Phase Focused Electron Beam-Induced Etching (LP-FEBIE)

Whether in the gas or liquid phase, focused electron beam-induced processing can only reach its full technological potential if nanostructures can be created and modified via both additive and subtractive patterning of materials (Figure 14.1). Since etch products should be volatile at room temperature, relatively few suitable precursors exist for FEBIE, limited in their elemental composition to hydrogen, oxygen, nitrogen, the halogens, and the noble gases. The most widely used etchant in gas-phase FEBIE is xenon difluoride (XeF_2), being one of the few noble-gas compounds stable at room temperature [2]. Gas- or vapor-assisted etching under electron beam irradiation has been reported for Si [68, 69] and silicon oxide, nitride, and carbide [70]; Ge [71, 72]; graphene

[73], diamond, and carbon nanotubes; alumina [74, 75], GaAs [76] (including AlGaAs) and GaN; TaN; Cr and chromium oxide; W; Ti [77, 78], photoresist and PMMA; see Refs. 2, 3, 9, and 79 for overviews and additional references. On the other hand, FEB-induced etching using bulk liquid reactants is currently in its infancy: thus far, LP-FEBIE has been reported only for Cu and silicon nitride, which we discuss below.

14.5.1 Silicon Nitride (Si_3N_4, $Si_{x>3}N_4$)

One of the earliest studies of gas-assisted electron beam-induced etching of silicon nitride was reported in 1979 by Coburn and Winters [80], who observed linear-with-time removal of Si_3N_4 only in the concurrent presence of the etchant gas (XeF_2) and the electron beam. More recently, advances in gas-assisted FEBIE of Si_3N_4 [70, 75, 81–83] have been spurred by the technological promise of artificial nanopores in inert membranes as platforms for biophysics experiments and biotechnology applications, such as the detection and sequencing of biomolecules [84]. Liquid-phase FEBIE of silicon nitride was first reported in 2013 [85], and some results are shown in Figure 14.10.

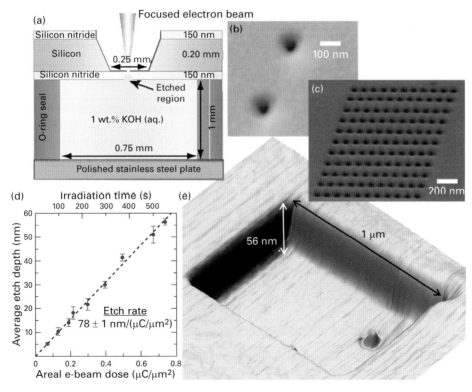

Figure 14.10. (a) Cross-sectional schematic of closed liquid cell for LP-FEBIE of silicon nitride (1 wt% KOH, 20 keV, ~1.5 nA). *Ex situ* tilted-view SEM images of holes etched into 100 nm thick, low stress Si_xN_y membrane (from SPI Supplies®) at (b) 300 nm lateral spacing and (c) 100 nm pitch dense array (2 nC/dot). (d) Etch rate (78 nm per µC/µm²) for larger pits in 150 nm thick, low stress Si_xN_y membrane, and (e) AFM image of 56 nm deep etch pit (240 scans at 2.5 s per scan).

A home-built stainless-steel liquid cell (Figure 14.10a), with a commercial low stress Si_xN_y ($x/y > 3/4$) membrane window, was filled with an aqueous solution of potassium hydroxide (KOH: 1 wt%), and the patterns were exposed in an electron beam lithography tool; after LP-FEBIE, the etched side of the membrane was inspected *ex situ* by SEM and AFM. Figure 14.10b, c shows SEM images of both sparse and dense single-pixel etched holes with diameters of ~50 nm. Etching larger rectangles (~1 μm²) revealed a linear relationship between etch depth and irradiation dose, plotted in Figure 14.10d, with an etch rate of 78 ± 1 nm per μC per μm², which corresponds to a volume removal efficiency of ~80 nm³/pC. Previous work with FEBIE of silicon nitride using XeF_2 [81] and H_2O [75] in gas-phase processes yielded lower efficiencies of ~10 nm³/pC and ~2 nm³/pC, respectively. However, some care must be taken in the comparison because processing conditions and patterns differ significantly in the three cases. An AFM scan of the deepest etch from Figure 14.10d is displayed in Figure 14.10e; the total exposure time for this pit was 600 seconds. In a control experiment using similar electron beam exposures but with an aqueous solution of potassium chloride (KCl: 1.3 wt%) as the liquid reactant, no etching was detected by *ex situ* SEM (not shown). Therefore, the mechanism for LP-FEBIE of silicon nitride likely involves a two-step process of Si_xN_y oxidation and subsequent dissolution in KOH. It is worth bearing this etching process in mind when using liquid cells with Si_xN_y membranes for experiments that include KOH.

14.5.2 Copper

Gas-assisted, focused ion beam-induced etching of Cu is a standard tool in integrated circuit editing and debugging (e.g. see Ref. 86). Electron beam-induced etching of Cu is desirable for improved resolution and selectivity with reduced damage; however, effective gas-phase processes remain elusive because of the lack of volatile Cu etch

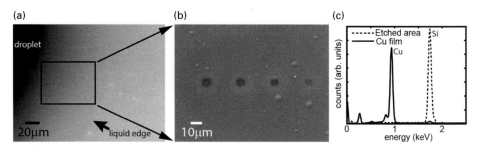

Figure 14.11. Liquid-phase focused electron beam-induced etching of 150 nm thick Cu film sputtered on Si. The liquid solution contained sulfuric acid and sodium dodecyl sulfate, the sample was held at 3 °C, and the chamber had a water-vapor background pressure of 770 Pa. The primary beam energy was 30 keV, with a beam current of 0.55 nA and a dose of 670 mC/cm². (a) *In situ* ESEM image showing liquid droplet, thin liquid film spreading over Cu film, and etched region. (b) *Ex situ* ESEM image of 5 μm × 5 μm squares etched into Cu film; liquid thickness decreases from left to right. (c) XEDS measurement of third square from the left in (b); XEDS confirmed that Cu was completely etched through for all but the right-most square in (b).

products at practical temperatures. Liquid-phase etching offers a promising alternative because there are a number of water-soluble Cu compounds. A good candidate etchant for open systems is H_2SO_4 because it has a low vapor pressure [87], does not spontaneously etch pure Cu, and offers soluble Cu sulfate as an etch product. Figure 14.11 shows a recent demonstration [88] that sputtered Cu films can be etched controllably in aqueous solutions of H_2SO_4 and sodium dodecyl sulfate (a surfactant) using an environmental SEM. The temperature (3.0 °C) and background water vapor pressure (770 Pa) imply an equilibrium H_2SO_4 concentration of ~3 M [89]. Etch rates appeared to increase with liquid layer thickness suggesting that reactions or mass transport in the bulk solution influence etching. Micrometer-scale features could be etched with excellent selectivity to the underlying silicon, but some etching was also observed in proximity to the target features. Efforts are currently underway to better control liquid layer thickness and solution composition to improve etch resolution and edge definition.

14.6 Mechanisms for LP-FEBIP

Although the interaction between an electron beam and liquid precursors results in observable surface modification, the fundamental mechanism is still under investigation. Mechanism possibilities include the generation of hydrated electrons as a reactive species and the potential for the electron beam to act as a driver of electrochemistry.

Hydrated (solvated) electrons (e_h^-) are well-known particles in the field of pulse radiolysis. While pulse radiolysis uses much higher beam energies (MeV) compared to SEM (typically keV), it is likely that SEM produces hydrated electrons in a similar manner, as discussed in Chapter 4. A hydrated electron is a result of water molecules reorienting to screen the charge of a thermalized electron, which may be produced by the primary beam or excited secondary electrons. Hydrated electrons have a strong reduction potential (−2.7 V), which enables tantalizing possibilities for chemical reactions. The data shown in Figure 14.8 for bimetallic alloys [14] show that LP-FEBID is well correlated to hydrated electron kinetics. However, parallel hydrated electron reactions do not adequately explain the resulting composition from LP-FEBID of CdS or PbS. It is also possible that radiolysis products, such as OH, H_2O_2, and e_h^- radicals, could act as intermediaries for secondary chemical reactions. While the presence of these species has not been experimentally confirmed during LP-FEBIP experiments, it is generally considered that they likely play a role in the process; experiments involving scavenger species could address this point. Further discussion regarding the production of various radiolysis species and the role of the hydrated electron in reducing metal ions in aqueous solutions can be found in Chapter 7.

Alternatively, the electron beam can be considered as the cathode in an electrochemical cell, with the sample (at ground) as the anode. This formalism is suggested by Randolph et al. [37], whereby the liquid injector needle can be biased as a reference electrode. The mechanism is supported by evidence of deposition in regions with conductive vs. insulating paths to ground. The electron beam has also been used to induce reactions within an electrochemical cell [19]. While these mechanisms are not mutually exclusive, the exact mechanism is likely complicated and requires further investigation, such as the observation, or lack thereof, of key reactive species during electron beam irradiation.

14.7 Outlook

Focused electron beam-induced processing in liquids has introduced a broad range of new reactants and mechanisms that may ultimately impact nanoscale rapid prototyping, nano-to-microscale interfacing, and device and circuit edit. The most promising early results have centered on higher-purity deposition of metals, deposition of new materials with no known gas-phase precursors, the etching of materials with soluble but non-volatile byproducts, and etching with high selectivity between materials. These opportunities are accompanied by a number of challenges. Most practical applications require bulk substrates, and maintaining high resolution requires that thin liquid layers be introduced, controlled, and exchanged in partial vacuum or in liquid cells with narrow membrane-to-substrate gaps. In addition, the mechanisms responsible for deposition and etching – radiation-chemical, electro-chemical, or otherwise – must be better understood in order to control composition and selectivity, maximize resolution, minimize proximity effects, optimize morphology, and adapt deposition to varied substrates. Progress will certainly require extensive modeling efforts across a range of length and time scales. Better understanding of these mechanisms will also impact liquid cell electron microscopy by clarifying, and perhaps mitigating, undesirable beam effects. Ultimately, the promise and challenge of liquid-phase FEBIP both lie in controlling radiation- and electro-chemical processes at interfaces in nanoscale volumes. As a result, novel and surprising experiments will continue to raise interesting questions for the foreseeable future.

Acknowledgements

Some previously unpublished results provided by the authors were based in part upon work supported by the National Science Foundation under Grant No. CMMI-1125998.

References

1. W. F. van Dorp and C. W. Hagen, A critical literature review of focused electron beam-induced deposition. *J. Appl. Phys.*, **104** (2008), 081301.
2. I. Utke, P. Hoffmann and J. Melngailis, Gas-assisted focused electron beam and ion beam processing and fabrication. *J. Vac. Sci. Technol. B*, **26** (2008), 1197–1276.
3. S. J. Randolph, J. D. Fowlkes and P. D. Rack, Focused, nanoscale electron beam-induced deposition and etching. *Crit. Rev. Solid State Mater. Sci.*, **31** (2006), 55–89.

4. A. Botman, J. J. L. Mulders and C. W. Hagen, Creating pure nanostructures from electron beam-induced deposition using purification techniques: a technology perspective. *Nanotechnology*, **20** (2009), 372001.
5. K. Furuya, Nanofabrication by advanced electron microscopy using intense and focused beam. *Sci. Technol. Adv. Mater.*, **9** (2008), 014110.
6. M. H. Song and K. Furuya, Fabrication and characterization of nanostructures on insulator substrates by electron beam-induced deposition. *Sci. Technol. Adv. Mater.*, **9** (2008), 023002.
7. S. W. Lee and R. M. Sankaran, Direct writing via electron-driven reactions. *Mater. Today*, **16** (2013), 117–122.
8. N. Silvis-Cividjian and C. W. Hagen, *Electron Beam-Induced Nanometer-Scale Deposition* (San Diego, CA: Academic Press, 2006).
9. I. Utke, S. Moshkalev and P. Russell, *Nanofabrication Using Focused Ion and Electron Beams: Principles and Applications* (Oxford; New York: Oxford University Press, 2012).
10. T. Takahashi, Y. Arakawa, M. Nishioka and T. Ikoma, Selective growth of GaAs wire structures by electron beam-induced metalorganic chemical vapor-deposition. *Appl. Phys. Lett.*, **60** (1992), 68–70.
11. P. A. Crozier, J. Tolle, J. Kouvetakis and C. Ritter, Synthesis of uniform GaN quantum dot arrays via electron nanolithography of D_2GaN_3. *Appl. Phys. Lett.*, **84** (2004), 3441–3443.
12. R. C. Che, M. Takeguchi, M. Shimojo, W. Zhang and K. Furuya, Fabrication and electron holography characterization of FePt alloy nanorods. *Appl. Phys. Lett.*, **87** (2005), 223109.
13. M. Winhold, P. M. Weirich, C. H. Schwalb and M. Huth, Superconductivity and metallic behavior in $Pb_xC_yO_\delta$ structures prepared by focused electron beam-induced deposition. *Appl. Phys. Lett.*, **105** (2014), 162603.
14. M. Bresin, A. Chamberlain, E. U. Donev *et al.*, Electron beam-induced deposition of bimetallic nanostructures from bulk liquids. *Angew. Chem. Int. Ed.*, **52** (2013), 8004–8007.
15. M. Bresin, B. R. Nadimpally, N. Nehru, V. P. Singh and J. T. Hastings, Site-specific growth of CdS nanostructures. *Nanotechnology*, **24** (2013), 505305.
16. M. Bresin, N. Nehru and J. T. Hastings, Focused electron beam-induced deposition of plasmonic nanostructures from aqueous solutions. In *Proc. SPIE 8613, Advanced Fabrication Technologies for Micro/Nano Optics and Photonics VI* (2013), p. 861306.
17. X. Chen, L. H. Zhou, P. Wang *et al.*, A study of electron beam-induced deposition and nano device fabrication using liquid cell TEM technology. *Chinese J. Chem.*, **32** (2014), 399–404.
18. X. Chen, L. H. Zhou, P. Wang, C. J. Zhao and X. L. Miao, A study of nano materials and their reactions in liquid using in situ wet cell TEM technology. *Chinese J. Chem.*, **30** (2012), 2839–2843.
19. M. den Heijer, I. Shao, A. Radisic, M. C. Reuter and F. M. Ross, Patterned electrochemical deposition of Cu using an electron beam. *APL Mater.*, **2** (2014), 022101.
20. E. U. Donev and J. T. Hastings, Liquid-precursor electron beam-induced deposition of Pt nanostructures: dose, proximity, resolution. *Nanotechnology*, **20** (2009), 505302.
21. E. U. Donev and J. T. Hastings, Electron beam-induced deposition of Pt from a liquid precursor. *Nano Lett.*, **9** (2009), 2715–2718.
22. E. U. Donev, G. Schardein, J. C. Wright and J. T. Hastings, Substrate effects on the electron beam-induced deposition of Pt from a liquid precursor. *Nanoscale*, **3** (2011), 2709–2717.
23. J. M. Grogan, N. M. Schneider, F. M. Ross and H. H. Bau, Bubble and pattern formation in liquid induced by an electron beam. *Nano Lett.*, **14** (2014), 359–364.
24. T. Hoshino and K. Morishima, Electron beam direct processing on living cell membrane. *Appl. Phys. Lett.*, **99** (2011), 174102.

25. E. Jensen, C. Kobler, P. S. Jensen and K. Molhave, In-situ SEM microchip setup for electrochemical experiments with water based solutions. *Ultramicroscopy*, **129** (2013), 63–69.
26. N. Kolmakova and A. Kolmakov, Scanning electron microscopy for in situ monitoring of semiconductor-liquid interfacial processes: electron assisted reduction of Ag ions from aqueous solution on the surface of TiO_2 rutile nanowire. *J. Phys. Chem. C*, **114** (2010), 17233–17237.
27. T. Kraus and N. de Jonge, Dendritic Au nanowire growth observed in liquid with transmission electron microscopy. *Langmuir*, **29** (2013), 8427–8432.
28. Y. Liu, X. Chen, K. W. Noh and S. J. Dillon, Electron beam-induced deposition of silicon nanostructures from a liquid phase precursor. *Nanotechnology*, **23** (2012), 385302.
29. Y. Liu, K. P. Tai and S. J. Dillon, Growth kinetics and morphological evolution of ZnO precipitated from solution. *Chem. Mater.*, **25** (2013), 2927–2933.
30. K. W. Noh, Y. Liu, L. Sun and S. J. Dillon, Challenges associated with in-situ TEM in environmental systems: the case of silver in aqueous solutions. *Ultramicroscopy*, **116** (2012), 34–38.
31. L. E. Ocola, A. Joshi-Imre, C. Kessel et al., Growth characterization of electron beam-induced silver deposition from liquid precursor. *J. Vac. Sci. Technol. B*, **30** (2012), 06FF08.
32. G. Schardein, E. U. Donev and J. T. Hastings, Electron beam-induced deposition of Au from aqueous solutions. *Nanotechnology*, **22** (2011), 015301.
33. T. J. Woehl, J. E. Evans, L. Arslan, W. D. Ristenpart and N. D. Browning, Direct in situ determination of the mechanisms controlling nanoparticle nucleation and growth. *ACS Nano*, **6** (2012), 8599–8610.
34. J. M. Yuk, J. Park, P. Ercius et al., High-resolution EM of colloidal nanocrystal growth using graphene liquid cells. *Science*, **336** (2012), 61–64.
35. H. M. Zheng, R. K. Smith, Y. W. Jun et al., Observation of single colloidal Pt nanocrystal growth trajectories. *Science*, **324** (2009), 1309–1312.
36. E. U. Donev, N. Nehru, G. Schardein et al., Recent advances in liquid-phase electron beam-induced deposition: characterizing growth processes and optical properties. *Microsc. Microanal.*, **17** (2011), 438–439.
37. S. J. Randolph, A. Botman and M. Toth, Capsule-free fluid delivery and beam-induced electrodeposition in a scanning electron microscope. *RSC Adv.*, **3** (2013), 20016–20023.
38. M. Bresin, A. Botman, S. J. Randolph, M. Straw and J. T. Hastings, Liquid phase electron beam-induced deposition on bulk substrates using environmental scanning electron microscopy. *Microsc. Microanal.*, **20** (2014), 376–384.
39. T. Tsuda, S. Seino and S. Kuwabata, Au nanoparticles prepared with a room-temperature ionic liquid-radiation irradiation method. *Chem. Commun.*, **44** (2009), 6792–6794.
40. P. Roy, R. Lynch and P. Schmuki, Electron beam-induced in-vacuo Ag deposition on TiO_2 from ionic liquids. *Electrochem. Commun.*, **11** (2009), 1567–1570.
41. A. Imanishi, M. Tamura and S. Kuwabata, Formation of Au nanoparticles in an ionic liquid by electron beam irradiation. *Chem. Commun.*, **44** (2009), 1775–1777.
42. A. Imanishi, S. Gonsui, T. Tsuda, S. Kuwabata and K. Fukui, Size and shape of Au nanoparticles formed in ionic liquids by electron beam irradiation. *Phys. Chem. Chem. Phys.*, **13** (2011), 14823–14830.
43. N. de Jonge, Introduction to special issue on electron microscopy of specimens in liquid. *Microsc. Microanal.*, **20** (2014), 315–316.
44. N. de Jonge, and F. M. Ross, Electron microscopy of specimens in liquid. *Nat. Nanotechnol.*, **6** (2011), 695–704.

45. S. Thiberge, O. Zik and E. Moses, An apparatus for imaging liquids, cells, and other wet samples in the scanning electron microscope. *Rev. Sci. Instrum.*, **75** (2004), 2280–2289.
46. D. R. Ciarlo, Silicon nitride thin windows for biomedical microdevices. *Biomed. Microdevices*, **4** (2002), 63–68.
47. N. A. Stelmashenko, J. P. Craven, A. M. Donald, E. M. Terentjev and B. L. Thiel, Topographic contrast of partially wetting water droplets in environmental scanning electron microscopy. *J. Microsc. Oxford*, **204** (2001), 172–183.
48. A. Botman, J. J. L. Mulders, R. Weemaes and S. Mentink, Purification of Pt and Au structures after electron beam-induced deposition. *Nanotechnology*, **17** (2006), 3779–3785.
49. R. M. Langford, T. X. Wang and D. Ozkaya, Reducing the resistivity of electron and ion beam assisted deposited Pt. *Microelectron. Eng.*, **84** (2007), 784–788.
50. J. F. Lin, J. P. Bird, L. Rotkina and P. A. Bennett, Classical and quantum transport in focused-ion beam-deposited Pt nanointerconnects. *Appl. Phys. Lett.*, **82** (2003), 802–804.
51. L. Penate-Quesada, J. Mitra and P. Dawson, Non-linear electronic transport in Pt nanowires deposited by focused ion beam. *Nanotechnology*, **18** (2007), 215203.
52. T. Tao, J. S. Ro, J. Melngailis, Z. L. Xue and H. D. Kaesz, Focused ion beam-induced deposition of Pt. *J. Vac. Sci. Technol. B*, **8** (1990), 1826–1829.
53. K. A. Telari, B. R. Rogers, H. Fang *et al.*, Characterization of Pt films deposited by focused ion beam-assisted chemical vapor deposition. *J. Vac. Sci. Technol. B*, **20** (2002), 590–595.
54. N. W. M. Ritchie, Spectrum simulation in DTSA-II. *Microsc. Microanal.*, **15** (2009), 454–468.
55. N. W. M. Ritchie, Using DTSA-II to simulate and interpret energy dispersive spectra from particles. *Microsc. Microanal.*, **16** (2010), 248–258.
56. A. Folch, J. Servat, J. Esteve, J. Tejada and M. Seco, High-vacuum versus "environmental" electron beam deposition. *J. Vac. Sci. Technol. B*, **14** (1996), 2609–2614.
57. T. Brintlinger, M. S. Fuhrer, J. Melngailis *et al.*, Electrodes for carbon nanotube devices by focused electron beam-induced deposition of Au. *J. Vac. Sci. Technol. B*, **23** (2005), 3174–3177.
58. T. A. Green, Au electrodeposition for microelectronic, optoelectronic and microsystem applications. *Gold Bull.*, **40** (2007), 105–114.
59. V. Friedli, I. Utke, K. Molhave and J. Michler, Dose and energy dependence of mechanical properties of focused electron beam-induced pillar deposits from $Cu(C_5HF_6O_2)_2$. *Nanotechnology*, **20** (2009), 385304.
60. Y. Ochiai, J. Fujita and S. Matsui, Electron beam-induced deposition of Cu compound with low resistivity. *J. Vac. Sci. Technol. B*, **14** (1996), 3887–3891.
61. R. R. Kunz and T. M. Mayer, Electron beam-induced surface nucleation and low-temperature decomposition of metal-carbonyls. *J. Vac. Sci. Technol. B*, **6** (1988), 1557–1564.
62. A. Chamberlain, E. U. Donev, C. B. Samantaray *et al.*, Electron beam-induced deposition of transition metals from bulk liquids: Ag, Cr, and Ni. In 55th International Conference on Electron, Ion, and Photon Beam Technology and Nanofabrication (Las Vegas, NV, 2011).
63. D. Spoddig, K. Schindler, P. Rodiger *et al.*, Transport properties and growth parameters of PdC and WC nanowires prepared in a dual-beam microscope. *Nanotechnology*, **18** (2007), 495202.
64. M. Anbar and P. Neta, A compilation of specific bimolecular rate constants for the reactions of hydrated electrons, hydrogen atoms and hydroxyl radicals with inorganic and organic compounds in aqueous solution. *Int. J. Appl. Radiat. Isot.*, **18** (1967), 493–523.

65. D. Hayes, O. I. Micic, M. T. Nenadovic, V. Swayambunathan and D. Meisel, Radiolytic production and properties of ultrasmall cadmium sulfide particles. *J. Phys. Chem.*, **93** (1989), 4603–4608.
66. M. H. Wu, H. J. Zhong, Z. Jiao, Z. Li and Y. F. Sun, Synthesis of PbS nanocrystallites by electron beam irradiation. *Coll. Surf. A*, **313** (2008), 35–39.
67. J. E. Evans, K. L. Jungjohann, N. D. Browning and I. Arslan, Controlled growth of nanoparticles from solution with in situ liquid transmission electron microscopy. *Nano Lett.*, **11** (2011), 2809–2813.
68. P. Roediger, G. Hochleitner, E. Bertagnolli, H. D. Wanzenboeck and W. Buehler, Focused electron beam-induced etching of silicon using chlorine. *Nanotechnology*, **21** (2010), 285306.
69. P. Roediger, H. D. Wanzenboeck, G. Hochleitner, E. Bertagnolli and W. Buehler, Focused electron beam-induced etching of silicon by chlorine gas: negative effects of residual gas contamination on the etching process. *J. Appl. Phys.*, **108** (2010), 124316.
70. A. A. Martin and M. Toth, Cryogenic electron beam-induced chemical etching. *ACS Appl. Mater. Inter.*, **6** (2014), 18457–18460.
71. P. Roediger, M. Mijic, C. Zeiner et al., Local, direct-write, damage-free thinning of germanium nanowires. *J. Vac. Sci. Technol. B*, **29** (2011), 06FB03.
72. P. Roediger, H. D. Wanzenboeck, G. Hochleitner and E. Bertagnolli, Crystallinity-retaining removal of germanium by direct-write focused electron beam-induced etching. *J. Vac. Sci. Technol. B*, **29** (2011), 041801.
73. D. Fox, A. O'Neill, D. Zhou et al., Nitrogen assisted etching of graphene layers in a scanning electron microscope. *Appl. Phys. Lett.*, **98** (2011), 243117.
74. T. Bret, B. Afra, R. Becker et al., Gas assisted focused electron beam-induced etching of alumina. *J. Vac. Sci. Technol. B*, **27** (2009), 2727–2731.
75. P. S. Spinney, D. G. Howitt, R. L. Smith and S. D. Collins, Nanopore formation by low-energy focused electron beam machining. *Nanotechnology*, **21** (2010), 375301.
76. A. Ganczarczyk, M. Geller and A. Lorke, XeF_2 gas-assisted focused-electron beam-induced etching of GaAs with 30 nm resolution. *Nanotechnology*, **22** (2011), 045301.
77. J. H. Noh, J. D. Fowlkes, R. Timilsina et al., Pulsed laser-assisted focused electron beam-induced etching of titanium with XeF_2: enhanced reaction rate and precursor transport. *ACS Appl. Mater. Inter.*, **7** (2015), 4179–4184.
78. F. J. Schoenaker, R. Cordoba, R. Fernandez-Pacheco et al., Focused electron beam-induced etching of titanium with XeF_2. *Nanotechnology*, **22** (2011), 265304.
79. M. Toth, Advances in gas-mediated electron beam-induced etching and related material processing techniques. *Appl. Phys. A*, **117** (2014), 1623–1629.
80. J. W. Coburn and H. F. Winters, Ion-assisted and electron-assisted gas-surface chemistry: important effect in plasma-etching. *J. Appl. Phys.*, **50** (1979), 3189–3196.
81. M. Yemini, B. Hadad, Y. Liebes, A. Auner and N. Ashkenasy, The controlled fabrication of nanopores by focused electron beam-induced etching. *Nanotechnology*, **20** (2009), 245302.
82. Y. Liebes, B. Hadad and N. Ashkenasy, Effects of electrons on the shape of nanopores prepared by focused electron beam-induced etching. *Nanotechnology*, **22** (2011), 285303.
83. P. A. Crozier, Nanoscale oxide patterning with electron-solid-gas reactions. *Nano Lett.*, **7** (2007), 2395–2398.
84. C. Dekker, Solid-state nanopores. *Nat. Nanotechnol.*, **2** (2007), 209–215.

85. E. U. Donev, C. B. Samantaray, M. Bresin and J. T. Hastings, Recent advances in liquid-phase e-beam-induced processing: silicon nitride etching and palladium deposition. In 39th International Conference on Micro and Nano Engineering (London, 2013), p. O-FEBIP-04.
86. Y. Drezner, Y. Greenzweig and A. Raveh, Strategy for focused ion beam compound material removal for circuit editing. *J. Vac. Sci. Technol. B*, **30** (2012), 011207.
87. A. Jaeckervoirol, J. L. Ponche and P. Mirabel, Vapor-pressures in the ternary-system water nitric-acid sulfuric-acid at low-temperatures. *J. Geophys. Res. Atmos.*, **95** (1990), 11857–11863.
88. M. Bresin and J. T. Hastings, Etching of Cu using liquid reactants and a focused electron beam. In International Conference on Electron, Ion, and Photon Beam Technology and Nanofabrication (Washington, D.C., 2014).
89. M. Massucci, S. L. Clegg and P. Brimblecombe, Equilibrium vapor pressure of H_2O above aqueous H_2SO_4 at low temperature. *J. Chem. Eng. Data*, **41** (1996), 765–778.

15 Liquid Cell TEM for Studying Environmental and Biological Mineral Systems

Michael H. Nielsen and James J. De Yoreo

15.1 Introduction

Nucleation and growth of crystals is a pervasive phenomenon in environmental, geochemical, and biological environments. Mineralization in soils is closely coupled to the activity of microbial systems [1, 2]. Moreover, soil mineral surfaces appear to influence the residence time of organics produced by microbes as well as plant matter [3], thereby impacting the cycling of carbon on a vast scale. The connection between mineralization, biota, and climate is perhaps more pronounced in the oceans, where marine organisms ranging from single cell phytoplankton to shell-forming mollusks impact seawater chemistry on a global scale through the deposition of calcium carbonate ($CaCO_3$) [4]. These vast deposits, which are revealed in stark relief in places such as the White Cliffs of Dover, UK, represent one of the largest terrestrial reservoirs of carbon and record the interaction between Earth and biota all the way back through the Cambrian period. In mammalian systems, directed nucleation and growth of crystals by organic matrices is a central process in the formation of mineralized tissues, both functional, as in the case of bones and teeth, and pathological, as with kidney stones and arterial deposits [5]. The production of nanoparticles, whether through biomineralization or abiotic processes, has a wide range of impacts on the environment, including promotion of aerosol nucleation in the atmosphere [6] and immobilization or remobilization of heavy metals in effluent streams [2]. The reaction of minerals to environmental changes, as in the hydration of clay particles [7, 8], represents a related problem that, while decoupled from the processes of mineral formation, still depends on similar interfacial reactions.

These natural examples of crystal nucleation and growth and their widespread impacts have driven numerous scientific and technological activities. Biomedical researchers are addressing challenges related to disease of mineralized tissues, both through stimulation of tissue remineralization in the case of dental caries or broken bones, and through inhibition of mineral formation in the case of various pathologies. Geochemists are investigating ways to inject supercritical CO_2 into underground geological reservoirs to induce carbonate mineralization for sequestering CO_2 from man-made sources. Environmental scientists are seeking to understand how changing temperatures and wetting cycles during a period of climatic change will impact carbon cycling, while climate scientists are working to understand how to correctly include the action of aerosols in predictive climate models. Chemical engineers are working to develop chemical additives that prevent scaling in pipes associated with both extraction of fossil fuels and generation of geothermal power. Finally, taking inspiration from the exquisite control over crystallization exhibited by the biological world, materials scientists are attempting to understand the controlling processes in order to develop biomimetic approaches to synthesis of functional materials. In all these cases, experimental methods that can follow both nucleation and subsequent growth in real time are critical to elucidating the physical principles and chemical controls that underlie these processes. In this chapter, we review recent efforts to utilize liquid cell TEM for this purpose.

15.2 Mechanisms of Mineral Formation

Mechanisms of mineral formation remain poorly understood due in part to a lack of experimental techniques that allow for direct observations at relevant length and time scales. Early events in the establishment of a growing nucleus, including initial cluster formation, ion or cluster attachment events, and transformation of intermediate phases, occur at nanometer length scales, rendering *in situ* optical and X-ray based imaging and spectroscopic methods of limited value. For surface-mediated nucleation where *in situ* scanning probe microscopy can be utilized, these processes can be captured with adequate spatial resolution. However, they often occur on time scales that are short compared to the time scale required for image collection, leading to uncertainty about initial formation events and the occurrence of short-lived intermediate phases [9]. Moreover, because surfaces often dramatically alter the free energies and kinetic barriers that control the location, rate, and mineral phase of nucleation [10, 11], methods that can minimize surface effects are desirable. Finally, for investigations of the role played by organic matrices, such as amelogenin [12], collagen [13], lipid vesicles [14], and globular phases of macromolecules [15], where mineral formation is likely to occur within the matrix, scanning probe techniques that can only look at surfaces are of little use.

The paucity of tools that allow for direct, *in situ* investigations into mineralization mechanisms has resulted in alternative approaches being utilized for understanding these processes. *Ex situ* imaging [16], spectroscopic [17], and other analytical tools [18] have provided snapshots at different time points of mineralization processes, while *in situ* techniques lacking adequate spatial and/or temporal resolution have yielded indirect evidence of pathways [9, 19]. The lack of direct experimental data has given rise to different theories of nucleation that run counter to the model embodied in classical nucleation theory, first proposed by Gibbs [20]. According to classical nucleation theory, *unstable* density fluctuations inherent in a solution at finite temperature give rise to nucleation through monomer-by-monomer attachment events that, by chance, create a cluster whose dimensions exceed the size of a critical nucleus, at which point continued growth is energetically favorable. In the case of calcium carbonate crystallization, analytical ultracentrifugation and titration data have been used to suggest the existence of a *stable* multi-ion species of "pre-nucleation clusters", which aggregate into an amorphous precursor phase before transformation into one of the crystalline phases [21]. Cryogenic transmission electron microscopy (cryo-TEM) data that revealed clusters in solutions were cited as supporting evidence [22]. However, the static data could not be used to determine whether the clusters were indeed stable or were transient species resulting from the density fluctuations expected in classical nucleation theory, nor could they be used to establish a link between the observed clusters and the nuclei that eventually appear. NMR data were used to propose a metastable, bicarbonate-rich dense liquid precursor phase [23] which forms prior to nucleation of the solid phase, and molecular dynamics simulations predicted the formation of highly dynamic clusters [24] followed by a liquid–liquid phase separation occurring through spinodal decomposition [25] at moderate supersaturations. Formation of solid mineral phases was then inferred to occur through dehydration of the dense liquid droplets to form amorphous calcium carbonate (ACC). Furthermore, numerous calcium carbonate crystallization studies have documented the formation of ACC as the first bulk phase [26], which then transforms into one of the crystalline phases. However, neither *in situ* X-ray studies [19, 27] nor *ex situ* electron microscopy [22] or X-ray spectroscopy [17] could determine whether the observed transformation was the result of a dissolution–reprecipitation process or direct phase transformation.

Observations on other systems have reinforced the idea that mineral nucleation pathways are both complex and involve aggregation of clusters and transformation of precursor phases. For example, when applied to calcium phosphate solutions held at high supersaturation, cryo-TEM provided direct evidence for the formation of a polymeric network of nanometer-size amorphous particles [28]. These gave rise to the final crystalline phase through a series of chemical, structural, and morphological transformations. Atomic force microscopy data reinforced the idea that the initial amorphous phase was formed from particles through an energetic analysis based on nucleation rate data. However, once again none of these techniques could provide direct evidence for particle formation and subsequent aggregation, leaving the results open to other interpretations.

Part of the reason for the seemingly complicated landscape in these mineral systems is the existence of multiple crystalline polymorphs with similar solubilities and relatively large interfacial energies. Due to the latter, in order to study nucleation of these materials on a manageable experimental time scale, the solution must be driven to high supersaturation. This results in multiple phases and pathways becoming thermodynamically available. For example, in the $CaCO_3$ system, where the interfacial free energy is over 100 mJ/m^2, even when the concentration is raised to the solubility limit of the amorphous phase so that all four of the common polymorphs are stable with respect to the dissolved state, the free energy barrier to forming calcite, which is the most stable phase, is still on the order of 100 kT [9]. Hence pathways can be complex. However, mineral systems are not the only systems with hierarchical and/or multiple pathways to the final crystalline state. Organic systems are notorious for exhibiting both multiple polymorphs and two-step nucleation pathways [29]. This behavior is also prevalent in proteins, even though they have low interfacial energies, because conformational disorder and tendencies towards aggregation often lead to condensation prior to crystallization due to kinetic limitations on ordering [30, 31].

The mechanisms of single crystal growth, post-nucleation, are likewise poorly understood. While many systems exhibit growth through monomer-by-monomer attachment, as predicted by well-developed classical theories, particle-based mechanisms are now recognized as important and perhaps dominant in a range of biological, environmental, and synthetic systems. The first discovery of single crystal growth by a particle-based process involved the attachment of TiO_2 crystals on specific crystallographic faces [32], giving rise to the term "oriented attachment" or "OA". OA has since been inferred as responsible for the growth of many materials, particularly semiconductor nanoparticles and nanowires exhibiting complex topologies, such as branched, chain-like, or tetrapod-shaped structures. However, particle-based growth mechanisms can also include nanoparticle attachment with random or somewhat misaligned orientations, followed by relaxation that eliminates the defects to produce a well-ordered single crystal. In still other cases, one observes the attachment of particles, composed of a phase – often disordered – that is either metastable or is stable only at the nanoscale, onto the surface of a crystal composed of the stable bulk phase [33, 34]. Attachment is then followed by recrystallization of the nanoscale phase to create a single crystal. Thus the final products of particle-based growth are rarely indicative of the pathway by which the crystals formed and, indeed, may contain no evidence of the pathway. The overwhelming majority of observations used to infer particle-based pathways have been based on *ex situ* techniques.

These and many other recent studies, though sometimes controversial or inconclusive, imply that there are numerous potential nucleation pathways and growth mechanisms that may contribute to the mineralization process [35]. This is illustrated in Figure 15.1. In order to develop a definitive picture of mineral formation and truly understand the physics underlying these "non-classical" mechanisms, it is essential to know which processes are active. Moreover, all pathways are dynamic responses to gradients in interaction potentials, changes in free energy, and barriers separating initial and final states; thus, observations of rates provide an opportunity to quantify the energetics of nucleation and growth. Consequently, the ability provided by liquid cell TEM to directly image crystal formation and growth while simultaneously gaining information on phase and crystallographic orientation is an invaluable asset in advancing our knowledge of mineralization.

15.3 Liquid Holder Design

As discussed thoroughly in Chapter 2, there now exist a variety of approaches for liquid cell TEM. Systems incorporating suspended silicon nitride membranes on microfabricated silicon wafers, to yield fluid cells with static or flowing liquid layers, have been used to acquire the data presented below. Two different experimental approaches using a fluid flow holder are shown schematically in Figure 15.2. Static cells (described in

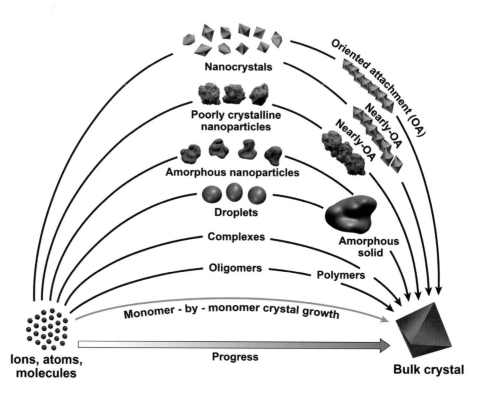

Figure 15.1. A plethora of nucleation and growth mechanisms have been proposed and/or observed for mineralization from solution. Formation mechanisms can rarely if ever be inferred from the final product, highlighting the importance of *in situ* observations of formation. This figure is reproduced from Ref. 35 and used with permission by AAAS.

detail in Chapter 2) are typically filled with undersaturated solutions (or low supersaturation solutions if the kinetics are slow enough) and provided with an external driving force to initiate the process of interest. This stimulus can be either the electron beam itself [36], an electrical bias [37], or temperature modulation [38]. Flow cells are generally filled with water or, in the case of multiple inlet designs, one reactant, which is then displaced by supersaturated solutions or mixed with combined reagent streams (Figure 15.2a) flowing in from one or multiple inlets, respectively. While static cells often require an external driving force to initiate the reaction, they have an advantage in the well-defined starting condition within the cell. By contrast, the flow patterns within, and mixing characteristics of flow cells and holders are largely unknown, leading to a degree of uncertainty as to the solution composition in the imaging area, particularly in the case of multiple inlet flow holders. These latter systems do, however, provide the capability of mixing multiple reagents to create supersaturated solutions and drive nucleation and growth processes in the absence of other external stimuli. Furthermore, they can be used to passively diffuse gas into a liquid droplet contained within the fluid cell (Figure 15.2b), broadening the experimental approaches feasible with such devices.

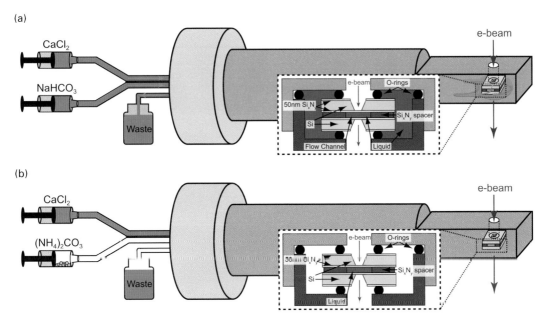

Figure 15.2. (a) Schematic of dual-inlet liquid flow holder configured for mixing $CaCO_3$ precursor solutions, with inset showing a side-view schematic of the sample region. Two separate inlets connect syringes to fluid lines that combine into a single solution stream shortly upstream of the liquid cell. Solution flows through a channel in the liquid cell and around the outside of the cell, exiting the holder through a third line to a waste container. (Inset) The liquid cell enclosed in the pocket of the holder. O-rings seal the liquid environment from the vacuum of the TEM column. (b) Dual-inlet flow holder configured for diffusing gaseous decomposition products of $(NH_4)_2CO_3$ into Ca^{2+}-rich droplets in the liquid cell, with inset showing the sample region in detail. The cell is pre-wet with a droplet of $CaCl_2$ solution during assembly. A syringe containing $(NH_4)_2CO_3$ powder is connected to an empty fluid line, to allow $NH_3(g)$ and $CO_2(g)$ to diffuse through the system and into the $CaCl_2$ solution filling the channel between the windows of the liquid cell. The diffusion of $CO_2(g)$ into the Ca^{2+}-rich solution continually increases the supersaturation of solid $CaCO_3$ phases, creating a driving force for precipitation. (a) Reproduced from Ref. 39 and used with permission by AAAS.

15.4 Calcium Carbonate Formation Pathways

Figure 15.2a shows how a dual inlet liquid cell TEM holder can be used to study formation pathways of $CaCO_3$ [39]. Reagent streams of $CaCl_2$ and $NaHCO_3$ flow separately into the holder and are combined into a single liquid stream shortly before reaching the liquid cell. Calcium carbonate is of interest due to the issues laid out in Section 15.1, with its complexity arising from the existence of multiple crystalline phases and amorphous states and the suggested existence of potential additional precursors. While calcite is the energetically stable anhydrous phase at standard temperature and pressure, aragonite and ikaite are the stable anhydrous and hydrated crystalline phases, respectively, at other conditions [40]. Vaterite is a metastable

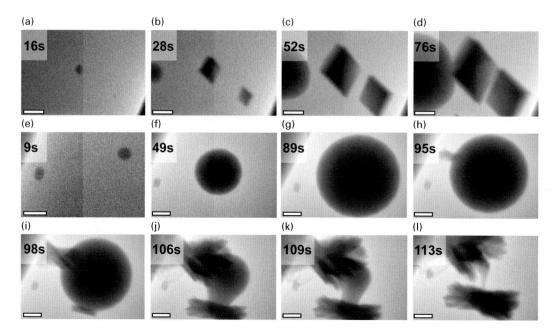

Figure 15.3. Direct and indirect formation pathways. (a–d) Concurrent formation of calcite rhombohedra and vaterite with no apparent precursor phase. (e) Initial formation of amorphous particle on the right side of the frame, followed by a period of growth (f, g). (h) Nucleation of aragonite on or near the surface of the amorphous particle leads to the growth of the crystalline phase at the expense of the amorphous precursor (i–l). Scale bars are 500 nm. Adapted from Ref. 39 with permission by AAAS.

anhydrous crystalline phase across the temperature–pressure phase space, as is the hydrated crystal monohydrocalcite. Furthermore there exist one [26] or multiple [18] metastable amorphous states. This multitude of phases has stimulated a number of investigations into the phase pathway or pathways taken to form the final crystals.

Our investigation into calcium carbonate formation has shown the existence of multiple pathways, both single and multi-step. Representative examples are shown in Figure 15.3. The first image sequence shows the formation and growth of calcite rhombohedra in the center and lower right of the frame. Concurrent to this is the formation of another particle, roughly circular in projection, at the left of the image. We could not obtain diffraction data for this particle but structure is evident from contrast differences in the particle. As such there is some ambiguity as to this particle's phase, as both ACC and vaterite have been observed to have similar appearance. However, the amorphous phase has been widely observed to dissolve in the presence of calcite and, in our experiments, is rapidly consumed when in contact with other phases, while in this sequence the two phases can be seen to grow post-contact. Thus it is likely that the second phase seen in this sequence is vaterite. The second image series shows the initial formation and growth of the amorphous phase. The particle grew to a diameter of a few micrometers, then briefly shrank before the appearance of two aragonite bundles at or just below the surface of the particle. The aragonite then grew

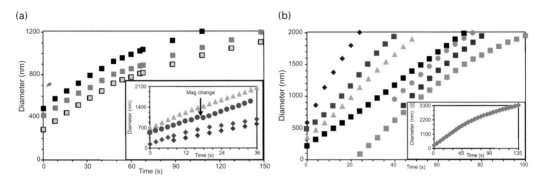

Figure 15.4. CaCO₃ post-nucleation growth rates. Lateral growth measurements of ACC (a) and vaterite (b) particles throughout post-nucleation growth show linear growth, indicating surface kinetics are the controlling factor for particle growth. Adapted from Ref. 39 and used with permission by AAAS.

at the expense of the ACC, while the two phases maintained physical contact throughout the transformation. This and other such observations demonstrate that, during multi-step nucleation pathways incorporating an amorphous precursor, the crystal forms through direct transformation rather than via complete ACC dissolution and re-precipitation of the solute as the crystal.

With the uncertainty regarding flow and mixing profiles in the liquid holder, the question arises whether the liquid streams are actively flowing through the fluid cell at all, or if instead the solutes are simply diffusing into the imaging region. Figure 15.4 shows measurements of lateral growth rates of multiple phases. In numerous experiments growth was linear, indicating that it was limited by surface kinetics rather than diffusive transport [41]. Some experiments showed a single growth rate over the entire course of observation, while others showed two distinct linear regions with the second rate slower than the first. A possible explanation for this phenomenon is a reduction in solute entering the liquid cell, perhaps due to nucleation events occurring upstream of the imaging region. This same phenomenon could account for the slight shrinkage of the ACC prior to aragonite formation (Figure 15.3g, h), although an alternate explanation would be the expulsion of water from a hydrated amorphous phase prior to the formation of the secondary crystalline phase.

Figure 15.5 summarizes the pathways observed: direct formation from solution, with no evidence of a precursor phase, of all three anhydrous crystalline phases and the amorphous phase, as well as multi-step pathways incorporating an amorphous or crystalline precursor. An ACC precursor was found to lead to the formation of vaterite and aragonite but, significantly, not to calcite. Although the latter pathway has long been assumed to exist, this liquid cell TEM data adds to the body of evidence that this process is unlikely and, indeed, has never been directly observed. Lastly, transformation of initially formed aragonite into calcite was observed. Phase assignment was made, when possible, by switching between imaging and diffraction modes throughout the nucleation process. However, oftentimes diffraction data were not obtained due to transformations occurring while operating the TEM in imaging mode, or because the

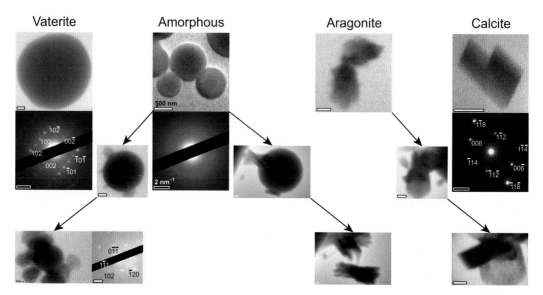

Figure 15.5. Formation pathways of CaCO$_3$. The amorphous phase and three anhydrous crystalline phases form directly from solution. Nucleation of vaterite and aragonite on existing amorphous particles leads to direct transformation from amorphous to crystal in these multi-step pathways. Crystal to crystal transformation is also observed in the case of calcite forming on aragonite and growing at the expense of the initially formed phase. Scale bars are 500 nm and 2 nm^{-1} for images and diffraction data, respectively. Elements of this figure are reproduced from Ref. 39 and used with permission by AAAS.

fluid layer was thick enough to degrade the diffraction signal beyond use. In those cases, phase was inferred based on the distinct morphologies exhibited by each phase.

With the caveat that the precise solution conditions in the imaging area are unknown, these formation pathways were found to exist across a range of supersaturations. As the solute concentrations were raised, ACC became increasingly predominant and more pathways were observed. These findings are consistent with the expectations of classical nucleation theory that as the energy barriers are lowered through increasing supersaturation, multiple phases and pathways should become increasingly accessible due to the high interfacial free energy and large free energy barrier to nucleation in this system, as discussed above. However, while the results contradict models that allow only for crystallization pathways involving initial formation of ACC from either pre-nucleation clusters or dense liquid droplets, they provide no evidence for or against the existence of either species, because the resolution is inadequate to see clusters of the size envisioned by the pre-nucleation cluster model and, when ACC precursors are observed, the TEM images do not definitively reveal whether they are solid or liquid.

15.5 Nucleation within an Organic Matrix

Organic matrices are well known to greatly influence the formation of biominerals in living organisms. In order to tease out aspects of biological control over mineralization,

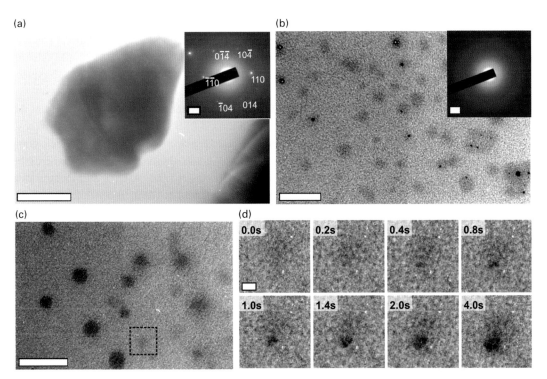

Figure 15.6. CaCO$_3$ nucleation within an organic matrix. (a) Formation of crystals with no apparent precursor phase, from non-PSS solution; diffraction spots (inset) identify crystals as vaterite. (b) The first CaCO$_3$ phase in PSS-bearing solutions forms within the organic matrix; the amorphous nature of ACC is seen in diffraction data (inset). (c) PSS binds with calcium to form a globular phase of Ca-PSS complexes prior to CaCO$_3$ formation. (d) Sequence of images following the rapid formation of an ACC particle within a Ca-PSS globule in the boxed region of (c). Scale bars are 1 μm (a), 100 nm (b, c), 10 nm (d), and 2 nm^{-1} (a, b insets). Adapted from Ref. 15 and used with permission by Nature Publishing Group.

nucleation of CaCO$_3$ in the presence of an organic additive can be studied using a liquid flow holder in a non-standard setup [15]. The liquid cell and one inlet are filled with a calcium-bearing solution, and the gaseous products of ammonium carbonate powder are diffused in through the other inlet. This experimental approach is often used in CaCO$_3$ precipitation studies, as the evolving CO$_2$ gas diffuses into the solution to continuously raise supersaturation. This procedure enabled observations of the effect on mineralization due to the presence of the organic additive polystyrene sulfonate (PSS).

In the purely inorganic calcium-bearing solution, vaterite was found to nucleate on the nitride membranes under these conditions (Figure 15.6a), with no indication of a precursor phase. This is similar to some of the observations reported in Section 15.4, albeit with the vaterite crystals adopting different morphologies. However, when PSS was introduced into the calcium solution, different behavior was observed. First, in the absence of the supersaturation-inducing gas, globules formed on the nitride windows (Figure 15.6c). These were found through independent calorimetric, spectroscopic, and ion-selective electrode measurements to be a globular phase [42] of Ca-PSS complexes formed through the binding of calcium to PSS, with approximately 56% of all calcium bound in this phase. Thus, the introduction of the organic matrix to the simple calcium solution produced locally concentrated regions of solute available for subsequent mineralization.

Upon introduction of the gaseous CO_2 to create supersaturated conditions within the fluid cell, particles rapidly nucleated within the Ca-PSS globules (Figure 15.6d). These particles rapidly grew to consume the Ca bound within the globules, at which point growth terminated. Diffraction data collected under these conditions identified the particles within the organic matrix as ACC (Figure 15.6b). After a sufficiently long exposure to the gas, vaterite crystals were observed to form independently of the organic matrix. The results show that through binding much of the calcium in solution into Ca-PSS complexes, the organic matrix biases the kinetics of the system towards the formation of ACC. The PSS globules provide chemical environments that are then amenable to ACC stabilization. Thus, the study provides an example of mineral nucleation directed by a chemical additive binding ions from solution to produce regions of high concentration. This supports a previously proposed model for nucleation of the biomineral nacre in mollusk shells [43, 44].

15.6 Particle-Based Crystallization

As discussed in Section 15.1, post-nucleation growth can occur via monomer-by-monomer (or perhaps cluster-cluster) attachment, as appears to be the case in the $CaCO_3$ experiments described above, or through particle-based mechanisms whereby fully formed primary particles coalesce to form larger scale crystals. Although the possibility that these particle-based mechanisms included OA has been recognized since the pioneering work of Penn and Banfield [32], proof that this process was at work was not obtained prior to the development of liquid cell TEM, which provided the ability to determine the phase and orientation of constituent particles throughout the growth process [45–47]. While those studies revealed growth by OA, other TEM studies have shown that particle-based growth can occur in the absence of particle co-alignment, for both crystalline and amorphous particles [46, 48]. In addition, a number of low resolution liquid cell TEM studies of various nanoparticle systems have shown coalescence to play a role in crystal development, but lacked the resolution to determine whether or not OA was the primary mechanism during the attachment events (see Refs. 36, 49, and 50 for examples, and discussion in Chapter 9).

Imaging at resolutions sufficient to distinguish crystal lattice fringes has been particularly useful for understanding which mechanisms contribute to particle-mediated growth. Such resolution has not been demonstrated with flow cells. However, static cells, using both Si-based microfabricated chips and graphene films, have enabled lattice resolution in a number of studies [45–47]. In our experience with silicon-based cells, there are two possible ways to achieve this resolution. The first and most useful results from inward bowing of the thin electron-transparent windows so that the

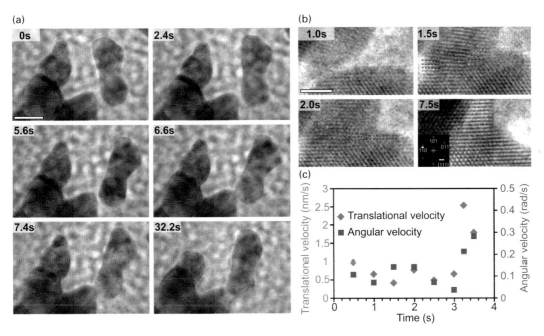

Figure 15.7. Oriented attachment of iron oxyhydroxide nanoparticles. (a) Two neighboring particles remain separate until the smaller particle rotates into a matching orientation, at which point the small particle jumps into contact with the larger particle and the interface grows. (b) Formation of dislocations from near-perfect attachment, followed by translation of defects out to the surface resulting in perfect lattice match. (c) Plot of angular and translational velocities during an OA event, showing the acceleration during jump to contact. Scale bars are 10 nm (a) and 2 nm (b). (a) Reproduced from Ref. 51. (b, c) Reproduced from Ref. 46, with permission by Cambridge University Press and AAAS, respectively.

membrane separation is below about 100 nm. However, bowing is random and, for many cells, the windows either end up touching so that there is no fluid in the imaging volume or they bow outward so that the separation is over one micrometer and the resolution is poor. The second way to achieve lattice resolution is to image with high electron beam current in a single location. Radiolysis effects will often cause movement of the fluid from the imaging region leaving a thin meniscus that progressively thickens as one moves away from this region.

Using cells for which the bowing resulted in a thin liquid layer through much of the cell, iron oxyhydroxide attachment events were observed using a static liquid cell filled with an $FeCl_3$–KH_2PO_4 solution [46], and found to exhibit OA. Figure 15.7a shows a sequence of images capturing this process. The smaller particle comes into close proximity with the larger crystal but does not attach. Instead, it rotates until it finds an orientational match, and then snaps into contact with the other particle. However, co-alignment is not perfect at the moment of contact and results in creation of lattice defects at the interface. These move rapidly to the surface and are removed from the lattice. The end result is a perfect lattice match across the contact region. In other such events, two particles bump into each other multiple times without coalescing, until rotation brings them into crystallographic registry and they then snap together to form either a twinned interface or a continuous orientation.

High resolution analysis of the lattice fringes permits a detailed look at these attachment events. Figure 15.7b highlights the dynamics of a representative attachment event, where defects are initially generated through the contact of two particles. These edge dislocations move rapidly to the right and within seconds exit the lattice, leaving behind a defect-free interface. The interface then grows through ion-by-ion addition to reduce the curvature of the coalesced particle. Analysis of the Fourier transform of the interface shows that the attachment event and subsequent transport of transient defects out of the lattice results in true crystallographic alignment between the joined particles.

The temporal and spatial resolutions afforded by liquid cell TEM allow for accurate determination of particle translational and rotational velocities during these attachment events. Velocity measurements from a typical OA observation are shown in Figure 15.7c. Both velocities are constant until shortly before attachment, at which point they both exhibit acceleration as the particles jump into contact with each other. This acceleration indicates the existence of an attractive potential as the driving force for OA. Using the velocity measurements, the magnitude of the attraction can be calculated. Under the assumption that the force is Coulombic in nature, it was calculated that the magnitude of charge inducing this force for these particles is of order one unit charge per particle.

While the iron oxide particles exhibit OA behavior in their growth mechanisms, other systems show different behavior. For example, liquid cell TEM observations of calcium carbonate solutions containing a $CaCO_3$-binding protein show that particle growth occurred through ACC attachment onto a crystal followed by crystallization of the amorphous particle [51]. This sequence of events is captured in the images shown in Figure 15.8. Lattice fringes identify the crystalline nature of the largest particle, whereas neither fringes nor structure in the contrast of the particle above it suggest crystallinity. Over time the two particles float into close proximity and attach, followed by the rapid appearance of lattice fringes across the point of contact indicating the crystallization of the previously amorphous particle. This stands in contrast to the behavior in the absence of an organic. Densely packed groups of distinct $CaCO_3$ crystals that form through the decomposition of ammonium carbonate display no evidence of particle attachment events during observation. These phenomena highlight the importance that additives can play in altering the formation and growth mechanisms utilized by a given mineral system.

15.7 Conclusions and Future Applications

The data reviewed in this chapter highlight the utility of liquid cell TEM in understanding formation mechanisms in mineral systems. High resolution imaging of iron oxide and calcium carbonate solutions has provided clear evidence for the existence of multiple formation pathways including single-step nucleation and growth of crystalline phases, multi-step mineral formation marked by direct transformation of initial

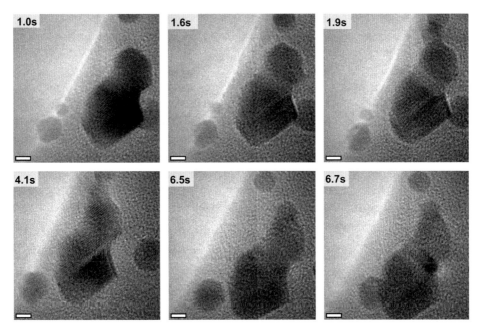

Figure 15.8. Influence of organic additive on $CaCO_3$ particle attachment. In the presence of SM-50, a $CaCO_3$-binding protein, ACC particles attach to crystalline particles, followed by crystallization of the amorphous constituent. Scale bars are 2 nm. Adapted from Ref. 51 and reprinted with permission by Cambridge University Press.

amorphous precursor particles, and particle-based growth via OA as well as random aggregation of amorphous nanoparticles onto secondary crystals. The observation of Ca-PSS globule formation and subsequent nucleation of ACC only within the globules under conditions that otherwise lead to vaterite formation has demonstrated the capability of liquid cell TEM in observing both soft macromolecular matrices and hard mineral constituents, revealing the potential role of ion binding within macromolecular matrices in controlling nucleation. Most importantly, the continual structural and morphological evolution observed throughout mineralization in all three of these examples brings into relief the reality that one cannot know what process led to mineral formation simply from looking at a final crystal. Liquid cell TEM has therefore become an invaluable tool in understanding both how minerals form and how to manipulate the mineralization process. Continued development of the technique will broaden its utility to enable investigations of an increasing number of materials systems and phenomena. Of particular interest would be the ability to run experiments in the liquid cell across a range of temperature and pressure. Furthermore, *in situ* diagnostic measurements of solution properties such as pH and composition are critical to many quantitative studies of mineral nucleation and growth of high relevance to the geochemical, biomineral, and environmental science communities.

Looking forward, some specific applications of liquid cell TEM in the fields of environmental and biological mineralization seem obvious. First is investigation of a wider range of mineral systems in order to compare and contrast pathways and identify underlying principles that determine which pathway a given system will follow. Second is to introduce natural macromolecular matrices such as collagen, amelogenin, and polysaccharides, to understand the mechanisms by which they direct nucleation with control over mineral phase, location, and crystallographic orientation. Third is to explore the impact of surface chemistry and topography on nucleation. In this case, the silicon nitride membranes must be patterned using organic films such as functionalized self-assembled monolayers, or through focused ion beam sputtering to form physical patterns. Observations of the locations and rates of mineral formation can be used to quantify the effects on the energetic parameters controlling nucleation [11]. Finally, the important problems in soil [52], aerosol [53], and climate science discussed in the introduction can be uniquely addressed using *in situ* liquid cell TEM and there is little doubt that, in all of these areas, new insights into mechanisms of mineralization will emerge.

Acknowledgements

This research was supported by the U.S. Department of Energy, Office of Basic Energy Sciences, Division of Chemical Sciences, Geosciences, and Biosciences at Pacific Northwest National Laboratory and Lawrence Berkeley National Laboratory and by the National Science Foundation under grant DMR-1312697. Work conducted at Lawrence Livermore National Laboratory (LLNL) was performed under the auspices of the U.S. Department of Energy by LLNL under Contract DE-AC52-07NA27344. Pacific Northwest National Laboratory is operated by Battelle for the U.S. Department of Energy under Contract DE-AC05-76RL01830.

References

1. T. J. Beveridge, Role of cellular design in bacterial metal accumulation and mineralization. *Annu. Rev. Microbiol.*, **43** (1989), 147–171.
2. J. F. Banfield and H. Z. Zhang, Nanoparticles in the environment. In J. F. Banfield and A. Navrotsky, eds., *Nanoparticles and the Environment*, Reviews in Mineralogy & Geochemistry **44** (Mineralogical Society of America, 2001) pp. 1–58.
3. M. W. I. Schmidt, M. S. Torn, S. Abiven *et al.*, Persistence of soil organic matter as an ecosystem property. *Nature*, **478** (2011), 49–56.
4. A. H. Knoll, Biomineralization and evolutionary history. In P. M. Dove, J. J. DeYoreo and S. Weiner, eds., *Biomineralization*, Reviews in Mineralogy & Geochemistry **54** (Mineralogical Society of America, 2003) pp. 329–356.
5. H. A. Lowenstam and S. Weiner, *On Biomineralization* (New York: Oxford University Press, 1989).
6. C. Hoose and O. Mohler, Heterogeneous ice nucleation on atmospheric aerosols: a review of results from laboratory experiments. *Atmos. Chem. Phys.*, **12** (2012), 9817–9854.

7. S. B. Hendricks, R. A. Nelson and L. T. Alexander, Hydration mechanism of the clay mineral montmorillonite saturated with various cations. *J. Am. Chem. Soc.*, **62** (1940), 1457–1464.
8. G. Sposito, N. T. Skipper, R. Sutton *et al.*, Surface geochemistry of the clay minerals. *Proc. Natl. Acad. Sci. USA*, **96** (1999), 3358–3364.
9. Q. Hu, M. H. Nielsen, C. L. Freeman *et al.*, The thermodynamics of calcite nucleation at organic interfaces: classical vs. non-classical pathways. *Faraday Discuss.*, **159** (2012), 509–523.
10. A. J. Giuffre, L. M. Hamm, N. Han, J. J. De Yoreo and P. M. Dove, Polysaccharide chemistry regulates kinetics of calcite nucleation through competition of interfacial energies. *Proc. Natl. Acad. Sci. USA*, **110** (2013), 9261–9266.
11. L. M. Hamm, A. J. Giuffre, N. Han *et al.*, Reconciling disparate views of template-directed nucleation through measurement of calcite nucleation kinetics and binding energies. *Proc. Natl. Acad. Sci. USA*, **111** (2014), 1304–1309.
12. P. A. Fang, J. F. Conway, H. C. Margolis, J. P. Simmer and E. Beniash, Hierarchical self-assembly of amelogenin and the regulation of biomineralization at the nanoscale. *Proc. Natl. Acad. Sci. USA*, **108** (2011), 14097–14102.
13. F. Nudelman, K. Pieterse, A. George *et al.*, The role of collagen in bone apatite formation in the presence of hydroxyapatite nucleation inhibitors. *Nat. Mater.*, **9** (2010), 1004–1009.
14. C. C. Tester, R. E. Brock, C. H. Wu *et al.*, In vitro synthesis and stabilization of amorphous calcium carbonate (ACC) nanoparticles within liposomes. *CrystEngComm*, **13** (2011), 3975–3978.
15. P. J. M. Smeets, K. R. Cho, R. G. E. Kempen, N. A. J. M. Sommerdijk and J. J. De Yoreo, In situ TEM shows ion binding is key to directing $CaCO_3$ nucleation in a biomimetic matrix. *Nat. Mater.*, **14** (2015), 394–399.
16. J. Rieger, T. Frechen, G. Cox *et al.*, Precursor structures in the crystallization/precipitation processes of $CaCO_3$ and control of particle formation by polyelectrolytes. *Faraday Discuss.*, **136** (2007), 265–277.
17. J. R. I. Lee, T. Y. J. Han, T. M. Willey *et al.*, Structural development of mercaptophenol self-assembled monolayers and the overlying mineral phase during templated $CaCO_3$ crystallization from a transient amorphous film. *J. Am. Chem. Soc.*, **129** (2007), 10370–10381.
18. A. V. Radha, T. Z. Forbes, C. E. Killian, P. Gilbert and A. Navrotsky, Transformation and crystallization energetics of synthetic and biogenic amorphous calcium carbonate. *Proc. Natl. Acad. Sci. USA*, **107** (2010), 16438–16443.
19. P. Bots, L. G. Benning, J. D. Rodriguez-Blanco, T. Roncal-Herrero and S. Shaw, Mechanistic insights into the crystallization of amorphous calcium carbonate (ACC). *Crys. Growth Des.*, **12** (2012), 3806–3814.
20. J. W. Gibbs, On the equilibrium of heterogeneous substances. *Trans. Connect. Acad. Arts Sci.*, **3** (1876), 108–248; (1878), 343–524.
21. D. Gebauer, A. Volkel and H. Colfen, Stable prenucleation calcium carbonate clusters. *Science*, **322** (2008), 1819–1822.
22. E. M. Pouget, P. H. H. Bomans, J. A. C. M. Goos *et al.*, The initial stages of template-controlled $CaCO_3$ formation revealed by cryo-TEM. *Science*, **323** (2009), 1455–1458.
23. M. A. Bewernitz, D. Gebauer, J. Long, H. Colfen and L. B. Gower, A metastable liquid precursor phase of calcium carbonate and its interactions with polyaspartate. *Faraday Discuss.*, **159** (2012), 291–312.
24. R. Demichelis, P. Raiteri, J. D. Gale, D. Quigley and D. Gebauer, Stable prenucleation mineral clusters are liquid-like ionic polymers. *Nat. Commun.*, **2** (2011), 590.

25. A. F. Wallace, L. O. Hedges, A. Fernandez-Martinez *et al.*, Microscopic evidence for liquid-liquid separation in supersaturated $CaCO_3$ solutions. *Science*, **341** (2013), 885–889.
26. L. Brecevic and A. E. Nielsen, Solubility of amorphous calcium carbonate. *J. Cryst. Growth*, **98** (1989), 504–510.
27. J. Rieger, J. Thieme and C. Schmidt, Study of precipitation reactions by X-ray microscopy: $CaCO_3$ precipitation and the effect of polycarboxylates. *Langmuir*, **16** (2000), 8300–8305.
28. W. J. E. M. Habraken, J. H. Tao, L. J. Brylka *et al.*, Ion-association complexes unite classical and non-classical theories for the biomimetic nucleation of calcium phosphate. *Nat. Commun.*, **4** (2013), 1507.
29. D. Erdemir, A. Y. Lee and A. S. Myerson, Nucleation of crystals from solution: classical and two-step models. *Accounts Chem. Res.*, **42** (2009), 621–629.
30. O. Galkin, K. Chen, R. L. Nagel, R. E. Hirsch and P. G. Vekilov, Liquid-liquid separation in solutions of normal and sickle cell hemoglobin. *Proc. Natl. Acad. Sci. USA*, **99** (2002), 8479–8483.
31. S. Chung, S. H. Shin, C. R. Bertozzi and J. J. De Yoreo, Self-catalyzed growth of S layers via an amorphous to-crystalline transition limited by folding kinetics. *Proc. Natl. Acad. Sci. USA*, **107** (2010), 16536–16541.
32. R. L. Penn and J. F. Banfield, Imperfect oriented attachment: dislocation generation in defect-free nanocrystals. *Science*, **281** (1998), 969–971.
33. C. Frandsen, B. A. Legg, L. R. Comolli *et al.*, Aggregation-induced growth and transformation of beta-FeOOH nanorods to micron-sized alpha-Fe_2O_3 spindles. *CrystEngComm*, **16** (2014), 1451–1458.
34. J. Baumgartner, A. Dey, P. H. H. Bomans *et al.*, Nucleation and growth of magnetite from solution. *Nat. Mater.*, **12** (2013), 310–314.
35. J. J. De Yoreo, P. U. P. A. Gilbert, N. A. J. M. Sommerdijk *et al.*, Crystallization by particle attachment in synthetic, biogenic, and geologic environments. *Science*, **349** (2015), aaa6760.
36. H. M. Zheng, R. K. Smith, Y. W. Jun *et al.*, Observation of single colloidal platinum nanocrystal growth trajectories. *Science*, **324** (2009), 1309–1312.
37. M. J. Williamson, R. M. Tromp, P. M. Vereecken, R. Hull and F. M. Ross, Dynamic microscopy of nanoscale cluster growth at the solid-liquid interface. *Nat. Mater.*, **2** (2003), 532–536.
38. M. H. Nielsen, J. R. I. Lee, Q. N. Hu, T. Y. J. Han and J. J. De Yoreo, Structural evolution, formation pathways and energetic controls during template-directed nucleation of $CaCO_3$. *Faraday Discuss.*, **159** (2012), 105–121.
39. M. H. Nielsen, S. Aloni and J. J. De Yoreo, In situ TEM imaging of $CaCO_3$ nucleation reveals coexistence of direct and indirect pathways. *Science*, **345** (2014), 1158–1162.
40. J. L. Bischoff, J. A. Fitzpatrick and R. J. Rosenbauer, The solubility and stabilization of ikaite ($CaCO_3 \cdot 6H_2O$) from 0–25 °C: environmental and paleoclimatic implications for thinolite tufa. *J. Geol.*, **101** (1993), 21–33.
41. A. A. Chernov, *Modern Crystallography III*. Springer Series in Solid-State Sciences (Berlin: Springer, 1984).
42. O. Trotsenko, Y. Roiter and S. Minko, Conformational transitions of flexible hydrophobic polyelectrolytes in solutions of monovalent and multivalent salts and their mixtures. *Langmuir*, **28** (2012), 6037–6044.
43. L. Addadi, J. Moradian, E. Shay, N. G. Maroudas and S. Weiner, A chemical model for the cooperation of sulfates and carboxylates in calcite crystal nucleation: relevance to biomineralization. *Proc. Natl. Acad. Sci. USA*, **84** (1987), 2732–2736.

44. F. Nudelman, B. A. Gotliv, L. Addadi and S. Weiner, Mollusk shell formation: mapping the distribution of organic matrix components underlying a single aragonitic tablet in nacre. *J. Struct. Biol.*, **153** (2006), 176–187.
45. J. M. Yuk, J. Park, P. Ercius *et al.*, High-resolution EM of colloidal nanocrystal growth using graphene liquid cells. *Science*, **336** (2012), 61–64.
46. D. S. Li, M. H. Nielsen, J. R. I. Lee *et al.*, Direction-specific interactions control crystal growth by oriented attachment. *Science*, **336** (2012), 1014–1018.
47. H.-G. Liao, D. Zherebetskyy, H. Xin *et al.*, Facet development during platinum nanocube growth. *Science*, **345** (2014), 916–919.
48. H. G. Liao, L. K. Cui, S. Whitelam and H. M. Zheng, Real-time imaging of Pt_3Fe nanorod growth in solution. *Science*, **336** (2012), 1011–1014.
49. L. R. Parent, D. B. Robinson, T. J. Woehl *et al.*, Direct *in situ* observation of nanoparticle synthesis in a liquid crystal surfactant template. *ACS Nano*, **6** (2012), 3589–3596.
50. T. J. Woehl, J. E. Evans, L. Arslan, W. D. Ristenpart and N. D. Browning, Direct *in situ* determination of the mechanisms controlling nanoparticle nucleation and growth. *ACS Nano*, **6** (2012), 8599–8610.
51. M. H. Nielsen, D. S. Li, H. Z. Zhang *et al.*, Investigating processes of nanocrystal formation and transformation via liquid cell TEM. *Microsc. Microanal.*, **20** (2014), 425–436.
52. A. Fukami, K. Fukushima, N. Kohyama, Observation technique for wet clay minerals using film-sealed environmental cell equipment attached to high-resolution electron microscope. In R. Bennett *et al.*, eds., *Microstructure of Fine-Grained Sediments* (New York: Springer, 1991) pp. 321–331.
53. K. Adachi, E. J. Freney and P. R. Buseck, Shapes of internally mixed hygroscopic aerosol particles after deliquescence, and their effect on light scattering. *Geophys. Res. Lett.*, **38** (2011), L13804.

16 Liquid STEM for Studying Biological Function in Whole Cells

Diana B. Peckys and Niels de Jonge

16.1 Introduction

Life happens in liquid water. The structure and properties of water play a vital role in the secondary and tertiary conformation of proteins, nucleic acids, and membranes. Thus, ever since the invention of electron microscopy, a wish has been to image cells in their native liquid environment as one can do with light microscopy. Various approaches have been developed [1–3] but the spatial resolution has mostly been disappointing. Studying biological functions at the nanoscale in whole cells in their native liquid environment has become possible only recently using a new approach involving scanning transmission electron microscopy and nanoparticles of high atomic number [4–6]. These nanoparticles, used as specific protein labels or as trafficking markers, provide sufficient contrast to obtain nanometer resolution on whole cells in a liquid layer [7]. This method entirely avoids the drying and sectioning associated with traditional EM sample preparation into plastic embedded sections, or the delicate preparation into cryogenic sections. Almost as simple as the preparation of cells for fluorescence microscopy, liquid STEM can be used to study tens of cells within only a few hours [8, 9] – about a factor of 50 faster than is possible using conventional electron microscopy. The images can be easily correlated with light microscopy recordings of the same cells [10], and unstained cells that are alive at the onset of imaging can be examined [11,12].

In this chapter, we provide an overview of the technique of liquid STEM as applied in two promising modes: at 200 keV electron energy using a microfluidic chamber in a STEM [13], and at 30 keV in an environmental SEM using a STEM detector [9, 14]. We describe the requirements, functioning, and equipment for these techniques, and their applications in cell biology. We expect that liquid STEM in both modes will develop in the near future into a mature analysis method used in modern cell biology, nano-biotechnology, and biomedical research, for example, to study cancer-related questions.

16.2 Liquid STEM Technology

16.2.1 Requirements, Benchmarks, and Advantages

High spatial resolution alone is not enough to study protein function in cells. Eight other important requirements are that the technique should [6, 15]

- Image in liquid water
- Image whole cells
- Image single objects (not arrays of objects)
- Be time-resolved
- Be intracellular
- Be three-dimensional
- Be reproducible, and
- Have easy sample preparation.

For existing techniques, it is interesting that the spatial resolution appears to be inversely proportional to the number of fulfilled requirements. Cellular function is often studied with fluorescence microscopy using cells with genetically encoded fluorescent protein tags [16, 17], but even the practically achievable resolution of ~20 nm of super-resolution fluorescence microscopy [18–23] is not sufficient to resolve the individual subunits of most protein complexes driving most cellular processes. The use of Förster resonance energy transfer (FRET) to study the stoichiometry of protein complexes is a matter of debate, since the FRET signal rapidly decreases for distances larger than 10 nm. The FRET signal is thus not necessarily sensitive to fluorescent labels attached to the subunits of a complex, but would rather detect interactions between neighboring protein complexes positioned back-to-back [24, 25]. FRET is also an indirect method. Certain advanced techniques, such as fluorescence bleaching microscopy [26] and fluorescence cross-correlation spectroscopy [27], measure protein clustering, but these are also indirect methods and typically involve cells with low concentrations of genetically engineered membrane proteins, not the high concentrations of endogenously expressed receptors in naturally occurring cells that are not genetically modified.

Using TEM, protein structure and cellular ultrastructure are commonly examined [28] as thin plastic embedded sections, frozen sections [28, 29] at the edges of frozen cells [30] or at fractures [31] of cellular samples, or in cells that are sectioned in a serial process [32]. But in all cases the native liquid state is lost and whole cells are cut into sections. Most importantly, these TEM studies involve ultrastructural information of a few exemplary samples, and typically only of thin sections of cells. Antibody labeling is commonly used, but is restricted in localization precision to 30 nm by the sizes of the antibodies [33] so that discrimination of clustering into protein complexes is not accurate. X-ray microscopy seems promising [34] but its growth in the scientific community seems to be hampered by the lack of table-top X-ray sources of high brightness.

Liquid STEM provides a different type of information because it detects protein positions in intact cells in a liquid state. It aims at studying many tens of cells, but typically does not provide much information about the ultrastructure. Once one realizes how simple it is to study cells directly in liquid, and recognizes the need to study proteins in intact cells and examine large numbers of cells, the value of this novel technology becomes clear. New experimental strategies are needed. Instead of providing detailed information about the cellular ultrastructure or the protein structure in a laborious electron microscopic study, liquid STEM focuses on a specific biological process, thereby limiting the amount of information collected per experiment. Similarly to fluorescence microscopy, liquid STEM uses labels to tag a specific subset of membrane proteins. The selected targets are then studied with high resolution in an intact cell in liquid, and the experiment is repeated for many cells and under varied conditions.

Liquid STEM presents a paradigm change in electron microscopy of cells. Its key benefits are:

- It adds a new level of analytical characterization possibilities, providing a different type of information compared to standard electron microscopy [4, 6]. It presents the unique capability of studying protein complexes at the single molecule level in the context of intact cells in liquid [35].
- A resolution better than 4 nm is possible on tagged proteins in whole eukaryotic cells in liquid [4, 9].
- Studies involving more than a hundred cells are possible [8].
- Correlative light microscopy and liquid STEM is readily possible [10, 11, 36].
- Live cell liquid STEM is feasible for "snap shot" images [6, 11, 12].
- Rapid sample preparation and study of selected proteins is similar to the well-established methods of fluorescence microscopy, but the results have much higher resolution [4, 9, 36].

16.2.2 Principles of Operation

Chapters 2–4 have described the two different experimental schemes that have been developed to study liquid samples, open and closed cell systems [3]. The closed systems, based on a pair of electron-transparent membranes that enclose a thin liquid layer [37], or a single membrane that encloses liquid in a capsule [38], are readily adapted to study whole biological cells. The first step is to grow eukaryotic cells on microchips with membranes, then prepare them, for example by incubating with specific labels, fixing, or keeping alive [6, 39]. Then, for STEM experiments, the microchip with the biological cells plus a counter chip with a spacer are assembled in their sample holder to form a microfluidic chamber [13]. This is then imaged with STEM (Figure 16.1a), typically at 200 keV beam energy. The typical liquid layer thickness for studying enclosed eukaryotic cells is 6 μm. Liquid can be flowed through the microfluidic chamber to provide nutrients to the cells, and to remove radicals, free electrons, and heat during electron microscopic investigation. Because STEM is so sensitive to atomic number, Au nanoparticles or high Z quantum dots can be detected above the background signal from the thick but low Z water and biological matter [4, 7, 10]. Although microfluidic cells can also be imaged using a STEM detector in a standard SEM [40], the biological samples, with their few-μm water thickness, would produce poor quality images at the low SEM accelerating voltage. On the other hand, the assembled microfluidic chamber in its sample holder (or a single microchip) can easily be examined with light microscopy of high numerical aperture [10, 11].

Other methods, described in Chapter 5, allow imaging of whole cells in SEM. A sample in a dish, which is open to ambient air and thus accessible for experiments, can be imaged from below through a thin membrane window using backscattered electrons [41];

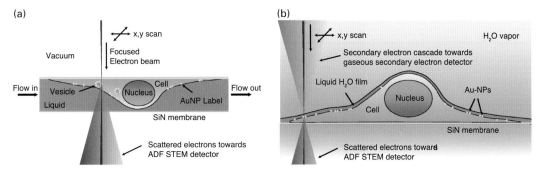

Figure 16.1. Concepts of liquid STEM of whole cells. Cells are grown on a supporting silicon nitride membrane. Target membrane proteins are labeled with Au nanoparticles and the labels are imaged in live or fixed cells. (a) A microfluidic chamber, formed by two microchips with Si_xN_y windows, fully encloses the cells for STEM imaging. (b) Using ESEM the labeled and fixed cells on a microchip are kept in a saturated water vapor atmosphere, while a thin layer of water covers them during imaging. From Ref. 6.

alternatively a microfluidic chamber may be imaged in an SEM within which an optical lens is integrated [42]. The high kV STEM technique described above, and the technique described below, should provide higher resolution for thick biological samples.

Open systems, as described in Chapter 3, expose the liquid to the vacuum but with pressure and temperature adjusted to the vapor pressure of the liquid to achieve equilibrium between liquid and vapor. The most commonly used technology is variable pressure or environmental SEM (ESEM) [43, 44] with a gaseous secondary electron detector placed above the sample [44]. But for whole cell studies, we instead use ESEM with so-called WetSTEM detection [14], achieving a few nanometers resolution on nanoparticles using an integrated STEM detector [9, 14]. In this technique (Figure 16.1b), cells in liquid are kept at 3 °C in a sample stage equipped with a Peltier cooling element, and the vacuum level in the sample chamber is carefully adjusted around 740 Pa, gradually evaporating water to achieve a thin liquid layer on the sample. Gold nanoparticles (AuNPs) used as specific protein markers on whole cells can be imaged with a spatial resolution of 3 nm [9], and a similar resolution can be obtained for quantum dots [35]. The large sample chamber volume available in SEM allows auxiliary equipment to be integrated, such as a light microscope with high numerical aperture [45], or a 90° tilting stage [46].

16.2.3 Equipment

Figure 16.2 illustrates equipment required for the experiments. It is not necessary to purchase a dedicated *in situ* microscope for high voltage STEM observations, only a specimen holder. ESEM-STEM of course requires an ESEM with a STEM detector. In both high voltage STEM and ESEM-STEM, the sample is placed on a silicon microchip supporting the Si_xN_y window [13, 47]. It is possible to achieve at least 3 nm resolution in ESEM [9] and atomic resolution at 200 keV using a 50 nm Si_xN_y membrane [48].

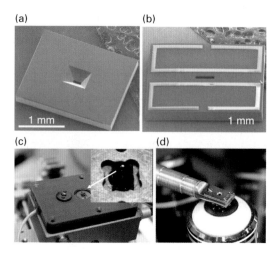

Figure 16.2. Equipment for liquid STEM. (a, b) Microchip imaged in SEM from the back (a) and front (b). The window is flanked by SU8 spacer. Charging effects distort the image at the spacer. With permission from Ref. 13. (c) Peltier-cooled ESEM-STEM stage (Quanta 400 FEG ESEM, FEI, USA) on which a microchip (inset) is held by a copper ring (left side) in a copper holder with openings above and below. An integrated STEM detector is located a few millimeters below (not visible). (d) For correlative microscopy, the sample holder tip (Protochips) is positioned on a water immersion lens (1.0 numerical aperture) of a fluorescence microscope (Nikon, Japan). With permission from Ref. 11.

The window is rectangular with a shortest dimension of 50 μm, balancing a maximal field of view with strength and stiffness, as discussed in Chapter 2, and sufficiently robust for handling [39].

The microfluidic chamber, formed by assembling two microchips, requires a chip separation of typically 4–6 μm, controlled using polystyrene microbead spacers deposited at the corners of the chip [4] or lithographically defined [13] (Figure 16.2b). This allows liquid to flow between the microchips and provides the required height to enclose an adherent eukaryotic cell. If the experiment does not require flow, then a sealed device can be used as well [5, 47, 49]. For STEM, two microchips are placed into the tip of a specimen holder [50]. For ESEM-STEM, a single microchip [9] is placed on a modified Peltier stage containing the STEM detector below the sample (Figure 16.2d).

16.2.4 Correlative Light and Electron Microscopy

For many biological experiments it is highly desirable to examine a sample containing many cells with fluorescence microscopy prior to liquid STEM studies. This correlative fluorescence microscopy and liquid STEM can be accomplished in two ways. The microchip with cells can simply be imaged with fluorescence microscopy prior to assembly in the liquid flow holder or before imaging with ESEM-STEM. An elegant solution for imaging samples within the microfluidic chamber can be accomplished using a water immersion lens with a working distance of around 2 mm, available with a relatively high numerical aperture of 0.9–1.0. The holder tip is placed on the lens and a droplet of liquid between the Si_xN_y membrane and the lens forms the immersion medium (Figure 16.2d) [11]. The spatial correlation between the images recorded with both modalities is determined from the relative positions of the image with respect to the corner of the Si_xN_y window [10].

16.3 Studying Membrane Proteins in Whole Cells in Liquid

16.3.1 The Epidermal Growth Factor Receptor (EGFR)

As a first experiment to demonstrate the capabilities of liquid STEM, we have studied the epidermal growth factor receptor (EGFR) on COS7 cells. EGFR is a transmembrane receptor [51, 52] playing a critical role in the pathogenesis and progression of many different types of cancer [53]. The EGFR is activated when epidermal growth factor (EGF) binds to it. The activated EGFR forms a homo dimer with a second EGFR [54], initiating a signal transduction to the cytosolic side of the receptor [55]. An important question is under which conditions and in which cellular regions dimerization occurs [52, 56]. The length scale at which EGFR dimerization needs to be studied is exemplified in Figure 16.3. If EGFR is activated by EGF that is attached to 10 nm AuNPs via a short streptavidin-biotin chain [4, 57], the molecular model of Figure 16.3 shows that the signature of dimerization is a center-to-center distance between two labels peaking at 20 nm.

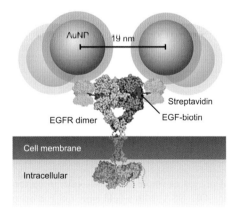

Figure 16.3. Molecular model of the EGFR dimer labeled with biotinylated EGF and streptavidin-conjugated AuNPs. The distance between the centers of the two AuNPs was estimated as ~19 nm for a tightly packed structure, although up 29 nm is possible by rotating the nanoparticle position on the streptavidin. The drawings were derived from CPK models of the 1stp (streptavidin), 1EGF (EGF), 1NQL, 2JWA, 1M17, 1IVO, and 2GS6 (EGFR) structures as found in the RCSB Protein Databank, created by Jmol Version 12.2.15, apart from biotin which was derived from RCSB Ligand Explorer Version 1.0. From Ref. 9.

16.3.2 Imaging Labeled EGFRs with Liquid STEM

To perform the experiment, COS7 cells were grown on microchips coated with poly-L-lysine to promote cell adherence, incubated with the labels, and fixed with glutaraldehyde [4, 39]. This protocol avoids the traditional preparation steps described in Section 16.2.1 [28] with their high risk of perturbing the original state of the cells. Instead, the protocol is comparable with that for fluorescence microscopy [39], where chemical fixation is used to preserve biological structures during exposure to the intense light beam needed to detect the fluorophores [17, 58]. The sample was then enclosed in the microfluidic chamber for STEM and images were recorded while maintaining a continuous flow of buffer. Figure 16.4a shows the edge of a COS7 fibroblast cell that was incubated for 5 minutes with the label. The liquid STEM spatial resolution of 4 nm was sufficient to distinguish adjacent labels, dimers, and larger clusters (see inset). The cellular material is visible as light blue shapes. The randomly distributed localization of the EGFRs over the cellular surface is consistent with this time window of incubation [57]. To test if the labeled EGFRs were still functional after the binding of the nanoparticle label, we examined the cells for endocytosis of labeled EGFRs, a process that naturally follows after the EGFR is activated [59]. The cells were incubated for 10 minutes with EGF-gold, washed, incubated an additional 15 minutes in buffer, then fixed. Figure 16.4b shows circular clusters of labels, consistent with clustering in internalized

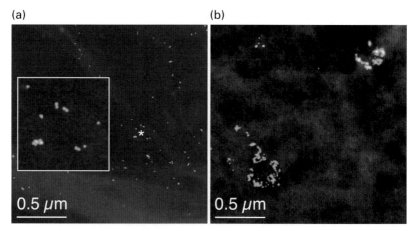

Figure 16.4. Liquid STEM images of whole eukaryotic cells in buffer solution obtained in a CM200 S/TEM (FEI, Hillsboro, OR, USA) with a field emission gun operated at 200 keV. (a) Color-coded image recorded at the edge of a fixed COS 7 cell. Au EGFR labels are visible as yellow spots and the cellular material as light blue over the dark blue background. The inset shows monomers, dimers, and a larger cluster at the location marked *. The image size was 1024 × 1024 pixels, the pixel size 2.9 nm, the pixel dwell time 20 µs and the total exposure time 20 s. The liquid thickness was 5.7 µm. (b) Image of a cell fixed after an additional chase (incubation) time of 15 min, following the 5 min EGF nanoparticle labeling, to allow the endocytosis of labeled EGFR. The image shows two endocytotic vesicles with labeled EGFRs, recorded at pixel size 4.4 nm. The liquid thickness was 8.6 µm. Images used with permission, modified with color, from Ref. 4.

endosomes [57]. The rounded shape in the top right corner is blurred indicating a different vertical position in the sample.

The images shown in Figure 16.4 were the first recorded of whole cells in their native liquid state obtaining nanoscale resolution [4] sufficient to recognize the individual constituents of protein complexes and the molecular factors driving cellular function. The experiments show how cellular function can be probed if samples are prepared with different ligand incubation times, other chemical substances are added to change the functional state of the cells, or labels for other proteins are added.

16.3.3 Spatial Resolution and Radiation Damage

The experiments in Figure 16.4 were recorded at an electron dose of 7×10^2 $e^-/Å^2$, only an order of magnitude larger than used for imaging frozen cells [60], and an order of magnitude smaller than the dose used for STEM on conventional thin sections [61]. Signs of radiation damage, such as displacement or shifting of the EGFR-bound AuNPs in subsequently recorded images, only appeared after several image exposures. The sample fixation thus provided sufficient stability to the sample, and the liquid flow presumably also helped to reduce the damaging effects from electron beam-induced radicals, free electrons, and heat [4].

The liquid thicknesses in Figure 16.4 were in the micrometers range, as measured from the total number of electrons scattered towards the annular dark field detector. The spatial resolution achieved is remarkably high, and is due to the atomic number (Z) contrast sensitivity of STEM [7, 62, 63]. It would not be possible to achieve nanoscale resolution at these thicknesses with TEM. Furthermore, although it has been suggested that Brownian motion should blur the images, this is not seen experimentally; instead, nanoparticles in close proximity to a membrane move several orders of magnitude slower than would be expected in a bulk liquid [64–66]. Indeed, atomic resolution has been reported for TEM in liquid [67, 68]. Nanoparticle labels appear to be sufficiently immobilized in live or fixed cellular structures to allow nanoscale microscopy [4, 11].

16.3.4 Imaging EGFR with ESEM-STEM

Whereas liquid STEM preserves the whole cell in a micrometers-thick buffer solution, there are certain cases where the experiment can be simplified, as shown in Figure 16.5a. Here, labeled receptors in the outer regions of the cells are imaged with ESEM-STEM [9]. The cells were grown and labeled on a microchip and were imaged at a location close to the edge of the cell while still covered by a thin layer of water, as shown in Figure 16.1b. This was achieved by evaporating the water layer (by control of temperature and pressure) down to the required thickness. The precipitation of salt crystals was avoided through an additional step in sample preparation, i.e. removal of the salts of the saline buffer solution by washing with pure water. In Figure 16.5b, many labels are visible, some are monodispersed, several are grouped as pairs and a couple of clusters are present as well.

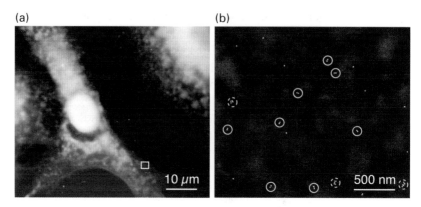

Figure 16.5. ESEM-STEM images of a whole, fixed A549 cell in hydrated state, obtained in a Quanta 400 FEG (FEI, Hillsboro, OR, USA). (a) Overview dark field ESEM-STEM image (pixel size 87 nm) showing parts of three cells. Thin cellular regions appear in grey, the thicker areas, such as the central part with the nucleus, appear in white. With permission from Ref. 9. (b) Image (pixel size 2.7 nm) recorded at a location close to the edge of one cell (rectangle in (a)) showing individual AuNPs, dimers (circles), and larger clusters (dashed circles). From Ref. 6.

The electron dose of one image amounted to 5×10^2 e$^-$/Å^2 (note that an incorrect value of the dose was published in Ref. 9). Recording a few images from one area did not lead to visible sample damage, as measured from the absence of shifts of the nanoparticles. Compared to use of the backscatter detector, the STEM detector is particularly efficient for label detection in terms of the resolution–dose balance [41, 45].

The total processing time including sample preparation, ESEM-STEM, and data analysis is as rapid as fluorescence microscopy for a skilled scientist. In our pilot study we examined 15 whole cells of two different types (A549 and COS7). The positions of 1411 labels were automatically determined and used for a statistical analysis [9]. It was found that the most probable center-to-center distance between pairs of labels amounted to 19 nm, consistent with the distance predicted from the model (Figure 16.3).

A key advantage of whole cell liquid STEM or ESEM-STEM is the capability to study the spatial distribution of membrane proteins at the nanoscale. This information is difficult if not impossible to obtain via conventional electron microscopy techniques using sectioning, because a section almost never cuts exactly through the plane of the plasma membrane, and the membrane is in any case typically curved. Membrane sheet or freeze fracture techniques [28, 69] are more time-consuming, lose the location of the analyzed membrane patch within the cellular context, and cannot image internalized nanoparticles. Considering the ease of conducting experiments on whole cells with ESEM-STEM, and the demonstrated feasibility of studying thousands of labels on tens of cells, we expect this to mature into a key analysis method for membrane proteins.

16.3.5 Correlative Fluorescence Microscopy and Liquid STEM

A practical advantage of microchips is the ease of performing correlative light and electron microscopy. For this purpose it is beneficial to use quantum dots instead of AuNPs as labels. Quantum dots with a high Z core of CdSe/CdZnS [70] have excellent

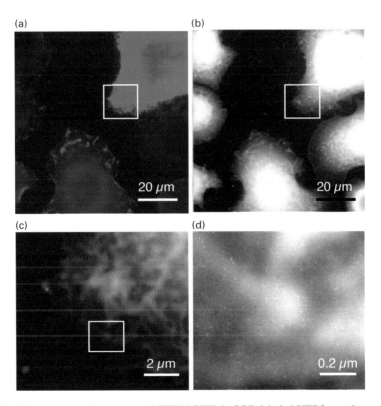

Figure 16.6. Correlative fluorescence microscopy and ESEM-STEM of QD labeled HER2 membrane proteins in SKBR3 breast cancer cells. (a) Overview fluorescence image of several cells demonstrating the heterogeneity in HER2 levels. (b) Overview ESEM-STEM image. (c) ESEM-STEM image of the boxed region in (a) and (b). Membrane ruffles are the brighter elongated shapes. (d) High resolution ESEM-STEM image of the marked area in (c). The locations of individual HER2s are indicated by the QDs, found predominantly as pairs and in smaller clusters. They reside preferentially on the ruffles. From Ref. 35.

fluorescence characteristics (very high brightness, non-bleaching), and also provide strong STEM contrast. Figure 16.6a shows fluorescence microscopy of HER2s in breast cancer cells [35] labeled with streptavidin conjugated quantum dots using a protocol [10, 36, 71] similar to that used for gold labeling. This type of image tests if the labeling experiment has been successful, identifies cellular labeling patterns, and allows search for possible inter- or intracellular differences. It was thus possible, for example, to select thin regions at the edge of the cell, or membrane ruffles, and differentiate between cell regions of higher and lower receptor expression.

Subsequent STEM is accomplished using either liquid STEM at 200 keV with higher spatial resolution [10] or ESEM-STEM. Figure 16.6b shows an ESEM-STEM image recorded at the same magnification as for the fluorescence image providing a spatial reference as needed to correlate the data. ESEM-STEM is then used for recording a higher-resolution image (Figure 16.6c). Several structural features of the cell can be recognized in the ESEM-STEM image, such as the shape and the edges of the cell, and membrane ruffles. Further electron microscopy, Figure 16.6d, is then performed at a higher magnification. Clustering and the stoichiometric distribution of the HER2 in monomers and dimers can be determined from this image. These observations are relevant for a better understanding of the mechanisms behind cancer cell growth and spreading [35]. Correlative fluorescence microscopy with liquid STEM is particularly useful for the study of membrane receptors, because information about assembly into protein complexes and nanoscale spatial information can be correlated directly with the cellular context obtained from fluorescence microscopy.

16.4 Live Cell Liquid STEM

16.4.1 Designing Live Cell Experiments

The STEM experiments described so far have been carried out on fixed cells. Chemical fixation should preserve the stoichiometry and spatial distribution of the membrane proteins [72], implying that nanoscale information can be obtained on the spatial distribution of labeled receptors, and cellular function can be studied by examining differences in these distributions between cells and between cellular regions. However, capturing images of unfixed, pristine cells would allow exciting direct studies of cellular function, as well as study of samples that do not withstand fixation, and measurement of the effects of fixation. Such experiments are possible in the microfluidic chamber but not in ESEM-STEM, as it requires the removal of salt. It is feasible to image the biological material of pristine cells at a resolution of several tens of nanometers [11], while nanometer resolution can be achieved using labels [6].

Some comments should be made up front about the prospects of studying cellular dynamics at the nanoscale. Live cell light microscopy is sometimes used to follow cellular processes, in which changes in the shape of cells or the locations of areas with tagged proteins are visible in a movie. It does not seem likely that analogous electron microscopy will be possible because the cells will probably not survive a series of images recorded over a time span of typically several tens of seconds to several minutes. The cells will most likely start necrosis, or at least apoptosis (programmed cell death), after the first STEM image, if we consider that the electron dose for a single image is six orders of magnitude above that for reproductive-cell death [73]. Processes that appear to take place in live cells [74] most probably reflect motions of material from dead cells. Furthermore, for live cells the microfluidic chamber must be large enough to contain the cells without damaging by squeezing. Even light microscopy does not seem an option to record live cell movies at the nanometer scale because the light intensity needed would lead to rapid cell death. In fact, it seems unlikely that movies containing nanoscale spatial information over at least several seconds can be recorded with either an electron beam, X-rays, or with light, since the obtained spatial resolution typically scales with the third power of the dose of any type of radiation [75].

Clever experiments must be designed to follow changes in the functional state of cells at the nanoscale. One option is to combine live cell fluorescence microscopy at intermediate resolution with the correlative recording of "snapshot" STEM images to elucidate the local and temporal stoichiometry of labeled protein complexes [4, 10]. The spatial correlation is readily obtained because drying and sectioning of the cells is avoided. The temporal correlation is determined by the speed of fixation, or by the speed of the change from light microscopy to electron microscopy, which can be optimized towards several tens of seconds [76]. It also appears feasible to record a series of images of a sample containing several cells, irradiating each cell only once [6]. This would involve inducing the cells to be in the same initial state and assuming a synchronous response to the stimulus. Perhaps some biochemical processes can be followed in real time in thin regions of cells during an interval over which the beam has not damaged the cell too much.

16.4.2 Pristine Yeast Imaging

The capability to study live cells with liquid STEM was tested first on yeast cells [11]. Yeast belongs to the eukaryotic cells but has a cell wall and is thus mechanically more stable than a COS7 cell. Figure 16.7a, b show light microscopy images of live *S. pombe* cells enclosed in the microfluidic chamber, acquired using an optical lens of high numerical aperture (Figure 16.2d). The viability of the cells was verified using the FUN-1 fluorescence marker. A few minutes after recording these images, liquid STEM images were recorded (Figure 16.7c) at a spatial resolution of about 30 nm through a liquid thickness of 4 μm and using an electron dose of 0.2 $e^-/Å^2$. This dose is below the limit for structural damage of native biological material in liquid [77]. Several structural

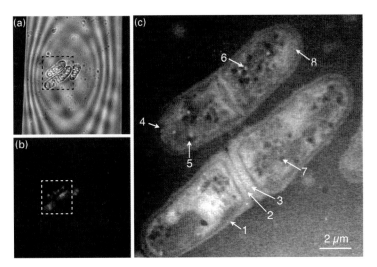

Figure 16.7. Light microscopy and liquid STEM of live, wild-type *S. pombe* yeast cells in buffer solution. (a) Phase contrast image of cells enclosed in the microfluidic chamber. (b) The corresponding fluorescence image shows that all cells accumulated FUN-1 dye and emitted a punctuated red fluorescence, the typical signal of living yeast cells. (c) Liquid STEM image acquired from the same cells, kept fully hydrated and still alive at the onset of STEM imaging. Numbered arrows indicate (1) cell wall, (2) primary septum, (3) secondary septum, (4) cell membrane invagination, (5) lipid droplet, (6) peroxisome, (7) unclassified vesicle, (8) AuNP. The color is an overlay of the red channel of the fluorescence image (b). Reprinted from Ref. 11.

features are visible, such as the cell wall, the septa, lipid droplets, and peroxisomes. The cells were alive at the onset of electron beam scanning but fluorescence images recorded afterwards revealed a leaking cell wall, so we must conclude that the cells were not viable afterwards. Liquid STEM can thus be used to study these pristine cells alive at the onset of imaging, bridging the capabilities of light microscopy and electron microscopy of cryogenic specimens.

16.4.3 Live COS7 Cell Imaging

A protocol was developed to keep typical cultured human and animal cells, such as COS7 cells, alive in the microfluidic chamber at room temperature with liquid flow for several hours [6, 12]. The cellular viability was tested for COS7, V79 hamster cells, and T98 cells, with the following criteria [6]:

- They flattened and outstretched on poly-L lysine coated microchips.
- They did not show signs of necrosis when maintained in the microfluidic device.
- No signs of apoptosis were observed.
- Additionally, for COS7, after enclosure in the microfluidic device, the device was opened and the microchip with cells was further incubated under cell growth conditions. Cell division was observed to take place after 39 hours.

It was thus concluded that cells from these three cell lines, and probably from many other cell lines with similar dimensions, can be kept healthy in the microfluidic chamber. However, when the viability of live COS7 cells was monitored after electron beam exposure of 0.4 $e^-/\text{Å}^2$, using the calcein AM dye (which provides a fluorescent signal in cells with an intact plasma membrane) [6], this inevitably led to cell death within 5–45 minutes. It was not possible to find a remedy by supplementing the cell culture medium with various antioxidants and/or oxygen scavengers. Interestingly, the degree of radiation damage, i.e. the speed of cell death, depended on the type of cellular region. Exposing the nucleus led to an immediate loss of the fluorescent signal, while the fluorescence was maintained for tens of minutes if a single image were recorded in the peripheral cellular region. Moreover, irradiation of one cell did not perturb the viability of adjacent cells.

Despite the loss of viability after recording a STEM image, it is still possible to obtain nanoscale information on the pristine cells. Individual gold labels attached to EGFRs can be imaged in unfixed COS7 cells [6]. Particle uptake can also be studied, as will be described in Section 16.5 below.

We can thus formulate the hypothesis that the first STEM image contains information about the ultrastructure of the live cell, that several additional images recorded at the outer regions of the same cell may possibly also contain live-cell information, and that adjacent cells can be studied independently. Future studies should address the question of how soon the cell initiates necrosis or apoptosis, and specifically how soon structural damage appears as a function of electron dose and type of irradiated cellular region.

16.5 Gold Nanoparticle Uptake Studied in Whole Cells

16.5.1 Liquid STEM of Nanoparticle Uptake in Pristine Cells

The study of AuNP uptake and fate in live cells provides another example of biological experiments that can be addressed with liquid STEM [12]. Figure 16.8 shows the results of such an experiment. COS7 cells were grown on microchips, serum-protein starved for 2 hours, then incubated for 2 hours at 37 °C in the CO_2 incubator with a solution containing serum-protein-coated AuNPs of 30 nm diameter [78]. The serum-protein coating prevents particle agglomeration in saline solutions and favors endocytotic uptake [79]. Subsequently, the samples were rinsed and incubated for 24 hours in cell culture medium, then enclosed in the microfluidic chamber and imaged in pristine state with liquid STEM at 200 kV. In Figure 16.8, a large intracellular cluster contains vesicles with AuNPs. The liquid was so thick (10 µm) that a contrast reversal occurred and the vesicles with AuNPs appeared black even though the annular dark field detector was used to record these images. The nanoparticle-filled vesicles had round or oval shapes with average diameter 0.26 ± 0.05 µm, measured on over 60 vesicles [12]. Individual AuNPs, enhanced by image processing (see the inset of Figure 16.8), show a remarkable distribution. One would expect that the cargo in an approximately spherical endosome would result in a density gradient towards the center of the vesicle. The more homogeneous density suggests that the AuNPs adhere to the vesicle membrane [12]. This indicates that the protein corona presumably remained intact because its dissociation from the AuNPs would result in a so-called "flocculation" phenomenon [80]. Over 25 images, the vesicles averaged 57 ± 26 nanoparticles.

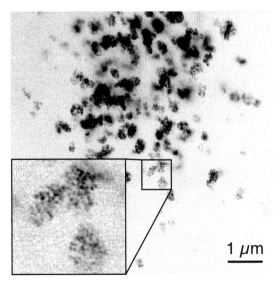

Figure 16.8. Liquid STEM images of live COS7 cells in a microfluidic chamber. 24 hours after incubation with serum-protein coated 30 nm AuNPs, the particles were found in a dense, three-dimensional cluster of vesicles. Inset is a processed image of the indicated region where individual AuNPs can be discerned within the vesicles. With permission from Ref. 12.

This nanoscale analysis of whole cells exhibits several important advantages over conventional TEM [81], where sectioning is problematic because vesicles are cross sectioned at a random plane, nanoparticles may be removed from the sample, and a section may not contain the entire cell and its cargo [82, 83]. Quantitative studies are possible with liquid STEM because the up-taken cargo is preserved in the sample.

16.5.2 ESEM-STEM of Nanoparticle Uptake in 145 Cells

The advantage of ESEM-STEM is its capability for studying many cells within a short period. Figure 16.9 shows results of a study where 145 whole cells were imaged to study AuNP uptake in the A549 human lung cancer cell line. The cells were subjected to uptake protocols with 10, 15, or 30 nm diameter AuNPs with adsorbed serum proteins. After 20 min, 24 h, or 45 h, the cells were fixed and imaged whole in a thin layer of liquid water with ESEM-STEM. The fast preparation and imaging of this many whole cells allowed nanoscale data collection within an exceptionally short time, only ~80 h. Analysis of 1041 nanoparticle-filled vesicles showed that the lysosomes increased their average size by 80 nm when uptaking 30 nm diameter nanoparticles, compared to lysosomes of cells incubated with 10 and 15 nm diameter nanoparticles.

16.6 Comparison with Cryo-TEM

It is important to benchmark electron microscopy of specimens in liquid with those in amorphous ice. The following aspects are of relevance:

- *Type of information.* Whole cell liquid STEM provides different information compared to that obtained via cryo-TEM. Conventional EM can provide detailed information about ultrastructure from a small set of samples, while liquid STEM can be used to obtain information on a selected set of labeled proteins or nanoparticles, similar to the results from fluorescence microscopy but with higher resolution. Ideally, a particular scientific question is addressed with many types of microscopy, each resolving a piece of the puzzle within the complexity of biology.
- *Number of cells studied.* Cryo-TEM is typically used to study a few selected specimens, usually thin sections of a cell, and it is not practical to study several tens of cells. Liquid STEM is capable of studying tens to hundreds of whole cells to provide quantitative data and statistics about a given nanoscale property over many cells.
- *Correlative microscopy.* Correlative light microscopy and cryo-TEM is possible but requires dedicated equipment [29, 84], and a complex specimen workflow. Moreover, it remains to be seen if live cell fluorescence microscopy combined with cryo-TEM is feasible. Cells grown on microchips can readily be imaged with fluorescence microscopy either in live state or fixed, then fairly easily imaged with liquid STEM in the microfluidic chamber or with ESEM-STEM. The sample workflow is similar to standard fluorescence microscopy.

Biological Function in Whole Cells 349

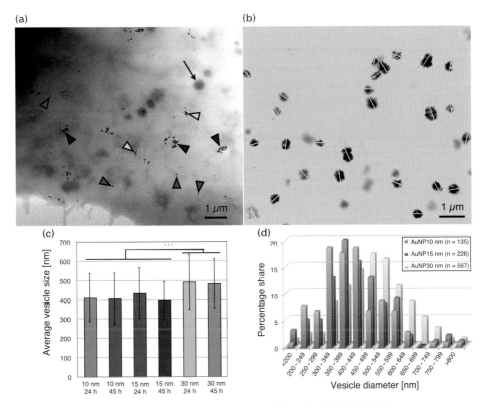

Figure 16.9. Quantitative study of AuNP containing and storing vesicles. (a) ESEM-STEM image showing an A549 cell at an early stage after exposure to serum protein coated 30 nm AuNPs (5 min pulse and 15 min chase). The particles, appearing as small black dots, adhere and cluster at the membrane or have just been up-taken into early endosomes. Single or paired AuNPs are highlighted with purple triangles. Clusters <200 nm and >200 nm in size are highlighted with light and dark blue, respectively. A laminar body (LB), typical for alveolar type II cells, is indicated with the arrow. (b) ESEM-STEM image of vesicular AuNP storage at later stages of the cellular uptake (2 h pulse and 22 h chase) depicting the procedure to measure the sizes of the vesicles across the largest diameter. Vesicles that appeared too blurred, due to their remote location from the focal plane, were not measured. (c) Bar graph of average vesicle sizes found for 10, 15, and 30 nm AuNPs and for different chase times. Blue: 10 nm (24 h, $n = 232$; 45 h; $n = 18$); red: 15 nm (24 h, $n = 161$; 45 h, $n = 63$); green: 30 nm (24 h, $n = 468$; 45 h, $n = 99$). After 24 and 45 h, vesicles with 30 nm AuNPs were significantly larger than those with 10 or 15 nm AuNPs. (d) Size distribution histogram of vesicles filled with AuNPs of different size. Green: 30 nm diameter, $n = 569$; blue: 10 nm diameter, $n = 250$; red: 15 nm diameter, $n = 224$. From Ref. 8.

- *Structure.* It should be stressed that, for unfixed specimens, high resolution electron imaging in liquid may generally open new insights into the native structure of cells. The structure and the special properties of water play a vital role in the secondary and tertiary conformation of proteins and nucleic acids, and lipid membranes. If the water structure changes by solidification, the macromolecular conformations can change [85] and thus cellular structures and properties may differ too.

- *Dynamics*. For unfixed samples, the dynamic aspect of liquid water is important. Functional proteins and other cellular components may not necessarily resemble static three-dimensional structures derived from X-ray crystallography, nuclear magnetic resonance, or single-particle TEM tomography. Instead, the functional state may involve continuous changes in the protein structure. However, in spite of its importance, the study of dynamics is obviously limited by radiation damage.
- *Radiation damage*. Electron beam-induced radiation damage propagates differently in liquid water, compared to ice. In ice, broken atomic bonds do not immediately lead to a loss of structure because the static matrix will hold the structure in place. However, liquid has advantages in that the radicals have a chance to move away by diffusion [4], and heat is dissipated better than in ice [86]. It will be interesting to learn which particular configurations are best imaged in water and which in ice [87, 88].

16.7 Conclusions and Outlook

Liquid STEM performed with the microfluidic chamber or with ESEM-STEM presents a new microscopy modality for studying cellular function at the molecular level. Nanometer spatial resolution is combined with much of the functionality of fluorescence microscopy. It should be kept in mind that the information provided is different from that obtained from cryo-TEM of conventional thin sections [29]. Liquid STEM does not provide detailed information about the ultrastructure, but rather focuses on labeled proteins or up-taken nanoparticles. In common with fluorescence microscopy, the experiment must be designed to address the spatial and temporal distribution of a certain molecular species in the cell. It is readily possible to image through a whole cell, to study several tens of cells, and to obtain statistics of many thousands of labeled proteins or nanoparticles. Correlative fluorescence microscopy and liquid STEM can be easily accomplished by moving the microchips with cells between both microscopes. Live cell imaging is possible using the microfluidic chamber.

The current status of the technology leaves much room for improvement. Microchips and sample holders are still far from optimal. The Si_xN_y membrane often bends too much, is still too thick, and too small. The holder is often difficult to load, and in practice the liquid gap cannot be set as precisely as the manufacturers suggest. It would be interesting to develop a liquid system comprising a large chamber for the central, nucleus-containing parts of a cell and a very thin imaging chamber [89] to allow higher resolution imaging of thinner cellular regions. Correlative fluorescence microscopy on a process occurring in a cell with subsequent STEM "snapshot" imaging [4] still needs to be demonstrated. Future studies may also include different types of high Z labels, such as nanoparticles of different shapes, sizes, and materials, and genetically encoded labels [90–92] for the parallel multiplexed imaging of different protein species.

We expect that liquid STEM will be used by different biological groups in the future, when a microscopy technique is needed with a better resolution than super-resolution light microscopy [22] to resolve protein complex stoichiometry. Scientists now using conventional electron microscopy may find it useful to study the distribution of a protein complex in many cells, while obtaining detailed information about the local organelle environment via cryo-TEM in a few cells. Membrane proteins can readily be studied on whole cells, while membrane sheet or freeze fracture techniques [28] are more difficult to apply and do not allow for correlation with light microscopy images of the intact cells. Nano-toxicological studies would benefit from the capability to study the up-taken cargo quantitatively without the need for sectioning [93]. New insights in the functioning of growth factor receptors in cancer cells have already been gained [35] and many more studies will follow. All in all this new microscopy method for cells, now in its early phase, has great potential for improvement and a broad range of important applications.

Acknowledgements

We thank E. Arzt for his support through INM. Research in part supported by the Leibniz Competition 2014.

References

1. D. F. Parsons, V. R. Matricardi, R. C. Moretz and J. N. Turner, Electron microscopy and diffraction of wet unstained and unfixed biological objects. *Adv. Biol. Med. Phys.*, **15** (1974), 161–270.
2. D. F. Parsons, Structure of wet specimens in electron microscopy. *Science*, **186** (1974), 407–414.
3. N. de Jonge and F. M. Ross, Electron microscopy of specimens in liquid. *Nat. Nanotechnol.*, **6** (2011), 695–704.
4. N. de Jonge, D. B. Peckys, G. J. Kremers and D. W. Piston, Electron microscopy of whole cells in liquid with nanometer resolution. *Proc. Natl. Acad. Sci. USA*, **106** (2009), 2159–2164.
5. D. B. Peckys, G. M. Veith, D. C. Joy and N. de Jonge, Nanoscale imaging of whole cells using a liquid enclosure and a scanning transmission electron microscope. *PLoS One*, **4** (2009), e8214.
6. D. B. Peckys and N. de Jonge, Liquid scanning transmission electron microscopy: imaging protein complexes in their native environment in whole eukaryotic cells. *Microsc. Microanal.*, **20** (2014), 346–365.
7. N. de Jonge, N. Poirier-Demers, H. Demers, D. B. Peckys and D. Drouin, Nanometer-resolution electron microscopy through micrometers-thick water layers. *Ultramicroscopy*, **110**, 1114–1119 (2010).
8. D. B. Peckys and N. de Jonge, Gold nanoparticle uptake in whole cells in liquid examined by environmental scanning electron microscopy. *Microsc. Microanal.*, **20** (2014), 189–197.
9. D. B. Peckys, J. P. Baudoin, M. Eder, U. Werner and N. de Jonge, Epidermal growth factor receptor subunit locations determined in hydrated cells with environmental scanning electron microscopy. *Sci. Rep.*, **3** (2013), 2626.
10. M. J. Dukes, D. B. Peckys and N. de Jonge, Correlative fluorescence microscopy and scanning transmission electron microscopy of quantum-dot-labeled proteins in whole cells in liquid. *ACS Nano*, **4** (2010), 4110–4116.

11. D. B. Peckys, P. Mazur, K. L. Gould and N. de Jonge, Fully hydrated yeast cells imaged with electron microscopy. *Biophys. J.*, **100** (2011), 2522–2529.
12. D. B. Peckys and N. de Jonge, Visualization of gold nanoparticle uptake in living cells with liquid scanning transmission electron microscopy. *Nano Lett.*, **11** (2011), 1733–1738.
13. E. A. Ring and N. de Jonge, Microfluidic system for transmission electron microscopy. *Microsc. Microanal.*, **16** (2010), 622–629.
14. A. Bogner, G. Thollet, D. Basset, P. H. Jouneau and C. Gauthier, Wet STEM: a new development in environmental SEM for imaging nano-objects included in a liquid phase. *Ultramicroscopy*, **104** (2005), 290–301.
15. N. de Jonge, R. Sougrat, D. B. Peckys, A. R. Lupini and S. J. Pennycook, 3-Dimensional aberration corrected scanning transmission electron microscopy for biology. In T. Vo-Dinh, ed., *Nanotechnology in Biology and Medicine-Methods, Devices and Applications* (Boca Raton, FL: CRC Press, 2007) pp. 13.11–13.27.
16. J. Lippincott-Schwartz, E. Snapp and A. Kenworthy, Studying protein dynamics in living cells. *Nat. Rev. Mol. Cell. Biol.*, **2** (2001), 444–456.
17. J. B. Pawley, *Handbook of Biological Confocal Microscopy*, 2nd edn. (New York: Springer, 1995).
18. K. I. Willig, S. O. Rizzoli, V. Westphal, R. Jahn and S. W. Hell, STED microscopy reveals that synapthotagmin remains clustered after synaptic vesicle exocytosis. *Nature*, **440** (2006), 935–939.
19. S. W. Hell, Far-field optical nanoscopy. *Science*, **316** (2007), 1153–1158.
20. E. Betzig, G. H. Patterson, R. Sougrat *et al.*, Imaging intracellular fluorescent proteins at nanometer resolution. *Science*, **313** (2006), 1642–1645.
21. M. Bates, B. Huang, G. T. Dempsey and X. Zhuang, Multicolor super-resolution imaging with photo-switchable fluorescent probes. *Science*, **317** (2007), 1749–1753.
22. J. Lippincott-Schwartz and S. Manley, Putting super-resolution fluorescence microscopy to work. *Nat. Meth.*, **6** (2009), 21–23.
23. S. Herbert, H. Soares, C. Zimmer and R. Henriques, Single-molecule localization super-resolution microscopy: deeper and faster. *Microsc. Microanal.*, **18** (2012), 1419–1429.
24. D. W. Piston and G. J. Kremers, Fluorescent protein FRET: the good, the bad and the ugly. *Trends Biochem. Sci.*, **32** (2007), 407–414.
25. C. M. Warren and R. Landgraf, Signaling through ERBB receptors: multiple layers of diversity and control. *Cell. Signal.*, **18** (2006), 923–933.
26. S. R. Needham, M. Hirsch, D. J. Rolfe *et al.*, Measuring EGFR separations on cells with ~10 nm resolution via fluorophore localization imaging with photobleaching. *PLoS One*, **8** (2013), e62331.
27. P. Liu, T. Sudhaharan, R. M. Koh *et al.*, Investigation of the dimerization of proteins from the epidermal growth factor receptor family by single wavelength fluorescence cross-correlation spectroscopy. *Biophys. J.*, **93** (2007), 684–698.
28. A. Hoenger and J. R. McIntosh, Probing the macromolecular organization of cells by electron tomography. *Curr. Opin. Cell Biol.*, **21** (2009), 89–96.
29. L. F. Kourkoutis, J. M. Plitzko and W. Baumeister, Electron microscopy of biological materials at the nanometer scale. *Annu. Rev. Mater. Res.*, **42** (2012), 33–58.
30. O. Medalia, I. Weber, A. S. Frangakis *et al.*, Macromolecular architecture in eukaryotic cells visualized by cryoelectron tomography. *Science*, **298** (2002), 1209–1213.
31. K. Fujimoto, Freeze-fracture replica electron microscopy combined with SDS digestion for cytochemical labeling of integral membrane proteins: application to the immunogold labeling of intercellular junctional complexes. *J. Cell Sci.*, **108** (1995), 3443–3449.

32. A. J. Bushby, K. M. P'Ng, R. D. Young *et al.*, Imaging three-dimensional tissue architectures by focused ion beam scanning electron microscopy. *Nat. Protoc.*, **6** (2011), 845–858.
33. L. H. Bergersen, J. Storm-Mathisen and V. Gundersen, Immunogold quantification of amino acids and proteins in complex subcellular compartments. *Nat. Protoc.*, **3** (2008), 144–152.
34. C. A. Larabell and K. A. Nugent, Imaging cellular architecture with X-rays. *Curr. Opin. Struct. Biol.*, **20** (2010), 623–631.
35. D. B. Peckys, U. Korf and N. de Jonge, Local variations of HER2 dimerization in breast cancer cells discovered by correlative fluorescence and liquid electron microscopy. *Sci. Adv.*, **1** (2015), e1500165.
36. D. B. Peckys and N. de Jonge, Studying the stoichiometry of epidermal growth factor receptor in intact cells using correlative microscopy. *J. Vis. Exp.* (2015). Epub. 2015/09/19.
37. M. J. Williamson, R. M. Tromp, P. M. Vereecken, R. Hull and F. M. Ross, Dynamic microscopy of nanoscale cluster growth at the solid-liquid interface. *Nat. Mater.*, **2** (2003), 532–536.
38. S. Thiberge, A. Nechushtan, D. Sprinzak *et al.*, Scanning electron microscopy of cells and tissues under fully hydrated conditions. *Proc. Natl. Acad. Sci. USA*, **101** (2004), 3346.
39. E. A. Ring, D. B. Peckys, M. J. Dukes, J. P. Baudoin and N. de Jonge, Silicon nitride windows for electron microscopy of whole cells. *J. Microsc.*, **243** (2011), 273–283.
40. J. M. Grogan and H. H. Bau, The nanoaquarium: a platform for in situ transmission electron microscopy in liquid media. *J. Microelectromech. Sys.*, **19** (2010), 885–894.
41. H. Nishiyama, M. Suga, T. Ogura *et al.*, Atmospheric scanning electron microscope observes cells and tissues in open medium through silicon nitride film. *J. Struct. Biol.*, **169** (2010), 438–449.
42. N. Liv, I. Lazic, P. Kruit and J. P. Hoogenboom, Scanning electron microscopy of individual nanoparticle bio-markers in liquid. *Ultramicroscopy*, **143** (2014), 93–99.
43. D. J. Stokes, Recent advances in electron imaging, image interpretation and applications: environmental scanning electron microscopy. *Phil. Trans. R. Soc. Lond. A*, **361** (2003), 2771–2787.
44. D. L. Stokes, *Principles and Practice of Variable Pressure/Environmental Scanning Electron Microscopy (VP SEM)* (New York: Wiley, 2008).
45. N. Li, A. C. Zonnevylle, A. C. Narvaez *et al.*, Simultaneous correlative scanning electron and high-NA fluorescence microscopy. *PLoS One*, **8** (2013), e55707.
46. K. Masenelli-Varlot, A. Malchere, J. Ferreira *et al.*, Wet-STEM tomography: principles, potentialities and limitations. *Microsc. Microanal.*, **20** (2014), 366–375.
47. N. de Jonge, D. B. Peckys, G. M. Veith *et al.*, Scanning transmission electron microscopy of samples in liquid (liquid STEM). *Microsc. Microanal.*, **13** (2007), 242–243.
48. R. Ramachandra, H. Demers and N. de Jonge, Atomic-resolution scanning transmission electron microscopy through 50 nm-thick silicon nitride membranes. *Appl. Phys. Lett.*, **98** (2011), 93109.
49. K. L. Liu, C. C. Wu, Y. J. Huang *et al.*, Novel microchip for in situ TEM imaging of living organisms and bio-reactions in aqueous conditions. *Lab Chip*, **8** (2008), 1915–1921.
50. K. L. Klein, I. M. Anderson and N. de Jonge, Transmission electron microscopy with a liquid flow cell. *J. Microsc.*, **242** (2011), 117–123.
51. U. Coskun and K. Simons, Cell membranes: the lipid perspective. *Structure*, **19** (2011), 1543–1548.
52. A. Arkhipov, Y. Shan, R. Das *et al.*, Architecture and membrane interactions of the EGF receptor. *Cell*, **152** (2013), 557–569.

53. N. Normanno, A. De Luca, C. Bianco et al., Epidermal growth factor receptor (EGFR) signaling in cancer. *Gene*, **366** (2006), 2–16.
54. J. Schlessinger, Signal transduction by allosteric receptor oligomerization. *Trends Biochem. Sci.*, **13** (1988), 443–447.
55. A. Ullrich and J. Schlessinger, Signal transduction by receptors with tyrosine kinase activity. *Cell*, **61** (1990), 203–212.
56. N. F. Endres, R. Das, A. W. Smith et al., Conformational coupling across the plasma membrane in activation of the EGF receptor. *Cell*, **152** (2013), 543–556.
57. D. S. Lidke, P. Nagy, R. Heintzmann et al., Quantum dot ligands provide new insights into erbB/HER receptor-mediated signal transduction. *Nat. Biotechnol.*, **22** (2004), 198–203.
58. K. A. Tanaka, K. G. Suzuki, Y. M. Shirai et al., Membrane molecules mobile even after chemical fixation. *Nat. Meth.*, **7** (2010), 865–866.
59. J. R. Glenney, Jr., W. S. Chen, C. S. Lazar et al., Ligand-induced endocytosis of the EGF receptor is blocked by mutational inactivation and by microinjection of anti-phosphotyrosine antibodies. *Cell*, **52** (1988), 675–684.
60. A. Hoenger and C. Bouchet-Marquis, Cellular tomography. *Adv. Protein Chem. Struct. Biol.*, **82** (2011), 67–90.
61. A. A. Sousa, A. A. Azari, G. Zhang and R. D. Leapman, Dual-axis electron tomography of biological specimens: extending the limits of specimen thickness with bright-field STEM imaging. *J. Struct. Biol.*, **174** (2011), 107–114.
62. H. Demers, N. Poirier-Demers, D. Drouin and N. de Jonge, Simulating STEM imaging of nanoparticles in micrometers-thick substrates. *Microsc. Microanal.*, **16** (2010), 795–804.
63. T. Schuh and N. de Jonge, Liquid scanning transmission electron microscopy: nanoscale imaging in micrometers-thick liquids. *C. R. Phys.*, **15** (2014), 214–223.
64. E. A. Ring and N. de Jonge, Video-frequency scanning transmission electron microscopy of moving gold nanoparticles in liquid. *Micron*, **43** (2012), 1078–1084.
65. E. R. White, M. Mecklenburg, B. Shevitski, S. B. Singer and B. C. Regan, Charger nanoparticle dynamics in water induced by scanning transmission electron microscopy. *Langmuir*, **28** (2012), 3695–3698.
66. A. Verch, M. Pfaff and N. de Jonge, Exceptionally slow movement of gold nanoparticles at a solid/liquid interface investigated by scanning transmission electron microscopy. *Langmuir*, **31** (2015), 6956–6964.
67. J. M. Yuk, J. Park, P. Ercius et al., High-resolution EM of colloidal nanocrystal growth using graphene liquid cells. *Science*, **336** (2012), 61–64.
68. J. E. Evans, K. L. Jungjohann, N. D. Browning and I. Arslan, Controlled growth of nanoparticles from solution with in situ liquid transmission electron microscopy. *Nano Lett.*, **11** (2011), 2809–2813.
69. B. F. Lillemeier, J. R. Pfeiffer, Z. Surviladze, B. S. Wilson and M. M. Davis, Plasma membrane-associated proteins are clustered into islands attached to the cytoskeleton. *Proc. Natl. Acad. Sci. USA*, **103** (2006), 18992–18997.
70. J. McBride, J. Treadway, L. C. Feldman, S. J. Pennycook and S. J. Rosenthal, Structural basis for near unity quantum yield core/shell nanocrystals. *Nano Lett.*, **6** (2006), 1496–1501.
71. D. B. Peckys, V. Bandmann and N. de Jonge, Correlative fluorescence and scanning transmission electron microscopy of quantum dot-labeled proteins on whole cells in liquid. *Meth. Cell Biol.*, **124** (2014), 305–322.
72. K. A. Tanaka, K. G. Suzuki, Y. M. Shirai et al., Membrane molecules mobile even after chemical fixation. *Nat. Meth.*, **7** (2010), 865–866.

73. L. Reimer and H. Kohl, *Transmission Electron Microscopy: Physics of Image Formation* (New York: Springer, 2008).
74. E. S. Pohlmann, K. Patel, S. Guo et al., Real-time visualization of nanoparticles interacting with glioblastoma stem cells. *Nano Lett.*, **15** (2015), 2329–2335.
75. J. C. H. Spence, *High-Resolution Electron Microscopy*, 3rd edn. (Oxford: Oxford University Press, 2003).
76. A. V. Agronskaia, J. A. Valentijn, L. F. van Driel et al., Integrated fluorescence and transmission electron microscopy. *J. Struct. Biol.*, **164** (2008), 183–189.
77. V. R. Matricardi, R. C. Moretz and D. F. Parsons, Electron diffraction of wet proteins: catalase. *Science*, **177** (1972), 268–270.
78. D. J. Siegwart, A. Srinivasan, S. A. Bencherif et al., Cellular uptake of functional nanogels prepared by inverse miniemulsion ATRP with encapsulated proteins, carbohydrates, and gold nanoparticles. *Biomacromol.*, **10** (2009), 2300–2309.
79. B. D. Chithrani, A. A. Ghazani and W. C. Chan, Determining the size and shape dependence of gold nanoparticle uptake into mammalian cells. *Nano Lett.*, **6** (2006), 662–668.
80. N. A. Bright, B. J. Reaves, B. M. Mullock and J. P. Luzio, Dense core lysosomes can fuse with late endosomes and are re-formed from the resultant hybrid organelles. *J. Cell Sci.*, **110** (1997), 2027–2040.
81. R. Tantra and A. Knight, Cellular uptake and intracellular fate of engineered nanoparticles: a review on the application of imaging techniques. *Nanotoxicology*, **5** (2011), 381–392.
82. M. I. Glavinovic, M. L. Vitale and J. M. Trifaro, Comparison of vesicular volume and quantal size in bovine chromaffin cells. *Neuroscience*, **85** (1998), 957–968.
83. C. Brandenberger, C. Muhlfeld, Z. Ali et al., Quantitative evaluation of cellular uptake and trafficking of plain and polyethylene glycol-coated gold nanoparticles. *Small*, **6** (2010), 1669–1678.
84. A. Sartori, R. Gatz, F. Beck et al., Correlative microscopy: bridging the gap between fluorescence light microscopy and cryo-electron tomography. *J. Struct. Biol.*, **160** (2007), 135–145.
85. B. L. Gilmore, S. P. Showalter, M. J. Dukes et al., Visualizing viral assemblies in a nanoscale biosphere. *Lab Chip*, **13** (2013), 216–219.
86. H. Zheng, S. A. Claridge, A. M. Minor, A. P. Alivisatos and U. Dahmen, Nanocrystal diffusion in a liquid thin film observed by in situ transmission electron microscopy. *Nano Lett.*, **9** (2009), 2460–2465.
87. T. J. Woehl, K. L. Jungjohann, J. E. Evans et al., Experimental procedures to mitigate electron beam induced artifacts during in situ fluid imaging of nanomaterials. *Ultramicroscopy*, **127** (2013), 53–63.
88. U. M. Mirsaidov, H. Zheng, Y. Casana and P. Matsudaira, Imaging protein structure in water at 2.7 nm resolution by transmission electron microscopy. *Biophys. J.*, **102** (2012), L15–17.
89. C. Mueller, M. Harb, J. R. Dwyer and R. J. Dwayne Miller, Nanofluidic cells with controlled pathlength and liquid flow for rapid, high-resolution in situ imaging with electrons. *J. Phys. Chem. Lett.*, **4** (2013), 2339–2347.
90. X. Shu, V. Lev-Ram, T. J. Deerinck et al., A genetically encoded tag for correlated light and electron microscopy of intact cells, tissues, and organisms. *PLoS Biol.*, **9** (2011), e1001041.
91. G. Gaietta, T. J. Deerinck, S. R. Adams et al., Multicolor and electron microscopic imaging of connexin trafficking. *Science*, **296** (2002), 503–507.
92. C. Risco, E. Sanmartin-Conesa, W. P. Tzeng et al., Specific, sensitive, high-resolution detection of protein molecules in eukaryotic cells using metal-tagging transmission electron microscopy. *Structure*, **20** (2012), 759–766.
93. R. Tantra and A. Shard, We need answers. *Nat. Nanotechnol.*, **8** (2013), 71.

17 Visualizing Macromolecules in Liquid at the Nanoscale

Andrew C. Demmert, Madeline J. Dukes, Elliot Pohlmann, Kaya Patel, A. Cameron Varano, Zhi Sheng, Sarah M. McDonald, Michael Spillman, Utkur Mirsaidov, Paul Matsudaira, and Deborah F. Kelly

17.1 Introduction: The Critical Need for Imaging Dynamic Events in Life Sciences

Macromolecules are heavily utilized worldwide in industry and in healthcare. Modern uses may include the production of next generation fuel cells, water filtration membranes, drug delivery systems, and human vaccines. Despite the fact that molecular polymers are exploited in the manufacture of nearly all substances, our current knowledge of how macromolecules work is still quite limited. A common goal in the biomedical community is to understand the fundamental properties of macromolecules to develop emerging technologies, such as new therapies to treat diseases. To date, characteristic information on macromolecules comes primarily from morphological studies performed in a fixed, non-native environment using electron microscopy. However, developing novel tools to visualize the dynamic behavior of macromolecules could provide a multi-dimensional framework to observe biological mechanisms at the nanoscale.

Transmission electron microscopes are ideal instruments to visualize the intricate details of molecular entities at atomic resolution. However, in order to survive the high vacuum system of a TEM, samples are typically embedded in thin films of vitreous ice, sugars, resins, or heavy metal salts [1–3]. Naturally, these procedures yield only static snapshots of dynamic processes. Early attempts to maintain liquid specimens in a TEM were performed by Parsons and colleagues in the 1970s [4, 5]. At the time, these experiments were limited by the thick, diffusing liquid that could not easily be controlled within the sample chamber. This challenge posed a major roadblock to implementing liquid imaging techniques using TEM. It is only recently that further experiments to observe liquid specimens for molecular imaging have been attempted.

Currently, there remains a critical need for innovative imaging platforms to illuminate the dynamic behavior of macromolecules in a native liquid environment. Here, we discuss how liquid cell TEM can address this problem and potentially complement cryo-EM methodologies. In doing so, we review recent results obtained in the study of biological macromolecules for use as model systems to peer into the world of active nanomachines at unprecedented resolution.

17.2 Recent Technical Advances: How Liquid Cell TEM Can Address This Critical Need

The use of liquid cell imaging platforms can transform our ability to visualize the structural details and functional capacities of biomedically relevant macromolecules at nanometer resolution. We recently tested the resolution capacity of the liquid imaging platform by examining the atomic features of therapeutic gold nanorods in solution [6]. Aliquots of polyvinyl pyridine (PVP)-Au nanorods (Nanopartz, Inc.) having dimensions of 50 nm × 20 nm were diluted in Milli-Q water (0.17 mg/ml) and loaded into a liquid chamber with 50 nm thick Si_xN_y windows, one of which was engineered with a 150 nm spacer. The Au lattice spacing (2.04 Å) visible in the images demonstrates the feasibility of resolving atomic information under fully hydrated conditions within Si_xN_y liquid cells [6]. Dynamic information is also accessible. The Au rods initially diffused to evade the beam then returned to fill the same liquid region, or were swept up in migrating liquid fronts, similar in appearance to a "nanoscale tidal wave" (Figure 17.1). These wavefronts served to concentrate the nanorods into a central location as the liquid retreated. We posit similar wavefronts may concentrate Au nanorods in tumor micro environments and contribute to the collective heating process achieved during therapeutic interventions such as photothermal ablation used to treat cancer [6]. With these new tools in hand, we can now use liquid cell TEM technology to image unlabeled biological assemblies in liquid.

17.3 The Affinity Capture Technique to Tether Unlabeled Biological Complexes onto Si_xN_y

Figure 17.1 demonstrated that long-range movements can persist on the level of micrometers while imaging samples in liquid. As described in Chapter 8, we would expect motion due to liquid movement or diffusion to limit our ability to discern the fine details inherent in biological macromolecules. As a means to control long-range diffusion in our liquid samples, we adapted the affinity capture technique, previously developed to enhance cryo-EM specimen preparation [7, 8]. In this section we discuss the implementation of the affinity capture procedure for use with Si_xN_y substrates, and apply this technique to study active viral processes in liquid using microchips with integrated microwells.

17.3.1 Capturing and Visualizing Ribosomal Assemblies in Solution

The affinity capture system is used in cryo-EM specimen preparation procedures to tether biological entities to carbon-coated EM grids. As the surface of cleaned Si_xN_y chips is naturally hydrophobic, we can functionalize the microchips with a tunable surface coating prior to assembling the microfluidic chamber [9, 10]. This coating contains nickel-nitrilotriacetic acid (Ni-NTA), a metal-chelated substrate that binds with

Figure 17.1. TEM imaging of mobile Au nanorods in liquid. Selected images indicate the micrometer-scale movements of Au rods in solution. "Nanoscale tidal waves" are the dark fronts emerging from the top left corner in panel 2. These waves pass through the imaging plane and concentrate the rods upon recession. Red arrows provide a reference landmark in the images. Images were recorded in a Poseidon Liquid Stage (Protochips, Inc.) using a FEI Tecnai Spirit BioTwin TEM (FEI Company) equipped with a FEI Eagle 2k HS CCD camera (30 μm pixel size) and operating at 120 kV under low dose conditions (0.01–0.03 electrons/nm^2). Scale bar is 100 nm. Adapted from Ref. 6.

high affinity to histidine (His)-tagged recombinant proteins. Figure 17.2a–c shows a first test of this system on His-tagged ribosomal assemblies produced in bacteria. To accomplish this, we overexpressed His-tagged protein subunits (rpl3) of the 50S ribosome in *E. coli* (BL21 strain), lysed the bacteria, and flowed the cleared lysate into a fluidic chamber having 150 nm spacers. As the sample flowed through the device while inside the TEM column, intact His-tagged ribosomal complexes were captured by the Ni-NTA-coated microchips (Figure 17.2d, e). These RNA-containing protein complexes exhibited dark contrast due to their density in solution, consistent with images of ribosomal complexes prepared in vitreous ice [7]. Figure 17.2f shows the control experiment where ribosomal assemblies do not appreciably accumulate on microchips lacking the Ni-NTA coating. By uniquely tethering the tagged ribosomes to the microchip surface we limited their long-range diffusion in solution and could image the complexes throughout the process. These results provided the first glimpse of individual macromolecules in liquid using TEM [9].

Although this provided the necessary proof-of-concept results and technical foundation for imaging individual biological assemblies in solution, the ribosomal assemblies lacked strong structural features in the images, limiting further single-particle image processing routines. This lack of detail in the images can be attributed to the inability to control the liquid thickness. We now discuss a means to circumvent this limitation.

17.3.2 Microwell Chip Design for Use with Biomedical Reagents

Figure 17.3a illustrates the concept used to improve image quality for biological assemblies: microchips with integrated microwells etched into the Si_xN_y membranes (Figure 17.3a). In the center of the 200 nm thick membrane, an array of 10 × 10 μm microwells is formed by etching to leave only ~30–50 nm Si_xN_y. Each microwell can therefore potentially contain a liquid layer 150 nm thick, although depending upon

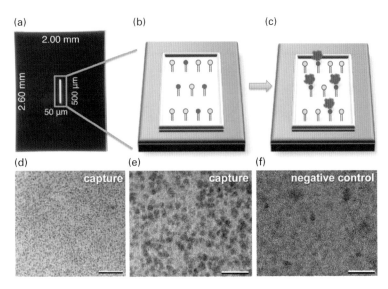

Figure 17.2. The production and use of affinity capture devices. (a) Commercially available Si_xN_y microchips were functionalized with lipid monolayers (b) containing Ni-NTA head groups (red circles) that interact with His-tagged protein targets (red complexes) (c). (d) Low magnification view of His-tagged ribosomes shows an abundance of captured complexes. Scale bar is 500 nm. (e) High magnification view of captured ribosomal assemblies shows native contrast in solution but lacks fine structural features. Scale bar is 100 nm. (f) Ribosomal assemblies do not readily accumulate on microchips that lack the Ni-NTA coating (negative control). Scale bar is 100 nm. Images were recorded using a FEI Tecnai Spirit BioTwin TEM equipped with a FEI Eagle 2k HS CCD camera (30 μm pixel size) and operating at 120 kV under low dose conditions (0.01–0.03 electrons/nm^2). Adapted from Ref. 9.

experimental procedures, the liquid can on occasion become even thinner or dry [11]. In general, the use of microchips with integrated microwells appeared to consistently control the thickness of the liquid layer and possibly reduce bulging of the microchip system while in the TEM vacuum. Further experiments to better define these effects are in progress. Employing these microchips, we show real-time events involving magnetic-based nanoparticles used in biomedical MRI applications in Figure 17.3b and liposomal-based drug delivery vehicles in Figure 17.3c [11]. Results from these initial experiments proved encouraging for extending the technique to biologically active macromolecules.

Another biomedical application for liquid cell TEM involves the real-time imaging of nanoparticles delivered to human cancer cells, in particular, for monitoring uptake and export processes involving Au nanoparticle reagents [12]. Glioblastoma stem cells (GSCs) are a rare population of cells that lurk within primary tumors. They are thought to give rise to recurrent tumors following conventional treatments. New treatment options for cancer cells and in particular GSCs include the use of nanoparticle-based therapies for targeted delivery. Developing technologies to evaluate these therapies at the molecular level may contribute much-needed insight on therapeutic efficacy.

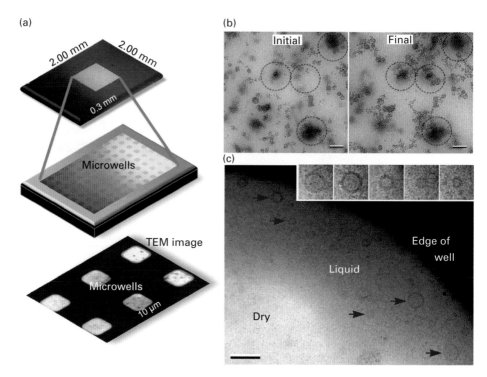

Figure 17.3. The use of Si_xN_y microchips containing integrated microwells to examine biomedical reagents. (a) Schematic and TEM image to illustrate the dimensions of the microwell-containing microchips used to form the base chip of the liquid chamber that fits within the holder. (b) TEM images acquired during magnetite-containing nanoparticle diffusion in liquid. Scale bar is 100 nm. (c) Liposome-based drug delivery vehicles exhibiting native contrast in hydrated microwell-containing microchips. Scale bar is 250 nm. Inset panels (200 nm in width) indicate individual liposomes in solution. Images were recorded using an FEI Eagle 2k HS CCD camera having a pixel size of 30 μm at a dose of less than 0.01 electron/nm². The camera was integrated into an FEI Tecnai Spirit BioTwin TEM operating at 120 kV using the Protochips, Inc. Poseidon system in a static configuration. Adapted from Ref. 11.

To address this issue, we engineered a toolkit for real-time imaging of Au nanoparticle interactions with individual GSCs [12]. We enriched for and tethered GSCs uniquely expressing the NOTCH1 protein receptor to Si_xN_y microchips with integrated microwells. The tethered cells were treated with PVP-encapsulated Au nanorods (Nanopartz Inc.). Mobility of the Au nanorods is visible at the edge of the cell membrane as well as within the cells in Figure 17.4.

In combination with cytotoxicity studies, it appears from these images that Au nanorods can penetrate the cell membrane and likewise be exported by the cells when used within the concentration ranges we tested. This is important technically as well as biologically, as GSCs are known to evade conventional drug therapies by exporting the drugs out of the tumor cells, and here we see this mechanism at

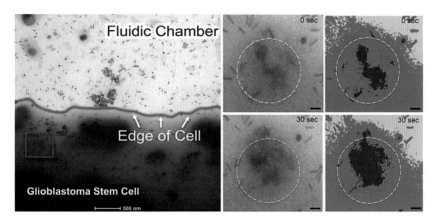

Figure 17.4. Gold nanorods effectively penetrated patient-derived GSCs (GS9-6/NOTCH1+ line) within 20 minutes. The nanorods were diluted to 0.17 mg/ml or 0.017 mg/ml in Milli-Q water and treated cells were imaged at 30 s intervals in different regions within the individual cells (red square). Contour plots showing the highest statistical differences show mobile nanorods (white dashed circles) within the tumor cells. Gradients of blue signify the degree of change in electron scattering events. Images were recorded under low dose conditions of <0.01 electron/nm^2 using a FEI Tecnai Spirit BioTwin TEM operating at 120 kV. Scale bars in side panels are 50 nm. Adapted from Ref. 12.

the nanoscale. Investigations aimed at monitoring drug delivery conditions to prevent drug export from GSCs would greatly contribute to the design of more effective treatments [12].

17.4 Correlative Nanoscale Imaging: What Information Can We Learn from Combining Liquid Cell TEM and Cryo-EM?

TEM liquid cells developed to image dynamics in liquids for materials science and physics problems are not ideal for imaging unstained biological material. Two critical modifications are necessary (Figure 17.5). First, the water thickness must be reduced from 0.2–10 μm to 100–200 nm, similar to the thickness of vitreous ice used in cryo-EM for single particle and tomography applications. At this thickness, the contrast in the liquid cell image should be similar to that in cryo-EM because the densities of water and vitreous ice are similar. Reducing liquid thickness further improves contrast in the image, but does not permit larger structures to be examined. Second, scattering from the windows must be minimized. Thinner windows provide improved resolution but are more fragile [13]. New materials, perhaps large, defect-free graphene windows, are required to resolve these deficiencies in liquid cell designs, and are discussed in Chapter 19. Ongoing work will improve our understanding of the differences that may arise when examining biological complexes in liquid and in vitreous ice, as both media represent near-native hydrated environments. Here, we discuss the current state of macromolecular dynamics that can be

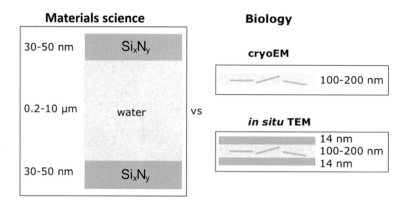

Figure 17.5. Nanoscale imaging in materials science and for unlabeled biostructures. Schematic representation of sample differences for materials science (left) and biological applications (right), employing either cryo-EM (top right) or liquid cell TEM (bottom right).

observed in cryo-EM and liquid cell EM, using viral assemblies and the actin cytoskeleton as model systems.

17.4.1 Case Study: Active Rotavirus Assemblies in Liquid and in Ice

Viral pathogens remain a global threat to human health and development. Hence, it is important to invent new technologies aimed at understanding the molecular intricacies of host–pathogen interactions. There is a fundamental gap in our understanding of how RNA viruses transcribe their genomes, which is a key step to host cell infection. Real-time *in situ* imaging may provide significant insights to explain dynamic viral processes that cannot be explained by static imaging or biochemical methodologies alone. Rotavirus (RV) is an important pediatric gastrointestinal pathogen that we can use as an experimental system to study the structural dynamics of viral nanomachines. Here, we discuss the imaging of transcriptionally active RV particles in liquid using the Si_xN_y microchips containing integrated microwells introduced in Section 17.3.2 above.

Upon invading a host cell, infectious triple-layered RV virions shed their outer capsid proteins (VP4 and VP7) leaving behind double-layered particles (DLPs) (Figure 17.6, step 1). Within the DLP cores, the viral RNA genome is transcribed into mRNA segments that become extruded from the particle and deposited into the host cell cytoplasm (Figure 17.6, step 2*). These mRNA segments are then translated into viral proteins by host ribosomes (Figure 17.6, step 3*). In this regard, new therapeutic targets may be designed against active DLPs that extrude their mRNA viral segments (step 2*) to initiate a hostile takeover of cellular ribosomal machinery (step 3*).

In developing useful assays to observe viral activities, we first need to produce transcriptionally competent DLPs. We therefore purified simian DLPs (strain SA11-4F) from monkey kidney MA104 cells [14], detecting and analyzing the protein composition of the purified DLPs using SDS-PAGE and silver staining (Figure 17.7a,

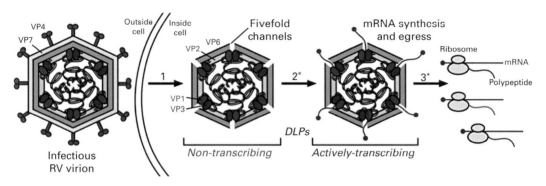

Figure 17.6. RV transcription within a host cell. The infectious RV virion sheds its outer VP4-VP7 layer during host cell entry to form a DLP (step 1). The viral RNA genome is actively transcribed by VP1 (pink) and VP3 (purple) in the interior of the DLP then extruded into the cytoplasm of the host cell (step 2*). Viral mRNA transcripts are translated into viral proteins by host ribosomes (step 3*). *denotes steps involved in viral hijacking of host cellular machinery.

left). This showed that the DLPs comprise four proteins (VP1, VP2, VP3, and VP6) and that they lack VP4 and VP7. To verify transcriptional competence, the purified DLPs were subjected to *in vitro* mRNA synthesis in the presence of [^{32}P]-ribonucleotides for 30 minutes at 37 °C. Specifically, each reaction contained DLPs, the ribonucleotides needed to synthesize nascent mRNA including adenosine triphosphate (ATP), and [^{32}P]-labeled ribonucleotides (Figure 17.7a, +ATP). Negative control reactions contained all transcription cocktail components except ATP (Figure 17.7a, −ATP). Radiolabeled viral mRNA products were only detected in the reactions containing the complete transcription cocktail; no radiolabeled mRNA products were seen in the reaction lacking ATP, as ATP was required by the viral assemblies to perform the synthesis reaction. This functional analysis therefore confirms that purified DLPs can be enzymatically activated.

In our attempts to visualize transcribing DLPs, viral reaction mixtures were prepared as above and added onto Ni-NTA-coated Si$_x$N$_y$ microchips with integrated microwells that were decorated with His-tagged protein A and polyclonal antibodies (IgGs) raised against the VP6 capsid protein (Figure 17.7a, schematic). The resulting TEM images showed dynamic attributes of the RV pathogen in liquid at 3 nm resolution (Figure 17.7b). We could also distinguish individual strands of molecules emerging from numerous DLPs. These strands had dimensions and shapes consistent with viral mRNAs transcripts (Figure 17.7b, 1–4, right panels). No such strands were identified in images of reactions that lacked ATP. We used the RELION software package to calculate 3D structures of the DLPs from a single EM image while enforcing icosahedral symmetry (Figure 17.7c). In comparing the 3D structures, the interiors of the DLP core structures revealed dynamic rearrangements during mRNA synthesis [11]. Based on these results, we anticipate ongoing work will permit the correlation of changes in 3D structures with real-time recording of transcribing DLPs [15]. As such, combining liquid cell imaging of dynamic behavior with high resolution cryo-EM structural

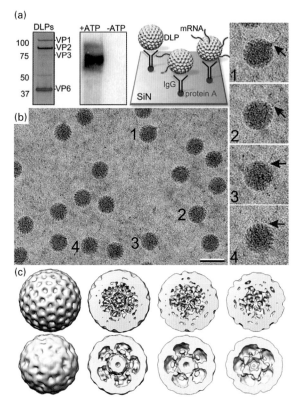

Figure 17.7. Liquid cell TEM of transcribing DLPs. (a) Purified DLPs comprise VP1, VP2, VP3, and VP6, as indicated on a silver-stained SDS-PAGE gel (left panel). Upon the addition of ribonucleotides including ATP, and [^{32}P]-labeled ribonucleotides, DLPs transcribe their RNA genome, leading to accumulation of mRNA transcripts, but in the absence of ATP no mRNA is detected (middle panel). The right panel shows a schematic representation of the active DLPs tethered to Si_xN_y microchips using protein A/IgG adaptors, preventing diffusion but not eliminating Brownian motion. (b) Images of transcribing DLPs in liquid recorded using a FEI Spirit Bio-Twin TEM equipped with a LaB_6 filament and operating at 120 kV under low dose conditions (~0.005 electrons/nm^2). Single-stranded mRNA emerges from the viral capsids (1–4). Scale bar is 100 nm. (c) 3D structures of active DLPs show differences in capsid protein arrangements (blue and yellow) and within their internal features during RNA synthesis. Adapted from Ref. 11.

insights may create new opportunities to develop therapeutic interventions based on protein capsid changes that occur during RNA synthesis.

It is important to compare the liquid imaging experiment described above with *ex situ* observations made using cryo-EM to further improve our understanding of transcribing DLPs in a frozen-hydrated state. Cryo-EM was used to image transcribing DLPs under low dose conditions (~0.05 electrons/nm^2). Individual DLPs were selected from the images using the PARTICLE software package (http://www.image-analysis.net/EM/, freely distributed through the SBGrid Consortium) and imported into the RELION software package, which computes 3D reconstructions using Bayesian inference [16].

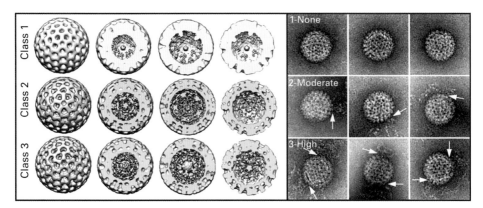

Figure 17.8. Ordered density in the interior of actively transcribing DLPs. Three distinct cryo-EM image reconstructions (Classes 1, 2, and 3; left panels) were determined from images of DLPs incubated in complete transcription cocktail for 30 min at 37 °C. Negatively stained DLP particles are shown at right with similar Class 1, 2, or 3 reconstructions. The transcriptional activities of the particles (none, moderate, or high) were interpreted based upon the amount of surrounding mRNA. These results suggest that Class 3 particles, which are robustly transcribing, have the most ordered internal density.

Based on statistical likelihood comparisons of the experimental images with respect to the reference model in Ref. 17, RELION defined three different reconstructions (Classes 1, 2, and 3) within the merged image stack (Figure 17.8, left panels). The Class 1 reconstruction exhibits the highest degree of order in the outer protein capsid layer, but little ordered density in the particle interior. In comparison, the ordered features in the protein capsid were somewhat diminished in the Class 2 reconstruction, but this structure exhibited stronger features throughout the cross sections, with prominent density at the apex of each fivefold axis. External order was somewhat diminished in the Class 3 reconstruction, however, this structure displayed the highest degree of order in the particle interior with strong densities emanating at the fivefold axes and continuing through each section.

Cryo-EM visualizes the presence of RNA transcripts, so it is possible to correlate a DLP's structure (i.e. Class 1, 2, or 3) to its transcriptional activity at the time of flash-freeze. Figure 17.8 (right panels) shows representative particles from the RELION output files. Few to no RNA transcripts are seen in the vicinity of the DLPs making up the Class 1 reconstruction, indicative of non-transcribing particles. DLPs in the Class 2 reconstruction had moderate levels of surrounding RNA, which typically emanated from a single general location on each particle, suggestive of particles ending a single round of synchronous transcription. DLPs constituting the Class 3 reconstruction contained several mRNA strands in their vicinity, which seemed to emerge from at least two positions on the DLP outer capsid, indicative of actively transcribing particles. Overall, these data suggest that non-transcribing particles (Class 1) are highly ordered on the outside, but quite disordered on the inside. In contrast, actively transcribing particles (Class 3) are more disordered on the outside (perhaps as a result of "messy" surface RNA strands), but show highly ordered density on the particle interior.

Fundamentally, these results support the idea that a dramatic structural rearrangement occurs within the DLP interior during the process of transcription.

Does the same relationship between protein capsid structure and mRNA production hold in the liquid cell observation? Upon comparing 3D reconstructions calculated for particles embedded in ice (Figure 17.8) with those enclosed in liquid (Figure 17.7), we observe that protein capsid rearrangements varied in accordance with mRNA production, and changes in the external features of the capsid proteins were inversely proportional to structural rearrangements in the interior of the DLP. The combination of cryo-EM and liquid cell imaging can reveal biologically meaningful insights to relate high resolution cryo-imaging with the direct visualization of dynamic viral systems. We anticipate that ongoing efforts to connect these two disciplines may have an important impact towards advancing our fundamental understanding of virus transcription. Global efforts to attenuate the detrimental health effects of viral pathogens may directly benefit from novel technologies that bridge molecular imaging modalities to provide a new view of these exquisite viral nanomachines.

17.4.2 Case Study: Acrosomal Process in Liquid and in Ice

The versatile actin cytoskeleton powers cell and tissue movements. One class of cell motility is the acrosome reaction of some marine invertebrate sperm [18]. When sperm contact the investing layers surrounding an egg, a 50 μm long finger of membrane, the acrosomal process, extends from the sperm head in about 5 seconds. In the horseshoe crab sperm this extension generates an impulse of 2 nN s [19], providing sufficient pressure, 1.6 MPa, to puncture the investing layers. These mechanical steps permit the acrosomal process to fuse with the egg membrane and activate the fertilization reaction of the egg. A crystalline bundle of actin filaments provides structural rigidity to the membrane process during extension (Figure 17.9). In the unactivated sperm, the actin bundle is preformed and coiled around the base of the sperm head. Interestingly,

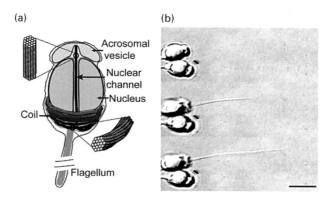

Figure 17.9. The versatile cytoskeletal arrangements in the acrosomal process. (a) The bundle of actin filaments coil around the base of the cell and extend through the nucleus. (b) At fertilization, the acrosomal process extends to contact the egg membrane. Scale bar is 5 μm. Reprinted from Ref. 19.

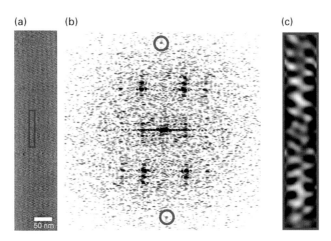

Figure 17.10. Liquid cell TEM analysis of the 80 nm diameter acrosome. (a, b) The corresponding Fourier transform of the image and (c) a reconstruction of a filament in the bundle (right) are consistent with cryo-EM images and structures. Adapted from Ref. 13.

the actin filaments in the bundle are over-twisted, thus building mechanical strain into the coiled bundle. The motive force for extension can be explained by the release of torsional strain when the coil is converted into a straight bundle. The change in twist of actin filaments can only be detected from electron micrographs of the acrosomal bundle.

A first step in characterizing the dynamics of this process is to visualize the acrosomal bundle in liquid. However, this is not possible by electron microscopy for several reasons. First, as discussed in Chapter 16, the highly energetic electron beam causes cell death. Second, it is thought that beam damage will prevent visualization of the structure at room temperature [20]. To explore these limitations and visualize the actin bundle structure from the coil to the straight bundle, we used liquid cell TEM [13]. Images of the acrosome taken under electron dose conditions identical to cryo-EM imaging of acrosomes in vitreous ice show the characteristic crystalline appearance of the bundle and the zig-zag-like structure of the filaments (Figure 17.10). The corresponding Fourier transform of the image shows the characteristic row and layer lines from a hexagonally packed bundle of actin filaments. The 2.7 nm meridional spot (circled) identifies the extent of the information in the image (see Chapter 8 for different methods to measure resolution). This information limit is identical to prior analysis of 120 kV cryo-EM images.

We use the time dependence of the ordering to evaluate the effects of electron beam-induced radiation damage (Chapter 7) in limiting spatial and temporal resolution in these observations. Figure 17.11 shows the fall-off of intensity in resolution measured for bundles in liquid room temperature water. Intensity is integrated over the spots in the diffraction pattern, and is shown for two separate frequency ranges: 2.7–5 nm periodicity (corresponding to the spacing between subunits in the filament) and >5 nm periodicity (corresponding to information about the filament and packing in the bundle). A surprising and controversial observation is that the fall-off of intensity of the spots

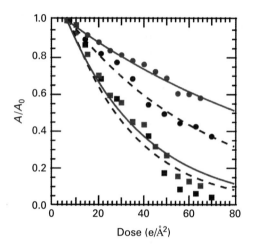

Figure 17.11. Fall-off of intensity as a function of electron dose for integrated intensities at 2.7–5 nm (squares) and >5 nm (circles) resolution shells and at 298 K (red) and 98 K (blue). Adapted from Ref. 13.

representing 2.7–5 nm periodicity is similar at room and liquid nitrogen temperatures (Figure 17.11). In contrast, prior experiments have documented that radiation damage is much faster at room temperature than at cryo temperatures [20]. The most significant source of damage from the electron beam is radiolysis of water and decomposition into H_2 gas, hydroxyl ions, and free radicals [21, 22]. Of the radiolysis products generated, gas formation is clearly limiting because it creates a void between the windows and leads to drying artifacts. However, negligible gas is produced at low dose conditions both at room temperature and at liquid nitrogen temperature [23]. The effects of the other radiolysis products are discussed in Chapter 7. Based on these experiments, imaging dynamic events in liquid at the molecular level should be feasible within a rapid time frame, and in particular with the use of the technologies discussed in Section 17.5 below.

17.5 New Directions: Use of Direct Electron CMOS Detectors to Acquire "Molecular Movies" of Fundamental Processes

To study real-time dynamics at the molecular level requires improvements to the liquid cells, cameras, and EM optics. Fortunately, several significant advances have been made commercially available in the past two years. The first is high spatial and temporal resolution cameras based on direct electron CMOS detectors. Second, phase plates have been developed to transform the contrast generated by low amplitude objects from a sine wave dependence to a cosine wave dependence, enhancing image contrast [24]. Lastly, new designs of liquid cells with larger and stronger windows will enable objects to be imaged more easily. The combined application of these three technologies will significantly reduce the time to capture images and thus the cumulative electron dose.

The latest generation of direct detection devices (DDDs) enables detection of incident electrons without the performance-limiting scintillator and fiber-optic coupling found in traditional charge-coupled devices (CCDs). DDDs, such as the DE-12 (Direct Electron, LP), are designed for direct detection of primary electrons from 60 keV to 1 MeV and are compatible with a wide range of instrument configurations. The major benefit of DDDs is the ability to capture "real-time movies" at significantly higher frame rates than CCDs. There is no dead time between frames in a DDD, unlike a CCD, enabling live streaming of movies with constant exposure to the electron beam. Additionally, smaller pixel designs and performance over broad spatial frequencies allows use of lower magnification for a larger field of view while retaining high resolution information. By contrast, CCDs are typically limited to a few frames per second (fps) and require binning of the image, restricting the field of view. DDDs also feature significantly increased sensitivity to incident electrons even with very low doses (<0.01 e$^-$/nm^2 per frame), especially helpful when dealing with radiation-sensitive biological samples.

Fundamental biological processes stand to benefit from the combination of DDDs and liquid cell holders that have been adopted by many materials science researchers. While biological specimens can be tethered within a liquid cell, specimens are still able to move due to fluid dynamic or beam-induced motion. This reduces the resolution from a typical CCD, but the higher frame rates available from DDDs can capture details that might otherwise have been blurred. The structures of hydrated, native-state biological specimens that can be obtained provide useful complements to cryo-EM studies.

Looking ahead, the large field of view and high frame rates of DDDs provide ideal conditions to observe biological samples in water, perhaps even permitting the imaging of viral particles that are actively transcribing RNA in real time at high resolution. The technique of liquid cell TEM is poised to open new frontiers in biological research related to the study of viral assembly, drug/vaccine interaction, and nanoparticle properties that were once thought impossible to observe at the nanoscale.

Acknowledgements

The authors acknowledge the following funding sources that supported this work: R21AI113402 NIAID/NIH (DFK, SMM) and R01CA193578NCI/NIH (DFK).

References

1. S. De Carlo and J. R. Harris, Negative staining and cryo-negative staining of macromolecules and viruses for TEM. *Micron*, **42** (2011), 117–131.
2. J. Dubochet *et al.*, Cryo-electron microscopy of vitrified specimens. *Q. Rev. Biophys.*, **21** (1988), 129–228.
3. N. Unwin and R. Henderson, The structure of proteins in biological membranes. *Sci. Am.*, **250** (1984), 78–94.

4. D. F. Parsons, V. R. Matricardi, R. C. Moretz and J. N. Turner, Electron microscopy and diffraction of wet unstained and unfixed biological objects. *Adv. Biol. Med. Phys.*, **15** (1974), 161–270.
5. D. F. Parsons, Structure of wet specimens in electron microscopy: improved environmental chambers make it possible to examine wet specimens easily. *Science*, **186** (1974), 407–414.
6. M. J. Dukes, B. W. Jacobs, D. G. Morgan, H. Hegde and D. F. Kelly, Visualizing nanoparticle mobility in liquid at atomic resolution. *Chem. Commun.*, **49** (2013), 3007–3009.
7. D. F. Kelly, P. D. Abeyrathne, D. Dukovski and T. Walz, The Affinity Grid: a pre-fabricated EM grid for monolayer purification. *J. Mol. Biol.*, **382** (2008), 423–433.
8. D. F. Kelly, D. Dukovski and T. Walz, A practical guide to the use of monolayer purification and affinity grids. *Methods Enzymol.*, **481** (2010), 83–107.
9. K. Degen, M. Dukes, J. R. Tanner and D. F. Kelly, The development of affinity capture devices: a nanoscale purification platform for biological in situ transmission electron microscopy. *RSC Adv.*, **2** (2012), 2408–2412.
10. B. L. Gilmore, S. P. Showalter, M. J. Dukes *et al.*, Visualizing viral assemblies in a nanoscale biosphere. *Lab Chip*, **13** (2013), 216–219.
11. M. J. Dukes, R. Thomas, J. Damiano *et al.*, Improved microchip design and application for in situ transmission electron microscopy of macromolecules. *Microsc. Microanal.*, **20** (2014), 338–345.
12. E. S. Pohlmann, K. Patel, S. Guo *et al.*, Real-time visualization of nanoparticles interacting with glioblastoma stem cells. *Nano Lett.*, **15** (2015), 2329–2335.
13. U. M. Mirsaidov, H. Zheng, Y. Casana and P. Matsudaira, Imaging protein structure in water at 2.7 nm resolution by transmission electron microscopy. *Biophys. J.*, **102** (2012), L15–17.
14. E. A. Ring, D. B. Peckys, M. J. Dukes, J. P. Baudoin and N. de Jonge, Silicon nitride windows for electron microscopy of whole cells. *J. Microsc. Oxford*, **243** (2011), 273–283.
15. A. Cameron Varano, A. Rahimi, M. J. Dukes *et al.*, Visualizing virus particle mobility in liquid at the nanoscale. *Chem Commun.*, **51** (2015), 16176–16179.
16. S. H. Scheres, A Bayesian view on cryo-EM structure determination. *J. Mol. Biol.*, **415** (2012), 406–418.
17. X. Zhang, E. Settembre, C. Xu *et al.*, Near-atomic resolution using electron cryomicroscopy and single-particle reconstruction. *Proc. Natl. Acad. Sci. USA*, **105** (2008), 1867–1872.
18. L. G. Tilney, Actin filaments in the acrosomal reaction of Limulus sperm: motion generated by alterations in the packing of the filaments. *J. Cell. Biol.*, **64** (1975), 289–310.
19. J. H. Shin, B. K. Tam, R. R. Brau *et al.*, Force of an actin spring. *Biophys. J.*, **92** (2007), 3729–3733.
20. R. H. Wade, The temperature-dependence of radiation-damage in organic and biological materials. *Ultramicroscopy*, **14** (1984), 265–270.
21. R. D. Leapman and S. Q. Sun, Cryoelectron energy-loss spectroscopy: observations on vitrified hydrated specimens and radiation-damage. *Ultramicroscopy*, **59** (1995), 71–79.
22. M. A. Aronova, A. A. Sousa and R. D. Leapman, EELS characterization of radiolytic products in frozen samples. *Micron*, **42** (2011), 252–256.
23. S. Yakovlev, M. Misra, S. Shi and M. Libera, Specimen thickness dependence of hydrogen evolution during cryo-transmission electron microscopy of hydrated soft materials. *J. Microsc. Oxford*, **236** (2009), 174–179.
24. R. Danev and K Nagayama, Transmission electron microscopy with Zernike phase plate. *Ultramicroscopy*, **88** (2001), 243–252.

18 Application of Liquid Cell Microscopy to Study Function of Muscle Proteins

Haruo Sugi, Shigeru Chaen, Tsuyoshi Akimoto, Masaru Tanokura, Takuya Miyakawa, and Hiroki Minoda

18.1 Introduction: Our Motivation for Liquid Cell Microscopy of Muscle Contraction

The structure of muscle fibers and the mechanism by which they contract have been a central interest of biologists for many years. Understanding the molecular mechanism of muscle contraction requires a microscopic view, visualizing the dynamics in an environment that permits control of the process. Our interest and research in muscle contraction has encompassed over 40 years. As early as 1986, we became aware that the liquid cell (at the time, also called the hydration chamber or gas environmental chamber) in a transmission electron microscope could be a direct way to approach the mysteries of muscle contraction. After a number of trials and errors, we were able to record myosin head movement in living muscle myosin filaments that retained their physiological function [1–5]. In this chapter, we first give an overview of the progress of research work on muscle contraction at the molecular level, and explain our motivation to perform "physiological experiments" using liquid cell electron microscopy. We then describe methods developed during the course of this research, and report dynamic myosin head movement induced by adenosine triphosphate (ATP) in living muscle myosin filaments. The methods that enable visualization of the functioning of biomolecules such as myosin and actin can be applied in research fields other than life sciences, and we anticipate that the techniques described in this chapter will be adopted more widely.

18.1.1 The Sliding Filament Mechanism and the Structure of Muscle Filaments

A vertebrate skeletal muscle has a complex, hierarchical structure. As illustrated in Figure 18.1, each muscle contains hundreds of muscle fibers (Figure 18.1a), which show periodic cross striations arising from thousands of well-aligned myofibrils having the same periodic cross striations (Figure 18.1b). The cross-striation pattern consists of the A-band, which is protein dense, and the I-band, which is less dense (Figure 18.1c, d). Although the striation pattern has been known since the nineteenth century, its functional meaning remained a mystery until the monumental work of Hugh Huxley and Jean Hanson.

We summarize Huxley and Hanson's sliding filament mechanism [6, 7] in Figures 18.2 and 18.3. The two main proteins that constitute muscle, actin and myosin,

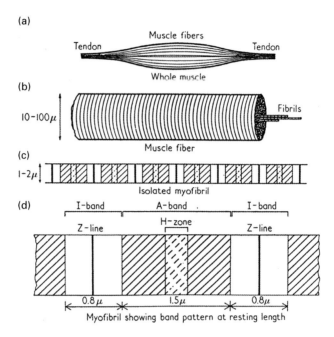

Figure 18.1. Diagram showing structure of vertebrate skeletal muscle with terms defined in the text.

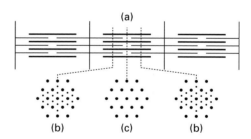

Figure 18.2. Arrangement of thick myosin filaments and thin actin filaments in vertebrate skeletal muscle myofibrils: (a) longitudinal view; (b, c) cross sections showing the hexagonal array of the filaments.

exist in muscle in the form of separate filaments, referred to as actin and myosin filaments (Figure 18.2a). The two filaments form a hexagonal array, in such a way that a myosin filament is surrounded by six actin filaments (Figure 18.2b, c). Myosin filaments are located in the A-band, while actin filaments originate from the Z-disc (or Z-band) at the center of the I-band, and extend between the myosin filaments in the A-band. Finally, a number of projections extend laterally from a myosin filament towards the neighboring six actin filaments (Figure 18.3d). In the sliding filament mechanism, the length of the actin and myosin filaments remains unchanged when a muscle is made to contract or is stretched. Instead, muscle contraction results from relative sliding between actin and myosin filaments.

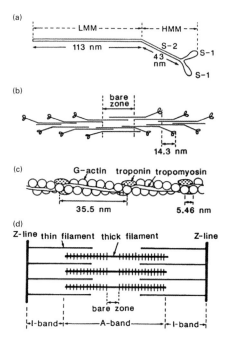

Figure 18.3. Structure of myosin and actin filaments and their arrangement in a sarcomere: (a) myosin molecule; (b) myosin filament consisting of myosin molecules; (c) actin filament; (d) sarcomere. Details are provided in the text. From Ref. 14.

In order to understand how the sliding takes place, it is necessary to describe the structure of each filament in more detail. Figure 18.3 illustrates how the actin and myosin molecules are organized within their respective filaments. A myosin molecule (molecular weight around 500 000) consists of two parts, shown in Figure 18.3a. A long rod (light meromyosin, labeled LMM in Figure 18.3a) is attached to the rest of the molecule (heavy meromyosin, labeled HMM) consisting of a short rod (subfragment-2, labeled S-2), and two heads (subfragment-1, S-1). Figure 18.3b shows how these molecules are arranged within a myosin filament. LMM aggregates to form the myosin filament backbone, which is polarized in opposite directions on either side of the center of the filament. HMM with its two S-1 heads (also called myosin heads or cross-bridges) extends laterally towards the neighboring actin filaments. Therefore, the direction of protrusion of HMM from myosin filaments reverses on moving across the center of the myosin filament. Actin filaments are shown in Figure 18.3c. They consist primarily of two helical strands of globular actin molecules, and also contain two regulatory proteins, troponin and tropomyosin.

The region between two adjacent Z-lines is regarded as the structural (and functional) unit of muscle fibers or myofibrils, and is called the sarcomere (Figure 18.3d). In the central region of the A-band, myosin filaments have no myosin head projections (as described above and in Figure 18.3b). This is called the bare region. It should be noted that, during muscle contraction, each sarcomere shortens as a result of the sliding of actin filaments into the A-band, where myosin filaments are located, and that the direction of sarcomere shortening changes sign across the bare region.

18.1.2 The Essential Role of Myosin Heads in Producing Sliding between Filaments

Muscle is an engine, utilizing ATP as fuel; in other words, muscle acts as a machine, converting chemical energy derived from hydrolysis of ATP [8]. Since both the actin binding site and the ATPase site are located in the two myosin heads [9], it is believed that the myosin heads play an essential role in muscle contraction. Concerning the question, *What makes actin and myosin filaments slide past each other?*, Huxley [10] presented an ingenious hypothesis that the filament sliding is caused by cyclic attachment and detachment between myosin heads extending laterally from myosin filaments and corresponding myosin-binding sites on actin filaments.

Due to the difference in axial periodicity between actin and myosin filaments, the cyclic actin–myosin interaction takes place asynchronously. This is shown schematically in Figure 18.4. In the upper diagram, a myosin head (located on the left) first attaches to a site on an actin filament. Then, the attached head changes its configuration to produce a unitary sliding between the filaments (middle diagram). The direction of filament sliding is such that the sarcomere shortens. After causing the unitary filament sliding, the myosin head detaches from the actin filament (lower diagram). It is generally believed that hydrolysis of one ATP molecule is coupled with each cycle of actin–myosin interaction. A more theoretical hypothesis was put forward in Ref. 11.

After the publication of the hypothesis that cyclic actin–myosin interactions produce muscle contraction, extensive studies were made in order to prove the conformational changes and movements of myosin heads that should be associated with ATP hydrolysis. These studies involved a range of experimental methods, including muscle mechanics, investigation of mechanical response of muscle fibers to applied length and force perturbation, time-resolved X-ray diffraction, chemical probes attached to myosin heads to detect myosin head movement, electron microscopy studies on rapidly frozen muscle

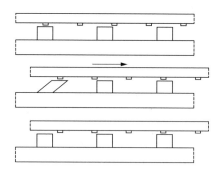

Figure 18.4. A proposed attachment–detachment cycle between a myosin head extending from the myosin filament and the corresponding site on the actin filament. From Ref. 10. Reprinted with permission from AAAS.

fibers, and crystallographic studies to detect nucleotide-dependent structural changes of myosin heads (S-1) detached from isolated myosin molecules. However, no definite consensus was reached concerning the amplitude of the myosin head movement (for reviews, see Refs. 12–16).

Thus, for many years, it was our dream to perform "physiological experiments" in which muscle contraction could be observed directly using the liquid cell. Fortunately, we were able to make our dream come true, as will be described in the following sections.

18.2 Experimental Methods for Recording Myosin Head Movement

The observation and measurement of myosin head movement requires a combination of experimental techniques. These include the preparation of the biological material, taking care to ensure that its physiological functioning is preserved; labeling the filaments so that movements can be quantified; sealing the filaments in a liquid cell that allows both the hydrated environment and the microscope resolution to be maintained; and the ability to deliver ATP into the liquid cell so that the contraction process can be triggered. Furthermore, it is essential to understand the dose limits that enable observation without damage to the filaments. The experimental complexities are considerable, but nevertheless solvable to produce useful information.

18.2.1 Carbon Sealing Film for the Liquid Cell

The experiments described here make use of a liquid cell that is sealed with a carbon film. This cell was developed by a pioneer of liquid observations in the TEM, Dr Akira Fukami of Nihon University. The carbon film successfully insulated hydrated, living biological specimens from the vacuum of the electron microscope [17]. Published in 1965, this work significantly predates the silicon nitride films used today. We found this sealing film to be suitable for experiments on myosin filaments. To our great pity, Dr. Fukami passed away in 2004, when our research was still at its preliminary stage.

As described in Chapters 2 and 8, the spatial resolution and the contrast of electron micrographs obtained using a liquid cell are improved if the sealing film thickness can be reduced. Preliminary studies made in Fukami's laboratory indicated that, to obtain a special resolution better than 1 nm, the sealing film thickness should be 15–20 nm. On the other hand, the robustness of the sealing film to the pressure difference becomes worse as the window area increases. According to this principle, a carbon sealing film covering a circular aperture of 50 μm diameter would need to be around 100 nm in thickness in order to maintain the pressure difference. This provides a technical paradox against preparing a carbon sealing film that is suitable for the experiments.

Fukami and Adachi [17] overcame this problem by the use of plastic microgrids made from a high molecular organic compound, cellulose acetobutylate. Figure 18.5 shows examples of microgrids. Microgrids with small holes (Figure 18.5a) or non-uniform holes (Figure 18.5b) were unsuitable, but those with fairly uniform holes of 5–8 nm diameter were suitable for electron microscopic observation of myosin filaments

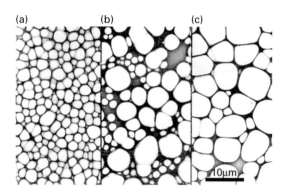

Figure 18.5. Photomicrographs of plastic microgrids with holes of (a) small diameter, (b) non-uniform diameter, and (c) nearly uniform diameter.

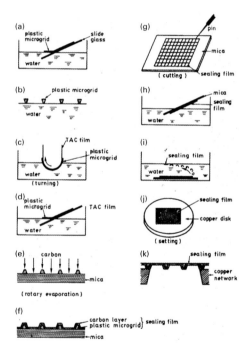

Figure 18.6. Procedures to prepare carbon sealing film supported by microgrids as described in the text. From Fukushima (1988) Ph.D. thesis.

(Figure 18.5c). Figure 18.6 illustrates the procedure to prepare a carbon sealing film over the microgrid. A plastic microgrid on a glass slide is put onto a water surface (a), so that the microgrid bars (with trapezoidal cross section) float on the surface with shorter edge down (b). The microgrids are inverted by means of a triacetylcellulose (TAC) membrane, and again put onto the water surface (c, d). The inverted microgrids are then transferred to a mica surface and a thin layer of carbon is evaporated (e, f). The carbon sealing film thus prepared on the mica surface is cut into rectangular pieces of appropriate size, and put into water (g–i). The microgrids with carbon sealing film separate from the mica and float on the water surface. Finally, the pieces are placed onto a copper grid so that the apertures on the copper grid are covered by the carbon (k).

In addition to the carbon insulating film described above, silicon nitride film is now commercially available and can be obtained at thicknesses of around 20 nm. It is interesting to note that the idea of using multiple, smaller window areas with a thinner film can be applied to silicon nitride also, as described in Chapter 2.

A second factor influencing spatial resolution is thickness of water covering the specimen. In our experiments, the cell is not completely filled with water. Instead, a water thickness of below 250 nm was used to obtain suitable spatial resolution. A sensitive image recording setup is essential to avoid dose-limited resolution, and is described in Section 18.2.3 below.

18.2.2 The Critical Electron Dose for Physiological Function of Muscle Proteins

The strong effect of the electron beam on water and on materials surrounded by water is well known and has been described in Chapter 7. For biomolecules such as myosin, physiological function is gradually impaired by electron beam irradiation. Fukami's group [18] determined the critical dose using isolated muscle myofibrils, as shown in Figure 18.7. At a total incident electron dose below 5×10^{-4} C/cm^2, all myofibrils within the field of view shortened in response to ATP. At higher dose, the ATP-induced myofibril contraction disappeared in a nearly all-or-none manner, even though the myofibrils exhibited no appreciable change in their appearance. The critical electron dose to impair function of myosin heads was also confirmed by our later experiments, with respect to ATP-induced movement of myosin heads in myosin filaments.

Based on these results, observation and recording of muscle proteins was made with total incident electron dose $<5 \times 10^{-4}$ C/cm^2. In order to fulfill this condition during

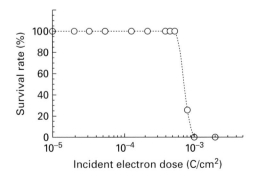

Figure 18.7. Relation between total incident electron dose and survival rate of muscle myofibrils mounted in the liquid cell. From Ref. 18. Reprinted by permission of Oxford University Press.

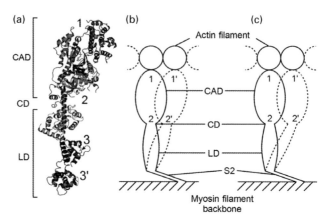

Figure 18.8. (a) Structure of myosin head showing the catalytic domain (CAD), converter domain (CD), and lever arm domain (LD). Attachment regions of antibodies 1, 2, and 3 are indicated by numbers 1, 2, and 3, 3′ respectively. (b, c) Diagrams showing the mode of myosin head power stroke in the isometric condition; (b) is at standard ionic strength and (c) at low ionic strength. From Ref. 22.

the course of experiments, samples were observed at extremely low intensities, below 5×10^{-13} A/cm² at the fluorescent screen, and the magnification was limited to 10 000×. The beam intensity was thus $5 \times 10^{-13} \times (10\,000)^2 = 5 \times 10^{-5}$ A/cm² at the sample. Immediately after focusing, the beam was blanked until the time of recording. Skill and patience are required in observing and focusing the sample, indicating that our work was largely a struggle to protect proteins against electron beam damage.

18.2.3 Synthetic Myosin Filaments with Position-Marked Heads

Synthetic myosin filaments were used to record myosin head movement in response to ATP. These filaments are prepared by polymerization of isolated native filaments in a low ionic strength solution [2]. They have length up to 3 μm and diameter up to 200 nm and consist of a 1:1 mixture of myosin rod (LMM) and whole myosin molecules.

The contrast of myosin filaments is too low to allow measurements, therefore it was necessary to position-mark individual myosin heads, as in Chapter 16 where low contrast cell components are labeled with heavy particles. For myosin filament labeling, we use antibodies that are specific to a region of the myosin molecule. Figure 18.8a shows the myosin head structure schematically. It consists of the catalytic (CAD), converter (CD), and lever arm (LD) domains. We use two different antibodies to two different regions of the myosin head; one (antibody 1) binds with a junctional peptide between the 50K and 23K fragments of the myosin heavy chain, located at the distal part of the myosin head catalytic domain, while the other (antibody 2) binds with a reactive lysine residue located in the myosin head converter domain [19]. In addition to these antibodies, we prepared another antibody (antibody 3), which binds with two

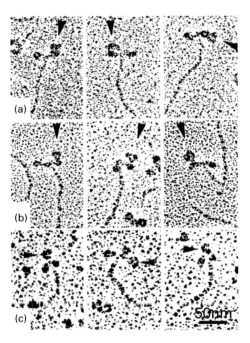

Figure 18.9. Rotary shadowed electron micrographs of antibody-1–myosin (a), antibody-2–myosin (b), and antibody-3–myosin (c) complexes. Triangular-shaped antibodies (IgG) are indicated by arrowheads. Reprinted from Ref. 3, Copyright 2011, with permission from Elsevier.

DTNB (or regulatory) light chains in the myosin head lever arm domain. The approximate binding regions of the three antibodies are indicated in Figure 18.8a. The actual binding locations to the three different regions of the myosin molecule are proved electron microscopically by examining negatively stained myosin molecules to which antibodies are attached (Figure 18.9). It can be seen that triangular-shaped antibodies 1 and 2 attach around the tip of the myosin head, while antibody 3 attaches to the proximal region of the myosin head.

Finally, colloidal Au particles of diameter 20 nm were coated with protein A, which serves as a glue to connect the particles to various proteins, and were attached to individual myosin heads via antibody 1, 2, or 3. Typical images of spindle-shaped myosin filaments with position-marked myosin heads are shown in Figure 18.10. Care was taken to distribute the position-marked myosin heads rather sparsely, so that individual head positions can be distinguished.

18.2.4 Image Recording and Analysis

The imaging plate (IP) system (JEOL PIX system) is a high sensitivity image recording system that has been used worldwide, although slow CCD and direct detection camera systems are now more popular. Each IP record is divided into 12 000 000 pixels (2.5×2.5 nm). For our liquid cell microscopy experiments, the IP allowed image

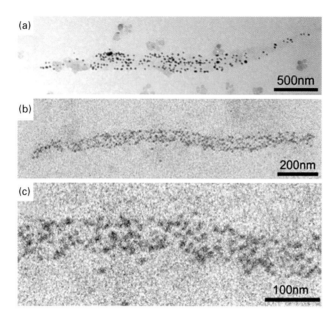

Figure 18.10. (a, b) Spindle-shaped bipolar myosin filaments with a number of Au particles on them. (c) Enlarged view of myosin filament shown in (a).

recording at 10 000 × with 0.1 s exposure time. Under these conditions, the number of electrons reaching each pixel is at most 7–8. Due to the poor statistics, the image of each Au particle consisted of 20–50 dark pixels with variable gradation. Contrast enhancement and binarization procedures allow particles to be identified by their nearly circular shape, then selected for analysis [2]. The center-of-mass position was determined to within ~0.6 nm. Different images were aligned via the position of natural markers such as bright spots on the carbon film [1, 2]. The shift in the center-of-mass position of each particle could then be calculated, and this value was taken as the amplitude of myosin head movement [1, 2].

18.2.5 Application of ATP to Myosin Filaments

A key element of *in situ* experiments addressing biological function is the ability to supply the fuel, ATP, to the biological material to power its physiological dynamics. We have found that iontophoretic application of ATP can be successful in the liquid cell [20]. Figure 18.11 illustrates how ATP is applied to myosin heads in myosin filaments in the liquid cell attached to a JEOL JEM 2000EX 200 kV electron microscope.

The liquid cell itself is a small cylindrical compartment of diameter 2.0 mm and height 0.8 mm. The upper and lower windows are made of copper films, each with nine 0.1 mm apertures that are sealed with the carbon film. The specimen, with thickness 50–250 nm, is placed on the lower sealing film and covered by a thin layer of the experimental solution, containing 25 mM KCl, 5 mM $MgCl_2$, 20 mM PIPES

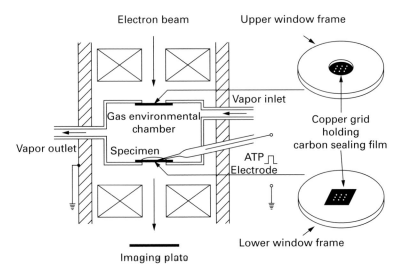

Figure 18.11. Diagram of the liquid cell showing the ATP-containing microelectrode. From Ref. 14.

(piperazine-1,4-bis(2-ethanesulfonic acid), and 0.1 mM DTT (dithiothereitol), at pH = 7.0. The thickness of the liquid layer was kept constant by ensuring that the air above the sample was saturated with water vapor at 60–80 Torr and 26–28 °C, constantly circulated with a flow rate of 0.1–0.2 liter/min. The thickness of this water vapor layer above the specimen was ~0.6 mm.

The liquid cell contains a glass capillary microelectrode containing 100 mM ATP (resistance 15–20 MΩ), whose tip is immersed in the experimental solution. ATP is applied to the specimen by passing a current pulse (10 nA for 1 s) from an electronic stimulator to the ATP electrode through a current clamp circuit [20]. Charged ATP ions carrying the current leave the electrode and reach the myosin filaments by diffusion through the experimental solution. The total amount of ATP released from the electrode is ~10^{-14} mol per pulse. Thus, assuming the volume of the experimental solution is ~10^{-6} ml, the ATP concentration around the specimen is estimated to be ~10 µM.

We believe that, although the iontophoretic application of charged molecules to a sample is widely used by physiologists, this technique has not been well known outside that field. We expect that iontophoretic techniques will have other applications in liquid cell microscopy experiments, where a controlled dose of a molecule is to be applied to a sample to trigger a process during observation.

18.3 ATP-Induced Movement of Individual Myosin Heads

18.3.1 Stability in Myosin Head Position in the Absence of ATP

When we first mounted myosin filaments on the carbon sealing film in the liquid cell, we felt anxious about their positional stability, in other words, how firmly they were

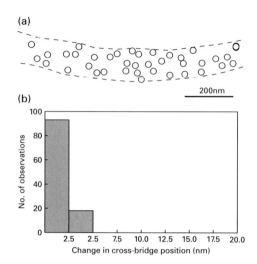

Figure 18.12. Stability of myosin head position in the absence of ATP. (a) Two superimposed IP records of the same filament showing stable positions of individual myosin heads. (b) Histogram of distance between myosin head position in (a). Among 120 different particles examined on three different image pairs, 93 particles exhibited no significant change in position ($D < 2.5$ nm), and 27 particles exhibited only small position changes (2.5 nm $< D <$ 5 nm). From Ref. 2.

fixed in position on the carbon sealing film. Positional stability is clearly necessary to interpret movements of individual myosin heads in response to ATP. IP records of the same myosin filament at intervals of 5–10 min, however, show no drift with time. A typical result is shown in Figure 18.12a, in which a circle of 20 nm diameter is drawn around the center-of-mass position of each selected particle image. The position of each particle, i.e. the position of each myosin head, remains almost unchanged. Figure 18.12b quantifies this statement through a histogram showing the distribution of changes in position. The myosin filaments are evidently fixed in position on the carbon surface, and, despite expected thermal fluctuations, the myosin head mean position, averaged over the exposure time of 0.1 s, remains almost unchanged.

The contraction model of A. F. Huxley [11] suggests that each myosin head has a definite equilibrium position, and the observed stability is consistent with this, suggesting that each head will fluctuate around its equilibrium position.

18.3.2 The Amplitude of Myosin Head Movement in Response to ATP

In Figure 18.13, we compare a filament before and after the iontophoretic application of ATP. The first IP record was taken 2–3 min before the current pulse to the electrode. Considering the time of diffusion of ATP from the ATP-containing microelectrode to myosin filaments, the second IP record was taken 40–60 s after the current pulse.

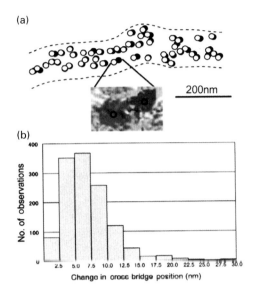

Figure 18.13. ATP-induced myosin head movement at one side of the bare region. (a) Comparison of myosin head position before (open circles) and during (filled circles) ATP application. The inset shows enlarged images of a particle before (red) and after (blue) application of ATP, with the center-of-mass position indicated. (b) Histogram of amplitude distribution of ATP-induced myosin head movement. From Ref. 2.

These data show clear signatures of the expected pattern of movement. In most cases, the myosin heads move in the same direction – consistent with the head movement expected at one side of the bare region. The histogram in Figure 18.13b, constructed from 1285 measurements on eight different parts of eight different myosin filaments, exhibits a peak at 5–7.5 nm. The average amplitude of ATP-induced myosin head movement (excluding values < 2.5nm) is 6.5 ± 3.7 nm (mean ± SD, $n = 1210$).

It is worth pointing out that we cannot determine the height of a given Au particle, since all particles appear in focus in the images. While the myosin heads located on the upper side of the filaments may move almost freely, those located at the lower side may be attached to the carbon film so their movement may be more or less limited. This may be the explanation of why a considerable proportion of myosin heads did not move appreciably or moved with small amplitude. If this interpretation is correct, the mean amplitude of ATP-induced movement in the myosin heads that are free to move would be greater than 7.5 nm.

18.3.3 Reversibility of ATP-Induced Myosin Head Movement

To create the cyclic interaction that produces muscle contraction, the recovery stroke should be the same in amplitude as, and opposite in direction to, the power stroke. To examine this reversibility, we recorded the same filament before ATP application, during ATP application, and after complete exhaustion of applied ATP. ATP exhaustion

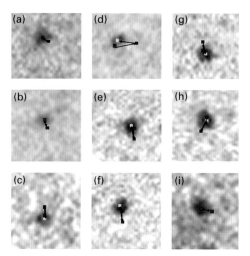

Figure 18.14. Examples showing sequential changes in position of pixels (2.5 × 2.5nm) where the center-of-mass position of an Au particle is located. Pixel positions before ATP application (red), during ATP application (blue) and after exhaustion of ATP (yellow) are shown in each frame. These pixel positions can be taken to represent sequential position changes of individual myosin heads. The direction of movement is indicated by arrows. From Ref. 2.

was facilitated by adding hexokinase and D-glucose to the experimental solution [20]. Figure 18.14 shows examples of sequential changes in the positions of nine different particles during this sequence.

It can be seen that individual myosin heads move on ATP application, and then return towards their initial position after exhaustion of the ATP. Heads for which the amplitude of the initial ATP-induced movement was small (~5 nm) returned almost exactly to their initial position. This may be taken to indicate reversibility of ATP-induced myosin head movement, being consistent with the idea that myosin head power and recovery strokes are the same in amplitude and opposite in direction. This point will be discussed in more detail below.

ATP-induced myosin head movement was no longer observed if the myosin head ATPase activity was inhibited by N-ethylmaleimide (NEM) [1, 2]. And myosin heads did not move in response to application of ADP (adenosine diphosphate), indicating that ATP-induced myosin head movement is actually coupled with ATP hydrolysis.

18.3.4 Reversal in Movement Direction across the Bare Region

We would expect myosin head movement to be reversed across the bare region at the center of myosin filaments (Figures 18.2 and 18.3). Despite extensive prior studies, no definite proof of this has been obtained. However, after a number of painstaking trials, we show a demonstration of this reversal in Figure 18.15. In response to ATP, individual myosin heads move away from, but not towards, the bare region, reflecting the reversed myosin head polarity across the bare region.

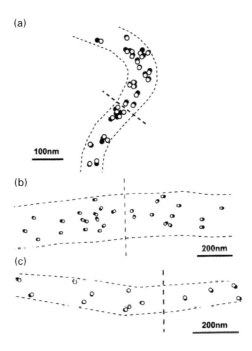

Figure 18.15. ATP-induced myosin head movement at the two sides of the myosin filament bare region (indicated by vertical broken lines). Open and filled circles represent positions before and during ATP application, respectively. Note that myosin heads move away from the myosin filament bare region. From Ref. 2.

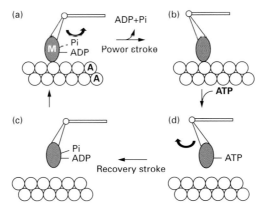

Figure 18.16. Diagram of the attachment–detachment cycle between a myosin head (M) and actin filament (A), coupled with ATP hydrolysis. From Ref. 2.

It was our great pleasure that H. E. Huxley, who discovered the sliding filament mechanism, was extremely happy with our finding. To our great sorrow, he passed away in 2013, leaving his monumental work in the history of muscle research.

A plausible attachment–detachment cycle between myosin heads and actin filaments, based on biochemical studies on ATP hydrolysis steps [21], is shown diagrammatically in Figure 18.16. First, the myosin head (M), in the form of M + ADP + Pi, attaches to the actin filament (Figure 18.16a), and performs a power stroke, coupled with release of ADP and Pi from the myosin head, to produce sliding between actin and myosin filaments (Figure 18.16a→b). After completion of the power stroke, M remains attached to actin until the next ATP comes to bind with M (Figure 18.16b). Upon binding with ATP, M detaches from A, and performs a recovery stroke, associated with reaction, MATP→M + ADP + Pi (Figure 18.16c→d). According to this diagram, the ATP-induced myosin head movement shown in the preceding sections corresponds to the recovery stroke (Figure 18.16c→d).

We have recorded the power stroke of myosin heads in the presence of actin filaments (Figure 18.16a→b), using a mixture of actin and myosin filaments, and describe the results in Section 18.3.6 below.

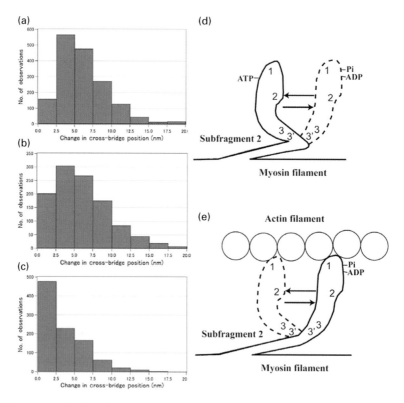

Figure 18.17. (a–c) Histograms showing the amplitude distribution of ATP-induced movement of individual myosin heads that are position-marked with antibodies 1, 2, and 3 respectively. (d, e) Diagrams of myosin head power and recovery strokes in the absence and presence of actin filaments, respectively. Approximate regions of attachment of antibodies 1, 2, and 3 are indicated by numbers 1, 2, 3, 3′, respectively.

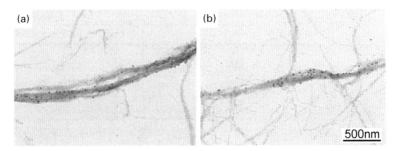

Figure 18.18. Low magnification conventional electron micrographs of an actin and myosin filament mixture. Myosin heads are position-marked with antibodies. From Ref. 22.

18.3.5 ATP-Induced Movement at Various Regions within Individual Myosin Heads

The three different antibodies described in Section 18.2.3 allow the recording of ATP-induced movement at the three different regions within individual myosin heads. The results are summarized in Figure 18.17. The three histograms (a–c) show the amplitude distribution of ATP-induced myosin head movement (in the absence of actin filaments) at the distal region of the myosin head catalytic domain, at the myosin head converter domain, and at the myosin head lever arm domain, respectively. The average movement was 6.14 ± 0.09 nm (mean $\pm$ SD, $n = 1692$), 6.14 ± 0.22 nm ($n = 1122$), and 3.55 ± 0.11 nm ($n = 981$), respectively. This data can be used to provide hints to the possible changes in configuration of myosin heads coupled with ATP hydrolysis (Figure 18.17d, e). Although we will skip further discussion of these results, as they are too specialized for general readers of this volume, it is clear that this type of site-specific data is extremely valuable in developing a detailed understanding of a physiological process.

18.3.6 ATP-Induced Myosin Head Power Stroke in the Presence of Actin Filaments

The experiments described in the preceding sections have addressed myosin motion in the absence of actin. For additional insight into the functioning of muscle it is important to measure the ATP-induced power stroke of individual myosin heads in the presence of actin filaments. This has been achieved only recently [22]. We prepared a hydrated mixture in which spindle-shaped myosin filaments were surrounded by actin filaments running approximately parallel with each other due to formation of rigor myosin head–actin linkages [23], Figure 18.18. As only an extremely small proportion of myosin heads in myosin filaments can be activated with iontophoretically applied ATP, individual ATP-activated myosin heads could only move by pulling adjacent elastic structures without producing gross sliding between actin and myosin filaments. The mode of performance of activated myosin heads in the filament mixture is therefore analogous to that of myosin heads in muscle with both ends fixed in position, i.e. the isometric condition in which a muscle generates force but does not shorten appreciably.

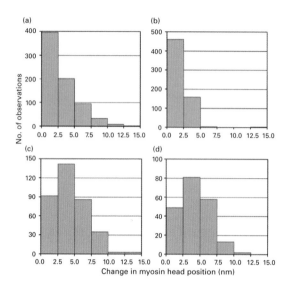

Figure 18.19. Histograms of amplitude distribution of ATP-induced myosin head power stroke: (a, b) at standard ionic strength (170mM) and (c, d) at low ionic strength (50 mM). In (a, c), individual myosin heads are position-marked with antibody 1 at the distal region of the myosin head catalytic domain. In (b, d) individual myosin heads are marked with antibody 2 at the myosin head converter domain. From Ref. 22.

As shown in Figure 18.19, the amplitude of the myosin head power stroke in this isometric condition was 3.3 ± 0.2 nm (mean $\pm$ SD, $n = 732$) at the distal region of the myosin head catalytic domain, and 2.5 ± 0.1 nm ($n = 613$) at the myosin head converter domain; at lower ionic strength of the experimental solution, the amplitude increased to 4.4 ± 0.1 nm ($n = 361$) and 4.3 ± 0.2 nm ($n = 305$) respectively. The clear increase in the amplitude of the myosin head power stroke is entirely consistent with our previous report that, at low ionic strength, the force generated by individual myosin heads in muscle fibers increases twofold [24]. Diagrams showing possible modes of the myosin head power stroke at standard and low ionic strengths are presented in Figure 18.8b and c.

18.4 Conclusions and Outlook

We have described a suite of experimental methods that are coupled with the use of liquid cells in the TEM and are extremely effective in studying dynamic structural changes of living, wet muscle proteins. The greatest advantage of these liquid cell microscopy experiments is that they enable us to record the behavior of individual myosin heads at high resolution, as they produce muscle contraction. Methods such as muscle mechanics, time-resolved X-ray diffraction, cryo-electron microscopy, and chemical probe experiments [12, 13] sample numerous myosin heads acting asynchronously. This provides averaged values of myosin head responses. Crystallographic and electron microscopic studies of the myosin head crystal and actin-S1 complex [15] were concerned only with static structures, and the results obtained are also static in nature. We believe that the use of liquid cell electron microscopy constitutes a breakthrough in the research field of life sciences, because it has proved the feasibility of measuring dynamic structural changes of individual biological macromolecules that retain their physiological function.

We emphasize that the methods developed for this particular process of muscle contraction can be used effectively for solving other mysteries in the field of biosciences. The expertise that is developing in liquid cell microscopy for studying dynamic processes in materials science, combined with the techniques described here for labeling, examining, and chemically dosing biomaterials, will contribute to a rapid advance in dynamic biological applications and will start to challenge other deep mysteries in the field of life sciences.

Acknowledgements

We would like to express our heartfelt thanks to Presidents Kazuo Ito, Terukazu Eto, and Yoshiyasu Harada of JEOL, Ltd for their generous support, which made our work possible. Our thanks are also due to Professors Kazuo Sutoh, Takeyuki Wakabayashi, and Eisaku Katayama of the University of Tokyo for their cooperation in this work.

References

1. H. Sugi, T. Akimoto, K. Chaen et al., Dynamic electron microscopy of ATP-induced myosin head movement in living muscle thick filaments. *Proc Natl. Acad. Sci. USA*, **94** (1997), 4378–4382.
2. H. Sugi, H. Minoda, Y. Inayoshi et al., Direct demonstration of the cross-bridge recovery stroke in muscle thick filaments in aqueous solution by using the hydration chamber. *Proc Natl. Acad. Sci. USA*, **105** (2008), 17396–17401.
3. H. Minoda, T. Okabe, Y. Inayoshi et al., Electron microscopic evidence for the myosin head lever arm mechanism in hydrated myosin filaments using the gas environmental chamber. *Biochem. Biophys. Res. Commun.*, **405** (2011), 651–656.
4. H. Sugi, H. Minoda, T. Miyakawa and M Tanokura, Electron microscopic recording of the cross-bridge power stroke in hydrated myosin filaments using the gas environmental chamber. *J. Muscle Res. Cell Motility*, **32** (2011), 34.
5. H. Sugi, Visualization and recording of the power stroke in individual myosin heads coupled with ATP hydrolysis using the gas environmental chamber. *J. Physiol. Sci. Japan*, **103** (2013), S53.
6. H. E. Huxley and J. Hanson, Changes in the cross-striations of muscle during contraction and stretch and their structural interpretation. *Nature*, **173** (1954), 973–976.
7. H. E. Huxley, The double array of filaments in cross-striated muscle. *J. Biophys. Biochem. Cytol.*, **3** (1957), 631–648.
8. R. C. Woledge, N. A. Curtin and E. Homsher, *Energetic Aspects of Muscle Contraction* (London and New York: Academic Press, 1985).

9. C. R. Bagshaw, *Muscle Contraction* (London: Chapman & Hall, 1993).
10. H. E. Huxley, The mechanism of muscular contraction. *Science*, **164** (1969), 1356–1366.
11. A. F. Huxley, Muscle structure and theories of contraction. *Prog. Biophys. Biophys. Chem.*, **7** (1957), 255–318.
12. R. Cooke, The mechanism of muscle contraction. *CRC Crit. Rev. Biochem.*, **21** (1986), 53–118.
13. M. G. Hibbard and D. R. Trentham, Relationships between chemical and mechanical events during muscular contraction. *Annu. Rev. Biochem.*, **15** (1986), 119–161.
14. H. Sugi, Molecular mechanism of actin-myosin interaction in muscle contraction. In H. Sugi, ed., *Muscle Contraction and Cell Motility*, Advances in Comparative & Environmental Physiology Vol. 12 (Berlin: Springer, 1992).
15. M. A. Geeves and K. C. Holmes, Structural mechanism of muscle contraction. *Annu. Rev. Biochem.*, **68** (1999), 687–728.
16. A. F. Huxley, Support for the lever arm. *Nature*, **396** (1998), 317–318.
17. A. Fukami and K. Adachi, A new method of preparation of a self-perforated micro plastic grid and its application. *J. Electron Microsc. (Tokyo)*, **14** (1965), 112–116.
18. H. Suda, A. Ishikawa and A. Fukami, Evaluation of the critical electron dose on the contractile activity of hydrated muscle fibers in the film-sealed environmental cell. *J. Electron Microsc. (Tokyo)*, **41** (1992), 223–229.
19. K. Sutoh, M. Tokunaga and T. Wakabayashi, Electron microscopic mapping of myosin head with site-directed antibodies. *J. Mol. Biol.*, **206** (1989), 357–363.
20. K. Oiwa, S. Chaen and H. Sugi, Measurement of work done by ATP-induced sliding between rabbit muscle myosin and algal cell actin cables in vitro. *J. Physiol. (London)*, **437** (1991), 751–763.
21. R. W. Lymn and E. W. Taylor, Mechanism of adenosine triphosphate hydrolysis by actomyosin. *Biochemistry*, **10** (1971), 4617–4624.
22. H. Sugi, S. Chaen, T. Akimoto *et al.*, Electron microscopic recording of myosin head power stroke in hydrated myosin filaments. *Sci. Rep.*, **5** (2015), 15700.
23. J. M. Squire, *The Structural Basis of Muscular Contraction* (New York; London: Plenum, 1981).
24. H. Sugi, T. Abe, T. Kobayashi *et al.*, Enhancement of force generated by individual myosin heads in skinned rabbit psoas muscle fibers at low ionic strength. *PLOS One*, **8** (2013), e63658.

Part III
Prospects

19 High Resolution Imaging in the Graphene Liquid Cell

Jungwon Park, Vivekananda P. Adiga, Alex Zettl, and A. Paul Alivisatos

19.1 Introduction to Graphene Liquid Cells: Advantages, Opportunities, and Fabrication Methods

Developments in electron microscopy of liquids provide visual insights into structure and function in physical and biological chemistry. The ultimate goal of liquid microscopy is to see materials in action with the greatest possible detail. Atomic resolution offers significant advantages in understanding phenomena associated with biological materials, electrochemical reactions, or nanoparticle growth in liquid environments. Conventional electron microscopy offers superb resolution. In liquid environments, atomic resolution initially eluded scientists and microscopists, even though diverse applications of the initial liquid cell design, based on silicon nitride, confirmed that liquid cell microscopy has the potential to achieve nanoscale resolution. The more recent development of the graphene liquid cell (GLC) now provides the chance to improve liquid cell resolution by using an atomically thin inert material, graphene, as the window material [1–6]. In this chapter, we outline the development of the GLC and its use in studying colloidal nanoparticles, bio-inorganic conjugated nanostructures, and protein molecules.

Graphene has emerged as a versatile membrane material for all types of TEM experiments [2, 7, 8] due to its electron transparency and favorable electrical, mechanical, and thermal properties. The strong van der Waals interaction between graphene sheets, together with graphene's impermeability to liquids and gases, makes it an ideal material to enable the trapping of small pockets of liquid containing a material of interest for TEM imaging [1]. The first generation design of GLCs can be used in a conventional TEM without modification or customization of either the TEM or the holder. The most elementary geometry, and that employed in the original GLC experiments [1], consists of two sheets of monolayer graphene arranged in a bubble-wrap or blister-like configuration, as illustrated in Figure 19.1. These GLCs were produced by transferring graphene grown by chemical vapor deposition (CVD) onto commercially available TEM grids by a direct transfer method [9], in which two gold TEM grids covered with a holey carbon film with well-defined periodic perforations (also known as Quantifoil®) are placed on CVD-grown graphene on copper, then the copper is removed using wet etching. The graphene on the grid is then transferred through several de-ionized water baths. After placing a drop of the solution of interest on one of the

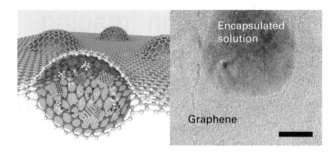

Figure 19.1. Conceptual diagram of the first generation GLC, where van der Waals interaction between two graphene sheets is used to trap the liquid, and a TEM image of the trapped liquid. The scale bar is 50 nm. Reproduced from Ref. 1 with permission from AAAS.

grids, the second grid is placed on top to create a sandwich structure where liquid is trapped in multiple blisters between the two graphene layers. Such blisters show circular shapes with a wide size range. Their diameter and thickness, as measured by atomic force microscopy, are tens of nanometers to micrometers. Blisters at the smaller end of the size range, typically <100 nm in diameter, have been used for atomic resolution imaging of liquid-borne metal nanocrystals [1], without the scattering and undesirable charging effects endemic to silicon nitride-based liquid cells.

The graphene has several key advantages when used in this way. The chemical and physical inertness of its surface helps in trapping biological samples without them being non-specifically adsorbed onto the blister walls. Unlike the artificial adhesion from the spacers in microfabricated liquid cells [10–21], the strong van der Waals attraction between the two graphene layers of the GLC minimizes contamination and leakage during sample preparation and observation. More importantly, for imaging delicate samples in liquid, the high electrical and thermal conductivities of graphene enable fast transfer of accumulated charges or heat due to the electron beam, and it is therefore a promising system to minimize radiation damage. Despite these advantages, controlling the size and shape (including aspect ratio) of the liquid chamber in these simple GLCs has been challenging. This difficulty is partly due to graphene's low bending rigidity, which often leads to irreversible collapse of the sheets onto each other giving a random overall thickness of the entrapped liquid packet. Similarly, the generation of bubbles by radiolysis (Chapter 7) can dynamically alter the overall size of the liquid packet in unpredictable ways. These limitations have led to the development of second generation GLCs.

Second generation GLCs (H. I. Rasool, private communication) are shown in Figure 19.2. They are made by transferring graphene onto highly regular prefabricated through-holes in thin (<200 nm) silicon nitride membranes. Strong adhesion between graphene and Si_xN_y results in size-controlled liquid packets trapped between the graphene sheets separated by the perforated Si_xN_y spacer. Perforated membranes are commercially available or can be fabricated with conventional photolithographic or focused ion beam techniques. Typical through-holes in Si_xN_y are in the size range of 100 nm to 1 μm. The perforated membranes can be pretreated with oxygen plasma

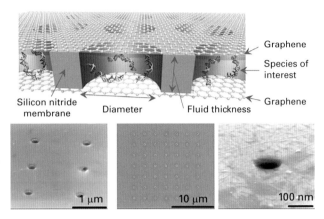

Figure 19.2. Schematic of liquid-filled, graphene-encapsulated perforations in a Si_xN_y membrane, and SEM images of graphene-covered perforations in Si_xN_y filled with liquid. Images courtesy Haider I. Rasool.

to make them hydrophilic. In one construction method, thin PMMA (20 nm) is spin-coated onto CVD-grown graphene/copper followed by wet etching of the copper in ammonium persulfate solution. The PMMA/graphene film is rinsed with deionized water, then transferred onto the wafer side of a perforated Si_xN_y membrane on a Si wafer. Finally, another graphene/PMMA film is floated in a solution of the liquid of interest, then transferred to the top surface of the Si_xN_y/graphene/PMMA substrate. To ensure sealing, drying must be done following the transfer. Finally, the PMMA film is removed using acetone/IPA. The result is a trapped liquid packet of pre-defined volume. However, this method also has its challenges. It is difficult to trap liquid/buffer solutions in small holes (<500 nm), but in large holes (1 μm or larger) the graphene sheets from both sides tend to collapse onto each other, resulting in problems similar to those discussed for the first generation design. Bubble formation from radiolysis can result in variations in pH [22, 23] and variable liquid thickness inside the cell, complicating the analysis of the data.

Some of these problems can be resolved by third-generation GLCs, called graphene liquid flow cells (GLFCs) because they employ controlled liquid flow (Figure 19.3) [24]. Liquid cells with flow capabilities have advantages when studying chemical reactions that require initiation on demand or continuous refreshing of the liquid environment. Using flow in microfabricated liquid cells with Si_xN_y windows can be difficult: the windows deform or bow under the pressure difference between flowing liquid and vacuum, creating liquid layers that are too thick for many kinds of nanoscale imaging [24]; the flow can be non-uniform; and bubble formation [22] also may cause unstable flow due to the membrane flexibility. The third-generation GLC adopts the beneficial aspects of flow established in Si_xN_y-based cells, but uses graphene technology to bypass some of the limitations.

The third-generation GLFC is not simply a conventional flow cell design with graphene replacing the Si_xN_y. The poor bending stiffness of graphene and difficulty in

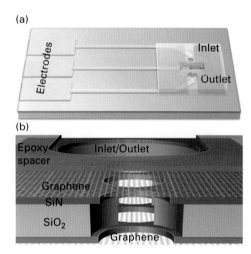

Figure 19.3. Conceptual design of a fully integrated nanofluidic channel that can be accommodated in existing electrochemical TEM holders. Two electrodes enable liquid flow through a nanofluidic channel formed by Si_xN_y membranes. The electrodes aid the motion of particles and molecules under the influence of an electric field using electrophoretic means. Also shown is a cross-sectional view of the flow cell with openings in the Si_xN_y window covered with graphene. From Ref. 24.

obtaining intact graphene sheets over large areas without pinholes makes it difficult to directly replace the Si_xN_y windows with graphene for flow cells. Instead, as shown in Figure 19.3, the flow region is a nanofluidic channel (1 μm wide) etched in a silicon dioxide layer (100 nm thick) sandwiched between thin (e.g. about 50 nm thick) Si_xN_y layers. These Si_xN_y layers have well-defined perforations (e.g. having dimensions of about 100 nm) sealed by few-layer graphene that serves as viewports for electron imaging. The small hole size in the Si_xN_y results in limited bowing of the graphene and hence controls the overall liquid thickness better. The channel in such a flow cell is capable of sustained liquid flow and can be electrically contacted for aiding the movement of particles in the liquid. These hybrid nanofluidic channels are expected to improve functionality and overall reliable resolution significantly.

19.2 Studying Growth Mechanisms in Atomic Detail by GLC-TEM

The first-generation GLC has enabled imaging of both growth and diffusion of nanoscale crystals. Below we highlight some of these results, illustrating the types of phenomena and data that are accessible through GLC experiments.

19.2.1 Nanocrystal Growth at Atomic Resolution

Transparency to the imaging electrons (which reduces background in the images) and surface inertness (which reduces crystal–wall interactions) enable us to study atomistic details of nanocrystal growth dynamics by recording high resolution *in situ* TEM movies in the GLC. Assuming that the graphene forms a conformal wrap over the encapsulated growth solution, we explore the sample at low magnification to search for liquid pockets whose diameter is <100 nm. This minimizes excessive liquid thickness and hence unwanted scattering. During imaging of the liquid packet, nanoparticles grow from solutions containing metal ions, as described in Chapter 9. The high resolution possible is illustrated by studies of Pt nanocrystal growth in GLCs obtained using aberration-corrected TEM [1]. Spherical and chromatic aberration correction contribute to the exceptionally high resolution and enhanced signal-to-noise ratio while the graphene membrane remains intact over the entire time period of observation.

Previous *in situ* studies by other types of liquid cell TEM and X-ray scattering have shown that nanocrystals in solution grow both by monomer addition and by frequent coalescence events [1, 19, 26 28]. The GLC allows direct observation of critically important features of the coalescence process at a higher resolution. Figure 19.4 shows

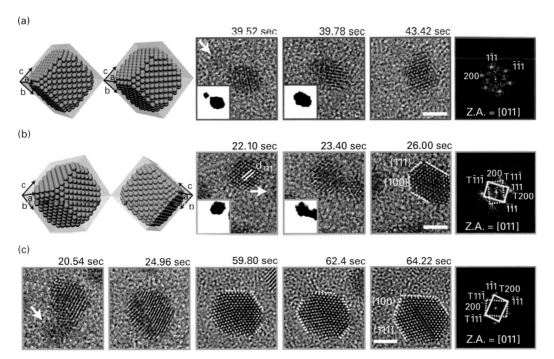

Figure 19.4. Atomic resolution image series obtained during Pt nanocrystal growth in a GLC in TEM. The images exhibit preferential coalescence along ⟨111⟩ directions which evolves into (a) a single crystalline face-centered cubic structure or (b) a twinned face-centered cubic structure. Incoming small nanocrystals are marked with white arrows. Black and white insets represent thresholded volume projections that clarify the nanocrystal shapes and positions. (c) Image series obtained during Pt nanocrystal shape evolution shows the straightening of the twin boundary and evolution towards a hexagonal shape consistent with the Wulff construction. FFT images for (a)–(c) correspond to the final frame of each sequence. Scale bars are 2 nm and zone axis is abbreviated as Z.A. Reproduced from Ref. 1 with permission from AAAS.

the beam-induced formation of Pt nanocrystals and subsequent crystal growth by coalescence. The crystallographic orientation relationship of the joining nanocrystals can be seen [1]. The most interesting finding is that most coalescence events proceed along one particular nanocrystal orientation, {111}. Once nanocrystals collide at {111} planes, they merge quickly, within the acquisition speed of one frame (0.26 s). Exposed {111} surfaces of a face-centered cubic crystal are known to have the lowest surface energy, therefore perhaps the lowest ligand coverage; incoming particles may thus experience minimal ligand obstruction. In this scenario, nanocrystals that contact at {111} planes quickly unify to minimize the total surface area and thus energy [29]. Contact that joins identically oriented crystals is shown in Figure 19.4a. Alternatively, contact that joins mirror {111} planes (Figure 19.4b and c) yields nanocrystals with a twin boundary. In addition, structural rearrangements of the nanoparticles are visible. These occur mainly by surface diffusion and rearrangement. Surface diffusion evolves the nanocrystal shape after coalescence from quasi-spherical to a hexagonal shape that minimizes surface energy, as expected from a Wulff construction (Figure 19.4c). Particle coalescence generating an interior crystal boundary, as observed in this study, may be the formation mechanism of the twin boundaries that are commonly observed in synthesized face-centered cubic metal nanocrystals [29].

19.2.2 Diffusion Dynamics of Nanocrystals

Along with the high resolution imaging of nanocrystal growth, GLC-TEM allows for the study of nanocrystal diffusion at high resolution without perturbation by the substrate. In other types of liquid cells with Si_xN_y windows, nanoparticle motion is perturbed by the strong potential of attraction to the windows. Nanoparticles of Au and Pt in non-graphene cells have been observed to execute complex non-Brownian stick-slip motion, possibly due to weak binding to one of the Si_xN_y windows [11, 19]. In contrast, the interaction of nanoparticles with graphene appears different, since growing Pt nanoparticles do not show stick-slip motion. Indeed, it is possible to discern colloidal Pt nanoparticles as small as 0.1 nm radius and to track their Brownian-like motion.

The motions observed in GLC experiments are much slower than what would be expected in an unconfined fluid with the normal bulk viscosity. Slow diffusion, in both Si_xN_y and graphene liquid cells, is discussed in more detail in Chapter 13. In the GLC, it is possible that sub-100 nm diameter liquid pockets may confine the particles significantly. But equally, it is possible that a few layers of solvent organic molecules near the substrate might retard particle dynamics locally. It may in fact be the case that particles in the center of any liquid cell are moving too quickly to be observed at the TEM frame rate, and only those particles that are confined between graphene windows in small liquid pockets move slowly enough to be tracked. The absence of stick-slip motion perhaps suggests a general slowing of diffusion, rather than surface interactions at particular locations. Thus it is probably the case that growth trajectories observed in GLCs may reveal events that can be hidden or distorted in the growth trajectories in non-graphene liquid cells.

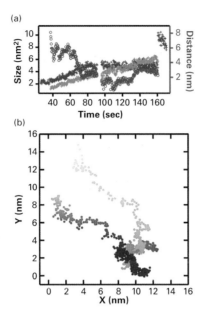

Figure 19.5. Direct observation of Pt nanocrystal growth and diffusion dynamics in the GLC. (a) The projected distance between two nanocrystals (red) and the projected sizes (blue and green) measured during a growth and coalescence sequence. (b) Two-dimensional projected position change of the two nanocrystals before coalescence. Blue and green colors correspond to the nanocrystals in (a). The color gradient represents time evolution from 40 s (bright) to 160 s (dark). Reproduced from Ref. 1 with permission from AAAS.

Since multiple liquid pockets with different sizes are formed in a GLC, a liquid pocket with optimal size can be selected for imaging, depending on the scale of the sample. By minimizing the extra amount of liquid over the sample, background scattering is reduced and contrast increased. This is important for obtaining quantitative data regarding non-perturbed dynamics. Such experiments unveil distinct correlated motion of two nanocrystals for a prolonged period before the particles join along the {111} orientation. In Figure 19.5a, b, blue and green trajectories show particle size, center-to-center distance, and position change during a coalescence event. Initially the freely moving nanocrystals draw close and the center-to-center distance rapidly decreases. This is followed by fluctuation of the center-to-center separation (here, between 1 and 6 nm) for the next 25 s. While this dynamic event proceeds, the nanocrystals also grow in size gradually by monomer addition. The sustained correlated motion in a confined regime exhibits three-dimensional behavior, collision, rolling, and relative sliding of nanocrystals over each other. Knowing the surface ligand length of 1 to 2 nm, these motions occur while the surface ligand layers touch and interleave [30]. After a prolonged period of this correlated motion, the center-to-center distance approaches the sum of the radii of the two nanocrystals and the nanocrystal lattices align, leading to coalescence.

19.3 Applications of GLC-TEM in Biological Studies

The GLC can be used to encapsulate water that contains biological materials. This allows exploration of the structure of biomolecules in the hydrated state, including their composition via analytical techniques, as well as evaluation of electron beam effects. It also allows recording of dynamic processes as biomolecules diffuse in water.

19.3.1 GLC for Studying Hydrated Proteins in STEM and EELS

Ferritin is a protein, 12 nm in diameter, that consists of a spherical protein shell (apo-ferritin) surrounding a 6 nm diameter core of hydrous ferric oxide. Ferritin molecules can be imaged in water in the GLC, and the results can then be compared with images obtained for dried molecules on a graphene substrate. Figure 19.6 shows high angle annular dark field and annular bright field STEM images of ferritin within a GLC, with lattice fringes of the ferritin core visible (Figure 19.6c). The resolution in this experiment even enables individual Fe atoms in water to be distinguished at the edge of the GLC (Figure 19.6d, e).

A key result of such studies is that beam-induced damage at 80 kV appears to be significantly reduced in the presence of liquid [3]. Furthermore, mass loss during EELS measurement is reduced if the sample is encapsulated by graphene [3]. In Figure 19.7, Fe from the cores and N and O from the protein shell are clearly resolved in EELS mapping of graphene-protected samples, whereas unprotected samples show significant mass loss of oxygen and nitrogen. The GLC-TEM appears to promise the potential of quantifying the charge state of Fe in the core in the presence of water via EELS fine structure (Figure 19.7c, d).

To minimize beam damage in these measurements, dose rate (rather than dose) appears to be critical in STEM imaging, if bubble formation is used as an indicator of beam-induced physical and chemical changes of the system [22]. It is therefore possible to define conditions for successful imaging of biological samples in STEM mode that minimize electron beam damage, by keeping the dose rate below a critical limit and decreasing the pixel dwell time. Presumably, below the threshold dose rate, the energy deposited by the incoming electrons can be dissipated in the GLC by the high thermal and electrical conductance of graphene and liquid [3].

Although the GLC therefore appears to be beneficial in imaging delicate biological samples, the question still remains how the environment in the GLC stabilizes the chemical bonding and assembly of macromolecules during irradiation. Direct structural observation of macromolecules in the GLC and comparison with structures obtained from methods such as cryo-EM and X-ray crystallography is needed to address this question quantitatively. The protection of the sample against mass loss by graphene hints at exciting future possibilities for the GLC in EELS measurements.

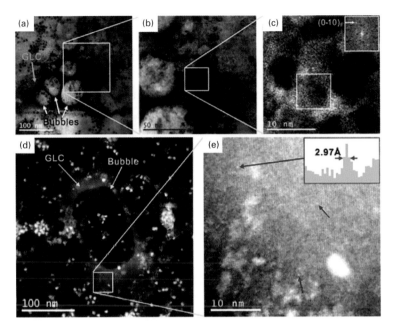

Figure 19.6. Annular bright field (a, b, c) and high angle annular dark field (d, e) STEM images of ferritin imaged in GLC TEM. Bubbles in (a), (b), and (d) were formed in advance using the Ronchigram mode to confirm the presence of liquid. Then, in STEM mode, electron dose rate and pixel dwell time were optimized to form no further bubbles during scanning. An inset in (c) shows FFT patterns of an area containing ferritin to confirm lattice spacing of the ferritin iron core (0.27 nm, consistent with ferrihydrite in the (0–10) orientation). Single Fe atoms are resolved in the liquid environment in (e) near the edge of a GLC. A line profile across the upper left atom is shown as an inset of (e), with each pixel corresponding to 0.99 Å. Reproduced from Ref. 3 with permission from John Wiley and Sons, Inc.

19.3.2 GLC as a Tool to Study Nano-Bio Conjugated Materials: Gold Dimers Conjugated by dsDNA

The motion and structure of double strand DNA (dsDNA) conjugated with Au nanocrystals has been studied in the GLC (Figure 19.8a) [2]. This DNA-assembled nanoconjugate integrates a biological molecule, the linker dsDNA, with inorganic material, two Au nanoparticles, which facilitates tracking of the structure in liquid due to its high contrast (Chapter 16). The moving Au-dsDNA nanoconjugates in liquid appear as clusters of dark circular shapes (Figure 19.8b). In particular, the dynamics of dimers, i.e. pairs of Au nanoparticles tethered by a single piece of dsDNA, can be distinguished from single Au particles, as well as from trimers, which are three Au nanoparticles connected in a linear configuration by two dsDNA bridges (discussed in Section 19.3.3 below). Measurement of the motion of a dimer allows one to infer the configuration and status of its DNA molecule under the electron beam.

A high electron dose is used continuously while imaging dimer motions. Reliable dynamics is observed for up to a few minutes, but then the initially conjugated Au

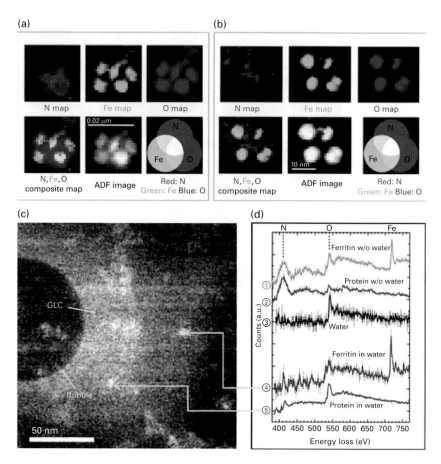

Figure 19.7. To demonstrate stable spectroscopic analysis of biological samples in a graphene sandwich, EELS maps with 1 nm resolution are shown for ferritin molecules (a) sandwiched between graphene sheets and (b) on monolayer graphene. The raw data are filtered using multivariate statistical analysis. The ferritin protein shell is clearly resolved in both cases and the Fe valence is identified as Fe^{3+} distributed across the whole core. But ferritin on graphene shows significantly reduced intensity for oxygen, presumably from sample degradation during imaging. (d) HAADF STEM image of a GLC containing ferritin. (d) EELS spectra of ferritin core, protein shell, or water, from the regions indicated in (c). Reproduced from Ref. 3 with permission from John Wiley and Sons, Inc.

particles are separated or the motion slows down. We assume that prolonged irradiation damages the molecular structure of dsDNA. The integrity of the graphene windows and encapsulated buffer solution can also be compromised after a prolonged time under strong irradiation. As was discussed in Section 19.3.1, control over the dose rate and pixel dwell time is important in order to observe dynamics for a long time without damaging the sample integrity too greatly.

With some understanding of the effect of the beam on the sample, we can compare the motion of various types of Au-dsDNA dimers in the same liquid

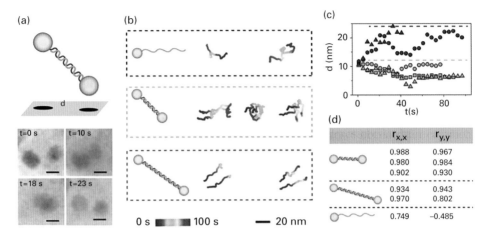

Figure 19.8. Gold nanoparticle dimers conjugated with double strand DNA. (a) Schematic of a dimer and its 2D projection, and representative TEM images of a rotating dimer showing different orientations. Scale bar 5 nm. (b) Various Au nanocrystal trajectories color coded with time in the same liquid pocket: individual particles (top, black box), dimers linked by 42 base pair dsDNA (middle, green box), and dimers linked by 84 base pair dsDNA (lower, purple box). (c) The fluctuation of projected interparticle distance with time for three 42 base pair Au dimers (green data points) and two 84 base pair Au dimers (purple data points). The dotted lines indicate the maxima of interparticle distances. (d) Measured Pearson's correlation coefficients showing the degree of linear correlation of x ($r_{x,x}$) and y ($r_{y,y}$) components of two trajectories. Au dimers conjugated by dsDNA strands show correlated motion for a prolonged time during imaging. Reproduced from Ref. 2 with permission from the American Chemical Society.

pocket. Figure 19.8c shows 2D projected trajectories for Au-single strand DNA, short Au-dsDNA-Au (42 base pair DNA), and long Au-dsDNA-Au (84 base pair DNA) encapsulated in one liquid pocket. Several aspects of the clustered trajectories of Au-dsDNA dimers confirm that the structural integrity of the dsDNA linkers is intact, holding adjacent nanocrystals over a prolonged period of imaging. The projected interparticle distance for each Au nanocrystal pair fluctuates but has a maximum that depends on the length of the dsDNA bridges (Figure 19.8c), while the trajectories of the paired particles show a high positive Pearson's correlation coefficient, with free particles showing low or even negative correlation (Figure 19.8d).

19.3.3 Real-Time Observation of 3D Configuration and Motions: dsDNA Conjugated Gold Nanoparticle Trimers

These types of experiments can be extended to more complex structures to provide an opportunity to track motion and configuration change in 3D [2]. For trimers, an iterative optimization method of frame-based capture of the continuous motion can be applied to extract both the rotational and translational motions (color-coded arrows and color-coded dots in Figure 19.9b, respectively) without *a priori*

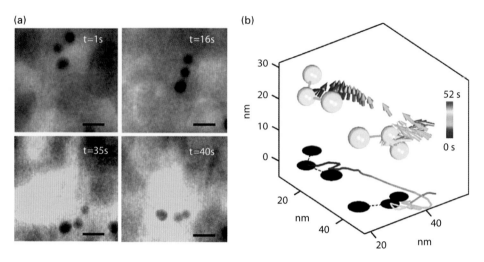

Figure 19.9. Three-dimensional configuration and motion of a Au trimer multiply conjugated with dsDNA. (a) Series of images of a trimer. The scale bar is 10 nm. (b) The reconstructed motion of the trimer in 3D: rotational (color-coded arrows) and translational (color-coded dots for the central nanocrystal). The two trimer diagrams (yellow spheres linked by green lines) show the configuration at $t = 1$ s and 40 s, respectively. Reproduced from Ref. 2 with permission from the American Chemical Society.

knowledge of the structural details of the trimers. The iteration method works by using the 2D projected interparticle distances of the two bridges of the trimer and their ratio to reconstruct the 3D orientation of the trimer in each frame. Repeating the same process for all frames, the standard deviation of the interparticle distances of the three Au particles calculated from the extracted 3D coordinates in each frame is used as a reference to fine tune the original structure of the trimer. Mathematical details of this method have been published elsewhere [2]. The principles of reconstructing 3D structure and motion from 2D projections can readily be extended to other structures in order to relate their 3D configuration and motion in solution to their function.

19.4 Future Directions

In this chapter we have summarized the recent developments in graphene liquid cell TEM and the use of the GLC in studying liquid sample dynamics at high resolution. The GLC has enabled the study of colloidal nanocrystal growth and diffusion processes with unprecedented resolution and without significant perturbation by substrate interactions. The GLC has also shown its worth as a platform for studying general protein samples in low dose mode and for carrying out electron spectroscopy, as well as *in situ* observation of the 3D dynamics of nanocrystal–DNA nanoconjugates.

We believe that the GLC provides multiple advantages over other imaging methods, suggesting new experimental capabilities that cannot be achieved otherwise. These are (i) ease of fabricating a liquid sample for TEM; (ii) compatibility with conventional TEMs and TEM holders; (iii) the opportunity for atomic resolution imaging of the sample; (iv) the chemical and physical inertness of the window material; (v) dissipation of accumulated heat and charge; (vi) enhanced sample protection while imaging; (vii) compatibility with versatile imaging and spectroscopy modes; and, most importantly, (viii) allowing direct observation of *in situ* dynamics at the sub-nanoscale.

However, GLCs also await technical improvement for reliable sample preparation and sophisticated *in situ* TEM experimentation. First of all, the amount of liquid that is encapsulated is not well controlled, especially in first generation GLCs. The efforts put into development of second and third generation GLCs will hopefully lead to more reliable *in situ* experiments. Fabricating functional components within the liquid enclosure, such as electrodes or heaters, can also be possible in principle, expanding the uses of GLC in high resolution TEM studies of reactive liquid environments.

A wide range of dynamics, interactions, and reactions await direct *in situ* GLC-TEM observation in materials sciences, physical chemistry, and biochemistry, correlating structure and development with reactions and functions in a realistic environment. The GLC-TEM satisfies requirements that might otherwise be considered incompatible. It provides high resolution imaging capability while maintaining a realistic liquid condition free from interactions with the windows. For X-ray analysis and EELS measurements, it minimizes unnecessary scattering and the strong core-loss signal usually associated with the presence of Si and N in non-graphene liquid cells. These requirements are not easily satisfied with other existing imaging tools, so we expect that the GLC will be applied to diverse dynamics that either have not been explored or need more detailed understanding. The few topics introduced above are the first steps in applying GLC-TEM for studying colloidal nanoparticle growth, diffusion dynamics at the nanoscale, hydrated biological specimens, and nano-bio interfaced materials. Following these initial studies, the next generation GLC devices we have introduced here will readily be applied to study diverse fluid samples as they evolve and function in time. Self-assembly at different length scales of atoms, nanocrystals, proteins, and macromolecules can benefit from GLC-TEM. Protein-to-protein interactions and macromolecule/polymer dynamics are examples where GLC-TEM will enable observation of delicate systems. Low dose imaging of protein molecules has already demonstrated the capacity of GLC TEM in studying such samples. The recent developments in direct electron detectors are expected to further improve imaging conditions by enhancing signal-to-noise ratio for even lower electron doses, suggesting promising directions for future research.

References

1. J. M. Yuk, J. Park, P. Ercius *et al.*, High-resolution EM of colloidal nanocrystal growth using graphene liquid cells. *Science*, **336** (2012), 61–64.
2. Q. Chen, J. M. Smith, J. Park *et al.*, 3D motion of DNA-Au nanoconjugates in graphene liquid cell electron microscopy. *Nano Lett.*, **13** (2013), 4556–4561.
3. C. Wang, Q. Qiao, T. Shokuhfar and R. F. Klie, High-resolution electron microscopy and spectroscopy of ferritin in biocompatible graphene liquid cells and graphene sandwiches. *Adv. Mater.*, **26** (2014), 3410–3414.
4. A. De Clercq, W. Dachraoui, O. Margeat *et al.*, Growth of Pt–Pd nanoparticles studied in situ by HRTEM in a liquid cell. *J. Phys. Chem. Lett.*, **5** (2014), 2126–2130.
5. N. Mohanty, M. Fahrenholtz, A. Nagaraja, D. Boyle and V. Berry, Impermeable graphenic encasement of bacteria. *Nano Lett.*, **11** (2011), 1270–1275.
6. P. Ericius, K. Kim, A. Zettl *et al.*, In-situ observations of Pt nanoparticle growth at atomic resolution using graphene liquid cells and Cc correction. *Microsc. Microanal.*, **18** (2012), 1096–1097.
7. R. R. Nair, P. Blake, J. R. Blake *et al.*, Graphene as a transparent conductive support for studying biological molecules by transmission electron microscopy. *Appl. Phys. Lett.*, **97** (2010), 153102.
8. Z. Lee, K. Jeon, A. Dato and R. Erni, Direct imaging of soft–hard interfaces enabled by graphene. *Nano Lett.*, **9** (2009), 3365–3369.
9. W. Regan, N. Alem, B. Alemán *et al.*, A direct transfer of layer-area graphene. *Appl. Phys. Lett.*, **96** (2010), 113102.
10. J. E. Evans, K. L. Jungjohann, N. D. Browning and I. Arslan, Controlled growth of nanoparticles from solution with in situ liquid transmission electron microscopy. *Nano Lett.*, **11** (2011), 2809–2813.
11. H. Zheng, S. A Claridge, A. M. Minor, A. P. Alivisatos, and U. Dahmen, Nanocrystal diffusion in a liquid thin film observed by in situ transmission electron microscopy. *Nano Lett.*, **9** (2009), 2460–2465.
12. T. J. Woehl, J. E. Evans, I. Arslan, W. D. Ristenpart and N. D. Browning, Direct in situ determination of the mechanisms controlling nanoparticle nucleation and growth. *ACS Nano*, **6** (2012), 8599–8610.
13. U. M. Mirsaidov, H. Zheng, Y. Casana and P. Matsudaira, Imaging protein structure in water at 2.7 nm resolution by transmission electron microscopy. *Biophys. J.*, **102** (2012), L15–L17.
14. H. Zheng, U. M. Mirsaidov, L.-W. Wang and P. Matsudaira, Electron beam manipulation of nanoparticles. *Nano Lett.*, **12** (2012), 5644–5648.
15. M. T. Proetto, A. M. Rush, M.-P. Chien *et al.*, Dynamics of soft nanomaterials captured by transmission electron microscopy in liquid water. *J. Am. Chem. Soc.*, **136** (2014), 1162–1165.
16. K. L. Jungjohann, S. Bliznakov, P. W. Sutter, E. A Stach and E. A. Sutter, In situ liquid cell electron microscopy of the solution growth of Au-Pd core-shell nanostructures. *Nano Lett.*, **13** (2013), 2964–2970.
17. U. M. Mirsaidov, H. Zheng, D. Bhattacharya, Y. Casana and P. Matsudaira, Direct observation of stick-slip movements of water nanodroplets induced by an electron beam. *Proc. Natl. Acad. Sci. USA*, **109** (2012), 7187–7190.
18. L. R. Parent, D. B. Robinson, P. J. Cappillino *et al.*, In situ observation of directed nanoparticle aggregation during the synthesis of ordered nanoporous metal in soft templates. *Chem. Mater.*, **26** (2014), 1426–1433.

19. H. Zheng, R. K. Smith, Y.-W. Jun et al., Observation of single colloidal platinum nanocrystal growth trajectories. *Science*, **324** (2009), 1309–1312.
20. H.-G. Liao, L. Cui, S. Whitelam and H. Zheng, Real-time imaging of Pt_3Fe nanorod growth in solution. *Science*, **336** (2012), 1011–1014.
21. J. Park, H. Zheng, W. C. Lee et al., Direct observation of nanoparticle superlattice formation by using liquid cell transmission electron microscopy. *ACS Nano*, **6** (2012), 2078–2085.
22. J. M. Grogan, N. M. Schneider, F. M. Ross and H. H. Bau, Bubble and pattern formation in liquid induced by an electron beam. *Nano Lett.*, **14** (2014), 359–364.
23. N. M. Schneider, M. M. Norton, B. J. Mendel et al., Electron–water interactions and implications for liquid cell electron microscopy. *J. Phys. Chem. C*, **118** (2014), 22373–22382.
24. V. P. Adiga, G. D. Dunn, A. P. Alivisatos and A. Zettl, Liquid flow cells having graphene on nitride for microscopy. US Patent Application No. US 20160042912 A1.
25. M. E. Holtz, Y. Yu, J. Gao, H. D. Abruña and D. A. Muller, In situ electron energy-loss spectroscopy in liquids. *Microsc. Microanal.*, **19** (2013), 1027–1035.
26. J. Polte, R. Erler and A. F. Thu et al., Nucleation and growth of gold nanoparticles studied via in situ small angle X-ray scattering at millisecond time resolution. *ACS Nano*, **4** (2010), 1076–1082.
27. M. Harada and E. Katagiri, Mechanism of silver particle formation during photoreduction using in situ time-resolved SAXS analysis. *Langmuir*, **26** (2010), 17896–17905.
28. J. Polte, T. T. Ahner, F. Delissen et al., Mechanism of gold nanoparticle formation in the classical citrate synthesis method derived from coupled in situ XANES and SAXS evaluation. *J. Am. Chem. Soc.*, **132** (2010), 1296–1301.
29. X. Lu, M. Rycenga, S. E. Skrabalak, B. Wiley and Y. Xia, Chemical synthesis of novel plasmonic nanoparticles. *Annu. Rev. Phys. Chem.*, **60** (2009), 167–192.
30. P. Schapotschnikow, R. Pool and T. J. H. Vlugt, Molecular simulations of interacting nanocrystals. *Nano Lett.*, **8** (2008), 2930–2934.

20 Analytical Electron Microscopy during *In Situ* Liquid Cell Studies

Megan E. Holtz, David A. Muller, and Nestor J. Zaluzec

20.1 Introduction

The driving goal for liquid phase microscopy is to understand the fundamental mechanisms and underlying processes in biological and materials science. A critical component of this is identifying the chemical state and/or elemental composition of the material being studied. Much of the recent progress in liquid cell electron microscopy has been in observing morphological and structural changes that occur on the subnanometer to micrometer scale; progress has been slower for techniques aimed at finding elemental and electronic structure information. The analytical electron microscope, or AEM, specializes in answering questions regarding composition (elemental and chemical) as well as electronic structures of materials, and can do so at the atomic scale for conventional crystalline samples in scanning transmission mode, and at lower resolution in conventional transmission electron microscopy mode. The AEM is often capable of two types of spectroscopic analysis methods: electron energy loss spectroscopy (EELS) and X-ray energy-dispersive spectroscopy (XEDS) [1]. These tools have recently been explored for liquid cell studies, providing elemental information on the micro- to nanoscale from XEDS and electronic structure information for dynamic processes in liquids from valence EELS. In this chapter, we discuss EELS and XEDS in liquid cell microscopy, and some of their achievements, strengths, and limitations.

20.1.1 Physical Principles and Information Available

A simple picture of the relevant physics for electron energy loss and X-ray spectroscopy begins with the incident electron beam striking a region of interest in the liquid cell and exciting bound electrons from occupied state to unoccupied states in the material [1]. After scattering from a bound electron state, the probing electron loses the amount of energy corresponding to the transition energy between the ground and excited states. The electron in the excited state then subsequently relaxes into a lower energy state, releasing the energy by either emitting a photon (such as an X-ray), by ejecting another electron (Auger transitions), or by heating the material (exciting lattice vibrations). The incident electron beam can also excite collective oscillations of valence electrons in the material, known as plasmons. Furthermore, the incident beam can also produce bremsstrahlung, radiation with a broad energy spread, as the electrons lose angular momentum and energy in the specimen.

The energy lost by the probing electron is measured by EELS, which provides elemental identification from the energy signatures of the primary core-level transitions (core-loss EELS). Additionally, EELS reveals the localized electronic state information from either the fine-structure features of the core edges or from lower energy valence-level transitions near the Fermi energy (valence EELS). Plasmons are a prominent feature of the valence EELS signal and generally have energies of 10–30 eV. EELS can measure the energy lost by these mechanisms with typical energy resolutions of 0.5–1 eV and as low as 0.05 eV when the instrument employs a monochromated electron source.

XEDS provides elemental identification from the measurement of characteristic energies of the X-rays emitted from the region of interest as a result of the decay transitions by electrons from their excited states, and can also detect bremsstrahlung radiation. The characteristic energies from single-electron core transitions provide elemental analysis and mapping, while bremsstrahlung is present as a continuum or background signal. Ultra-high energy resolution XEDS systems exist that can also provide chemical/valence state information. However, these have not been employed in liquid cell studies, principally due to their extremely low collection solid angles and thus low detection efficiency.

EELS and XEDS can both provide a measure of thickness of the whole sample, i.e. the sum of windows + liquid medium + material of interest. EELS does this by measuring the low loss scattering and applying Beer's law, as discussed in Section 20.3 [2, 3]. XEDS can derive similar information by noting that both the characteristic and the bremsstrahlung signals are linear in thickness and beam current to first-order approximation [4].

Both EELS and XEDS involve interfacing ancillary spectrometers to the electron optical column of the instrument used for imaging. In the case of EELS, a magnetic sector spectrometer is generally used, located after the objective lens, while for XEDS a solid state semiconductor detector is located in the objective lens area of the TEM/STEM or SEM, and thus in the immediate vicinity of the liquid cell.

20.1.2 Spectroscopic Imaging Modes

Spectroscopic imaging modes are possible using both EELS and XEDS. One method for creating spectroscopic images is to scan the incident beam across a region of the sample in STEM mode and collect an energy loss or X-ray spectrum at each point. The resulting hyperspectral data cube can then be post-facto analyzed and the integrated elemental signals used to create spectroscopic maps of the region of interest. This method allows the analyst the luxury of recording spectroscopic information from many points in the specimen, which can be interrogated post-measurement for subtle details that might escape initial observations. However, a major limitation of the point-by-point method for acquiring spectra is the amount of time needed to acquire a statistically relevant signal at each location.

An alternate mode for spectroscopic imaging exists for EELS, known as energy-filtered TEM (EFTEM). In this mode, an analyst operates the instrument using a relatively large illuminated region in TEM mode. After interacting with the sample, the forward-scattered electrons (which includes both elastic and inelastic signals) are directed into an imaging spectrometer, where a small range of electron energies is selected and subsequently used to create an image of the sample at energy losses chosen by the energy-selecting slit. These filtered images provide a means for spectroscopic imaging of large areas of the specimen, but only within fixed energy intervals. Mapping of elemental signals in EFTEM is achieved by recording images for "energy slices" lower and higher than the energies of characteristic loss events and then computationally processing to extract the intervening characteristic data [3]. EFTEM can also be used for purely elastic imaging, as discussed in Section 20.2.3.

In the limit of infinite amounts of time and data space, both of the aforementioned spectroscopic methodologies can, in principle, achieve identical results. Practical details dictate which mode is used for a particular experiment. EFTEM is often used for broad sweeping overviews, while hyperspectral imaging is used for detailed assessments using small focused probes.

20.1.3 AEM for Liquid Cell Samples

EELS and XEDS have been widely used in conventional (thin, solid) TEM samples in vacuum for decades. Although they have only recently been applied to studies in liquids [4–8], the results are promising. A brief comparison of XEDS and EELS in liquids in the AEM environment is given in Table 20.1. Particular challenges associated with liquid cell spectroscopy in the AEM are the degradation of the EELS signal in thick liquids, and the geometrical challenges of detecting XEDS signals from samples in liquid cell holders.

For EELS, the principal challenge for liquid cell studies is the rapid deterioration of the signal arising from multiple inelastic scattering events in the environment (sample + liquid medium + windows). Multiple scattering leads to compounded energy loss events which become increasingly difficult to interpret [3]. For regions of interest thicker than one mean free path of the electron in the material (~100–150 nm at 200 keV) [3], the singly scattered EELS signal becomes weaker. The thickness degradation of the EELS signal is exacerbated for higher-energy (core-loss) transitions because multiple scattering events of lower-energy valence signals create a large background, resulting in a strong decrease in signal-to-background ratio for the core losses. In contrast, because there are fewer lower-energy transitions below the valence EELS regime, valence EELS is resolvable in thicker samples than core-loss EELS [6]. Despite these limitations, EELS analysis can be performed successfully in thin liquids to obtain a wide variety of information about the sample and the liquid. The types of information obtainable and the thickness limitations of both core and valence EELS that have been explored will be discussed in Section 20.2.

While EELS in liquid media is fundamentally limited by the physics of multiple scattering, XEDS suffers a different technical problem, namely, signal detection.

Table 20.1 Comparison of XEDS and EELS as applied to liquid cell studies

	XEDS	EELS
Relevant physics	Incident electron excites core-level electron. Higher energy electron relaxes into core hole, releasing a characteristic X-ray.	Incident electron excites valence- or core-level electron or collective mode (i.e. plasmon), and incident electron loses a corresponding amount of energy.
What is measured	X-rays emitted from the region of interest in the liquid cell (sample + liquid + windows) dispersed by energy.	Energy distribution of electrons passing through the region of interest in the liquid cell (sample + liquid + windows).
Location of the detector	Proximate to the liquid cell in the objective lens area of the AEM, where a solid state detector collects from 0.1 to ~2 sR of the X-ray emission.	Remote to the liquid cell, after the projector lenses where an electron spectrometer collects 10–30 mrad of the central beam of the scattered electrons.
Kinds of information	Elemental identification, thickness of the region of interest.	Elemental identification, bonding environment, local electronic structure information, thickness of the region of interest.
Spatial resolution	Defined by interaction volume of the broadened electron beam in the liquid cell, the probe size, and the location of the region of interest in the liquid cell.	Defined by the broadened electron probe in the sample, by the probe size, and the location of the region of interest in the liquid cell.
Thickness dependence	Increased thickness can degrade spatial resolution, but generally not the XEDS signal. Highest spatial resolution when the region of interest is proximate to the electron entrance window.	Spectra degraded primarily by multiple inelastic scattering of the electron beam, resulting in multiple energy losses obscuring the signal after 3–6 mean free paths (~300–600 nm at 200 keV) for core and valence EELS respectively.
Beam dose and data acquisition times	Beam dose-intensive, requires long acquisition times.	Low loss: low dose, fast data acquisition for dynamic spectral imaging; high loss: large dose, longer data acquisition.
Higher sensitivity to:	$Z > 6$ and particularly medium to heavy elements, with limitations due to the elemental composition of the window and medium.	Lighter elements $Z < 30$, with limitations due to the elemental composition of the window and medium.
Spectroscopic imaging	In STEM mode, collect spectra as you scan the beam from pixel to pixel, and then form spectroscopic images by integrating the peaks corresponding to elements. In TEM mode, spectra from large regions are readily achievable, giving spatially averaged information.	In STEM mode, collect spectra as you scan the beam from pixel to pixel, and then form spectroscopic images by integrating the peaks corresponding to elements with appropriate background subtraction. In EFTEM mode, form image using electrons that have lost particular energies.
Strengths	Works in thick liquids, easy to map $Z > 6$ and heavier elements; modified liquid holders and optimized detector geometry required.	Beam-sensitive materials (valence EELS), rapid spectroscopic imaging (EFTEM); unmodified holders can be used.

The initial construction geometry of holders in TEM/STEM or SEMs for microfabricated liquid cells was focused on imaging and on protection of the instrument from liquid leakage into the vacuum system. Absent from these holder designs was an optimized line-of-sight path from the point of generation of the X-ray to an appropriate detector, making XEDS studies impossible in first-generation liquid cell holders [4, 6]. Recent holder modifications have enabled XEDS analysis by providing this line of sight path from the sample to the XEDS detector. Because X-rays interact very weakly with both the liquid medium and the encapsulating Si_xN_y windows, the XEDS signal is significantly less sensitive to the liquid thickness than EELS [4]. This makes XEDS an effective technique for elemental analysis in thicker liquids typical in many *in situ* studies [4, 8]. Holder designs for microfabricated liquid cells and the successes of XEDS studies in liquids will be discussed in Section 20.3.

While this chapter focuses mainly on microfabricated liquid cells with Si_xN_y windows, graphene liquid cells offer interesting opportunities for AEM. Because graphene liquid cells tend to be thin, they are promising for EELS studies which are limited by cell thickness. Graphene liquid cells, fabricated as described in Chapter 19, can be loaded into conventional TEM holders, which makes them promising for XEDS because conventional TEM holders have a significantly lower penumbra, hence less shadowing of the XEDS detector. EELS and XEDS are starting to be applied in the graphene liquid cell [9], and we anticipate further use in graphene liquid cell studies in the future. However, one potential limitation of the graphene cell is that graphene damages quickly at high operating voltage (above ~100 keV) due to knock-on damage, and spectroscopy typically requires a large beam dose, as we discuss in the next section.

20.1.4 Electron Beam Effects

During any microanalysis of samples that include liquids, the electron beam will not only interact with the sample, but also alter the chemical/elemental environment of the liquid or perhaps locally heat the region of interest (see Chapter 7). This interaction can range from minor to serious effects that can obfuscate observations. Analytical spectroscopies such as EELS and XEDS inherently require larger electron beam doses than imaging, since inelastic cross sections are smaller than elastic cross sections for most materials. The larger dose required for spectroscopy means that the beam interaction and subsequent irradiation effects can be correspondingly more severe. Both core-loss EELS and XEDS elemental mapping are dose-intensive methodologies requiring longer acquisition times to accumulate sufficient statistics, compared to other electron imaging methodologies. For beam-sensitive samples, valence EELS is nearly always possible, due to its larger scattering cross sections and thus higher intensity in the low loss scattering spectral regime, enabling spectroscopic imaging with relatively short acquisition times. To illustrate this, low loss valence EELS studies can be done in seconds, but statistically significant elemental maps involving core-loss EELS and/or XEDS can take minutes or longer. To minimize beam damage, lower resolution studies are likely to be effective. Thus, as with any TEM experiment, analysis must be done with radiolysis and other beam effects in mind.

20.2 Electron Energy Loss Spectroscopy

EELS can provide information about the compositional, electronic and dielectric properties of a material, if the specimen is sufficiently thin. Core-level EELS facilitates elemental identification as well as insight into the oxidation state and bonding environment from the fine-structure of the near edge profile [1]. If the thickness is less than one mean free path, λ, sub-nanometer to atomic resolution is possible for core-loss EELS [10–13]. Energy losses to the valence electrons can reveal the electronic structure of materials, such as the optical gap and electron density of the specimen. Recently, EELS has been used in liquid cell experiments to determine liquid presence and thickness [5, 6, 14]. In thin liquids, elements [6] and bonding states [15] have been identified. In moderately thick liquids, electronic structure information has been explored [6, 7]. Energy-filtered TEM (EFTEM) can track chemical changes in dynamic processes on the nanometer scale [7]. EFTEM can also improve imaging conditions in liquids by selecting only the elastically scattered electrons to reduce chromatic blur, in which the electrons of different energies come to different focal points (Chapter 21). Thus, the wealth of information provided by EELS makes it an attractive technique for exploring processes in thin ($t/\lambda < 2$) liquid layers.

20.2.1 Liquid Thickness Limitations on Core and Low Loss EELS

The primary challenge for EELS in liquids is that the signal is degraded in thick specimens ($t/\lambda > 1$), since the incident electron can lose energy from multiple inelastic scattering events which occur along the entire scattering path length of sample + surrounding liquid + cell windows. The effects of multiple scattering dominate the EELS spectra, as shown in Figure 20.1a where the multiply scattered plasmon peak becomes larger with increasing water thickness. Fortunately, when the sample composition is different from both the surrounding media and the cell windows, its EELS edges can remain distinct and are still detectable in thin liquid, as shown in Figure 20.1b for core-loss EELS of iron phosphate nanoparticles in a 180 nm thick liquid layer.

For any particular EELS edge, the effects of multiple scattering are manifest in two ways. First, the transition height of the edge will decrease with thickness since some signal from that edge may additionally lose energy by another scattering event. Second, the background from lower-energy scattering events will increase, further lowering the signal-to-background ratio. Thus, EELS features with higher energy (core-loss edges) degrade more rapidly with thickness since there are more lower-energy scattering events which convolve with the signal. Because the dominant energy loss feature in thick regions is the plasmon peak, core-loss edges above the plasmon energy are generally not observable in liquid media thicker than two or three inelastic mean free paths ($t/\lambda > 2$–3). This makes core-loss EELS for elemental identification only practical in liquids thinner than a few inelastic mean free paths. The oxygen-K edge is highlighted in Figure 20.1a and it is clear that the bulk plasmon peak in water overwhelms it in all but the thinnest ($t/\lambda < 3$) liquid layers. While core-loss EELS can identify

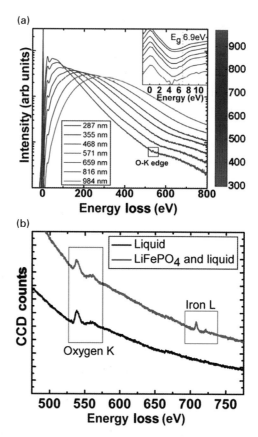

Figure 20.1. (a) STEM-EELS measurements through water as a function of thickness. The core-loss O-K edge is resolvable for the thinnest layer only, while the optical gap at 6.9 eV is resolvable out to 6.5 inelastic mean free paths, or ~650 nm (inset). (b) Core-loss EELS of LiFePO$_4$ nanoparticles in 180 nm of aqueous solution. The Fe-L edge is visible on the particle, while the O-K edge is visible in both the particle and the liquid. Adapted from Ref. 6.

elements present in the liquid up to three inelastic mean free paths, quantification of the amount of a particular element present and its oxidation state will be challenging in all but the thinnest cells ($t/\lambda < 1$). Because the Si$_x$N$_y$ windows themselves are each 50 nm thick ($t/\lambda \sim 0.5$ for 200 keV), elemental and chemical quantification from core-loss EELS in liquid media is likely to be difficult without using thinner membranes, such as graphene [16], discussed in more detail in Chapter 19.

The valence EELS features at and below the plasmon energy will become obscured when the fraction of the beam that has only scattered once is no longer detectable. This is governed by Poisson statistics. Since there are only modest amounts of signal in the very low loss regime, the background from this low intensity multiply scattered signal is low and thus the signal-to-background ratios are relatively high. The Poisson distribution dictates how much signal is present, by giving the likelihood P of n scattering events to occur in a specimen of thickness t/λ, where λ is the (inelastic) mean free path:

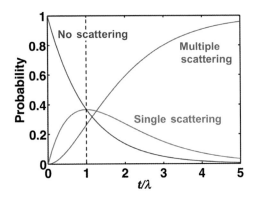

Figure 20.2. The Poisson distribution, which gives the probability of scattering n times for a given thickness in terms of the mean free path (λ). The probability for no scattering falls exponentially, and the probability for single scattering peaks at one mean free path. The sum of Poisson distributions for $n > 1$ shows the probability for multiple scattering, which grows as thickness increases. A microfabricated liquid cell with no liquid in it is roughly $t/\lambda = 1$ due to the thickness of the windows.

$P(n) = \frac{(t/\lambda)^n}{n!} e^{-(t/\lambda)}$. The probability for no scattering ($n = 0$), single scattering ($n = 1$), and multiple scattering (sum over $n > 1$) are shown in Figure 20.2. From Poisson's first-order distribution, we expect the fraction of the signal that scatters once will drop to 1% of the total signal at roughly 6.5 inelastic mean free paths. If the singly scattered signal is below the 1% level, we expect it to be difficult to resolve in the experimental EELS data. Indeed, the experimental valence spectra (inset of Figure 20.1a) have features that are resolvable for thickness up to roughly six inelastic mean free paths [6]. Thus, the thickness limitation for valence EELS is nearly twice as large as the limitation for core-loss EELS, because there is little lower-energy signal to obscure the valence EELS peaks. The more favorable thickness dependence and the large cross sections of valence EELS makes it an attractive technique for effective signal collection in thicker liquid volumes.

20.2.2 Applications of Electron Energy Loss Spectroscopy in Liquids

One of the first applications of EELS in a liquid was to study hydrated bacteria that were enclosed an environmental cell (Figure 20.3) [15]. The bacteria were metal-reducing and naturally encrusted in chromium, and EELS of the hydrated bacteria yielded spectra of the Cr-$L_{2,3}$ edge, which allowed fine structure analysis to determine the chromium oxidation state. Although this measurement was conducted in a gaseous environmental cell, rather than a liquid cell completely filled with water, EELS through the hydrated bacteria demonstrated that, given the water present in and around the bacteria, fine-structure analysis of core-loss edges could determine both oxidation and bonding information.

In liquid cells completely filled with liquid, one of the primary uses of EELS is the investigation of the liquid itself. In many situations, it is essential to determine whether

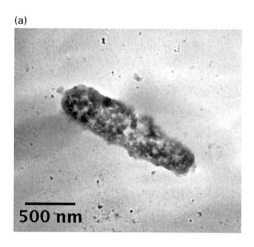

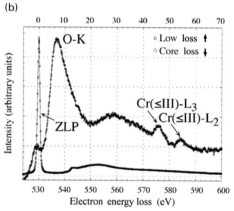

Figure 20.3. Micrograph and EELS of a metal-reducing bacterium, *Shewanella oneidensis*, naturally encrusted with chromium, imaged in an environmental cell containing humid air and water. EELS spectra show both the zero loss peak (ZLP) and O-K and Cr-$L_{2,3}$ core-loss spectra. Adapted from Ref. 15.

the liquid is even present (Chapter 8). For water, EELS spectra display a characteristic interband transition [5] shown in Figure 20.4. Additionally, the plasmon peak and the presence of multiply scattered or bulk plasmon peaks are often an indicator of liquid presence, since the mean free path in liquid is roughly 3 orders of magnitude smaller than the mean free path in gas. In many cases, a Si_xN_y membrane cell that is filled with gas (i.e. no liquid present) is approximately one inelastic mean free path thick, being composed of two Si_xN_y membranes each ~50 nm thick. The gas in the cell contributes a relatively small amount to the total scattering. Adding liquid increases the effective thickness of the sample. With the windows + liquid thicker than one mean free path, the probability for single scattering decreases as the first order ($n = 1$) Poisson distribution, so the first plasmon peak diminishes (Figures 20.2 and 20.4). Similarly, the probability for multiple scattering increases and the second and bulk plasmon peak intensities grow correspondingly.

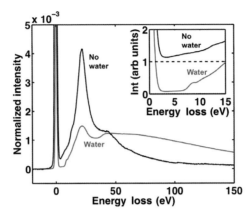

Figure 20.4. EELS spectra of a liquid cell before and during flow of water. The empty cell produces a signal from the two 50 nm Si_xN_y windows and the gas in between them, which is in total $0.9t/\lambda$ thick. The first plasmon is the dominant inelastic signal from the empty cell spectrum. The water-filled cell is $2.6t/\lambda$ thick so the scattering is dominated by the thick liquid layer, with absorption starting at the exciton peak at 6.9 eV (see inset). The plasmon peak is observed at 20.4 eV. The intensity of the first plasmon peak is lower with liquid than without due to multiple scattering, creating a bulk plasmon at higher energies. For the main figure, spectra were normalized by the total electron beam dose for the spectrum acquisition (i.e. the integrated intensity of the spectrum).

The composite thickness of the liquid cell (windows + liquid + material of interest) can then be determined using Beer's Law (or the $n = 0$ Poisson distribution), which predicts an exponentially decaying unscattered intensity $I = I_o\, e^{-t/\lambda}$ in the zero loss peak [2, 3]. By integrating the zero loss peak intensity (I) and comparing it to the integrated intensity over the entire spectrum (I_o), the thickness is measured in terms of the inelastic mean free path (t/λ). For thick specimens, it is important to record the EELS spectra out to high energies to capture most of the scattering (for specimens below $7t/\lambda$, acquiring up to 500 eV introduces a maximum error of 1%) [6]. Thickness determination using EELS and Beer's law has been used to map the bulging of the encapsulating Si_xN_y membranes due to the pressure difference, Figure 20.5.

Valence EELS has also been used to determine electronic structure properties of the liquid such as the plasmon energy, optical gap, and electron density [6]. Measurements on several liquids show that their valence electrons behave according to the free electron model (Figure 20.6) [6]. While this may seem surprising, it is not unexpected since many insulators also appear to follow the free electron model [3].

20.2.3 Energy-Filtered Transmission Electron Microscopy (EFTEM)

The discussion above concentrated mainly on spectroscopic aspects of EELS. We briefly turn now to its complementary imaging mode, EFTEM. For liquid studies, the EFTEM mode is particularly useful when applied to dynamic or time-resolved measurements. As outlined in Section 20.1.2, the EFTEM mode involves formation of

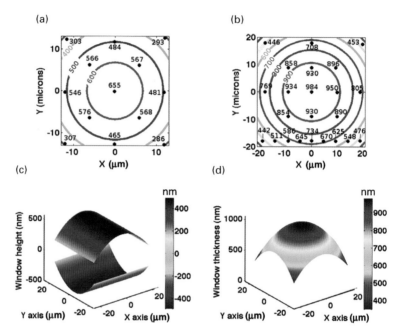

Figure 20.5. Measurement of the variations in liquid thickness due to window bulging. STEM-EELS measurements of thickness (points) with a parabolic fit (contours) are shown for (a) water between 25 μm wide windows with a 150 nm Au spacer and (b) ethylene glycol between 40 μm wide windows with a 500 nm SU-8 spacer. (c) A three-dimensional representation of the bulging of the two rotated rectangular Si_xN_y membranes, assuming each window bows out as an independent orthogonal parabola and (d) the total thickness between the windows from (c). The profile shown in (d) corresponds to the contour lines in (b). Adapted from Ref. 6.

images resulting from electrons that have lost a fixed amount of energy. The energy loss is selected by means of an operator-defined energy window within the optics of a suitably equipped imaging spectrometer [3]. In this operating mode, a large area integrating digital camera at the output of the imaging spectrometer is used to rapidly record and process electron scattering from discrete transitions selected by the analyst. Selection of high signal-to-background ratio energy windows can result in imaging times of seconds or less. This can be used for dynamic imaging of chemical processes for spectroscopic signals with large cross sections, such as the valence transitions.

Valence transitions are especially sensitive to electronic structure changes during *in situ* electrochemical or chemical reactions. For example, EFTEM has been used to study charging and discharging of a battery cathode material, $LiFePO_4$, where the electronic structure reveals the material's charge state (see Chapter 11 for more details). Figure 20.7 shows how the energy-selecting slit allows one to monitor the signal from the chemically active 5 eV EELS peak whose presence or absence identifies if the particles are the delithiated (charged) $FePO_4$ or lithiated (discharged) $LiFePO_4$. By comparing valence images to the elastic/zero loss image, one can verify that chemical (and not diffraction) contrast is changing [7]. In addition to observing chemical changes

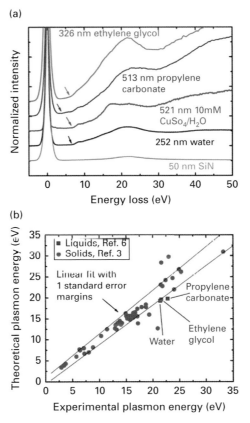

Figure 20.6. (a) Valence EELS of ethylene glycol, propylene carbonate, 10 mM of $CuSO_4/H_2O$, and water. Arrows point to the optical gap. The background spectrum from the membranes alone shows no dominant features on the same scale without the presence of liquid. (b) Theoretical free-electron estimates of the plasmon energy compared to experimental plasmon energy for solids [3] and measurements of the pure liquids. The solid lines are the one-sigma standard error margins of a linear fit. Adapted from Ref. 6.

in the particles, the lithium content in the liquid between particles was also observable in the 5 eV spectroscopic EFTEM image, which darkens on discharge as the solution near the particles is depleted of lithium. The spectroscopic fingerprint was identified by joint density functional theory to be $LiSO_4^-$, which has a peak at 6 eV, well within the 5 eV EFTEM window. Tracking changes in both the solid material and the solution gives useful information in analyzing reaction kinetics. In principle, concentrations of any dissolved species should be measurable if there is an accessible spectroscopic peak. Dynamic imaging of the specimen and solution is achievable, as long as the cross sections are large enough to provide rapid imaging.

In addition to mapping high signal-to-background transitions as illustrated above, EFTEM can also be used to improve image resolution in thick materials. By selecting the zero-loss (or elastic) signal, image blurring due to chromatic defocusing can be

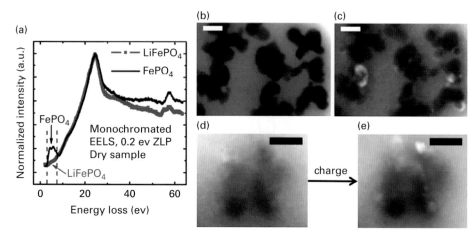

Figure 20.7. Spectroscopy of LiFePO$_4$ and its delithiated counterpart FePO$_4$. (a) Monochromated EELS, with an energy resolution of 0.2 eV, of a dry sample shows a peak at 5 eV for FePO$_4$ but not for LiFePO$_4$. EFTEM of LiFePO$_4$ in 0.5 M Li$_2$SO$_4$/H$_2$O with a 5 eV energy slit around (b) 0 eV, where the liquid dominates the signal and the particles look fairly homogeneous, showing no diffraction contrast, and (c) 5 eV, which highlights the FePO$_4$. In the electrochemical cell during cycling of the battery cathode, 5 eV EFTEM maps corresponding to the (d) discharged and (e) charged state showing bright delithiated regions of FePO$_4$ that formed during charging. Scale bars are 200 nm. Adapted from Ref. 7.

minimized. This is particularly important in thick ($t/\lambda > 2$) regimes. A key improvement we expect to see in future EFTEM experiments is discussed in Chapter 21: the correction of chromatic defocus in instruments equipped with combined spherical and chromatic aberration correctors. C_S+C_C correction will significantly enhance the signal intensity in all EFTEM imaging experiments [19], thus improving the overall performance of EFTEM. However, the use of C_C correctors in liquid cell studies has been limited, principally due to the small number of instruments with C_C correction installed.

20.3 X-ray Energy Dispersive Spectroscopy

Elemental spectroscopy through liquid media of moderate thicknesses typical of liquid cell microscopy studies is invaluable for applications such as corrosion in metallic systems, catalyst evolution during reactions, nanoparticle synthesis and degradation, as well as any situation where elemental identification is important. XEDS has an inherently better peak/background in thick media than EELS. Elemental spectroscopy as well as spatially resolved mapping by XEDS is being used for both qualitative and quantitative spectroscopy at resolutions ranging from the submicrometer to nanometer scale in liquids ranging from pure water to hydrocarbon solutions [4, 8, 17, 18].

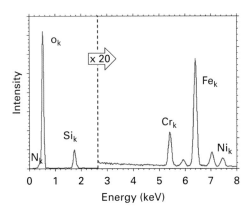

Figure 20.8. An XEDS spectrum taken from a TEM steel specimen fully immersed in liquid water in the AEM. Discernable in the spectrum is the strong X-ray line from water (O-K), followed by lower-intensity lines from the Si_xN_y window (Si-K, N-K), then the Cr-K, Fe-K, and Ni-K from the steel. The vertical dashed line marks a change in vertical scale by 20×. Note this spectrum was acquired with a modified liquid cell holder (Section 20.3.1).

As outlined earlier in this chapter, the principal signal used for XEDS spectral analysis is the characteristic X-ray emission due to transitions from excited states into ground states [1]. Figure 20.8 illustrates an XEDS data set taken of a TEM steel specimen fully immersed in liquid water. This spectrum demonstrates the straightforward ability to discern the nominal elemental composition of the steel (Cr, Fe, Ni), as well as the surrounding media (O from H_2O, and Si and N from the window) by means of characteristic X-ray emission peaks. XEDS is so useful for microanalysis in liquids because the energy of most characteristic X-ray lines is high enough that their detectability is not strongly affected by attenuation and multiple scattering processes in liquid cell experiments. This also means that under the appropriate conditions the quantification methodologies long ago established for XEDS in the AEM [1] can also be directly applied to liquid cell studies. Finally, and significantly, the new generation of solid state X-ray detectors commercially available today have collection solid angles that have been greatly increased over their predecessors (by a factor of ~4–5) and this enhances the ability to rapidly and efficiently measure X-ray signals [19]. Importantly, although many instrumental advances have been made for XEDS detection, it should be appreciated that EELS is a more efficient signal generation process compared with XEDS, and low atomic number elements having X-ray lines at energies less than 300 eV with meager X-ray fluorescence yields (such as lithium and beryllium) and valence electron transitions are impractical to measure by XEDS in liquid cell studies.

Recalling from Section 20.2.1 that the key parameter for assessing the applicability of EELS in liquid studies is the relative mean free path (t/λ), the corresponding parameter for XEDS is the X-ray absorption path length. This parameter is the physical distance along a line of sight path from the point of generation of an X-ray through any ensuing material to the semiconductor detector (Figure 20.9). This path length, d, varies based upon the location of the detector, the geometry/orientation of the liquid cell, and the location of the region of interest within the cell. The value and

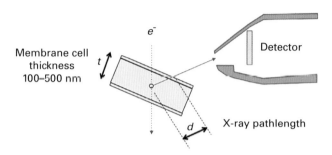

Figure 20.9. A schematic of the geometry of XEDS experiments in the liquid cell. The X-ray path length (d) is the distance travelled in the material of interest, liquid, and windows from the point of generation of the X-ray to the detector.

Table 20.2 Calculation of X-ray absorption as a function of path length for a range of characteristic X-ray lines by a Si_xN_y window and various amounts of liquid H_2O. As I/I_0 approaches unity the effects of X-ray absorption can be ignored.

X-ray line	Energy (keV)	μ/ρ in Si_xN_y $\rho \approx 1.33$ g/cm^3	μ/ρ in H_2O $\rho \approx 1$ g/cm^3	I/I_0 in Si_xN_y $d \approx 50$ nm	I/I_0 in H_2O $d \approx 150$ nm	I/I_0 in H_2O $d \approx 500$ nm	I/I_0 in H_2O $d \approx 1$ μm
Ni-K$_\alpha$	7.48	48.64	12.98	0.999	0.999	0.999	0.999
Ti-K$_\alpha$	4.51	197.3	56.22	0.998	0.999	0.997	0.994
S-K$_\alpha$	2.31	1267	390.6	0.991	0.994	0.981	0.962
Al-K$_\alpha$	1.48	655.2	1418	0.996	0.979	0.932	0.868
Na-K$_\alpha$	1.04	1767	3943	0.988	0.943	0.821	0.674
Ne-K$_\alpha$	0.85	3143	7050	0.979	0.899	0.703	0.494
F-K$_\alpha$	0.68	5878	13678	0.961	0.815	0.505	0.255
O-K$_\alpha$	0.52	12064	29376	0.923	0.644	0.230	0.053
C-K$_\alpha$	0.28	67557	176560	0.638	7.1×10^{-2}	1×10^{-4}	1×10^{-8}

attenuation that occur over the X-ray path length can vary significantly and will be determined by the experimental geometry established by the analyst. Minimizing d is an important part of any XEDS liquid cell experimental design, and thus requires understanding and careful setting up of experimental conditions, which include optimizing the microscope parameters and properly locating the XEDS detectors, the liquid cell holder, and the sample within the liquid cell itself. For reference, the gap between the Si_xN_y windows in current microfabricated liquid cells is typically 100–500 nm, and each window is ~50 nm thick.

The primary degradation process of XEDS signals as a function of path length is simple X-ray absorption and is described by an equation of the form:

$$I/I_0 = \exp\left[-\left(\frac{\mu}{\rho}\right)_{E_x} \rho d\right]$$

Table 20.2 compares this attenuation (I/I_0) for various characteristic energy (E_x) X-ray lines as a function of d for the mass absorption coefficients (μ/ρ) and densities (ρ) of water and Si_xN_y. To include attenuation in both the windows and liquid, calculate the product of the exponential attenuation (I/I_0) resulting from each material, as a function of location in the cell and the relative orientation of the holder and detector.

If the XEDS detector is positioned nearest to the entrance surface of the liquid cell (the most common configuration in the AEM) then the *worst* case geometry for XEDS analysis is when the region of interest (ROI) is located on the electron exit window of the cell, a situation that should be avoided by judicious experimental design. Alternatively, when the ROI is located on the electron entrance window this is the best-case condition for XEDS, since it minimizes the path length through the liquid, as well as facilitating the highest resolution STEM imaging (Chapter 8). However, having the specimen on the entrance surface is also the *least* favorable condition for conventional TEM imaging, resulting in substantial chromatic blurring [20]. This top/bottom/multiple scattering effect is a recurrent problem in all EM work involving thick materials. The use of EFTEM or chromatic aberration correction can partially alleviate the imaging issues in TEM mode (see Chapter 20), but will not mitigate problems in spectroscopy [20]. Referring to Table 20.2, which calculates the absorption terms for conditions that are typical (rather than worst or best case), one can see that a 50 nm Si_xN_y entrance window has negligible effect for X-rays with energies greater than O-K_α (~520 eV). Indeed, even the water has little effect on X-rays above ~1 keV until the path length approaches 1 μm. Higher energy X-rays ($E > 2$ keV) typical of the transition or noble metals should be even less affected than low energy ($E < 1$ keV) lines. But elements at or below C-K_α ($E \sim 280$ eV) show large absorption effects, making them difficult to analyze in liquid cells (actually, they are difficult even in vacuum). Higher density liquids will also generally introduce concomitant changes owing to their different mass absorption coefficients and densities. The X-ray path length limitations apply equally to the characteristic and bremsstrahlung signals. This is the reason that the deleterious effect of thickness in X-ray spectroscopy is substantially less than that due to multiple scattering in EELS, which is a result of a physically different process (see Section 20.2.1).

20.3.1 Liquid Thickness Limitations on X-ray Energy Dispersive Spectroscopy

As we documented in Table 20.2, the XEDS signal is generally not significantly degraded with increasing liquid thickness, since escaping X-rays with high enough energy are not strongly attenuated by the liquid or windows. Figure 20.10 directly compares the effects of thickness in the two complementary spectroscopies, EELS and XEDS, via their signal-to-background ratios. Two different thickness areas of the same NiO specimen are shown, normalized for comparison, with the EELS data scaled to the pre-O-K edge background and the XEDS data scaled to the Ni-K_α X-ray peak. In the thicker area, the EELS edges are degraded with low edge/background ratios due to multiple inelastic scattering events. In contrast, the XEDS signal-to-background is only slightly affected because both the signal and the background increase approximately linearly with thickness [1, 4]. Although the lower energy X-rays (O-K_α and Ni-L_α) are absorbed in the thicker areas, the signal is still clearly resolvable and does not overly complicate qualitative spectroscopy. This suggests that XEDS can be a more suitable

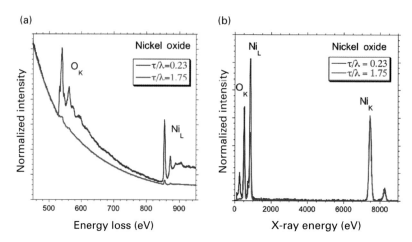

Figure 20.10. Comparison of normalized EELS (a) and XEDS (b) spectra recorded from two areas of the same NiO sample with mean free paths (t/λ) of 0.23 and 1.75 at 200 keV. Note the change in signal-to-background ratios with thickness. From Ref. 4.

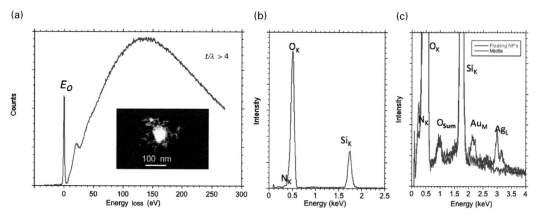

Figure 20.11. (a) EELS spectra taken from a thick layer of water illustrating the dominance of inelastic scattering. Inset, HAADF-STEM image of nanoparticles moving on the membrane surface in the water. (b) XEDS spectrum from the same area showing the O-K signal from the water as well as the weaker Si-K and N-K signals from the window. (c) Expanded scale XEDS spectra from the drifting nanoparticle array (blue) and a neighboring equivalently sized region (red) devoid of nanoparticles, with identical acquisition time, showing identification of the elemental species present. From Ref. 17.

technique for liquid cell studies than EELS when the sample is larger than two inelastic mean free paths; but for ultra-thin specimens ($t/\lambda < 0.3$), EELS almost always outperforms XEDS when comparable measurements are undertaken.

Figure 20.11 brings this point home by comparing EELS and XEDS data from an agglomeration of Ag/Au nanoparticles in a massively thick water window ($t/\lambda > 4$). In the low loss EELS spectrum (Figure 20.11a), the elastic peak is overshadowed by the inelastic signal and the mean energy loss occurs at nearly 150 eV; no core-loss edges are detectable due to multiple scattering. In the XEDS data (Figure 20.11b), the O-K peak from H_2O dominates the spectrum, followed by the Si-K and the barely discernable N-K buried in the low energy tail of the O-K peak. In this medium were sub-50 nm nanoparticles of Ag and Au, shown in the inset STEM HAADF image. These nanoparticles moved along the membrane surface so were manually tracked using a ROI window. It should be appreciated that nanoparticles that are truly freely floating in the liquid generally cannot be easily imaged nor can spectroscopy be easily accomplished, since their diffusion length is large on the temporal and spatial scale of the TEM image (Chapter 22); for example, the diffusion constant of Au nanoparticles in water is roughly 10^{-11} m^2/s, or 10^7 nm^2/s [21]. The particles that appear to be mobile in liquid cell experiments are generally weakly attached to the membrane windows but loose enough to move around. Figure 20.11c shows the accumulated spectra from the moving nanoparticle cluster, illustrating the detection of both Au and Ag X-ray signals, and the lack of these signals in a control spectrum. This example also highlights a significant problem for XEDS. Although we have gone to great lengths to optimize data collection, the spectral data collection rate of nanoparticles in liquid can be very sparse (<1 kilocounts per second) and thus the accumulated X-ray signal is both time and dose intensive, requiring long acquisition times (seconds to minutes) to acquire statistically reliable data. Unless the region of interest is relatively beam insensitive and is stationary (or drift corrected), XEDS can become challenging even in light of its excellent peak/background capabilities. Dose tolerance and sample stability should not be overlooked when using XEDS for characterization.

20.3.2 Liquid Cell Holders for X-ray Energy Dispersive Spectroscopy

While Figure 20.8 and 20.11 illustrated that XEDS can achieve elemental spectroscopy in liquids, the liquid cell holder designs were until recently not optimal. They blocked or shadowed the XEDS detector from the specimen, precluding any elemental analysis in most implementations. As described in Chapter 2, the liquid cell holders first introduced were optimized for microfabricated window chips (made from, say, 300 μm thick Si wafers) to seal in the liquid, protect the microscope vacuum, and provide the ability to form a transmission image. Although successful for imaging, their design has a major limitation with respect to XEDS. Specifically, the penumbra or shadow of a conventional holder for microfabricated liquid cells nearly completely blocks a line of sight path from the ROI in the liquid to the X-ray detector [19]. This is schematically illustrated in Figure 20.12a. Depending upon the geometry of the Si_xN_y membrane, the surrounding Si substrate, the holder sidewalls at its periphery, and the X-ray detectors, most of the X-ray emission from the probed area may not reach the detector, making XEDS analysis untenable. Fortunately, all of these issues can be mitigated by [4]

- careful redesign to minimize the penumbra of the holder (Figure 20.12b);
- optimizing the construction materials of the holder lid to minimize "systems peaks" through the use of low atomic number metals such as beryllium;

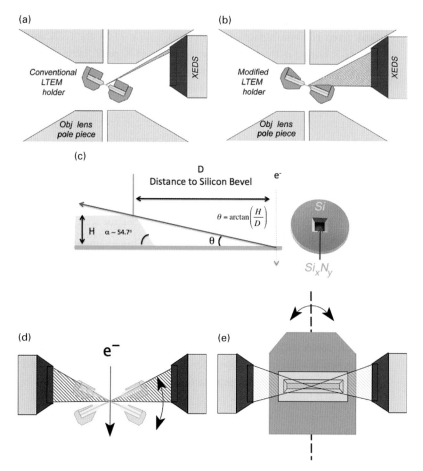

Figure 20.12. Penumbra of the conventional (a) and modified (b) membrane liquid cell holders. (c) Penumbra angle to Si wafer ⟨111⟩ face is dependent upon location and height of the side wall. (d) Tilting of holder relative to detector geometry to minimize penumbra. (e) View of the holder top-down (along direction of incident electrons). The long axis of slit-type windows should be aligned perpendicular to the tilt axis on a dual detector system. Cross hatched: acceptance solid angle of the XEDS. Red/dark grey, grey: XEDS detector, green/yellow: Si_xN_y /Si chips, blue: holder body, medium grey: holder lid, light grey: objective lens polepiece.

- configuring instruments with X-ray detectors having both large solid angles and large elevation angles within the constraints imposed by the objective lens polepieces;
- and optimizing the Si_xN_y/Si window chip to have wide windows and minimum Si thickness.

Once the holder is modified and detector geometry optimized, the limiting factor actually becomes the penumbra of the Si substrate used to support the Si_xN_y window. As described in Chapter 2, the walls supporting the window are ⟨111⟩ planes at 54.7° to the window. They shadow or block the detector, especially for small or long narrow

Si_xN_y windows and thick Si substrates. Indeed, the Si ⟨111⟩ face penumbra can block over 50% of the Si_xN_y viewing area even with an "optimized holder" (Figure 20.12c). H (the Si substrate thickness) and D (the distance between the region of interest and the sidewall) together define an effective cutoff angle θ. For instruments with multiple XEDS detectors, particularly those with detectors symmetrically arranged with respect to the holder tilt axis, the effect of the sidewalls can be mitigated by simply tilting the holder towards the conjugate detector (Figure 20.12d). However, the tilt needed may not be achievable depending on the value of θ and the microscope geometry. It is worth using thin Si substrates and long slit-shaped windows (which also reduce membrane bulging) with the long axis aligned perpendicular to the tilt axis, as illustrated in Figure 20.12e.

20.3.3 Applications of X-ray Energy Dispersive Spectroscopy in Liquids

Just as EELS can be used to identify the presence of liquids, similar results can be obtained using XEDS. Figure 20.13 compares spectra taken from liquid cells containing different liquids at atmospheric pressure [17]. Comparing the data set without liquid (i.e. cell filled with air) to the spectra of liquid-filled cells, one can immediately not only identify the presence and nominal composition of the liquids (and the windows) in the respective cells, but also monitor changes in the relative amount of liquid present versus time. The amount of liquid is monitored by taking the ratio of the integrated peak intensities of the liquid to that of the fixed (known) amount of silicon (i.e. the Si_xN_y window) and comparing with those ratios in full and empty cells. It is worth remembering that hydrogen and helium do not emit characteristic X-rays and thus an absolute elemental compositional analysis is not possible using XEDS unless standards are employed. In addition, we also note that even in vacuum small oxygen and chlorine peaks are typically detected from the windows used in TEM/STEM/SEM experiments. These peaks vary in magnitude with manufacturing process and supplier, but are always seen in "low stress" Si_xN_y films produced using any low pressure chemical vapor deposition process that employs dichlorosilane and ammonia for nitride film growth [22].

To illustrate the application of XEDS during nanoscale materials characterization in liquids, we turn next to a collection of hyperspectral images (Figure 20.14) taken from a liquid cell containing a stationary array of nanostructures [8]. HAADF STEM shows an assortment of nanorods, nanotubes, and nanoparticles (Figure 20.14a). Because of the complexity of structures and the surrounding liquid, the image alone is insufficient to identify uniquely the various materials via atomic number contrast. However, since the structures were effectively anchored (i.e. immobile) in the cell, long hyperspectral data collection was possible. Elemental maps reconstructed from the raw data (i.e. full X-ray spectra at each pixel) allowed mapping and identification of 40 nm gold particles (yellow), 30 nm diameter silver nanorods (light blue), palladium particles (green) decorating the carbon (red) nanotube (CNT), as well as the iron catalyst (purple) at the CNT center. Coating the edge of the CNT, one additionally resolves copper (orange), which was deposited by beam-induced precipitation from the aqueous

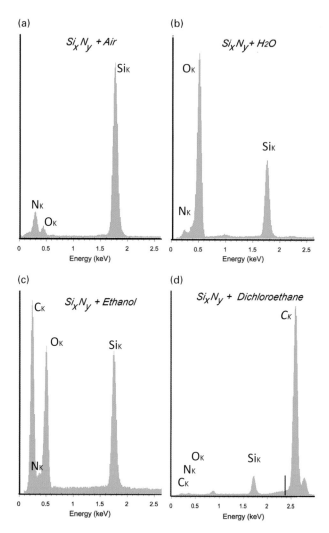

Figure 20.13. Elemental identification of liquids. A microfabricated liquid cell was used with a cell gap of 150 nm and 50 nm Si_xN_y windows, at 1 atmosphere of pressure, filled with (a) air, (b) water, (c) ethanol (C_2H_5OH), and (d) dichloroethane (CH_2ClCH_2Cl). The AEM was operated at 200 kV in TEM mode. From Ref. 17.

solution, which unintentionally was contaminated with Cu^+ ions. The smallest elementally resolved structures in this image are ~10–15 nm, and although this spatial resolution is currently not as good as the corresponding STEM-HAADF image in Figure 20.14a, it is sufficient to identify the species present.

Although this example demonstrates the resolution limits of XEDS in liquids, the more challenging experiments deal with controlled exposure to a liquid environment to drive chemical reactions in catalysts, or electrochemical deposition and dissolution as in corrosion (Chapter 12) or fuel cell and battery cycling (Chapter 11). These are difficult experiments as they entail fabricating well-controlled samples and inserting them in the

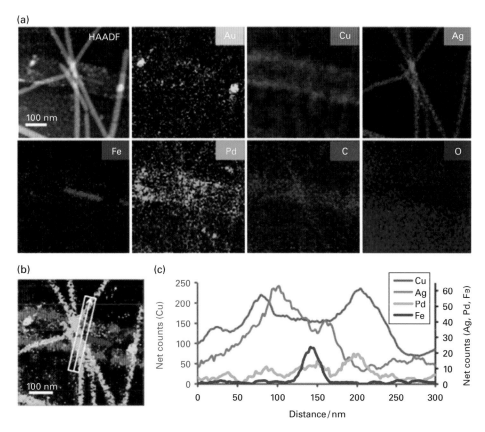

Figure 20.14. Hyperspectral composite (a, b) of nanoparticle "soup" in liquid water. Nominal spatial resolution ~10–15 nm. (c) Line profile from the region marked in (b). Adapted from Ref. 8.

liquid in a configuration that facilitates both imaging and spectroscopy. To date there has been little controlled spectroscopic work in this area, because of the difficulty in sample preparation. However, at least one approach has been demonstrated that employs a dual-beam FIB (FIB/SEM) [18, 23]:

- electropolishing of the target material into a conventional TEM specimen;
- identification of the ROI using TEM/STEM;
- transferring the TEM specimen to a FIB/SEM;
- identification of the identical ROI using STEM in the FIB/SEM;
- careful FIB cutting and lift out of the ROI (minimizing Ga ion implantation and Pt contamination);
- mounting of the FIB section on top of (but not touching) the top Si_xN_y window of the liquid cell;
- mounting and sealing the top and bottom chips in the liquid cell holder without breaking either window (generally using an overly safe cell gap of 500 nm);
- transfer of the liquid cell holder into the AEM with introduction of appropriate liquid.

Figure 20.15 illustrates a schematic of the experimental configuration of a FIB-cut section suspended between two windows, with AEM images in liquid of a meticulously sectioned 100 × 100 μm electron-transparent area of a steel specimen, oriented and attached to a slot window on the top chip then mounted into the holder. As one might appreciate, a mistake at any of the steps above can result in a ruined specimen and/or a broken window, and thus a diligent and unhurried approach is required. An even more challenging step is to carry this out in an electrochemical cell and also make the appropriate electrical connections [18] (see Chapters 10 and 11). Figure 20.16 shows a full corrosion hyperspectral experiment starting with a 304 steel specimen mounted in a dry (i.e. air-filled) cell, mapped in air to delineate the starting elemental distribution, and then mapped "wet", post corrosion. A MnS inclusion has completely dissolved after ~24 hours of exposure to water – during which time the beam was blanked to reduce beam-induced processes.

We finally comment on spatial resolution for XEDS. Although we have demonstrated that the degradation of the XEDS signal itself is not strongly thickness dependent, the spatial resolution of XEDS mapping does deteriorate with thickness since the electron probe diameter broadens as it propagates due to multiple elastic scattering. Since the probe diameter defines the region of generation of the X-ray signal, spatial resolution depends on the probe diameter as well as the region of interest within the liquid cell. Spatial resolution will be best when the region of interest is located at the electron entrance surface and worse at the exit surface; better resolution is achievable when the liquid is thinnest. Nevertheless, with stationary, irradiation-insensitive structures fully immersed in micrometer-thick liquid, relatively high spatial resolution measurements can be conducted.

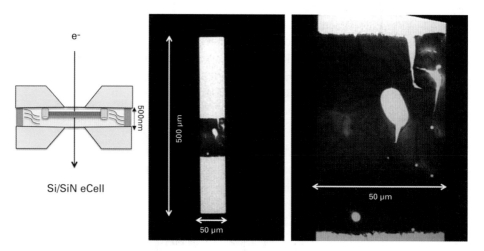

Figure 20.15. Illustration of the use of TEM and FIB/SEM to prepare and mount an electropolished sample into a liquid cell for combined XEDS/corrosion studies. (a) A schematic of the intended setup. (b, c) TEM images of thin FIB-cut sample mounted on the 50 μm × 500 μm × 50 nm thick Si_xN_y slot window prior to introduction of water. Adapted from Ref. 23 and 24.

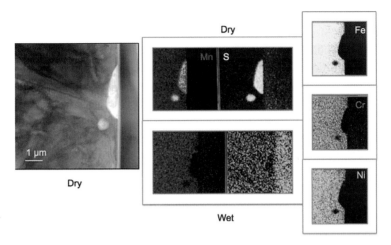

Figure 20.16. Hyperspectral images of dissolution of MnS inclusions in FIB-cut 304 steel in a liquid cell. Shown are the TEM and spectroscopic images prior to dissolution, with the cell filled with air (Dry), and post dissolution (Wet) in H_2O, where the two large MnS inclusions have dissolved leaving voids.

20.4 Summary

Electron energy loss spectroscopy and X-ray energy dispersive spectroscopy are challenging but also powerful techniques for elemental and chemical identification of materials suspended in liquids. EELS is degraded quickly by multiple scattering in the liquid. Nevertheless, core-loss EELS can determine composition and bonding if the sample is thinner than two to three mean free paths, while valence EELS can offer electronic structure information out to about six mean free paths. The larger cross sections for valence EELS provide a higher dose efficiency, making it more useful for rapid imaging. XEDS is significantly less affected by thick liquids and offers routine elemental identification in liquids for thicknesses significantly greater than used in EELS. EELS and XEDS are complementary and can be combined to elucidate submicrometer to nanoscale features.

Both EELS and XEDS require careful control of experimental conditions, and samples and liquids that are stable over the time scale of any spectroscopic measurements. In particular, careful attention must be paid to dose tolerance and radiation damage. Because the dose required for inelastic imaging is typically greater than the dose required for elastic imaging, the dose-limited resolution is worse for hyperspectral imaging compared to standard elastic imaging modes. Due to dose constraints, dynamic imaging with EELS and XEDS may be limited in materials and liquids that are beam sensitive. To avoid putting a large dose into the sample, it may be necessary to perform dynamic imaging with the elastic scattering signal and to intermittently interrogate the composition of the sample with core-loss EELS and XEDS. Achieving dynamic imaging or spectroscopy will rely on large cross sections to enable rapid acquisition times. Valence EELS often provides large cross sections that can be imaged quickly in EFTEM mode (~1–2 s per spectral image), but interpretation of the spectra can be ambiguous if there is little presumed knowledge about the system. For XEDS, the acquisition rates will continue to improve as X-ray detector solid angles become greater, but for statistically comparable signals, acquisition times of tens of seconds to minutes per hyperspectral image may be required for the foreseeable future.

For the successful application of core-loss EELS and/or XEDS in liquids, forming the thinnest liquid cell is paramount. One promising development is the use of graphene or ultra-thin Si_xN_y windows with thicker support bars (Chapter 2). Graphene films produce the least amount of scattering from the window, but because of graphene's flexibility, replacing Si_xN_y with graphene on standard silicon chips is likely to lead to window bulging (as well as being difficult to fabricate). The graphene liquid cell, fabricated as described in Chapter 19, is also promising for EELS and XEDS if beam damage can be controlled. An alternative way to reduce the effective thickness is to use higher operating voltages to 200–300 keV, which increases the mean free path of the electron in liquid and thus reduces the total amount of scattering. With these advances, we expect to see much more use of AEM in liquid cell experiments in the future.

Acknowledgements

This work was supported in part at Cornell University at the Energy Materials Center at Cornell (emc^2), an Energy Frontier Research Center funded by the U.S. Department of Energy, Office of Basic Energy Sciences under Award Number DESC0001086, in the electron microscopy facility of the Cornell Center for Materials Research (CCMR) with support from the National Science Foundation Materials Research Science and Engineering Centers (MRSEC) program (DMR 1120296).

The work at Argonne National Laboratory was supported by the U.S. Department of Energy, Office of Basic Energy Sciences under Contract Number DE-AC02-06CH11357, at the Electron Microscopy Center, Nanoscience and Technology Division at Argonne National Laboratory.

References

1. J. Goldstein, D. Joy, D. Maher, J. Silcox and N. J Zaluzec, *Introduction to Analytical Electron Microscopy* (New York: Plenum Press, 1979), Chapters 3, 4, 7, 9 and 10.
2. T. Malis, S. C. Cheng and R. F. Egerton, EELS log-ratio technique for specimen-thickness measurement in the TEM. *J. Electron Microsc. Tech.*, **8** (1988), 193–200.
3. R. F. Egerton, *Electron Energy-Loss Spectroscopy in the Electron Microscope* (New York: Plenum Press, 2011).
4. N. J. Zaluzec, M. G. Burke, S. J. Haigh and M. A. Kulzick, X-ray Energy-dispersive spectrometry during in situ liquid cell studies using an analytical electron microscope. *Microsc. Microanal.*, **20** (2014), 323–329.

5. K. L. Jungjohann, J. E. Evans, J. A. Aguiar, I. Arslan and N. D. Browning, Atomic-scale imaging and spectroscopy for in situ liquid scanning transmission electron microscopy. *Microsc. Microanal.*, **18** (2012), 621–627.
6. M. E. Holtz, Y. Yu, J. Gao, H. D. Abruña and D. A. Muller, In situ electron energy-loss spectroscopy in liquids. *Microsc. Microanal.*, **19** (2013), 1027–1035.
7. M. E. Holtz, Y. Yu, D. Gunceler et al. Nanoscale imaging of lithium ion distribution during in situ operation of battery electrode and electrolyte. *Nano Lett.*, **14** (2014), 1453–1459.
8. E. A Lewis, S. J. Haigh, T. J. A. Slater et al., Real-time imaging and local elemental analysis of nanostructures in liquids. *Chem. Commun.*, **50** (2014), 10019–10022.
9. J. M. Yuk, H. K. Seo, J. W. Choi and J. Y. Lee, Anisotropic lithiation onset in silicon nanoparticle anode revealed by in situ graphene liquid cell electron microscopy. *ACS Nano*, **8** (2014), 7478–7485.
10. M. Scheinfein, Electronic and chemical analysis of fluoride interface structures at subnanometer spatial resolution. *J. Vac. Sci. Technol.*, **4** (1986), 326.
11. P. E. Batson, Simultaneous STEM imaging and electron energy-loss spectroscopy with atomic-column sensitivity. *Nature*, **366** (1993), 727–728.
12. D. A. Muller, Y. Tzou, R. Raj and J. Silcox, Mapping sp(2) and sp(3) states of carbon at subnanometer spatial-resolution. *Nature*, **366** (1993), 725–727.
13. D. A. Muller, L. Fitting Kourkoutis, M. Murfitt et al., Atomic-scale chemical imaging of composition and bonding by aberration-corrected microscopy. *Science*, **319** (2008), 1073–1076.
14. K. L. Klein, N. de Jonge and I. M. Anderson, Energy-loss characteristics for EFTEM imaging with a liquid flow cell. *Microsc. Microanal.*, **17** (2011), 780–781.
15. T. L. Daulton, B. J. Little, K. Lowe and J. Jones-Meehan, In situ environmental cell-transmission electron microscopy study of microbial reduction of chromium(VI) using electron energy loss spectroscopy. *Microsc. Microanal.*, **7** (2001), 470–485.
16. J. M. Yuk et al., High-resolution EM of colloidal nanocrystal growth using graphene liquid cells. *Science*, **336** (2012), 61–64.
17. N. J. Zaluzec et al., X-ray and electron energy loss spectroscopy in liquids in the analytical S/TEM. *Microsc. Microanal.*, **20** (2014), 1518–1519.
18. S. Schilling, A. Janssen, X. L. Zhong, N. J. Zaluzec and M. G. Burke, Liquid in situ analytical electron microscopy: examining SCC precursor events for Type 304 stainless steel in H_2O. *Microsc. Microanal.*, **21** (2015), 1291–1292.
19. N. J. Zaluzec, Analytical formulae for calculation of X-ray detector solid angles in the scanning and scanning/transmission analytical electron microscope. *Microsc. Microanal.*, **20** (2014), 1318–1326.
20. N. J. Zaluzec, The influence of Cs/Cc correction in analytical imaging and spectroscopy in scanning and transmission electron microscopy. *Ultramicroscopy*, **151** (2015), 240–249.
21. K. Wong, C. Chen, K. Wei, V. A. L. Roy and S. M. Chathoth, Diffusion of gold nanoparticles in toluene and water as seen by dynamic light scattering. *J. Nanoparticle Res.*, **17** (2015), 153-1–153-8.
22. N. J. Zaluzec, When is Si_3N_4 not Si_3N_4? When it is a low stress SiN_x membrane window. *Microsc. Microanal.*, **21** (2015), 959–960.
23. X. Zhong et al., Novel hybrid sample preparation method for in situ liquid cell TEM analysis. *Microsc. Microanal.*, **20** (2014), 1514–1515.
24. S. Schilling, A. Janssen, M. G. Burke et al., In situ analytical election microscopy: imaging and analysis of steel in liquid water. *Proc. Intl. Microsc. Conf. 2014*, Prague (2014), Ed. P. Hozak, IT-7-O-2947.

21 Spherical and Chromatic Aberration Correction for Atomic-Resolution Liquid Cell Electron Microscopy

Rafal E. Dunin-Borkowski and Lothar Houben

21.1 Introduction

Compared with carrying out liquid cell experiments in an uncorrected electron microscope, spherical aberration correction in the TEM and STEM can be used to improve spatial resolution, as discussed in Chapter 8. Spherical aberration correction also improves the depth of field, allows the effects of image delocalization to be minimized, provides better precision in measurement of the positions of features of interest in the sample, and facilitates the quantitative comparison of experimental images with computer simulations in terms of absolute intensities rather than arbitrary contrast values. These advantages, however, should be balanced against the increased complexity and cost of an aberration-corrected instrument. It is also important to take into consideration the information that is desired from each particular experiment.

In this chapter, we begin by providing an introduction to the basics of spherical aberration-corrected imaging in the TEM. Very few liquid cell experiments have been carried out in aberration-corrected electron microscopes at the time of writing. We therefore present representative examples of the application of spherical aberration correction in both TEM and STEM to studies of more conventional (primarily inorganic) samples. When possible, we highlight specific aspects of the designs and outcomes of these experiments that are likely to be relevant for future liquid cell studies.

After considering spherical aberration correction alone, we introduce combined chromatic and spherical aberration correction in the TEM. This is currently only available on a handful of instruments but promises to improve spatial resolution further, when compared with the use of spherical aberration correction alone. The improvements are especially clear when studying beam-sensitive samples at lower microscope accelerating voltages, examining thicker samples and acquiring atomic-resolution energy-filtered TEM (EFTEM) images using wide energy-selecting windows and large objective aperture sizes. Chromatic aberration correction also promises to further improve the precision with which the positions of features of interest in the sample can be measured and to allow atomic spatial resolution imaging when using wider objective lens pole-piece gaps.

21.2 Spherical Aberration Correction in the TEM

High resolution TEM (HRTEM) involves the acquisition of images with a spatial resolution that is sufficient to separate single atomic columns. However, the interpretation of such images is not straightforward, as the recorded intensity is rarely a direct representation of the structure in the sample. Instead, an HRTEM image can be regarded as an interference pattern that is affected by both the interaction of the incident electrons with the sample and the transfer function of the microscope.

21.2.1 Spherical Aberration and the Contrast Transfer Function

The interaction of an incident electron wave with a TEM sample can be calculated by solving the relativistically corrected Schrödinger equation for the electron wavefunction $\Psi(r)$ in the crystal potential. In the phase object approximation, the atoms in a thin sample are described by a projected electrostatic potential that is constant in the direction of the incident electron beam. After passing through a sample of thickness t, the electron wavefunction can be written in the form:

$$\Psi(r) \approx \exp[i\phi(r,t)] = \exp[i\sigma V_P(r)t], \quad (21.1)$$

where r is a two-dimensional vector in the plane of the specimen, ϕ is the phase of the electron wave, σ is an interaction constant that depends on the microscope voltage (kV), and V_P is the projected electrostatic potential. In the weak phase object approximation, it is further assumed that the change in phase of the electron wave is small after passing through a very thin sample, resulting in the expression:

$$\Psi(r) \approx 1 + i\phi(r,t) = 1 + i\sigma V_P(r)t. \quad (21.2)$$

The effect of the microscope lenses on the recorded image intensity is described by modifying the exit plane wavefunction $\Psi(g)$ (now expressed in reciprocal space) using a phase factor $\exp[-i\chi]$ to give:

$$\Psi_i(g) = \Psi(g)\exp[-i\chi(g)], \quad (21.3)$$

where $g = \omega/\lambda$, ω is the scattering angle, and λ is the electron wavelength. Considering only the lowest order rotationally symmetric aberrations of a round imaging lens, the aberration function takes the form:

$$\chi(g) = 2\pi\left(\tfrac{1}{4}C_S\lambda^3 g^4 + \tfrac{1}{2}f\lambda g^2\right). \quad (21.4)$$

In Equation (21.4), f is the defocus and C_S is the spherical aberration coefficient of the objective lens. C_S is illustrated schematically in Figure 21.1 and describes the deviation of rays that travel through the outer part of the lens compared to near-axis rays. Spherical aberration cannot be avoided for a round electromagnetic lens [1].

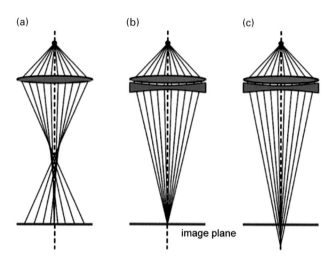

Figure 21.1. Schematic illustration of (a) positive spherical aberration, (b) no spherical aberration, and (c) negative spherical aberration.

For a very thin sample, the weak phase object approximation results in the following expression for the linear image intensity:

$$I_L(g \neq 0) \approx 2\sigma V(g) t \sin \chi(g), \tag{21.5}$$

from which it is apparent that optimum contrast is obtained when the coherent contrast transfer function (CTF) $\sin \chi = \pm 1$, i.e. when $\chi(\mathbf{g})$ is an odd multiple of $\pi/2$ for all values of $\mathbf{g}$. However, as $\sin \chi$ is in general a function that oscillates strongly with g, atomic columns that are arranged with spacing $d = 1/g$ are imaged as white dots only for selected spatial frequencies for which the CTF is close to 1. At spatial frequencies for which the CTF is close to -1, they are imaged as black dots, while they may be invisible if the spatial frequency corresponding to their spacing coincides with a zero in the CTF.

In order to approach ideal phase contrast transfer behavior, a defocus can be chosen that balances the g^2 and g^4 terms in the aberration function χ, allowing for a relatively broad frequency band to be transferred with a CTF close to -1. This defocus setting,

$$f_S \approx -\sqrt{\tfrac{4}{3} C_S \lambda}, \tag{21.6}$$

is known as Scherzer defocus [2]. The point resolution $d_S = 1/g_S$ is defined by the first zero crossing of the CTF at Scherzer defocus and is given by the expression:

$$g_S \approx \left(\tfrac{3}{16} C_S \lambda^3\right)^{-\tfrac{1}{4}}. \tag{21.7}$$

Typical values of point resolution for (non-aberration-corrected) commercially available medium accelerating voltage (200–400 kV) electron microscopes are in the range 0.24–0.17 nm.

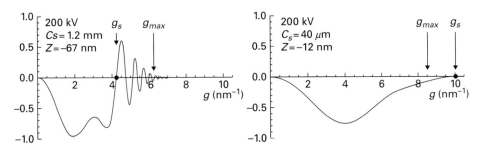

Figure 21.2. Partially coherent contrast transfer functions at Scherzer defocus calculated for a CM20 FEG TEM operated at 200 kV for (left) C_S = 1.2 mm, and (right) C_S = 0.04 mm. Z is the defocus and g_S and g_{max} are defined in the text.

The limited coherence of the electron source and electronic instabilities of the microscope have the additional effect of multiplying the coherent CTF by envelope functions, which result in a cutoff of the contrast transfer at high spatial frequencies. The spatial frequency at which the partially coherent CTF falls below a threshold defines the information limit g_{max} of the microscope, which is usually the spatial frequency at which the combined dampening envelope function drops by $1/e^2$. However, for an uncorrected field emission gun (FEG) TEM, the information limit g_{max} can be higher than the spatial frequency g_S corresponding to the point resolution, leading to strong oscillations in the partially coherent CTF and a decrease in the interpretability of images.

Figure 21.2 shows examples of the function $\sin \chi$ for a 200 kV TEM for two different values of the spherical aberration coefficient C_S. In an uncorrected electron microscope, at higher spatial frequencies the CTF oscillates rapidly up to the information limit. If C_S is reduced in a microscope that is equipped with a hardware spherical aberration corrector, then a broad transfer band extends up to the information limit of 0.125 nm, improving the point resolution significantly.

21.2.2 Delocalization Due to Spherical Aberration

As well as affecting interpretability, the oscillations of the transfer function also lead to the delocalization of information [3, 4], i.e. to the lateral displacement of high resolution image detail by a distance that depends on spatial frequency, defocus and C_S. For a homogeneous crystal with uniform thickness and translational symmetry, delocalization effects are not readily apparent. However, a discontinuity, such as an edge, locally destroys the symmetry.

Figure 21.3 shows that lattice fringes can be displaced significantly with respect to the true position of a particle. Delocalization effects are minimized in one image, despite the fact that C_S is not zero, because a single spatial frequency dominates the contrast. When a general sample is imaged in an uncorrected TEM, the effects of delocalization can usually only be minimized for a single spatial frequency at each defocus value.

If a hardware aberration corrector is not available, there is another method to overcome the effect of the strongly oscillating part of the CTF between the point

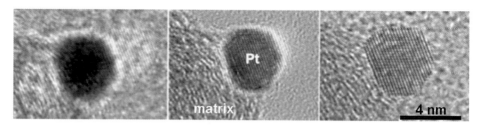

Figure 21.3. High resolution TEM images of a 5 nm Pt particle supported on graphitic C and imaged out of focus, recorded at 200 kV using a TEM that has a spherical aberration coefficient of 0.5 mm. The defocus step between successive images is 100 nm. Adapted from Ref. 4. © IOP Publishing. Reproduced with permission. All rights reserved.

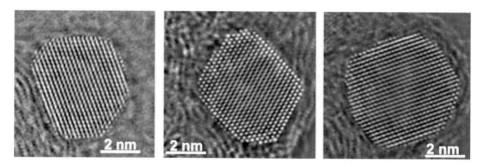

Figure 21.4. Phase images of three 5–6 nm Pt nanoparticles supported on graphitic C, obtained by applying exit wavefunction restoration to defocus series of high resolution TEM images. Adapted from Ref. 8. Reproduced with permission.

resolution and the information limit. A defocus series of typically 10–20 images is recorded from one region of the sample and software is used to numerically reconstruct the high spatial frequency components of the phase of the exit plane wavefunction [5–7]. Examples of the resulting phase images are shown in Figure 21.4.

Other similar techniques make use of multiple images recorded at different CTF settings to reconstruct the high frequency parts of the complex electron wavefunction and fill gaps in the frequency spectrum of a single image. All techniques require the sample not to change between successive images. Therefore, unless the microscope defocus can be changed rapidly and images can be acquired at a very fast frame rate, the application of exit plane wavefunction restoration to studies of materials imaged in liquid cells, in particular for real-time observations of dynamic processes, is expected to be challenging.

21.2.3 Correcting or Tuning C_S in TEM

The ability to tune the spherical aberration coefficient of the objective lens using a hardware spherical aberration corrector provides much more interpretable individual

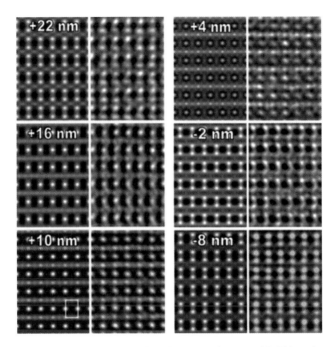

Figure 21.5. Comparison of an experimental defocus series of HRTEM images of SrTiO$_3$ oriented along [110] with simulated images. The experimental images were acquired using a C_S-corrected CM200ST FEG microscope with the value of C_S adjusted to −0.04 mm. Simulated images are shown on the left of each column, while experimental images are shown on the right. The defocus values of the simulated images are indicated. The thickness used in the simulations was 3.5 nm. Half of a SrTiO$_3$ cell projected along [110] is indicated by the white frame in the image simulated for a defocus of +10 nm. Unpublished results.

images. It is preferable to software reconstruction of the exit plane wavefunction from image series, particularly for beam-sensitive materials and for structures, such as those in liquids, that change rapidly over time.

The first truly successful demonstration of hardware spherical aberration correction in the TEM using multipole lenses was achieved in the late 1990s [9, 10]. Modern C_S-corrected transmission electron microscopes are now installed in hundreds of laboratories worldwide and are compatible with liquid cell TEM sample holders. By tuning the spherical aberration coefficient and other higher order aberrations in such instruments, it is possible to achieve optimum contrast transfer and a dramatic improvement in spatial resolution, visibility, and interpretability of individual images.

Figure 21.5 shows a comparison between an experimental defocus series of images of [110] SrTiO$_3$ recorded using a C_S-corrected Philips CM200ST FEG microscope and simulated images. A structure image predicted by simulations for a defocus of +10 nm is reproduced experimentally. Significantly, Figure 21.5 illustrates the fact that the recorded image intensity still depends sensitively on defocus (as well as on other aberrations). Comparisons of experimental images with computer simulations of image contrast are therefore usually still required for quantitative interpretation of C_S-corrected images.

The ability to tune the spherical aberration coefficient of the objective lens of a TEM using hardware provides a powerful method for recording interpretable images directly. In particular, the choice of a small negative value for C_S has been shown to result in strong contrast from columns of light atoms, which then appear bright [11]. In this so-called "negative C_S imaging" (NCSI) mode, paraxial rays that have travelled through the objective lens come to a focus ahead of outer rays. Columns of light atoms, such as O columns located adjacent to columns of heavy atoms in $Pb(Zr_{0.2}Ti_{0.8})O_3$, can then be resolved with improved contrast [12]. The ability to balance different symmetrical round aberrations in this way can be used to improve the precision of measurements of elemental composition and atom positions [13]. For a thin sample, the strong image contrast that is provided by the NCSI technique is highly sensitive to the scattering density accumulated along each atomic column. As a result, when applied in combination with image simulations, it is a valuable technique for measuring local variations in the occupancy of columns of atoms that have low nuclear charge, such as O in perovskites, as well as their positions with sub-5-pm precision [14]. A further major advantage of the NCSI technique for *in situ* experiments is its dose efficiency, since the reduced point spread and delocalization result in an optimal signal-to-noise ratio.

The significant improvement that hardware C_S correction provides in the ability to interpret recorded HRTEM image intensities quantitatively is illustrated in Figure 21.6, which shows a recent example of the experimental application of C_S-corrected imaging to recover the three-dimensional morphology and thickness of part of an MgO crystal with atomic precision from a single image recorded under NCSI conditions [15].

In the study illustrated in Figure 21.6, precise measurements of the imaging parameters of the microscope, in combination with prior knowledge of the crystal structure of the material, allowed quantitative comparisons of the recorded image intensity with computer simulations to be performed on an absolute scale. Three-dimensional information can be recovered from just one image because electron scattering is intrinsically a three-dimensional process. A statistical confidence check was used to establish the uniqueness of the best-fitting number of atoms in each column and its position along the optical axis of the microscope relative to neighboring atomic columns, thereby allowing the surface morphology of the crystal to be determined. In future studies, it may be possible to use a similar single-image approach to follow dynamic changes in the three-dimensional morphologies of individual nanoparticles in the presence of external stimuli such as high temperature or the presence of a reactive environment in the TEM.

21.2.4 Higher-Order Aberrations and Lifetime of the Corrected State

When applying hardware spherical aberration correction, precise control and correction of residual higher-order aberrations is essential to take full advantage of the improved spatial resolution for quantitative measurements. The complete range of coherent wave aberrations can be represented by a series expansion expressed as a function of either scattering angle or reciprocal distance. Sub-angstrom resolution imaging requires

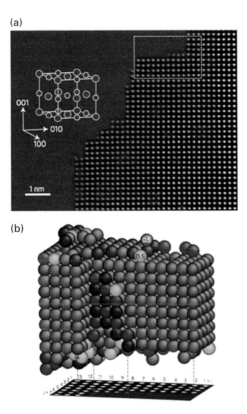

Figure 21.6. (a) C_S-corrected high resolution TEM image of an MgO crystal taken along the [001] direction under NCSI conditions. Side terraces are visible at the boundary between vacuum (left) and crystal (right). A perspective view of an MgO unit cell is shown on the left. (b) Atomically resolved view of the best-fitting three-dimensional atomic arrangement of Mg and O atoms in the region indicated in (a), based on a quantitative comparison between experimental and simulated image intensity across the region of interest. Red spheres indicate fully occupied Mg sites and blue spheres fully occupied O sites. In the surface layers, brown spheres indicate formally half-occupied Mg sites, while cyan spheres indicate formally half-occupied O sites. One possible explanation for the presence of formally half-occupied sites is the movement of surface atoms during acquisition of the image. Reproduced from Ref. 15 with permission.

both a knowledge of all significant wave aberrations up to sixth order in g and the stability of these aberrations over the time needed to acquire one or more images [16].

The decomposition of a typical aberration function into its components is illustrated in Figure 21.7, while Figure 21.8 shows an example of the time stability of the objective lens defocus measured for two different electron microscopes. After optimal alignment of a microscope, the typical lifetime of an optical state that preserves image detail up to the information limit is in most cases no longer than a few minutes [17].

The lifetime of the optical state of a microscope is, to a large extent, influenced by the thermal and environmental stability of the power supplies and the microscope lenses.

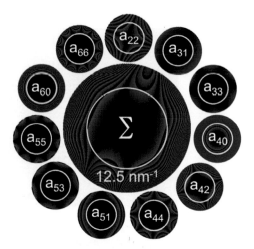

Figure 21.7. Representative aberration function (center) and its decomposition into series terms. The first index of each coefficient describes the power in g, while the second index describes the rotational symmetry. Positive values of the aberration function are depicted in red, negative values in blue. Sawtooth jumps occur at intervals of $\pi/4$. The yellow circle marks a value of $g = 12.5$ nm^{-1}, corresponding to a spatial resolution of 0.08 nm. Aberration-free imaging is achieved when the modulus of the total aberration function does not exceed $\pi/4$ within the circle that defines the resolution limit.

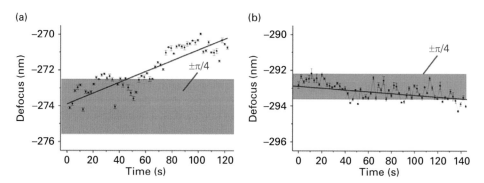

Figure 21.8. Measured time variation of the objective lens defocus over a period of approximately 2 minutes for (a) a 200 kV Philips CM200 TEM with a resolution of 0.12 nm and (b) a 300 kV FEI Titan TEM with a resolution of 0.08 nm. Both thermally induced long-term drift and short-term fluctuations caused by instabilities of the accelerating voltage and the objective lens current are observed. Unpublished data.

It can be optimized but not eliminated with today's electron optical lens designs. Therefore, continuous aberration control will also be required in future. The current state of the art is a lifetime of a few minutes at the information limit and 10–20 minutes at 1 Å resolution. For liquid cell experiments, the lifetime is likely to be much longer, since the attainable resolution, which is limited by the depth of field and multiple scattering in the sample, is worse than the optical resolution of the instrument, and since the lifetime scales approximately quadratically with resolution.

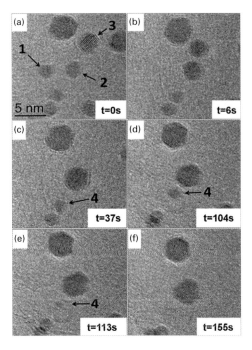

Figure 21.9. Frames taken from a sequence of HRTEM images of the sintering of a Au nanoparticle catalyst supported on BN and imaged in 130 Pa of H_2 at 410 °C at the indicated times in an environmental TEM. The images were recorded at 300 kV with the spherical aberration coefficient of the objective lens set to below 5 μm. The numbers refer to particles that undergo different sintering mechanisms. Between (a) and (b), particles 1 and 2 coalesce. Between (b) and (c), particle 3 coalesces. In contrast, particle 4 becomes smaller between (c) and (f), consistent with Ostwald ripening. Adapted from Ref. 18. Reproduced with permission.

The significant benefit of using C_S-corrected imaging to study dynamic processes in materials in real time is illustrated in Figure 21.9, in the form of a time sequence of images recorded during a gas phase sintering experiment [18]. Two primary sintering mechanisms have been proposed in such reactions, one based on the migration of particles over the support and coalescence when they are in close proximity, and the other based on Ostwald ripening, which involves mass transport from smaller to larger particles by the diffusion of atoms or molecular species. Figure 21.9 shows that both mechanisms can take place in closely adjacent regions of the same sample examined under the same conditions. Discerning such mechanistic details at the nanoparticle surfaces would have been much more difficult in an uncorrected electron microscope. Similar considerations are expected to apply to processes in liquids, such as the oriented attachment and growth experiments described in Chapters 9 and 15.

These examples serve to illustrate the significant benefit of spherical aberration correction of the TEM objective lens for atomic-resolution liquid cell studies, for which the acquisition of multiple images to improve spatial resolution or signal to noise is difficult even when using fast electron detectors. The spatial resolution of an image of a sample in liquid can be affected by a number of factors, including scattering from both the liquid and the enclosing membranes. Nevertheless, spherical aberration correction provides significant benefits for atomic-resolution imaging, not only in terms of spatial resolution but also as a result of reduced delocalization and improved visibility and interpretability of the recorded contrast. The fact that this information can be obtained from individual recorded images, rather than using an approach such as focal series restoration and subsequent numerical processing, is especially important for real-time studies of surface and interface structures that are changing during observation. Conversely, the disadvantage of hardware spherical aberration correction when compared to focal series restoration is that the full complex electron wavefunction is not recovered and so no post-processing is possible to refocus images that occur at different depths in the liquid.

21.3 Spherical Aberration Correction in STEM

When a scanned image of a sample is recorded in a STEM using either an on-axis bright field detector or a high angle annular dark field detector, the spatial resolution is related primarily to the diameter of the focused electron probe, which is in turn determined by parameters that include the diameter of the condenser aperture and the aberrations of the condenser lens system of the microscope. Correction of the spherical aberration of the condenser lenses is therefore crucial to obtain the highest spatial resolution.

An important advantage of the HAADF image formation process is that it can be regarded as incoherent if the inner semi-angle of the detector is much larger (typically at least 100 mrad) than the semi-angle of the condenser aperture that defines the convergence angle of the illumination (typically ~20 mrad). For a sufficiently thin sample, in which the effects of beam broadening and dynamic diffraction are negligible, the recorded intensity is approximately the convolution of the square of the probe function with the square of the sample transmission function. To a good approximation, it is then proportional to the atomic number raised to the power 1.7.

A further advantage of HAADF imaging is that electron energy loss spectroscopy can be performed simultaneously, permitting the local measurement of both low (valence) energy losses (i.e. plasmon excitations, inter- and intraband transitions) and higher energy losses that are associated with inner shell ionization of atoms within the material. As the total signal in an ionization edge is proportional to the number of excited atoms, EELS can be used for quantification of the local chemical composition, typically after subtraction of the background in the spectrum acquired at each probe position using a power law expression of the form:

$$I(\Delta E, A, r) = A \, \Delta E^{-r}, \tag{21.8}$$

where ΔE is the energy loss and A and r are fitting parameters [19].

Spherical and Chromatic Aberration Correction

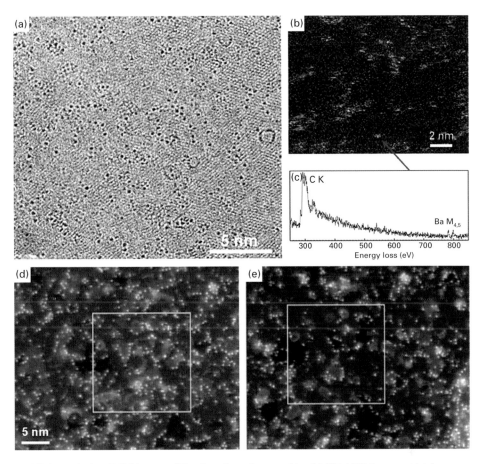

Figure 21.10. (a) High resolution TEM image of Ba-doped graphene oxide. (b) HAADF image and (c) corresponding EELS signal recorded from the indicated area during the acquisition of a spectrum image from the entire field of view. (d) and (e) HAADF images recorded before and after acquiring a spectrum image from the indicated area, showing strong bright contrast from individual Ba atoms, but also the effects of beam-induced damage both within and also to a lesser extent outside the scanned area. Adapted from Ref. 20. Reproduced with permission.

In order to illustrate the care that is required when applying aberration-corrected STEM HAADF imaging and EELS to a beam-sensitive material, as would be expected in a liquid cell experiment, Figure 21.10 shows a comparison of high resolution TEM and STEM images recorded from graphene oxide, a form of graphene whose surface is modified by the addition of functional groups [20]. This material is both sensitive to damage by the primary electron beam and highly prone to contamination during STEM imaging. As the functional groups show little contrast in HRTEM and STEM images, they were doped with Ba in order to reveal their positions. This sample required examination at low accelerating voltage (80 kV) to minimize beam damage and at elevated temperature (800 °C) to minimize contamination during imaging. Figure 21.10a shows an HRTEM image of the Ba-doped graphene oxide sample recorded at 80 kV in a C_S/C_C-corrected TEM. In this image, the individual black dots are thought to correspond to Ba atoms, whereas in HAADF images (Figures 21.10d and e) they are imaged as bright dots with high contrast. Unfortunately, for such a sample it was found that C_S/C_C-corrected EFTEM could not be used to form elemental maps using the Ba M edge at 781eV as a result of radiation damage to the sample over the required exposure time. Instead, spectrum images acquired using EELS (i.e. background-subtracted elemental maps formed from spectra acquired point-by-point as the electron beam was scanned across the specimen) were used to identify individual Ba atoms (Figure 21.10c). A significant advantage of spectrum imaging is that the recorded signal can be correlated directly with the HAADF image intensity (Figure 21.10b). However, Figures 21.10d and e illustrate the fact that the sample is still damaged and drifts during the acquisition of each image (over ~10 mins) and that great care is required both with experimental design for studies of electron beam sensitive materials using STEM imaging and spectroscopy and with subsequent image interpretation.

21.4 Chromatic Aberration Correction in the TEM

Chromatic aberration (C_C) correction is only available on a small number of instruments worldwide at the time of writing. It is generally pursued because of the improvement in resolution limit that it provides. However, for liquid cell microscopy additional benefits of C_C correction become important. These include atomic resolution analytical EFTEM, the opportunity for more space in the sample area for *in situ* experiments, high contrast at low voltages for unstained biological materials and high resolution field-free Lorentz microscopy. In this section we discuss the opportunities that C_C correction can provide in the context of liquid cell TEM.

21.4.1 Image Resolution

In Section 21.2 we showed that the instrumental resolution of a TEM improves appreciably as a result of coherent aberration correction. C_S correction improves the spatial resolution of an uncorrected microscope from ~100 λ to ~40 λ. In order to improve the information limit further, it is also important to minimize incoherent disturbances that affect image resolution. C_C correction promises to provide a further improvement in spatial resolution to better than ~20 λ and is especially effective at lower accelerating voltages [21–24].

The primary characteristic of a C_C-corrected TEM is that the defocus f of an image remains almost constant when the electron energy varies, either due to the energy spread of the electron gun or due to the inelastic scattering of electrons within the sample. The axial chromatic change in defocus Δf associated with an energy change ΔE is given by the expression:

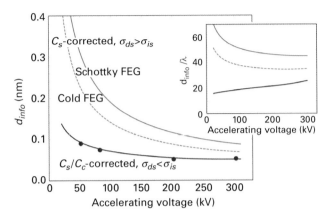

Figure 21.11. Information limit of a C_S-corrected TEM and a C_S/C_C-corrected TEM equipped with a Rose-Haider achroplanator, plotted as a function of kV. The defocus spread σ_{ds} resulting from the energy spread of the electron source determines the information limit of the C_S-corrected instrument. The cold field emitter has a better resolution than a Schottky field emitter because of a smaller energy spread. In the C_S/C_C-corrected case, the image spread σ_{is} originating from fluctuating lateral image shifts damps the high frequency transfer more strongly than the residual defocus spread due to lens instabilities. The solid curve is based on best parameter estimates between 80 kV and 300 kV and was extrapolated to lower voltages by using a linear scaling of σ_{is} with wavelength. Solid circles represent the experimental information limit measured on the PICO instrument in Jülich. The inset shows the information limit in units of wavelength. Unpublished results.

$$\Delta f(\Delta E) = C_C \frac{\Delta E}{E_0}, \qquad (21.9)$$

where C_C is the chromatic aberration coefficient of the objective lens and E_0 is the energy of the primary electron beam. The greatest benefit of C_C correction is at low accelerating voltages of a few tens of kV, at which the relative energy spread is much greater than at higher kV; C_C correction is a prerequisite for atomic resolution imaging at low kV. C_C correction is most important for atomic resolution studies of materials for which low kV provides a reduction in beam-induced damage. It is potentially of great interest for liquid cell experiments, especially for studies of thick liquid layers in which inelastic scattering would normally contribute significantly to blurring of image detail.

Figure 21.11 compares, as a function of kV, the information limit of a representative C_S-corrected TEM with that of the C_S/C_C-corrected PICO TEM in Jülich, which is equipped with a Rose-Haider quadrupole-octupole achroplanator [25–27]. After the chromatic focus spread associated with the energy spread of the electron gun has been reduced in an intermediate voltage electron microscope equipped with a C_C corrector, the most significant remaining influence on instrumental resolution is associated with parasitic noise-related factors that give rise to rapid time-dependent image displacements [28]. While mechanical instabilities, lens instabilities, and time-varying stray fields may contribute to the image spread, a more fundamental limitation is magnetic field noise from thermally driven currents in the conductive parts of the instrument that cause stochastic beam deflections [29].

Within the framework of validity of linear contrast transfer theory, the influence of defocus spread and image spread on the information limit can be described in terms of a spectral damping envelope that drops by $1/e^2$ at

$$\frac{1}{d_{\text{info}}} = \frac{\sigma_{\text{is}}}{\lambda \sigma_{\text{ds}}} \sqrt{2\left[\sqrt{1 + \left(\frac{\lambda \sigma_{\text{ds}}}{\pi \sigma_{\text{is}}^2}\right)^2} - 1\right]}, \qquad (21.10)$$

where d_{info} is the information limit, σ_{ds} is the root-mean-square defocus spread, and σ_{is} is the root-mean-square image spread [24]. C_C correction reduces the defocus spread of the PICO TEM in Jülich by more than a factor of 10, leaving image spread as the most significant contribution to the information limit. Resolutions of 50 pm at 200 kV and 90 pm at 50 kV have been achieved after reduction of defocus spread, with a remaining major limitation due to image spread.

21.4.2 Energy-Filtered TEM

In addition to the improved information limit and contrast transfer for fine image detail described above, C_C correction improves resolution when images contain a significant contribution from inelastic scattering over a large energy range. This occurs in energy-filtered TEM [23, 30, 31] or because of substantial inelastic scattering through thicker objects such as liquid cells [32].

EFTEM is described in detail in Chapter 20. When acquiring energy-filtered images, C_C correction allows large energy windows and large objective aperture sizes to be used without compromising spatial resolution significantly. The immediate advantage of a large energy window is that imaging is more dose efficient when substantial inelastic scattering is present. C_C correction is therefore expected to be highly beneficial for unstained biological samples, for tilt-series tomography when the projected sample thickness at high tilt angles exceeds the inelastic mean free path, and for liquid cell experiments in which the window material and thick liquid layer cause inelastic scattering. Furthermore, the improved signal-to-noise ratio facilitates the recording of EFTEM chemical maps on the atomic scale. In each of these examples, the energy-selecting window can be extended to tens of eV in width without compromising spatial resolution.

Figure 21.12 shows the instrumental resolution for EFTEM for both a C_S-corrected microscope and a C_S/C_C-corrected microscope plotted as a function of collection semi-angle. In the C_S-corrected microscope, the instrumental resolution is limited by chromatic aberration and the diffraction limit, or, at large collection angles, by residual aberrations. Inelastically scattered electrons that fill the collection angle β homogeneously cause a chromatic broadening of

$$r_C = C_C \frac{\Delta E}{E_0} \beta, \qquad (21.11)$$

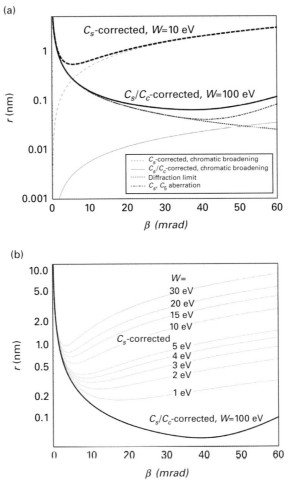

Figure 21.12. (a) EFTEM spatial resolution r plotted as a function of collection semi-angle β for a C_S-corrected instrument and a C_S/C_C-corrected TEM with field emission gun at 200 kV. The thick dashed line shows the total resolution of the C_S-corrected microscope for an energy-selecting slit width of 10 eV, whereas the thick solid line shows the total resolution of the C_S/C_C-corrected microscope for an energy-selecting slit width of 100 eV. The thin blue lines show the contribution of chromatic broadening, the grey dotted line the diffraction limit, and the red dot-dashed line the residual delocalization due to spherical aberration. In the example shown, a non-zero C_S is used to balance a higher-order spherical aberration C_5 of -4 mm for optimum phase contrast, with the reciprocal information limit β/λ. The total resolution is the sum of the residual delocalization and the chromatic broadening. (b) Total resolution plotted as a function of the width of the energy-selecting slit β. Unpublished results.

which limits the EFTEM spatial resolution in the C_S-corrected microscope to worse than a few angstroms even when the energy-selecting slit width and collection semi-angle are small. C_C correction reduces r_C by more than two orders of magnitude and provides sub-angstrom resolution when using the largest slit width and largest collection angle, as required to achieve a good signal-to-noise ratio when examining thick

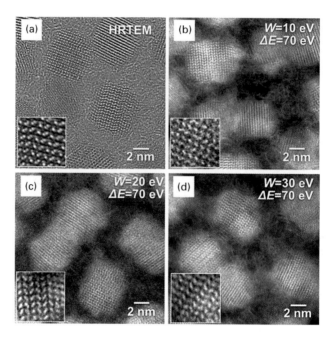

Figure 21.13. (a) High resolution and (b–d) EFTEM images of CdSe nanoparticles on a C support recorded at 80 kV in the C_S/C_C-corrected PICO instrument in Jülich. The EFTEM images were acquired using energy-selecting slit widths of 10, 20, and 30 eV with an exposure time of 20 s at an energy loss of 70 eV. They include contributions from plasmon losses and the Se M_{45} core loss. No objective aperture was used. Unpublished results.

materials or for high resolution EFTEM images, especially when the maximum tolerable dose is limited (Chapter 8).

Figure 21.13 shows that the spatial resolution in EFTEM images of CdSe nanoparticles is barely affected by the choice of energy-selecting slit width in a C_S/C_C-corrected instrument. The Cd–Se dumbbells with a spacing of 0.157 nm are resolved in each EFTEM image. It should, however, be noted that the preservation of elastic contrast can dominate atomic-resolution detail in such images, resulting in the need for comparisons with quantum mechanical image simulations for unambiguous interpretation [30, 31].

21.4.3 Thick Samples

When examining a thick sample in a C_S-corrected microscope using energy-filtered bright field TEM, the spatial resolution and depth of focus are usually optimized by reducing the chromatic broadening. This involves selecting an optimal objective aperture size, centering a narrow energy-selecting slit on the most probable energy loss that the primary electrons have suffered in the sample, and then refocusing the image to the defocus that corresponds to this most probable energy loss [33]. Figure 21.14 shows an example as applied to whole mount macrophage cells [32]. In contrast, in a C_S/C_C-corrected microscope, all electron energies and angles are in focus.

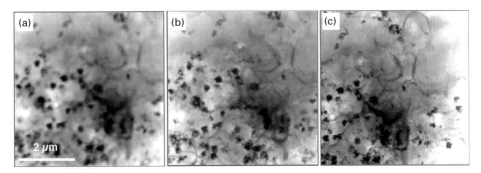

Figure 21.14. Bright field TEM images of a thick sample of a whole mount macrophage cell. (a) Energy-filtered C_S-corrected only. (b) C_C/C_S-corrected. (c) As in (a), but refocused to the most probable energy loss; (b) provides similar resolution to (c) but without excluding electrons from image formation and without the need to adjust the focus to a certain energy loss. Adapted in part from Ref. 32. Reproduced with permission.

Since inelastically scattered electrons typically lose energy without changing their direction significantly, we expect that C_C correction will allow a much larger fraction of incident electrons to be used to record high spatial resolution images of thick samples, compared to C_S correction and energy filtering alone. A further important advantage of C_C correction is that regions of the sample that have different thicknesses can all be imaged at an optimal defocus without the need to refocus to the most probable energy loss in the same field of view [32].

When a thick sample is imaged in TEM or STEM, plural elastic scattering, chromatic broadening, and defocus broadening all contribute to the loss of resolution [34]. As described in Chapter 8, in STEM, beam broadening by elastic scattering dominates and results in a distinct top–bottom effect, where objects at the top of the sample are resolved with the resolution of the unbroadened probe and those at the bottom with the resolution of the broadened probe [35–37]. In TEM, the chromatic error blurs the image detail of objects at any depth in the sample, in combination with multiple elastic scattering and the defocus error. The chromatic error is of minor importance in a C_S/C_C-corrected microscope, unlike in a conventional or C_S-corrected microscope. For the sample shown in Figure 21.14, the resolution was measured to vary between 2.5 nm and 4.5 nm between nanoparticles located at the bottom and the top of the sample, respectively, in a region of the sample of thickness 2.09 μm [32].

On the basis of reciprocity, the resolution loss due to beam divergence in STEM [38] is equivalent to depth of field and hence the effect of defocus in a C_S/C_C-corrected TEM. As chromatic broadening is no longer the limiting factor for the resolution in a C_S/C_C-corrected TEM image of a thick sample, a wide energy range of inelastically scattered electrons can be exploited. C_C correction therefore promises to be particularly beneficial for dose-optimized studies of beam-sensitive materials in liquid cells, especially when using lower accelerating voltages to attain higher contrast.

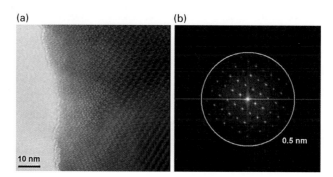

Figure 21.15. (a) C_C/C_S-corrected image of $Cs_{0.5}[Nb_{2.5}W_{2.5}O_{14}]$ [001] acquired at 300 kV in Lorentz mode with the objective lens turned off. (b) Fourier transform revealing 0.5 nm image detail. Unpublished results.

21.4.4 Wide Polepiece Gaps and Magnetic-Field-Free Imaging

A final important benefit of C_C correction results from the fact that combined C_S/C_C correction can be used to improve image resolution when using either a wide objective lens polepiece gap or a Lorentz lens. C_C correction of the Lorentz lens of a TEM allows ferromagnetic materials to be imaged with the conventional TEM objective lens switched off, i.e. in magnetic-field-free conditions, with a spatial resolution of better than 0.5 nm. Figure 21.15 shows a C_S/C_C-corrected Lorentz TEM image of $Cs_{0.5}[Nb_{2.5}W_{2.5}O_{14}]$ [001] acquired at 300 kV using the Titan PICO microscope in Jülich operated in Lorentz mode (with the objective lens turned off). The image contains 0.5 nm detail, suggesting that it may be possible to image magnetic fields in selected materials with close to atomic spatial resolution, as well as to acquire atomic-resolution images of ferromagnetic materials in liquids in magnetic-field-free conditions. For microscopes that have wide polepiece gaps, C_C correction promises to retain atomic spatial resolution while providing additional space for sophisticated holders, detectors, and other components around the sample.

21.5 Conclusions

In this chapter, we have discussed the benefits of C_S correction in TEM and STEM for liquid cell studies. These benefits include improved spatial resolution, visibility, and interpretability of images, as well as reduced delocalization of image contrast and smaller depth of field. We then described the potential benefits of combined C_S/C_C correction of the objective lens of the microscope. These benefits include an improvement in spatial resolution compared to the use of C_S correction alone, especially at low accelerating voltages, as well as the ability to study thick materials, to acquire atomic-resolution EFTEM images using large energy-selecting windows, and to improve spatial resolution in wide-polepiece-gap microscopes or microscopes that do not have an objective lens, such as those used for studies of magnetic materials.

C_S correction has been used in very few liquid cell experiments at the time of writing. However, its potential benefits are clear, especially in combination with the use of direct electron detectors and imaging energy filters. The future use of combined C_S/C_C correction also promises to allow new categories of experiments, involving thicker samples and the use of lower accelerating voltages for the examination of beam-sensitive materials at high spatial resolution.

Acknowledgements

We are grateful to Martina Luysberg, Juri Barthel, Andreas Thust, Knut Urban, Shaobo Mi, Chris Boothroyd, András Kovács, Joachim Mayer, Les Allen, Ben Forbes, Jörg Jinschek, Niels de Jonge, Jean-Pierre Baudoin, Lionel Cervera Gontard, Dogan Ozkaya, Thomas Hansen, and Maya Bar Sadan for ongoing discussions and contributions to this work.

References

1. O. Scherzer, Über einige Fehler von Elektronenlinsen. *Z. Phys.*, **101** (1936), 593–603.
2. O. Scherzer, The theoretical resolution limit of the electron microscope. *J. Appl. Phys.*, **20** (1949), 20–29.
3. W. Coene and A. J. Jansen, Image delocalisation and high resolution tranmission electron microscopic imaging with a field emission gun. *Scanning Microsc. Suppl.*, **6** (1992), 379–403.
4. L. Cervera Gontard, R. E. Dunin-Borkowski, M. J. Hÿtch and D. Ozkaya, Delocalisation in images of Pt nanoparticles. *J. Phys. Conf. Ser.*, **26** (2006), 292–295.
5. W. M. J. Coene, A. Thust, M. Op de Beeck and D. van Dyck, Maximum-likelihood method for focus-variation image reconstruction in high resolution transmission electron microscopy. *Ultramicroscopy*, **64** (1996), 109–135.
6. A. Thust, W. M. J. Coene, M. Op de Beeck and D. van Dyck, Focal-series reconstruction in HRTEM: simulation studies on nonperiodic objects. *Ultramicroscopy*, **64** (1996), 211–230.
7. C. Kisielowski, C. J. D. Hetherington, Y. C. Wang *et al.*, Imaging columns of the light elements carbon, nitrogen and oxygen with sub angstrom resolution. *Ultramicroscopy*, **89** (2001), 243–263.
8. L. Cervera Gontard, L.-Y. Chang, C. J. D. Hetherington *et al.*, Aberration-corrected imaging of active sites on industrial catalyst nanoparticles. *Angew. Chem.*, **46** (2007), 3683–3685.
9. M. Haider, H. Rose, S. Uhlemann *et al.*, A spherical-aberration-corrected 200 kV transmission electron microscope. *Ultramicroscopy*, **75** (1998), 53–60.
10. M. Lentzen, B. Jahnen, C. L. Jia *et al.*, High-resolution imaging with an aberration-corrected transmission electron microscope. *Ultramicroscopy*, **92** (2002), 233–242.
11. C. L. Jia, M. Lentzen and K. Urban, Atomic-resolution imaging of oxygen in perovskite ceramics. *Science*, **299** (2003), 870–873.
12. C. L. Jia, S. B. Mi, K. Urban *et al.*, Atomic-scale study of electric dipoles near charged and uncharged domain walls in ferroelectric films. *Nat. Mater.*, **7** (2008), 57–61.

13. C. L. Jia, L. Houben, A. Thust and J. Barthel, On the benefit of the negative-spherical-aberration imaging technique for quantitative HRTEM. *Ultramicroscopy*, **110** (2010), 500–505.
14. C. L. Jia, J. Barthel, F. Gunkel *et al.*, Atomic-scale measurement of structure and chemistry of a single-unit-cell layer of $LaAlO_3$ embedded in $SrTiO_3$. *Microsc. Microanal.*, **19** (2013), 310–318.
15. C. L. Jia, S.-B. Mi, J. Barthel *et al.*, Determination of the 3D shape of a nanoscale crystal with atomic resolution from a single image. *Nat. Mater.*, **13** (2014), 1044–1049.
16. J. Barthel and A. Thust, Aberration measurement in HRTEM: implementation and diagnostic use of numerical procedures for the highly precise recognition of diffractogram patterns. *Ultramicroscopy*, **111** (2010), 27–46.
17. J. Barthel and A. Thust, On the optical stability of high-resolution transmission electron microscopes. *Ultramicroscopy*, **134** (2013), 6–17.
18. T. W. Hansen, J. B. Wagner and R. E. Dunin-Borkowski, Aberration corrected and monochromated environmental transmission electron microscopy: challenges and prospects for materials science. *Mater. Sci. Technol.*, **26** (2010), 1338–1344.
19. R. F. Egerton, *Electron Energy-Loss Spectroscopy in the Electron Microscope* (New York: Springer, 2011).
20. C. B. Boothroyd, M. S. Moreno, M. Duchamp *et al.*, Atomic resolution imaging and spectroscopy of barium atoms and functional groups on graphene oxide. *Ultramicroscopy*, **145** (2014), 66–73.
21. J. Zach, Chromatic correction: a revolution in electron microscopy? *Phil. Trans. R. Soc. A*, **367** (2009), 3699–3707.
22. H. Rose, Future trends in aberration corrected electron microscopy. *Phil. Trans. R. Soc. A*, **367** (2009), 3809–3823.
23. B. Kabius, P. Hartel, M. Haider *et al.*, First application of C_C-corrected imaging for high-resolution and energy-filtered TEM. *J. Electron Microsc.*, **58** (2009), 147–155.
24. R. Leary and R. Brydson, Chromatic aberration correction: the next step in electron microscopy. *Adv. Imagi. Electron Phys.*, **165** (2011), 73–130.
25. M. Haider, P. Hartel, H. Müller, S. Uhlemann and J. Zach, Information transfer in a TEM corrected for spherical and chromatic aberration. *Microsc. Microanal.*, **16** (2010), 393–408.
26. H. Rose, Outline of an ultracorrector compensating for all primary chromatic and geometrical aberrations of charged-particle lenses. *Nucl. Instrum. Methods Phys. Res. A*, **519** (2004), 12–27.
27. H. Rose, Prospects for aberration-free electron microscopy. *Ultramicroscopy*, **103** (2005), 1–6.
28. M. Haider, H. Müller, S. Uhlemann *et al.*, Prerequisites for a Cc/Cs-corrected ultrahigh-resolution TEM. *Ultramicroscopy*, **108** (2008), 167–178.
29. S. Uhlemann, H. Müller, P. Hartel, J. Zach and M. Haider, Thermal magnetic field noise limits resolution in transmission electron microscopy. *Phys. Rev. Lett.*, **111** (2013), 046101.
30. K. W. Urban, J. Mayer, J. R. Jinschek *et al.*, Achromatic elemental mapping beyond the nanoscale in the transmission electron microscope. *Phys. Rev. Lett.*, **110** (2013), 185507.
31. B. D. Forbes, L. Houben, J. Mayer, R. E. Dunin-Borkowski and L. J. Allen, Elemental mapping in achromatic atomic-resolution energy-filtered transmission electron microscopy. *Ultramicroscopy*, **147** (2014), 98–105.

32. J. P. Baudoin, J. R. Jinschek, C. B. Boothroyd, R. E. Dunin-Borkowski and N. de Jonge, Chromatic aberration-corrected tilt series transmission electron microscopy of nanoparticles in a whole mount macrophage cell. *Microsc. Microanal.*, **19** (2013), 814–821.
33. L. Reimer and M. Ross-Messemer, Top–bottom effect in energy-selecting TEM. *Ultramicroscopy*, **21** (1987), 385–388.
34. L. Reimer and P. Gentsch, Superposition of chromatic error and beam broadening in TEM of thick carbon and organic specimens. *Ultramicroscopy*, **1** (1975), 1–5.
35. P. Gentsch, H. Gilde and L. Reimer, Measurement of the top–bottom effect in scanning transmission electron microscopy of thick amorphous specimens. *J. Microsc.*, **100** (1974), 81–92.
36. A. A. Sousa, M. F. Hohmann-Marriott, G. Zhang and R. D. Leapman, Monte Carlo electron-trajectory simulations in bright-field and dark-field STEM: implications for tomography of thick biological sections. *Ultramicroscopy*, **109** (2009), 213–221.
37. H. Demers, R. Ramachandra, D. Drouin and N. de Jonge, The probe profile and lateral resolution of scanning transmission electron microscopy of thick specimens. *Microsc. Microanal.*, **18** (2012), 582 590.
38. J. K. Hyun, P. Ercius and D. A. Muller, Beam spreading and spatial resolution in thick organic specimens. *Ultramicroscopy*, **109** (2008), 1–7.

22 The Potential for Imaging Dynamic Processes in Liquids with High Temporal Resolution

Nigel D. Browning and James E. Evans

22.1 Introduction

The last few years have seen a dramatic increase in the number of experimental studies in the scanning/transmission electron microscope that use liquid cells and holders, as described elsewhere in this book [1–11]. In all of these experiments, the spatial resolution that can be obtained in the images is limited either by the thickness of the liquid or by the motion of the structures being imaged. The limitation caused by thickness can perhaps be overcome by designing windows with reduced bowing (Chapter 2). In that case, motion blur becomes the main intrinsic limitation to spatial resolution and image interpretation. Overcoming motion caused by growth, mobility under external driving forces or Brownian motion all require the ability to image quickly. The more quickly the sample is imaged, the less the movement will be between and during each frame, and the higher the intrinsic spatial resolution will be in each frame.

Ideally, we would like to increase the speed of observation so that we can see individual atoms move. Using current technologies, the temporal resolution of images is limited by the beam current of the instrument (low beam current limits speed, as the images become increasingly noisy) and by the speed of the camera used in the experiment (the speed is again governed by signal/noise and also by data readout limitations). Even with the recent improvements afforded by direct electron detectors, it is hard to see how conventional methods in the S/TEM will improve the temporal resolution much beyond ~0.1 ms. However, the imaging speed can be increased significantly by moving away from the traditional field emission microscope, or by using non-standard imaging methodologies such as compressive sensing (CS). While there will always be cases where the liquid will be too thick for high resolution imaging (in the study of whole biological cells, for example [2]), an increase in the intrinsic speed of imaging opens up the possibility of studying fundamental atomic scale diffusion and structural order/disorder processes directly at the solid–liquid interface [3, 12].

In this chapter we discuss the hardware and software requirements necessary to increase temporal resolution to ~1 μs and faster. On the hardware side, we discuss the design and implementation of dynamic TEM (DTEM) [12], where the use of a photoemission source to create intense electron pulses can dramatically improve temporal resolution. From a software standpoint, CS [13] approaches to image acquisition and retrieval could provide a major boost in temporal resolution while at the same time helping to offset the data challenges that are associated with acquiring movies with high frame rates. As with conventional microscopy, to fully interpret images on the atomic scale, we need to understand what causes the contrast by performing image simulations. Uniquely for liquid cell experiments, we also need to quantify how electron dose, beam damage, and signal/noise control our ability to resolve processes on the critical spatial and temporal scales. Chapter 8 has discussed the simulation of images from the liquid stage [14]; here we describe the use of automated tracking algorithms to follow dynamic processes as they occur [15]. As frame rates increase, the only way to fully track transient dynamics will be through automated software – watching a movie at conventional frame rates, it would take over 8 hours to watch a 1 second movie acquired at 1 million frames/second! Finally, we discuss two areas of research that benefit from the liquid cell, electrochemical processes and structural biology, showing results obtained with conventional S/TEM and highlighting what could be achieved with an increase in spatio-temporal resolution.

22.2 Why Do We Need Better Temporal Resolution?

It is straightforward to see that, in general, the faster we acquire a single image and the more frames per second we can achieve in a movie, the higher the precision of the observation will be and the more transient structures we will be able to observe. But what temporal resolution and frame rate is optimal for studying processes in liquids? Designing a system to study irreversible processes over the temporal range of 1s to 1 fs (10^0–10^{-15}s) would be a daunting challenge (and physically impossible if we want atomic spatial resolution as well). Similarly, designing a system to observe a critical set of events that have a natural time scale of 1 millisecond with a microscope capable of 1 nanosecond temporal resolution and 10^9 frames per second creates a level of oversampling that makes the experiment untenable even if it could be performed. Fortunately, we can constrain the requirements through a basic understanding of the physical processes taking place at the solid–liquid interface.

As a first approximation, we can estimate the temporal resolution that we will need by looking at an atomic diffusion process (assuming that the required spatial resolution will be on the atomic scale). The density functional theory (DFT) simulations shown in Figure 22.1 [16] calculate the amount of time that it takes for Au atoms to hop along the surface of a TiO_2 (rutile) substrate. These simulations show that Au atoms can move very fast on a stoichiometric TiO_2 surface: an individual hop can take place in ~10 picoseconds (ps) at room temperature. However, if there is an oxygen vacancy on the surface, the hop time slows to ~1 millisecond (ms) since the Au atom is effectively pinned by the vacancy. If the cluster size is increased to either 2 or 3 Au atoms, then the hopping time on the stoichiometric surface is now reduced to ~1 ms to ~1 microsecond (μs) and the effect of the pinning vacancy is reduced. The hopping speed can therefore be controlled by the temperature and the surface chemistry. If we can obtain images in the temporal range of 1 μs or faster (with a correspondingly fast repetition rate) we can see all of the individual hops that take place on the surface, other than the

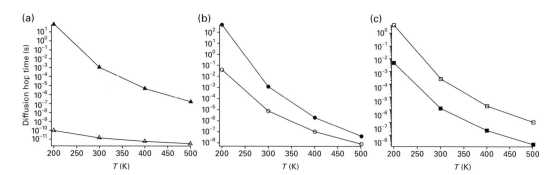

Figure 22.1. DFT simulations of the average hopping times for the diffusion of (a) single, (b) double, and (c) triple atom Au clusters on a TiO_2 surface. Open symbols are for a stoichiometric surface and closed symbols are for a surface containing oxygen vacancies. Note the different scales for the hop time. Reprinted from Ref. 16.

non-interacting single atoms. In images on this temporal range, the un-pinned individual atoms will "appear" at random locations far apart, allowing them to also be identified and viewed as a source of adatoms for the larger structures. Conversely, if we measure the speed of interaction and can observe differences in the speed for different particle sizes and shapes, we can infer the effect of different chemistries. While this example refers to atoms moving on the surface of a support under ultra-high vacuum conditions, it provides a maximum estimate of the level of temporal resolution that will be needed for diffusion experiments, since, intuitively, atoms at solid–solid and solid–liquid interfaces move slower than at a solid–vacuum interface.

While the simulations in Figure 22.1 show us the temporal resolution needed for atomic diffusion processes, they tell only half the story of what happens in the liquid cell. In cases where particles float freely in a liquid, we need to take Brownian motion into account. Figure 22.2 shows a bright field/dark field image pair where ferritin complexes appear distorted due to constrained Brownian motion [17]. Such distortions can be overcome by attaching the inorganic/biological material to the window of the liquid cell. Particles can be immobilized through covalent linkages, electrostatic effects or affinity tagging (Chapter 17) to minimize blur from Brownian motion. While attachment may sound ideal for improving attainable resolution, it has been well established that surface immobilization of proteins alters their native microenvironment and can cause conformational changes that perturb their functionality [18, 19]. Thus, maintaining most biological samples free-floating in solution would provide the most physiologically relevant view of their structure and dynamics, and faster imaging methods are needed to mitigate the effect of Brownian motion and other flow and beam charging effects for particles in solution.

Recent analysis using optical traps to measure the motion of microspheres immersed in liquid suggests that Brownian motion should cause jumps in the position of nanostructures by ~1 nm on a time scale of ~1–10 μs [20–22]. Using these parameters, we would expect the motion observed in Figure 22.2 to be several orders of magnitude larger than the measured ~5 nm displacement. However, we must remember that the

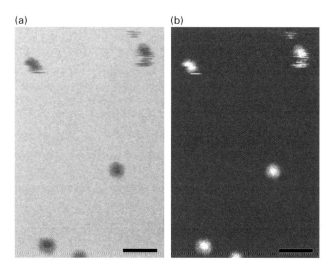

Figure 22.2. Liquid cell STEM of soluble proteins. (a) Bright field and (b) corresponding dark field STEM images of purified ferritin complexes with iron oxide core. Note the two particles at the top of each image that experienced Brownian motion or beam-induced charging during image rastering. Particles that appear stable may be tumbling end-over-end in solution. Scale bars are 25 nm.

liquid layer is thin. The ~500 nm liquid bounded by membranes is likely to display constrained Brownian motion. Whether we consider the full Brownian motion or some constraint imposed by thickness, it is clear from the available data that getting to the 1 µs or faster time scale will reduce the blurring effect to below atomic dimensions and hence show atomic diffusion processes. Note that other effects, such as particle charging or fluid flow [23], create motion in the liquid that is on a time/length scale that is slower than Brownian motion. Therefore, as long as we are in the 1 µs time regime for image acquisition, atomic resolution should be achieved.

As a final note on the need for increased time resolution, it is important to understand that liquid cell imaging is fundamentally different from conventional high resolution TEM. Atomic scale images of interfaces have long been used to provide insights into the structure–property relationships of materials; for example, observations of atomic scale intermixing at interfaces in semiconducting–oxide heterostructures helped to explain the unique electronic and magnetic properties of these systems [24, 25]. The development and application of the S/TEM techniques used in these and other studies, however, start from the expectation that the atoms in the structure do not move, as atomic motion would destroy the property of interest. As such, it is relatively straightforward to identify sample preparation damage or beam damage as the atoms start moving around. However, systems in liquids by definition involve atoms moving around, and the intrinsic mechanisms/functionality being studied are wholly dependent on the motion of atoms. For example, in Li-ion batteries the charge/discharge cycle involves the mobility of ions across the electrolyte–electrode interface [9]. As such, being able to control the experiment to the level where each observation is reproducible and free from specimen and beam-induced artifacts assumes a far greater importance than for high resolution microscopy (see the full discussion on resolution and image artifacts in Chapter 8). Achieving this level of dose control and interpreting the effects that are taking place is much easier with faster time resolution, as we can see the initiation of reactions rather than simply trying to backtrack from the outcome of the reaction.

22.3 Hardware/Software Developments for Fast Temporal Resolution

To achieve a temporal resolution of 1 μs or faster coupled with atomic spatial resolution, several developments have to be made in both the hardware and the software associated with image acquisition and analysis. For the hardware, the only means to get enough electrons into a microsecond pulse that is currently compatible with an electron microscope column is through photoemission [26]. Even though this provides enough electrons to form an image, the image will have low signal-to-noise ratio and so it is important that calibration, as described in Chapter 8, is performed for each experiment. Acquiring an individual frame on the microsecond time scale is only the first step, as analysis of dynamic events requires a sequence of frames. Assuming that a factor of 10 difference between the acquisition of a single frame and the time between individual frames will be sufficient, this gives a required frame rate of 100 000 per second, which is 60 or 2500 times faster than current direct electron and charge-coupled device (CCD) cameras respectively. Existing cameras have limitations due to signal to noise (which will be mostly overcome with pulsed beams, see below) and data rate. Reading out at 100 000 full frames per second (2048 × 2048 pixels at 16-bit depth) would give a data rate of ~1 TB/s. This implies some form of compression in either acquisition or storage. Compressive sensing offers one mechanism to accomplish this. Of course, if we can achieve these data rates, then analyzing the images becomes a challenge and we will need automated tracking algorithms to filter the data and identify critical events that take place.

22.3.1 Dynamic Transmission Electron Microscopy (DTEM)

The dynamic TEM (DTEM) is a relatively new technology that combines pulsed laser systems with the electron optics of a standard TEM [12, 27–36]. The initial work in this area was performed by Bostanjoglo and his collaborators at TU-Berlin [27, 28] working with a 1950s vintage TEM that achieved spatial resolution of ~100 nm. This work was extended by the group at Lawrence Livermore National Laboratory (LLNL), where a 1970s vintage instrument can obtain a spatial resolution of <10 nm with ~10 ns pulses [29–34]. In both of these instruments, the primary driving force was to move towards faster images; with the older electron optics, the spatial resolution was never expected to reach atomic dimensions. A project at Pacific Northwest National laboratory (PNNL) therefore aims to achieve atomic spatial resolution with 0.1–1 μs pulses by using a state-of-the-art double-corrected

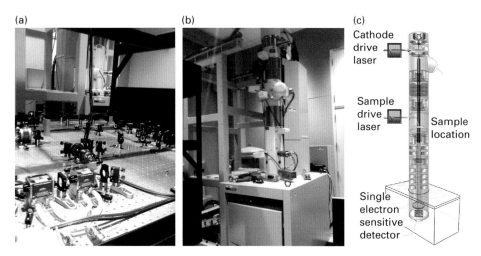

Figure 22.3. (a) The setup for the laser and (b) the DTEM at PNNL. (c) Schematic of the column and the implementation of *in situ* experiments.

JEOL 2200 as the base microscope [35, 36]. In the following discussion of time resolution and frame rates, we will assume these performance parameters.

The very high time resolution (1 µs–1 ns) in the DTEM is achieved by producing a short burst of electrons (up to 10^9 electrons) to illuminate the sample, coupled with a sensitive CCD/direct electron detector. The microscope works by irradiating a photocathode with a pulsed UV laser with photon energy greater than the target work function (Figure 22.3). Tantalum is typically used as the photocathode because of its combination of efficiency and stability [37]. A flux of electrons is then produced via photoemission with approximately the same time duration as the stimulating laser pulse. After this photoemission process, the microscope directs the emitted electron pulse in the traditional way (acceleration, focusing, magnification, detection, etc.). Images are obtained with approximately the same time resolution as the pulse duration and all normal TEM imaging modes are possible. If the photoemission pulse is synchronized with a second laser that stimulates the sample, both non-equilibrium and equilibrium *in situ* reactions can be studied in liquids with high temporal precision.

The goal in setting up the DTEM to study processes in liquids is to ensure that each electron pulse has enough electrons to form a complete image. This typically requires $\sim 10^9$ electrons per pulse. This "single shot" approach means that the process being studied need not be perfectly reversible (e.g. for chemical reactions, diffusion, and nucleation/growth events in liquids) as all the information is obtained from a single drive event. This provides advantages in the study of nucleation and growth compared to time-resolved 4D-EM techniques that are used to study perfectly reversible reactions on the femtosecond time scale [38–43]. However, the limitation of single shot imaging is that space-charge effects in the beam lead to a degradation of resolution: even with an optimized microscope source, column, and detector, the high current limits the overall temporal and spatial resolution of the instrument. Thus, single shot combined atomic resolution femtosecond imaging is unlikely ever to be possible. The key to using the single shot approach is therefore to optimize the components in the microscope to define the space-charge limited resolution. Calculations based on the best estimates of pulse and instrumentation characteristics suggest that space charge effects do not significantly limit the spatial resolution of the microscope for pulses as fast as 100 ns in duration [26]. These simulations predict the resolution seen with the LLNL DTEM and so provide a high level of confidence that we will be able to use the PNNL DTEM to study processes in liquids on the critical temporal scale to achieve atomic resolution in the images. Another important part of the design for the DTEM at PNNL is the in-column filter, which permits low energy core losses and plasmons to be observed, providing the only possible level of microanalysis that can exist on this time scale – energy-filtered TEM (EFTEM) [44].

A further advantage of the DTEM is that the trigger used to set the timing is a second laser that is incident on the sample. This second laser provides either heat or an optical stimulation that has the effect of initiating a reaction. Hence, whilst most conventional *in situ* liquid experiments have to turn on the reaction through resistive heating or through the electron beam damage effect itself (both of which are inherently slow processes as the heat/dose has to be ramped up), in the DTEM the experiment can be turned on with a nanosecond laser pulse. As this pulse can be up to 100 times faster than the pulse of electrons used to probe the reaction, we can essentially start the reaction from any point we want with the laser. In addition, by keeping the overall electron dose rate low, we can separate the reaction from the observation [4]. Figure 22.4 shows a nucleation and growth experiment using a liquid cell in the LLNL DTEM. The beam

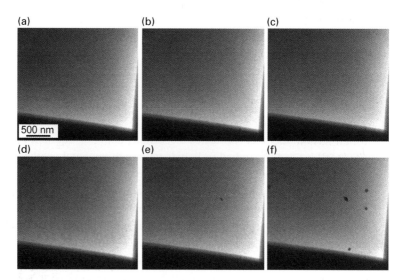

Figure 22.4. Controlled synthesis of nanoparticles using the DTEM sample drive laser. (a–c) Correlated image series of the same area of a 1:1000 dilution of 1:1 (Pb:S) solution illuminated solely with electron irradiation, or (d–f) with electron irradiation but following a single sample drive laser pulse at an energy of 0.6 µJ; (a) corresponds to time 0, while (b) and (c) are for 5 and 15 min respectively; (d, e, f) correspond to images 2 s, 2 min and 5 min after sample drive laser initiation, respectively. Scale bars represent 3 µm. Adapted from Ref. 4.

pulses cause no nucleation and growth until the laser is used to create the reaction. While the ~10 nm spatial resolution of the LLNL precludes observation of all the reaction products at high resolution, it does show the concept of the experiment that will be employed in the higher spatial resolution DTEM at PNNL.

22.3.2 Compressive Sensing

The DTEM described above has the potential to deliver individual images with a temporal precision in the images of ~0.01–1 μs. To make full use of this single frame precision, it is essential that multiple frames are recorded. Indeed, the LLNL instrument showed that careful calibration leads to insights from single images, but movies are more straightforward to interpret and much more likely to catch transient events [13]. Using an approach where consecutive pulses are directed onto different parts of a detector, we can build movies at standard frame readout speeds but where each frame actually contains a grid of 9 (3×3) to 64 (8×8) sub-frames (achieved by stepping across various sized cameras with minimum sub-frame sizes of 512×512 pixels). If a 4k × 4k direct electron detector could provide a full frame readout of 1 ms, such tiling could enable effective rates of 64 000 frames per second (fps). However, issues such as beam deflection efficiency and stability then become a limiting factor. To achieve the goal of 100 000 fps we will therefore likely need an alternate approach. As mentioned above, increasing data rates in existing cameras to 100 000 fps would require data transfer and storage at ~1 TB/s if we acquire images in the conventional way. This type of data rate is not going to be possible for many years at least. Compressive sensing (CS) is one way that the faster frame rates can be achieved without increasing the data transfer rates significantly. In CS, sensing and compression is combined in one operation, representing a fundamentally new approach for improving temporal resolution of an acquisition system. Temporal compressive sensing represents one of many variants of general compressive sensing and signal recovery, which has now been shown to have applicability in a number of scientific and technological areas. For example, in the field of electron microscopy alone, variants of CS have been successfully applied for reducing tilt artifacts in electron tomography [45] and enabling low dose STEM imaging [46]. For DTEM, a coded aperture (mask) could be mechanically/electrically translated during detector integration to induce modulation and compression of the signal. When recording in this way, the frame rate of the camera is not dictated by speed of the acquisition system, but the range of coded aperture translation during integration. The concept of coded aperture compressive temporal imaging (Figure 22.5) has been developed by researchers at Duke University, and their proof of concept performed on the optical bench has demonstrated an enhancement of frame rate by 14× [47]. Further calculations suggest that it should be possible to extend the frame rate by a factor of up to 100 [48], meaning that we would be able to get 10 000 frames per second from an existing camera now, with 100 000 frames per second in the near future. It should be noted that there is nothing to stop the compressive sensing approach being used for conventional microscopes. However, in this case the images will run out of electrons (i.e. be noise limited) before the limit in temporal resolution is reached.

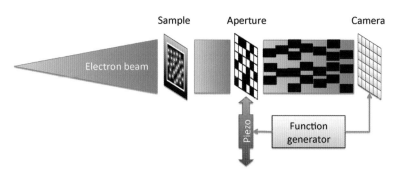

Figure 22.5. A schematic of the TEM setup for compressive sensing. After the beam passes through the sample, portions of it are occluded by the aperture. The occluded images are integrated together on the camera. Because each image has a different encoding, defined by the position of the aperture, they can be recovered by CS inversion.

22.3.3 Tracking Algorithms

Images acquired from the liquid cell often show nucleation and growth events. For one or two "representative" particles, it is possible to track size and shape evolution by hand from frame to frame using simple image analysis software. However, in many of the experiments that are being performed now, there can be hundreds of nucleating events, all of which potentially provide insights into the process taking place. To characterize and quantify the process fully, with statistics that are free from analysis bias (the user picking samples based on a known or unintentional bias), all particles have to be tracked using automated software codes. These codes are a research challenge in themselves and represent an area of image analysis that is rich with applications. While there are many approaches to solving the problem of automated tracking, the one that is needed for liquid cell microscopy must be capable of identifying particles in low signal environments and tracking the agglomeration and/or separation of these particles from one frame to the next. The approach must also have the potential to be used in a high performance computing environment (when the movie has 1 million frames, the analysis cannot take 1 second per frame). There are obviously opportunities here for further data compression by only storing frames where changes have occurred and recording the number of frames between such events in a metadata file. However, for the present discussion we focus on a set of codes that has already been implemented [15].

As an example we show a data association formulation and solution method for tracking a varying number of interacting targets that undergo birth, death, splitting, and merging [15]. The algorithm assumes the visual measurements from a video sequence are noisy, so there might be false and missed measurements (more likely to happen at the start when the contrast variations are smaller). An example of the use of this algorithm is given in Figure 22.6 for a movie [7] showing the growth of Ag nanoparticles by electron beam reduction. Ostwald ripening is one mechanism for solution-phase nanoparticle growth that dictates the dynamic and final properties

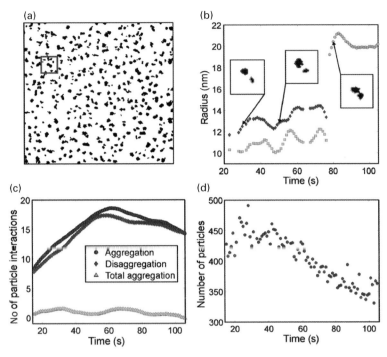

Figure 22.6. (a) Rendering of a BF-STEM image at $t = 105$ s, showing the morphology of individual particles/aggregates. (b) Growth of Ag nanoparticles by monomer attachment and aggregation. Insets show the particle morphologies at various time points before and after aggregation at $t = 78$ s. Data are filtered using a 5 s running filter to reduce noise. (c) Number of particle interactions as a function of time. The data are filtered using a 20 s running filter to reduce noise. (d) Number of nanoparticles in the ensemble as a function of time.

of the nanoparticle ensemble, such as average particle size and particle size distribution (PSD). While experiments have shown that direct microscopic observations of single nanoparticle growth trajectories can be interpreted quantitatively in terms of Ostwald ripening [5, 49], the *ensemble* properties of nanoparticle growth (agglomeration and coalescence) have not been tested against Ostwald ripening models. Using statistical clustering analysis to track and group all the different growth modes seen in the complete experiment, it was found that while the average nanoparticle radius scaling agrees with the classical model, the PSD is artificially broadened by abnormal aggregative growth of nanoparticles. Measurements of the local nanoparticle environments show that aggregative growth leads to depletion of the areas surrounding the abnormal particles, which further invalidates the mean field approach of the classical model (Figure 22.7). These observations of abnormal aggregative growth modes in nanoparticle ensembles highlight the need for complete analysis of each frame in the acquired movie to fully understand the process taking place. If we do achieve 0.1–1 μs temporal resolution with accompanying frame rates, this is the type of analysis that will be needed.

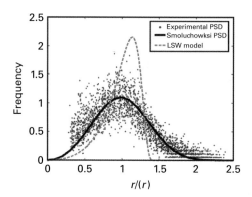

Figure 22.7. The scaled Ag nanoparticle PSD measured for all times for $t = 15-105$ s. The dashed blue curve is the LSW PSD (sum of squared errors SSE = 1550), and the solid black curve is the Smoluchowski PSD with $\langle \beta_{PSD} \rangle = \langle \beta \rangle = 0.31$ (SSE = 78). Nanoparticle radii are normalized to the respective mean nanoparticle radius for that time.

22.4 Materials and Biological Examples

As discussed in Section 22.2, increasing the imaging speed in liquids will automatically increase the spatial resolution as motion blur will be reduced. When thinking of where such capabilities can make an important impact, it is clear that it is the combined spatio-temporal resolution that is important. In materials processes such as nucleation and growth or corrosion [50], chemical processes such as catalysis [51], or biological processes such as self-assembly or conformational changes [52], it is important to know where the event takes place first. While a technique may have the spatial resolution to observe an atomic site, if the temporal resolution is not fast enough to catch the event, or the first event is subsequently swamped by other events, then that crucial first stage of the process will be missed. In virtually every aspect of materials and structural biology, knowing exactly where something happens first is a key to controlling the subsequent evolution. While there are many possible examples that can be discussed, here we focus on two areas where liquid cell microscopy combined with the spatio-temporal resolution of the DTEM will allow for unprecedented analysis. These areas are electrochemical processes for energy storage and dynamic processes in structural biology.

22.4.1 Electrochemistry

One of the main types of experiments for which the liquid cell is ideal involves electrochemical processes and, in particular, processes related to energy storage. In this case we restrict the discussion to batteries [53–58]. A key area concerns fundamental mechanisms in Li-ion/metal batteries [53, 54] and related Li-sulfur [55] and Li-air [56] systems. Open cell geometry [59], described in Chapter 3, allows direct study of Li diffusion and intercalation in Si, but not the formation of the main structural features in Li-ion batteries, i.e. the solid–electrolyte interphase layer and Li metal dendrites. As discussed in Chapter 11, understanding the initial formation of the SEI layer,

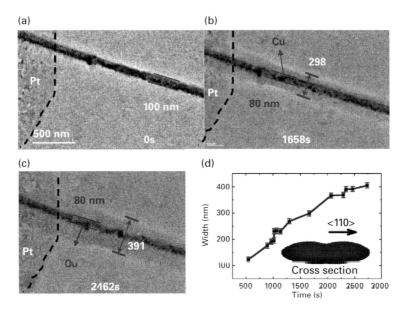

Figure 22.8. (a) TEM image showing the pristine state of a Cu-coated Si nanowire at 0 s. (b, c) Core-shell formation during lithiation shown at two times, 1658 s and 2462 s. (d) Change in width of the nanowire as a function of time. The maximum volume expansion is along the $\langle 110 \rangle$ direction.

how the anode surface, the electrolyte salts/solvent, and the presence of additives affect its thickness, and how it breaks to allow the formation of dendrites, is key to improving lifetime and safety in Li-ion batteries and other promising systems such as Na, Mg, or Ca. Although battery cycling is not generally considered a fast mechanism, the high spatio-temporal resolution afforded by DTEM will be essential for imaging in a liquid electrolyte to identify the initial stages or simply the location where the process happens first.

We can see how electrochemistry experiments will work in the DTEM by looking at results from conventional S/TEM. One of the first experiments that used a liquid cell to form a "real" battery (i.e. with an anode, cathode, and liquid electrolyte) clearly showed the volume expansion during lithiation of a Si nanowire acting as an anode material (Figure 22.8 [9, 60]). Here the nanowire was FIB-welded to a Pt contact in the liquid cell and a small piece of Li formed the counter electrode. More recently [61], Li deposition has been observed directly on Au or Pt electrodes. Lithium is less dense than the electrolyte material so appears bright in bright field images, but a small degree of alloying with the contact metal (Au or Pt) reverses the contrast, making Li easy to identify and distinguishing it from the SEI layer. After extended cycling of the anode, Figure 22.9 clearly shows this alloying, the formation of the SEI layer (bright line) and the presence of significant Li deposits on the electrode and "dead" Li away from it. In these observations, the dose rate was calibrated to be below the onset of damage to the electrolyte to ensure that what was being measured was purely an electrochemical effect [10]. STEM movies show the evolution of these features clearly, albeit with

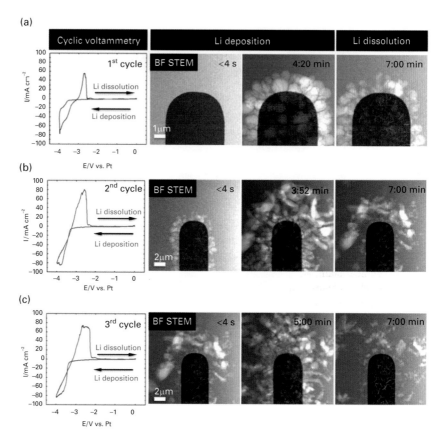

Figure 22.9. Bright field STEM images of Li deposition and dissolution at the interface between a Pt working electrode and the LiPF$_6$/PC electrolyte during (a) the first, (b) the second, and (c) the third charge/discharge cycles of an *operando* cell, with cyclic voltammetry for each cycle. From Ref. 61.

limited spatio-temporal resolution. The use of the DTEM in such experiments will provide a significant increase in the sensitivity of the observations.

22.4.2 Structural Biology

TEM exhibits the unique ability to visualize length scales relevant to life science research at nanometer resolution or better. However, as discussed in Chapters 16 and 17, it is limited to static observations of macromolecules. The vitrification of protein samples helps to optimally preserve the integrity of biological material [62], but the freezing process for high resolution cryo-electron microscopy prevents observation of dynamic phenomena. In an effort to overcome this limitation, time-lapse cryo-EM has been developed using laser flash photolysis, microfluidic mixing, or directed spraying of an aerosol onto a sample-coated grid to trigger a reaction just prior to plunging the grid into a volume of liquefied ethane [63–65]. Unfortunately, time-lapse cryo-EM remains

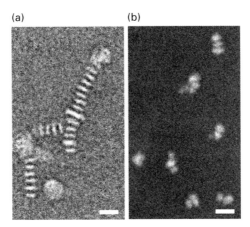

Figure 22.10. (a) Cryogenic negative stain TEM images of frozen and stained nanolipoprotein discs (NLPs). (b) Aberration-corrected dark field STEM image of NLPs fully hydrated, without stain, and suspended in a liquid cell. Note that the NLPs in (a) are arranged as long stacked chains whereas in (b) the NLPs stacks are shorter dimers and trimers. Scale bars represent 20 nm.

restricted to the time scale of the freezing process (~0.1 ms) and solely provides "snapshots" of processes. For real-time TEM observations of biological dynamics that occur on the µs or faster time scale, the samples need to be imaged in a hydrated state, i.e. in a liquid cell, that is coupled with a microscope capable of spatial and temporal resolutions on the order of 0.1–1 nm and 0.01–1 µs, as provided by the PNNL DTEM [36].

As a test of the versatility of imaging macromolecular biological samples within a liquid environment, consider nanolipoprotein discs (NLPs). NLPs have been proposed as a useful self-assembling platform for both functional and structural studies of membrane protein complexes [66]. Although NLPs should ideally be monodisperse in solution, direct imaging by negative stain TEM and cryo-EM shows discoidal stacks as well as monomeric NLPs (Figure 22.10a, b). In the liquid cell (Figure 22.10), NLPs in suspension at the same concentration used for the TEM images clearly aggregate into short stacks even though the samples have low protein and salt concentrations and were not subjected to drying. The relative ratio of monomeric to non-monomeric NLPs is approximately the same for each imaging method, but the stacks appear shorter in the liquid cell; the stacking association is clearly not an artifact caused by TEM sample preparation, but is likely due to an electrostatic interaction [66]. Although these images in the conventional STEM, as well as the images in Figure 22.2 of ferritin in solution, suffer from the effects of Brownian motion limiting resolution for particles not attached to the windows, the results show the benefits of imaging in the hydrated state for maintaining more native environments and protein interactions.

To establish whether the liquid cell in DTEM can potentially improve on cryo-EM, we need to compare the signal to noise in each case, since cryo-EM resolution is controlled by the tolerable dose to the sample. Figure 22.11 shows a simulation of

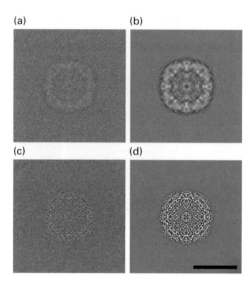

Figure 22.11. (a, b) Individual and averaged ($n = 100$) bright field cryo-EM simulations of apo-ferritin within an ice layer of 100 nm thickness and using a defocus of -500 nm with $C_S = 2.0$ mm. (c, d) Individual and averaged ($n = 100$) bright field liquid DTEM simulation of apo-ferritin between two 5 nm thick silicon nitride membranes and within a 50 nm fluid path length and with $C_S = 0.005$ mm. The simulations are optimized for defocus that maximizes intensity in the first Thon ring at 0.5 nm and results in a point resolution of 0.35 nm. Both images assume a total dose of 1000 electrons/nm^2 and a signal-to-noise ratio of 4%. Scale bar is 10 nm for all images.

apo-ferritin (the protein shell from the particles shown in Figure 22.2) under optimal cryo-EM conditions, and what would be expected for DTEM imaging with a liquid cell. These simulations show higher resolution for the DTEM. Of course, this high resolution would also be realized in an aberration-corrected field emission TEM, but only if motion effects were ignored, since these degrade the resolution during the ms data acquisition of the DTEM. Even for cryo-EM, the use of DTEM may also increase the resolution limit by reducing the effect of beam-induced motion of the sample (whether this motion is from charging or other mechanisms that occur on the ms to μs time scale). While direct electron detectors have mitigated some of the beam-induced charging effects, the potential to bypass this limitation altogether using pulsed beams may allow DTEM to push the frontiers of structural biology by further improving the attainable resolution and making atomic resolution cryo-EM a routine and standard technique for single particle and crystalline samples.

As stated previously, the DTEM is designed for pump-probe reactions that are initiated by a second laser. For biological samples, this means that the DTEM can photo-activate conformational changes through direct light stimulation or via ultraviolet photolysis of caged ligands, or stimulate reactions with a localized temperature jump from an infrared pulse. Other mechanisms for triggering reactions are also feasible, including creating localized regions of pH gradients and thermal and electrochemical regulation. This inherent flexibility for reaction initiation and synchronicity means that many different types of dynamics and compatible samples could be imaged with the DTEM, ranging from assembly/disassembly of microfilaments, pore gating of ion channels, conformational switching of enzymes and membrane fusion/fission events. Ref. 36 provides additional details on how such experiments can be designed.

22.5 Conclusions

The development and application of liquid cells and holders has opened up many new opportunities for research in materials science, chemistry, and biology. Using conventional and state-of-the-art aberration corrected S/TEM to perform the analyses has already led to significant advances in our understanding of the processes behind nucleation and growth and electrochemistry. However, there is an intrinsic limitation in the spatio-temporal resolution that can be obtained with these microscopes even for perfectly designed and constructed liquid cells. By moving to alternative sources for the TEM, in this case photoemission, it is possible to create electron pulses containing enough electrons to enable atomic resolution on the critical sub μs time scale that is important for phenomena such as diffusion processes at the solid–liquid interface. Design and implementation of a DTEM is already underway, as is the implementation of compressive sensing and automated tracking algorithms that will permit data to be quantified from high frame rate observations. As with conventional electron microscopy methods, the DTEM has the advantage over other high temporal resolution techniques in that it provides direct images of the process taking place; there is no uncertainty caused by the inversion from a spectrum or diffraction pattern. Given the success that has already been achieved, the future for studying fast dynamics in the liquid cell using DTEM looks certain to produce unique scientific results that can solve major challenges in materials synthesis, energy technologies and structural biology.

Acknowledgements

We would like to express our thanks to the many collaborators who worked with us on the design and implementation of the liquid holder used for the results that are shown and cited in this chapter: Katherine Jungjohann, Taylor Woehl, Lucas Parent, David Welch, Peony Wong, Po-Lin Chiu, Gavin Dutrow, Meng Gu, Patricia Abellan, Layla Mehdi, Joseph Patterson, Libor Kovarik, Chiwoo Park, Andrew Stevens, Nathan Gianneschi, Roland Faller, William Ristenpart, Chongmin Wang, and Ilke Arslan. The development of the *operando* stage was supported by the Chemical Imaging Initiative, a Laboratory Directed Research and Development Program at Pacific Northwest National Laboratory (PNNL). PNNL is a multi-program national laboratory operated by Battelle for the U.S. Department of Energy (DOE) under Contract DE-AC05-76RL01830. A portion of the research was performed using the Environmental Molecular Sciences Laboratory (EMSL), a national scientific user facility sponsored by the Department of Energy's Office of Biological and Environmental Research and located at PNNL. In addition, the electrochemical aspects of this work were primarily supported by Joint Center for Energy Storage Research (JCESR), an Energy Innovation Hub funded by the Department of Energy, Office of Science, Basic Energy Sciences. The development of the DTEM to study biological systems was supported by the National Institutes of Health under grant numbers S10RR025032 and 5RC1GM091755. The study of nucleation and growth phenomena was supported by the Department of Energy, Office of Science, Basic Energy Sciences under grant number DE-FG02-03-ER46057.

References

1. M. J. Williamson, R. M. Tromp, P. M. Vereecken, R. Hull and F. M. Ross, Dynamic microscopy of nanoscale cluster growth at the solid-liquid interface, *Nat. Mater.*, **2** (2003), 532–536.
2. N. de Jonge, D. B. Peckys, G. J. Kremers and D. W. Piston, Electron microscopy of whole cells in liquid with nanometer resolution. *Proc. Natl. Acad. Sci. USA*, **106** (2009), 2159–2164.
3. N. de Jonge and F. M. Ross, Electron microscopy of specimens in liquid. *Nat. Nanotechnol.*, **6** (2011), 695–704.
4. J. E. Evans, K. L. Jungjohann, N. D. Browning and I. Arslan, Controlled growth of nanoparticles from solution with *in situ* liquid transmission electron microscopy. *Nano Lett.*, **11** (2011), 2809–2813.
5. H. M. Zheng, R. K. Smith, Y. W. Jun *et al.*, Observation of single colloidal platinum nanocrystal growth trajectories. *Science*, **324** (2009), 1309–1312.
6. D. S. Li, M. H. Nielsen, J. R. I. Lee *et al.*, Direction-specific interactions control crystal growth by oriented attachment. *Science*, **336** (2012), 1014–**1018**.
7. T. J. Woehl, C. Park, J. E. Evans *et al.*, Direct observation of abnormal Ostwald ripening in nanoparticle ensembles caused by aggregative growth. *Nano Lett.*, **14** (2014), 373–378.
8. E. R. White, S. B. Singer, V. Augustyn *et al.*, In situ transmission electron microscopy of lead dendrites and lead ions in aqueous solution. *ACS Nano*, **6** (2012), 6308–6317.
9. M. Gu, L. R. Parent, L. Mehdi *et al.*, Demonstration of an electrochemical liquid cell for operando transmission electron microscopy observation of the lithiation/delithiation behavior of Si nanowire battery anodes. *Nano Lett.*, **13** (2013), 6106–6112.
10. P. Abellán, C. Park, B. L. Mehdi *et al.*, Probing the degradation mechanisms in electrolyte solutions for Li-ion batteries by in-situ TEM. *Nano Lett.*, **14** (2014), 1293–1299.
11. E. Sutter, K. L. Jungjohann, S. Bliznakov *et al.*, In situ liquid-cell electron microscopy of silver-palladium galvanic replacement reactions on silver nanoparticles. *Nat. Commun.*, **5** (2014), 4946.
12. J. S. Kim, T. B. LaGrange, B. W. Reed *et al.*, Imaging of transient structures using nanosecond in situ TEM. *Science*, **321** (2008), 1472–1475.
13. E. J. Candes, J. Romberg and T. Tao, Near-optimal signal recovery from random projections: universal encoding strategies? *IEEE Trans. Inform. Theory*, **52** (2006), 489–509.
14. D. A. Welch, R. Faller, J. E. Evans and N. D. Browning, Simulating realistic imaging conditions for in-situ liquid microscopy. *Ultramicroscopy*, **135** (2013), 36–42.

15. C. Park, T. J. Woehl, J. E. Evans and N. D. Browning, Minimum cost multi-way data association for optimizing large-scale multitarget tracking of interacting objects. *IEEE Trans. Patt. Anal. Mach. Intell.*, **37** (2015), 611–624.
16. N. Goldman and N. D. Browning, Gold cluster diffusion kinetics on stoichiometric and reduced rutile TiO_2 (110). *J. Phys. Chem.* C, **115** (2011), 11611–11617.
17. J. E. Evans, K. L. Jungjohann, P. C. K. Wong *et al.*, Visualizing macromolecular complexes with in-situ liquid transmission electron microscopy. *Micron*, **43** (2012), 1085–1090.
18. T. Kobayashi and K. Laidler, Kinetic analysis for solid-supported enzymes. *Biochim. Biophys. Acta*, **302** (1973), 1–12.
19. R. C. Rodrigues, C. Ortiz, A. Berenguer-Murcia, R. Torres and R. Fernandez-Lafuente, Modifying enzyme activity and selectivity by immobilization. *Chem. Soc. Rev.*, **42** (2013), 6290–6307.
20. B. Lin, J. Yu and S. A. Rice, Direct measurements of constrained Brownian motion of an isolated sphere between two walls. *Phys. Rev. E*, **62** (2000), 3909–3919.
21. S. Kheifets, A. Simha, K. Melin, T. Li and M. G. Raizen, Observation of Brownian motion in liquids at short times: instantaneous velocity and memory loss. *Science*, **343** (2014), 1493–1496.
22. P. S. Burada, P. Hanggi, F. Marchesoni, G. Schmid and P. Talkner, Diffusion in confined geometries. *ChemPhysChem*, **10** (2009), 45–54.
23. E. R. White, M. Mecklenburg, B. Shevitski, S. B. Singer and B. C. Regan, Charged nanoparticle dynamics in water induced by scanning transmission electron microscopy. *Langmuir*, **28** (2012), 3695–3698.
24. D. E. Jesson, S. J. Pennycook and J. M Baribeau, Direct imaging of interfacial ordering in ultrathin $(Si_mGe_n)_P$ superlattices. *Phys. Rev. Lett.*, **66** (1991), 750–753.
25. D. A. Muller, L. F. Kourkoutis, M. Murfitt *et al.*, Atomic scale chemical imaging of composition and bonding by aberration corrected microscopy. *Science*, **319** (2008), 1073–1076.
26. B. W. Reed, M. R. Armstrong, N. D. Browning *et al.*, The evolution of ultrafast electron microscope instrumentation. *Microsc. Microanal.*, **15** (2009), 272–281.
27. O. Bostanjoglo, High-speed electron microscopy. *Adv. Imag. Electron Phys.*, **121** (2002), 1211–1251.
28. O. Bostanjoglo and W. R. Horinek, Pulsed TEM: a new method to detect transient structures in fast phase-transitions. *Optik*, **65** (1983), 361–367.
29. T. B. LaGrange, M. Armstrong, K. Boyden *et al.*, Single shot dynamic transmission electron microscopy for materials science. *Appl. Phys. Lett.*, **89** (2006), 044105.
30. M. Armstrong, K. Boyden, N. D. Browning *et al.*, In-situ synthesis of nanowires in the dynamic TEM. *Ultramicroscopy*, **107** (2007), 356–367.
31. M. R. Armstrong, N. D. Browning, B. W. Reed and B. R. Torralva, Prospects for electron imaging with ultrafast time resolution. *Appl. Phys. Lett.*, **90** (2007), 114101.
32. M. L. Taheri, B. W. Reed, T. B. Lagrange and N. D. Browning, In-situ synthesis of nanowires in the dynamic TEM. *Small*, **4** (2008), 2187–2190.
33. B. W. Reed, T. LaGrange, R. M. Shuttlesworth *et al.*, Solving the accelerator-condenser coupling problem in a nanosecond dynamic transmission electron microscope. *Rev. Sci. Instrum.*, **81** (2010), 053706.
34. D. J. Masiel, T. LaGrange, B. W. Reed, T. Guo and N. D. Browning, Time resolved annular dark field imaging of catalyst nanoparticles. *ChemPhysChem*, **11** (2010), 2088–2090.
35. N. D. Browning, M. A. Bonds, G. H. Campbell *et al.*, Recent developments in DTEM. *Curr. Opin. Solid State Mater. Sci.*, **16** (2012), 23–30.

36. J. E. Evans and N. D. Browning, Enabling direct nanoscale dynamic observations of biological systems with DTEM. *Microscopy*, **62** (2013), 147–156.
37. B. L. Rickman, J. A. Berger, A. W. Nicholls and W. A. Schroeder, Intrinsic electron beam emittance from metal photocathodes: the effect of the electron effective mass. *Phys. Rev. Lett.*, **111** (2013), 237401.
38. V. A. Lobastov, R. Srinivasan and A. H. Zewail, Four-dimensional ultrafast electron microscopy. *Proc. Natl. Acad. Sci. USA*, **102** (2005), 7069–7073.
39. A. H. Zewail, 4D ultrafast electron diffraction, crystallography and microscopy. *Annu. Rev. Phys. Chem.*, **57** (2006), 65–103.
40. F. Carbone, O. H. Kwon and A. H. Zewail, Dynamics of chemical bonding mapped by energy resolved 4D electron microscopy. *Science*, **325** (2009), 181–184.
41. A. Yurtserver and A. H. Zewail, 4D nanoscale diffraction observed by convergent beam ultrafast electron microscopy. *Science*, **326** (2009), 708–712.
42. A. H. Zewail, 4D electron microscopy. *Science*, **328** (2010), 187–193.
43. O. H. Kwon and A. H. Zewail, 4D electron microscopy. *Science*, **328** (2010), 1668–1673.
44. F. Hofer, W. Grogger, G. Kothleitner and P. Warbichler, Quantitative analysis of EFTEM elemental distribution images. *Ultramicroscopy*, **67** (1997), 83–103.
45. R. Leary, Z. Saghi, P. A. Midgley and D. J. Holland, Compressed sensing electron tomography. *Ultramicroscopy*, **131** (2013), 70–91.
46. A. Stevens, H. Yang, L. Carin, I. Arslan and N. D. Browning, The potential for Bayesian compressive sensing to significantly reduce electron dose in high resolution STEM images. *Microscopy*, **63** (2014), 41–51.
47. G. R. Arce, D. J. Brady, L. Carin, H. Arguello and D. S. Kittle, Compressive coded aperture spectral imaging. *IEEE Signal Proces. Mag.*, **31** (2014), 105–115.
48. A. Stevens, L. Kovarik, X. Yuan, L. Carin and N. D. Browning, Applying compressive sensing to TEM video: a substantial frame rate increase on any camera. *Adv. Struct. Chem. Imag.*, **1** (2015), 10.
49. Y. Liu, K. Tai and S. J. Dillon, Growth kinetics and morphological evolution of ZnO precipitated from solution. *Chem. Mater.*, **25** (2013), 2927–2933.
50. M. T. Proetto, A. M. Rush, M. Chien *et al.*, Transmission electron microscopy of a synthetic soft material in liquid water. *J. Am. Chem. Soc.*, **136** (2014), 1162–1165.
51. P. L. Gai, Developments in in situ environmental cell high-resolution electron microscopy and applications to catalysis. *Topics in Catalysis*, **21** (2002), 161–173.
52. A. McPherson and D. Eisenberg, In R. Donev. ed., *Protein Structures and Diseases*, Advances in Protein Chemistry and Structural Biology (New York: Academic Press, 2011).
53. J. M Tarascon and M. Armand, Issues and challenges facing rechargeable lithium batteries. *Nature*, **414** (2001), 359–367.
54. J. B Goodenough and Y. Kim, Challenges for rechargeable Li batteries. *Chem. Mater.*, **22** (2010), 587.
55. X. L. Jie and L. F. Nazar, Advances in Li-S batteries. *J. Mater. Chem.*, **20** (2010), 9821–9826.
56. P. G. Bruce, S. A. Freunberger, L. J. Hardwick and J. M. Tarascon, Li-O_2 and Li-S batteries with high energy storage. *Nat. Mater.*, **11** (2012), 19–29.
57. P. Verma, P. Maire and P. Novak, A review of the features and analyses of the solid electrolyte interphase in Li-ion batteries. *Electrochem. Acta*, **55** (2010), 6332–6341.
58. J. Wen, Y. Yu and C. Chen, A review on lithium-ion batteries safety issues: existing problems and possible solutions. *Mater. Express*, **2** (2012), 197–212.

59. J. Y. Huang, L. Zhong, C. M. Wang *et al.*, In situ observation of the electrochemical lithiation of a single SnO_2 nanowire electrode. *Science*, **330** (2010), 1515–1520.
60. B. L. Mehdi, M. Gu, L. R. Parent *et al.*, In-situ electrochemical transmission electron microscopy for battery research. *Microsc. Microanal.*, **20** (2014), 484–492.
61. B. L. Mehdi, E. Nasybulin, J. Qian *et al.*, Observation and quantification of nanoscale processes in lithium batteries by operando electrochemical S/TEM. *Nano Lett.*, **15** (2015), 2168–2173.
62. R. M. Glaeser, K. Downing, D. DeRosier, W. Chiu and J. Frank, *Electron Crystallography of Biological Macromolecules* (Oxford: Oxford University Press, 2007).
63. J. Berriman and N. Unwin, Analysis of transient structures by cryomicroscopy combined with rapid mixing of spray droplets. *Ultramicroscopy*, **56** (1994), 241–252.
64. T. R. Shaikh, D. Barnard, X. Meng and T. Wagenknecht, Implementation of a flash-photolysis system for time-resolved cryo-electron microscopy. *J. Struct. Biol.*, **165** (2009), 184–189.
65. S. Subramanian and R. Henderson, Electron crystallography of bacteriorhodopsin with millisecond time resolution. *J. Struct Biol.*, **144** (1999), 2546–2562.
66. L. Zhang, J. Song, G. Cavigiolio *et al.*, Morphology and structure of lipoproteins revealed by an optimized negative-staining protocol of electron microscopy. *J. Lipid Res.*, **52** (2011), 175–184.

23 Future Prospects for Biomolecular, Biomimetic, and Biomaterials Research Enabled by New Liquid Cell Electron Microscopy Techniques

Taylor Woehl and Tanya Prozorov

23.1 Introduction

In this chapter we describe several research areas in biomolecular sciences and biomaterials where emerging liquid cell electron microscopy experiments have the potential for immense impact through the use of advanced imaging techniques. We outline possible future contributions in the areas of protein structural analysis, *in vivo* biomineralization, mesocrystals, and biomimetics. The chapter is written in a notional way to outline possible future avenues for liquid cell research: to this end, our discussion of future research assumes that technical challenges described in previous chapters are successfully resolved, and we can obtain the required spatial and temporal resolution. In some cases we discuss the difficulties that may be expected in performing the experiments we suggest, such as radiation damage or practical experimental design considerations. While this list is by no means exhaustive, we hope that these examples will stimulate the interest of a broad research community and advance the field of liquid cell electron microscopy. The unparalleled direct nanoscale visualization capabilities afforded by liquid cell electron microscopy have already had an enormous impact on research fields in physical and life sciences, and we believe it will continue to transform science in the years to come.

23.2 Visualizing Protein Structure in Liquid Water at High Resolution

X-ray crystallography and single-particle cryo-EM have both advantages and limitations for determining protein and molecular machine structure. The limitations can be used to shape approaches for the future development of liquid cell electron microscopy techniques for protein structure determination. In this section we discuss some possibilities. It appears that the methods used for structural determination via single-particle cryo-EM will translate well to liquid cell electron microscopy, albeit with a few technical caveats. We conclude this section with a discussion of possible obstacles for developing a robust liquid cell electron microscopy technique for determining protein structure.

23.2.1 State of the Art in Determining Protein Structure

Determining protein structure is essential for understanding biomolecular function, which facilitates drug discovery [1] and fundamental understanding of cellular function. X-ray protein crystallography [2] and single-particle cryo-EM [3] are two of the conventional techniques of choice for determining three-dimensional protein structure. Protein crystallography is the most prolific technique for structural biology and is capable of resolving biomolecular structures at resolutions up to 0.048 nm [4], with typical resolutions of 0.2–0.3 nm. However, this technique only works for proteins that can be crystallized, and the result is a static representation of the lowest energy conformation. On the other hand, single-particle cryo-EM is capable of determining structures of biomolecules that cannot be crystallized or are highly asymmetric, such as membrane proteins or heterogeneous macromolecules, albeit with a lower resolution typically on the order of 0.6–2 nm [3, 5, 6]. Cryo-EM involves analyzing thousands of images containing hundreds of copies of an identical biomolecule at random orientations to back-calculate its structure [3]. State-of-the-art single-particle imaging techniques employ direct electron detectors, image drift correction algorithms, and dedicated cryo-TEM instruments to achieve spatial resolutions up to 0.3 nm (Figure 23.1) [6].

The major factors limiting spatial resolution in cryo-EM are radiation damage and image drift [5]. Radiation damage leads to production of radicals, breaking of covalent

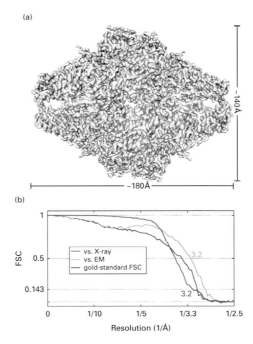

Figure 23.1. (a) Surface representation of the density map of β-galactosidase at a resolution of 0.32 nm derived by cryo-EM. The dimensions of the β-galactosidase tetramer are ~18 × 14 × 8.7 nm. (b) Fourier shell correlation (FSC) plots to estimate resolution of the maps shown in (a). Reproduced with permission from Ref. 6.

bonds, and mechanical stresses in the ice due to buildup of hydrogen [5]. These electron beam artifacts are mitigated using low cumulative electron doses (~1000 electrons/nm^2) [7] and, more recently, dose fractionation techniques made possible by direct electron detectors [5, 6]. As discussed in Chapter 8, a balance must be struck between electron dose and spatial resolution in order to preserve the macromolecular structure and achieve the highest possible resolution [7, 8]. With reports on near atomic-resolution cryo-EM structures, preferential radiation damage of certain protein functional groups has become an important artifact to consider as well [6].

23.2.2 Protein Structure Determination via Liquid Cell Electron Microscopy

As discussed in Chapters 16 and 17, cryo-EM imaging captures biomolecular structures in their nearly native environment, but the vitreous ice environment of the sample may introduce artifacts, for instance due to the use of cryo-protectants. It has also been shown that structural homogeneity may be artificially increased in vitrified samples compared to the native liquid state [9]. This drives interest in determining the structure of macromolecules in their native liquid environment.

Imaging at cryogenic temperatures has long been used to mitigate radiation damage of biological TEM samples [10]. It works because of the greatly reduced production rates of radicals at low temperatures [10]. However, the beam-induced radicals are effectively trapped in the frozen cryo-TEM sample, allowing rapid local radical buildup and sample damage even at low cumulative electron doses (>1000 electrons/nm^2). This can even lead to sample sublimation in the high vacuum environment of the TEM. Liquid cell EM may provide a route to increase the tolerable dose above the cryo-EM damage threshold (see Section 17.4.2), since the liquid environment permits diffusion of radicals, gases, and thermal energy away from the sample area, while simultaneously maintaining the liquid at atmospheric pressure [11–13]. The flow capabilities of commercial fluid cells offer an additional convective transport mechanism to continuously remove damaging radicals during imaging (cf. Section 16.2.2). As discussed in Chapter 17, protein crystals have been imaged with liquid cell TEM at 2.7 nm resolution with a cumulative electron dose of 3500 electrons/nm^2 [14].

By using the imaging and analysis techniques developed for cryo-EM (Section 23.2.1), liquid cell EM of biomolecular structures could become a power tool for structural biology. The single-particle cryo-EM reconstruction method [3] should translate well to liquid cell electron microscopy [9], albeit with a few caveats that are discussed in Chapter 17 and in Section 23.2.3. While protein crystallography and single-particle cryo-EM only permit static imaging of biomolecular structure, the dynamic and flow capabilities of liquid cell electron microscopy will add a temporal dimension to biomolecular imaging, possibly allowing determination of protein structure in four dimensions (three spatial and one temporal). Ultrafast imaging techniques such as DTEM (Chapter 22) could enable direct real-time visualization of protein structure reorganization or binding in response to external stimuli that could be applied using fluid flow or the pump-probe capabilities of the DTEM [15].

23.2.3 Obstacles for Determining Biomolecular Structure via Liquid Cell Electron Microscopy

Determining biomolecular structure in liquid water introduces additional technical obstacles and considerations. While biological material is frozen in place in cryo-EM samples, biomolecules in liquid water are free to translate and rotate by random Brownian motion. To image mobile biomolecular species, one of two strategies must be implemented: (1) immobilization of the structures onto the fluid cell window surfaces, or (2) high temporal resolution imaging with ultrafast microscopy techniques such as DTEM and direct electron detectors. For examples of these two strategies the reader is referred to Chapter 17 for specific binding of rotovirus particles and Chapter 22 on the prospects of DTEM imaging. While biomolecules may be more dose-tolerant in liquid water than in ice, radiation damage mechanisms must still be considered for determining biomolecular structure with liquid cell electron microscopy. Radiolysis experiments [11, 13, 16] and kinetic simulations [12] have begun to paint a picture of the damaging effects of radiolysis on various types of liquid cell samples, but it is important to establish standard damage thresholds similar to those existing for cryo-EM to faithfully interpret biomolecular structures determined via liquid cell electron microscopy.

23.3 Elucidating Fundamental Biomineralization Mechanisms via *In Vivo* Imaging

23.3.1 Biomineralization in Single Cell Organisms

Biomineralization of inorganic materials is well known to occur in a number of organisms, ranging from complex mammals to single cell organisms [17]. Chapter 15 describes several aspects of liquid cell TEM related to biomineralization studies. The mechanism of biomineralization in multi-cellular organisms is highly complex, so single cell organisms are often used as model systems. Examples include biomineralization of calcium carbonate in cyanobacteria [18], magnetite and greigite in magnetotactic bacteria [19], and silica in diatoms [20].

These organisms often form impeccable biomineralized structures, such as the nanoscale magnetosome magnetite crystals formed by magnetotactic bacteria (Figure 23.2), or the intricate silica shells of diatoms. As described in Chapter 15, nucleation and crystallization of biomineralized nanomaterials is often mediated by biomacromolecules in single cell organisms, such as membrane-bound proteins that bind iron in magnetotactic bacteria [22] or silaffin peptides that direct biosilica formation in diatoms [23]. Biomimetic syntheses employ these biomacromolecules to produce synthetic nanomaterials at near-ambient conditions [24], which could conventionally only be made at high temperature or pressure (see Section 23.5). Because of the benefits of biomimetic synthesis of nanomaterials, understanding the fundamental mechanisms of biomineralization in single cell organisms is important for a wide range of materials applications [24], including nanotechnology [25], functional materials [26], and mesostructured materials (see Section 23.6) [27].

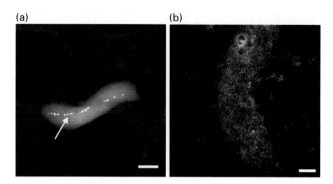

Figure 23.2. Biomineralized magnetite magnetosome nanocrystals in magnetotactic bacteria. (a) HAADF-STEM image of a glutaraldehyde-fixed magnetotactic bacterium containing a single chain of biomineralized magnetite nanocrystals (white arrow). Scale bar is 500 nm. (b) False colored EFTEM elemental map of a magnetotactic bacterium. Green represents oxygen while purple represents iron. Scale bar is 200 nm. Reproduced with permission from Ref. 21.

23.3.2 Characterization of Biomineralized Nanomaterials with TEM

TEM has been applied widely to characterize inorganic biomineralized nanomaterial structure and chemistry at high spatial resolution [28]. Cryo-TEM has been employed successfully to image whole single cell organisms, such as magnetotactic bacteria containing magnetite magnetosomes [29], as well as nanoscale biogenic materials [30]. However, as discussed above for protein studies, the vitreous ice environment of cryo-TEM samples may not be representative of the native hydrated state due to sample preparation and cryo-protectants, and cryo-TEM is also limited in temporal resolution. Liquid cell EM therefore has a role to play.

23.3.3 Towards Observations of Biomineralization in Single Cell Organisms *In Vivo*

23.3.3.1 Case Study: Magnetite Biomineralization in Magnetotactic Bacteria

Magnetotactic bacteria form magnetite and greigite nanocrystals with consistent species-specific morphologies and nearly perfect mineral crystal structures and faceting, which leads to well-defined magnetic properties that are often superior to those of non-biogenic nanocrystals [31]. The current picture of the magnetite biomineralization process begins with cellular uptake of soluble iron, followed by redox reactions, transport to the magnetosome vesicles, and finally nucleation and growth of mature magnetite nanocrystals [32]. Despite intense research on the mechanisms of magnetite biomineralization in magnetotactic bacteria, many aspects of the process remain unclear. The pathway for soluble iron uptake and transport is not well known due to a lack of *in vivo* observations of the biomineralization process. Iron binding occurs in the magnetosome membrane at some point in the process, likely directly prior to nanocrystal nucleation; however, the exact timeline of events is unclear. Likewise, specific proteins contained within the magnetosome membranes have been identified to be crucial in iron binding and controlling the size and morphology of the fully grown magnetite nanocrystals [22, 33]; however, their exact role in templating the nanocrystal nucleation and growth has not been elucidated.

Liquid cell electron microscopy is an ideal technique for visualizing biomineralized nanostructures in their native cellular environment, and has the potential to answer many of the unresolved questions posed for biomineralization in single cell organisms. In the following section, we highlight an approach for observing magnetite magnetosomes in their natural cellular environment via correlative liquid cell STEM and fluorescence microscopy [21]. While this example is specific to magnetotactic bacteria, a similar approach is expected to be applicable to other biomineralizing single cell organisms.

23.3.3.2 Correlative Fluorescence Microscopy and STEM of Magnetotactic Bacteria

Figure 23.3 shows the exciting possibilities that arise from correlative fluorescence microscopy and liquid cell STEM [21]. We imaged the magnetotactic bacterium *Magnetospirillum magneticum* strain AMB-1 containing nanoscale biomineralized magnetite using liquid cell STEM. Cells of *M. magneticum* were stained with a green fluorescent nucleic acid stain and a red nuclear counterstain to indicate cell viability via post-STEM fluorescence microscopy. The correlated STEM and fluorescence technique was implemented as follows: Initial fluorescence microscopy of the stained bacteria contained in the liquid cell was performed to identify viable bacteria. STEM imaging was then performed to visualize the biomineralized magnetite nanocrystals in their native cellular environment (Figure 23.3a). Post-STEM fluorescence imaging was then performed to verify that the STEM images contained viable bacteria (Figure 23.3b). The successful spatial correlation of individual bacteria between STEM and

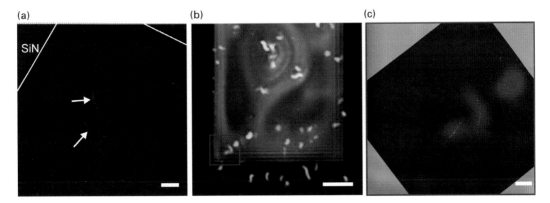

Figure 23.3. Correlative fluorescence and liquid cell STEM imaging of viable cells of *M. magneticum*. (a) False colored, background-subtracted HAADF-STEM image of two bacterial cells near the corner of the Si_xN_y window. The magnetosome chains appear in purple and are indicated with white arrows. (b) Post-STEM composite fluorescence image of the same fluid cell sample. (c) Correlated STEM and composite fluorescence image of the bacterial cells highlighted in the red box in (b). The scale bar is 1 μm in (a) and (c) and 10 μm in (b). Reproduced with permission from Ref. 21.

fluorescence microscopy revealed that the bacterial cell walls remained intact following STEM imaging (Figure 23.3c).

This study suggests new radiation damage criteria for imaging viable bacteria with liquid cell STEM. First, STEM imaging at electron doses below 100 electrons/nm^2 resulted in no detectable ultrastructural damage to the cells, suggesting that the biomineralized magnetite magnetosomes were imaged in an environment representative of the original live cell, even if the radiation dose altered or arrested cellular function. The imaging electron dose was under the established damage threshold for both nucleic and amino acids [10], further corroborating that the ultrastructure of the cells was unaltered. While the electron doses used in these experiments were orders of magnitude higher than those shown to cause reproductive death of *Escherichia coli* [34], the thin sample geometry and scanning electron probe likely decreased the amount of radiation absorbed by the cells.

23.3.3.3 Potential Future Studies of Biomineralization in Magnetotactic Bacteria

Liquid cell STEM imaging presents opportunities for future dynamic studies of biomineralization in magnetotactic bacteria, and biomineralizing single cell organisms in general. Biomineralization induction studies should be possible by introducing an iron source to iron-deficient bacterial cells via the flow capabilities of commercial liquid cells [21]. While it is almost certain that the biomineralization process will be altered due to radiation damage, the question remains whether damage incurred during liquid cell EM imaging will inhibit the bacteria's ability to biomineralize magnetite. One possible approach is to image several bacteria during biomineralization and capture each step of the process in a separate bacterium, such that each STEM image is of an initially live cell. Correlative liquid cell STEM and fluorescence microscopy could be used to establish the connections between biomolecular processes and biomineralization. For example, the localization of membrane proteins during magnetite biomineralization has been visualized using fluorescence microscopy [35]; correlation of fluorescence images of protein localization with liquid cell STEM images of the bacterial magnetosomes would aid in elucidating the role of protein localization and templating in the biomineralization process.

Further investigation of magnetite biomineralization could involve visualizing soluble iron uptake using the z-contrast capabilities of HAADF-STEM imaging. There have been several demonstrations that soluble ions in water can produce substantial HAADF contrast [36, 37]. Another route for visualizing iron uptake might involve STEM-EELS or STEM-XEDS spectral imaging of iron in bacteria; both of these techniques have experienced recent technological advancements that allow for fast spectral mapping of solution phase sample chemistry, as described in Chapter 20.

23.3.4 Establishing Tolerable Radiation Doses for *In Vivo* Liquid Cell Electron Microscopy

As discussed for proteins in Section 23.2.3, strict radiation dose limits must be set for live cell imaging in order to assess the validity of the information obtained from liquid cell microscopy. Quantitative methods of determining radiation dose response have not yet been applied to liquid cell electron microscopy. Future experiments may adapt the principles of radiochromic liquid dosimetry to liquid cell imaging, i.e. introduce radiochromic dyes to the liquid cell to generate a dose–response curve [38]. For now, we use *ex situ* dose limits to guide liquid cell experiments.

Dose limits for electron irradiation of live cells established *ex situ* involve, for example, irradiating bacterial pellets on wet filter paper [34] or use of cryogenic samples [10]. However, there are several reasons why the lethal electron doses obtained for common microbial cultures by these *ex situ* methods may not be applicable for liquid cell experiments. In cryo-TEM, radiation effects are reduced due to the temperature, but also highly localized due to slow diffusion in the vitreous ice [14]; in the liquid cell, diffusion in the water vs. ice environment can alter radiation damage thresholds greatly from the established cryo-TEM limits. Comparison between TEM and *ex situ* bulk samples is also not straightforward; the energy absorbed may be different in a thin sample that allows transmission of the majority of the irradiating electrons [21], and furthermore the density of the biological cells in *ex situ* radiation experiments is orders of magnitude larger than that in liquid TEM experiments [39]. Molecular level damage has been described in bacterial cultures with 10^{7-8} cells per ml exposed to high [40] and low [41] power pulsed laser irradiation, but the damage mechanisms may not be directly comparable with electron beam irradiation.

Further systematic experiments must be carried out to establish various radiation damage limits for live imaging of bacterial cells via liquid cell STEM or TEM. One of the challenges encountered in assessing bacterial viability is the difficulty in culturing irradiated bacterial cells, as some of these microorganisms can be *viable* but *not culturable*. De Jonge and co-workers are employing correlative fluorescence microscopy in systematic experiments to investigate the effect of radiation damage on eukaryotic cells imaged by liquid cell STEM, with results described in Chapter 17. A possible future route to further explore radiation damage of live cells could be to employ various fluorescence staining protocols to examine the function of efflux systems and other metabolic processes in microbes following irradiation at various cumulative electron doses.

Another approach is to combine *in situ* HAADF-STEM liquid cell imaging with fluorescence imaging in one instrument. This would establish direct visual correlation between the electron dose on the specimen and the fluorescence signal, and access cell viability.

23.4 Visualizing Electromagnetic Fields and Nanoparticle Interactions in Biomolecular Systems

23.4.1 Membrane Potentials and Cellular Signaling with Electric Fields

Biological systems are susceptible and responsive to electromagnetic fields and signaling. The ability to visualize field-induced interactions between individual protein molecules is of significant fundamental and practical interest to numerous biomedical applications [42–44]. Proteins undergo conformation changes and bind a variety of ions. The local electric fields present can be inferred to some extent with spectroscopic and computational approaches. Section 23.2 described how these biomacromolecules can be imaged with electron microscopy in liquid with high accuracy and spatial resolution, despite being composed predominantly of light elements. Visualization of electromagnetic fields in proteins *in situ* in liquid would reveal additional information about the dynamics of protein self-assembly, aggregation, and folding. Building an experimental charge map of the protein could, for example, help to visualize the C- and N-termini in amphiphylic proteins and their role in highly dynamic processes taking place in protein solutions, from stimuli response to specific binding and identification of binding sites. Mapping electric fields could be ultimately accomplished via liquid cell electron holography, as described below.

23.4.2 Electron Holography in Biological Samples

23.4.2.1 Common Challenges and Limitations Associated with Electron Holography in Biological Samples

Electron holography is most commonly used by physicists and materials scientists [45–47], while reports on its use for biological objects remain scarce [48]. The benefits of holography can be understood if we consider the incident electrons as a plane wave that experiences a phase shift when interacting with matter. In biological samples consisting predominantly of light elements, the induced phase shift is small, leading to a poor overall contrast in conventianal images. This necessitates the use of contrast-enhancing techniques such as negative staining with heavy metal compounds or imaging at high defocus [48]. Electron holography provides simultaneously the phase and amplitude of the "object wave" exiting the sample. It can detect minute phase shifts, eliminating the need for defocusing or staining techniques and permitting characterization of materials in their unaltered state [48]. Of the available electron holography techniques, the most commonly used are *in-line* electron holography (Fresnel imaging) and *off-axis*, or "sideband", electron holography [45]. In the in-line scheme, an object is placed into a divergent electron beam and the detector forms a pattern made up of the interference between the object wave and the reference (non-scattered) wave. In the off-axis scheme, the electron beam is split by means of a charged thin wire (biprism); one beam passes through the sample and the other through vacuum, and the two waves are recombined at the detector to produce an interference pattern [45].

The application of electron holography to biological materials has serious restrictions. First, compared to conventional electron microscopy techniques, holography requires higher doses. Next, to accumulate an adequate phase shift given the weakly interacting beam and the light sample atoms, samples should be thicker, which leads to more inelastic scattering and background noise. Furthermore, the region of interest must be in close vicinity to a hole (vacuum) to obtain a reference wave, which otherwise must be conducted through the substrate, again increasing the noise in the phase image. Finally, only a small field of view is effectively utilized due to the available area over which the beams can interfere in the normal operating mode.

23.4.2.2 A Positive Outlook for Liquid Cell Electron Holography

Even though biological matter is prone to radiation damage, a number of unstained biological specimens have been successfully studied with off-axis electron holography, including ferritin, tobacco mosaic virus, collagen bacterial flagellum, and protein S-layers [48]. Furthermore, magnetic fields due to the presence of biomineralized magnetite in magnetotactic bacteria have been visualized using off-axis electron holography [49–51]. In-line electron holography has been used to study DNA, bacteriorhodopsin, tobacco mosaic virus, and collagen [48], all imaged at nanometer resolution. Some biomimetic systems have proven resilient to beam damage under low dose imaging conditions. However, the problem of electron beam damage in liquid remains a serious obstacle in attaining the desired resolution while maintaining minimal disruption of the specimen (Chapter 8).

To address the experimental challenge associated with acquisition of the reference wave when using a liquid cell, we could for example design the liquid cell so as to include a series of voids (vacuum references) and liquid-filled spaces: the hologram utilizes a fairly small field of view, and a cell could be constructed in such a way as to designate a small area as a reference, ideally positioned within several micrometers of the region of interest. The holes in this designated area of the cell could be strategically microfabricated or drilled using a focused ion beam in places where membranes are attached or joined by posts. The answer to this challenge might be already within reach: on occasion, microfabricated cells are assembled in such a way that the windows are touching each other (i.e. a sealed pocket is formed, rendering them unsuitable for the liquid flow experiments). Assuming that the windows can be manufactured sufficiently thin, such an arrangement could be used to obtain the reference wave from the region with the double-thick Si_xN_y. The charging effects of Si_xN_y on the electron hologram, however, would have to be considered. Similarly, working with graphene liquid cells (cf. Chapter 19), the reference wave could be obtained from the region formed by the two graphene layers next to the liquid pocket. Furthermore, when studying nanometer-sized objects in liquid, it might be entirely possible to use the reference wave acquired from liquid and compare it to the "object wave" acquired from the region containing the suspended nanoparticles. Finally, the reference wave can be recorded in vacuum with a partially retracted holder, although this would be the least ideal approach.

Ferritin is a biological mineral–protein composite, and its iron hydroxide core produces a reasonable phase shift and notable contrast in electron holograms, as shown in Figure 23.4. Recent nanometer-resolution low energy electron holography of ferritin has succeeded in imaging a single protein molecule attached to a carbon nanotube suspended over a hole [52]. Imaging required an extended period of time without any sign of radiation damage. While replicating the exact experimental conditions could be problematic, it might be possible to utilize this kinetic hologram approach for the *in situ* analysis of free-floating ferritin in liquid [52]. An additional step suggested in this study involves combining holography with coherent diffraction imaging [52]. Since ferritin has been well characterized with electron

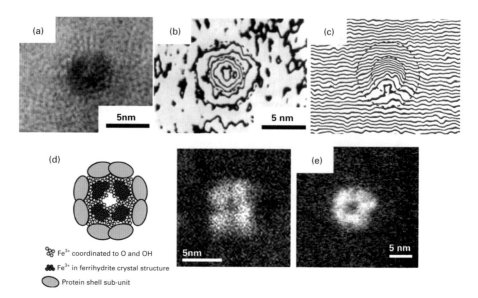

Figure 23.4. Observations of ferritin with electron microscopy: (a–c) Electron holography study of unstained ferritin. This is compared to (d) characterization of the ferritin molecule *ex situ* and (e) *in situ* in a liquid cell using HAADF-STEM. Reproduced with permission from (a–c) Ref. 53, (d) Ref. 54, and (e) Ref. 37.

holography, it could serve as an initial test of the feasibility of liquid cell electron holography.

Currently, the notion of *electron holography in liquid* as an experimental technique is deemed impractical. While it is possible to predict the phase shift values for a specific sample based on the inner potential of the sample and its thickness and compare these values to the reported phase detection limits [55], the experimental data are rarely available, and without new data the field struggles to progress. The technical difficulties, however, should not rule it impossible.

23.4.2.3 Visualization of Particle–Biomaterial Interactions

Chapter 16 has outlined the progress in observing nanoparticle uptake and binding to eukaryotic cells using liquid cell STEM. Incorporation of various magnetic nanoparticles into simpler systems involving biocompatible matrices, scaffolds, and vesicles is of relevance to a number of applications in the field of biotechnology. The majority of experiments performed to determine nanoparticle uptake properties employ *ex situ* cryo-EM analysis in an attempt to visualize the particle–biomacromolecule interactions. An example of magnetite nanoparticle loading into a vesicle bilayer is shown in Figure 23.5, where the mechanism for nanoparticle interaction with the vesicle is inferred based on the cryo-TEM images [56].

Assuming that the electron beam damage to nanoparticle–interface systems can be minimized, visualization of nanoparticle–interface interactions in liquid water

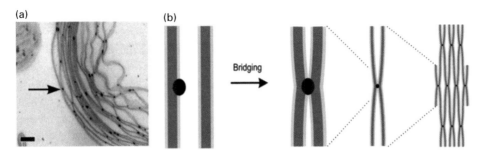

Figure 23.5. (a) Cryo-EM image of magnetite nanoparticles loaded into polyimide–polyethylene oxide (PI-PEO) vesicle bilayers with stable nanoparticle connections (arrow) formed between adjacent bilayers. (b) Schematics of nanoparticle-induced bilayer pairing. The proposed role of magnetite nanoparticles in bilayer bridging is inferred based on the static cryo-EM images. Scale bar 200 nm. Reproduced with permission from Ref. 56.

might provide important clues to the dynamic processes taking place. The cumulative electron dose in the liquid cell can be minimized by adapting cryo-EM "low dose" techniques, such as focusing on one region then moving to obtain the final image. Lowering the accelerating voltage might also be useful, with resolution being the obvious trade-off. To allow for lower accelerating voltages while still maintaining nanometer resolution in the TEM or STEM, the window thickness can be reduced to 10–20 nm.

There is another challenge associated with the characterization of magnetic particles in liquid. The magnetic structure in nanoparticles is difficult to determine with any technique other than TEM, due to the obvious need for high resolution, and *in situ* TEM characterization is crucial for understanding bulk magnetometry measurements [47, 57–60]. Another open question is the evaluation of magneto-hysteretic heating of an individual nanocrystal, which is particularly important due to the growing demand for materials in the field of magnetic fluid hyperthermia [61, 62]. Devising a means to control the movement of magnetic nanoparticles *in situ* in the liquid cell would allow one to observe the interactions of magnetic nanoparticles with each other and with macromolecular matrices in fluid, possibly while being magnetized. With the fabrication of a magnetizable liquid cell holder, it might be entirely possible to visualize nanoparticle interactions. Such a development, currently impractical due to geometric constraints around the liquid cell, might become possible by working closely with specialty TEM holder manufacturers. Visualization of nanoparticle response to external magnetic stimuli would provide fundamental tools for studying such interactions at the nanoscale, deemed impractical due to the lack of suitable characterization tools. Hologram noise will likely be detrimental for the image processing needed to extract magnetic information from the interference pattern. Additionally, controlling an applied magnetic field in the presence of the fields in the electron microscope poses a challenge. It will be necessary to utilize the expertise developed for *in situ* analysis of magnetic structures.

23.5 Biomimetics

Biomimetics utilizes our understanding of the biochemical and biomineralization processes found in nature to design and fabricate, in aqueous media under mild conditions, new functional materials and composites with specific properties. The use of proteins, peptides, viruses, polysaccharides, nucleic acids, and polymers as matrices, scaffolds, and templating agents ultimately permits molecular-level control over the new material. Working with these classes of bio-inspired materials presents a unique set of challenges, addressed in Chapter 15 and below in more detail.

Biomimetic syntheses are carried out *in vitro,* and the resultant nanomaterials are analyzed to infer the role the biomacromolecule plays in their formation. An idealized reaction schematic is often proposed based on the morphology of the product, even though the characterization employed does not provide sufficient information to define the actual process taking place during the reaction. Lacking the details tying the pieces of the proposed schematics together, such empirical models could lead to incorrect conclusions about the biomolecule-mediated nucleation and crystallization processes. This approach is not necessarily incorrect; it is the best that can be done with the techniques available. Liquid cell electron microscopy could therefore be a transformative technique for understanding the fundamental processes involved in biomimetic and biomacromolecule templated nanoparticle synthesis, by revealing the nanoscale dynamics that are lacking in the current proposed mechanisms.

23.5.1 Amphiphilic Block Copolymers and Hydrogels

Using self-assembly and hierarchical templating of a variety of supramolecular structures allows preparation of controllable complex nano- and mesostructures for technologically important applications that range from catalysis to optoelectronics and particle separation processes. The intricate network of macromolecule-produced cavities and restricted spaces acting as nanoreactors is thought to provide control over reagent delivery and formation of the functional nanoscale material [63–69]. Amphiphilic block copolymers and surfactants self-assemble into a variety of supramolecular structures that can be used as templates to prepare functional materials. A large number of minimal surface structures can be predicted by utilizing a molecular dynamics approach [63]. The model predictions can be tested using scattering techniques, but direct imaging of the structures is not usually possible. Liquid cell TEM permits direct nanoscale visualization of self-assembled macromolecular structures in the native fully hydrated state. An example is shown in Figure 23.6. It is worth pointing out that HAADF-STEM can image these low atomic number structures in liquid, even in the absence of staining or contrast enhancing agents [70]. Moreover, liquid cell microscopy allows for observation of the dynamics of self-assembly. Processes that be triggered by changes in pH or introduction of specific reagents to the reaction vessel can be replicated in a continuous flow liquid cell with relative ease. Thermoresponsive self-assembly, however, would require modification of the liquid cell holder, perhaps including a Peltier cooler,

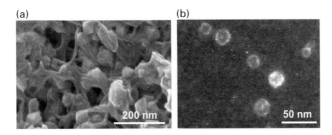

Figure 23.6. Direct visualization of self-assembled macromolecular structures used in biomimetic reactions. (a) Agarose gel imaged with low vacuum SEM. Reproduced with permission from Ref. 64. (b) HAADF-STEM image of micelles in Pluronic® F-127 block copolymer acquired in liquid, with the individual micelles visible due to the higher scattering compared to the surrounding liquid. Unpublished. Such hierarchical structures are usually characterized via scattering techniques or with cryo-TEM, but are visible in the liquid cell.

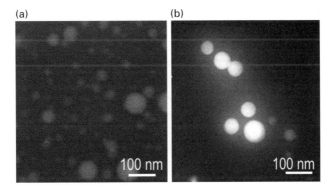

Figure 23.7. Iron-binding protein micelles: (a) Liquid cell HAADF-STEM image of hydrated micelles in buffer solution; (b) Z-contrast enhancement of iron-incubated protein micelles is attributed to binding and subsequent surface localization of iron. Reproduced with permission from Ref. 37.

to enable temperature control in a mild, physiological range. As the field progresses, temperature-controllable liquid cells are becoming a reality, as discussed in Chapter 6.

23.5.2 Visualization of Protein Domains

Functional domains of many biomineralization proteins that are amphipathic (having both hydrophilic and hydrophobic regions) can be visualized with the liquid cell. The initial assumption behind such an effort is that charge-bearing protein domains participate in simple ion binding. Figure 23.7 shows a recent example of successful visualization of C-termini in the acidic biomineralization proteins MamC and Mms6 self-assembled into macromolecular complexes [37, 70]. The incubation of these protein micelles with ferric chloride solution *in situ* led to a noticeable enhancement in the Z-contrast of the protein

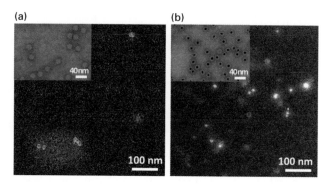

Figure 23.8. HAADF-STEM images of brome mosaic virus protein dimers acquired *in situ* with the liquid cell: (a) Empty protein dimer shells, (b) Au nanoparticles mixed with protein dimer. The insets show corresponding images of the negatively stained material obtained with cryo-EM. Adapted from Ref. 77 and unpublished data.

micelles, attributed to binding and subsequent surface localization of iron. Iron binding of the self-assembled bacterial recombinant acidic biomineralization protein, Mms6, in the solution phase provided the first visual indication of the specific surface interactions of self-assembled protein micelles with metal ions [37]. Since the iron binding took place preferentially on a static extended micellar surface, the observation was easy to interpret; however, it remains unclear whether iron binding to individual monomers of protein followed by dynamic self-assembly would produce a similar outcome.

The acidic C-termini of the biomineralization proteins are believed to be involved in cation binding and nucleation control, whereas N-termini are typically associated with the oligomerization of proteins, from microfibrils to viral capsid shells [71–76]. The visualization of the individual N-terminal domains might prove problematic; however, the self-assembled protein complexes can be visualized in liquid, as exemplified in Figure 23.8. Moreover, this method offers a rare opportunity to observe interactions between nanoparticles and protein complexes along with the formation of the resultant functional entities in real time. Unwanted aggregation of these proteins can be remedied by anchoring them into liposomes or lipid nanodiscs.

23.5.3 Biomacromolecule-Mediated Nanoparticle Nucleation and Growth: Biomimetic Magnetic Nanoparticles

Biomimetic approaches permit room-temperature fabrication of a variety of magnetic nanostructures with sizes and shapes not realizable via conventional syntheses [62, 78–81]. Different hypotheses have been proposed concerning the roles of biomolecular agents in nanoparticle nucleation and growth [82, 83]. The proteins, larger complexes, and membrane components reportedly promoting nucleation *in vivo* are presumed to expose positively charged residues to negatively charged crystal surfaces, while acidic components bearing negatively charged functionality are thought to act as nucleation inhibitors capable of stabilizing an amorphous structure through the coordination of iron [83, 84].

Several questions are critical for the understanding and rational design of biomimetic nanostructured magnetic materials: What are the steps involved in protein-mediated nanoparticle nucleation? Where does nucleation occur with respect to the template? How does the protein affect nucleation? What role does protein aggregation play in the templating process? Does the surface localization of the protein affect its templating ability? Recent reports bring the field closer to answering these fundamental questions [24], yet systematic *in situ* experiments at high spatial resolution are needed to reveal the specific nanoscale interactions controlling biomimetic syntheses.

The field of biomacromolecule-mediated particle growth would greatly benefit from development of reliable protocols aimed at visualizing biomimetic reactions in the liquid cell *in situ*. This would involve exercising high precision control over reagent delivery in the liquid cell along with modeling of the fluid flow in the confined nanometer-sized spaces, combined with experiments aimed at visualizing the movement of liquid in the cell.

23.6 Mesocrystal Formation

23.6.1 Mesocrystals and Current Understanding of Their Growth Mechanisms

Mesocrystals are superstructures with sizes on the order of hundreds of nanometers to several micrometers formed by mutual alignment of colloidal nanocrystal building blocks (Figure 23.9) [27]. The ordered superstructures can be one [85], two [86, 87], or three dimensional, and typically diffract as a single crystal due to the coordinated

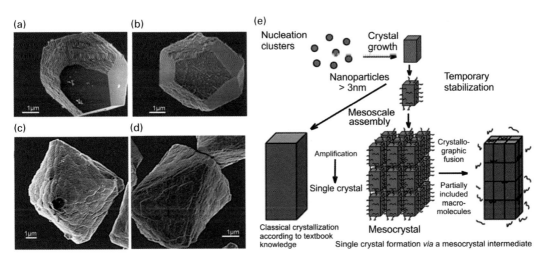

Figure 23.9. (a)–(d) SEM images of calcite mesocrystals produced using the ammonium diffusion method. In this method, calcite crystals precipitate from calcium ions in response to introduction of carbon dioxide and ammonium released from decomposition of ammonium carbonate. Reproduced with permission from Ref. 90. (e) Single-crystal formation via classical crystallization and mesocrystal assembly. Reproduced with permission from Ref. 27.

crystallographic arrangement of the constituent nanocrystals. These structures are important for applications that include functional materials [88], catalysis [89], and biomineralization systems [90]. Mesocrystals form by non-classical particle-mediated growth mechanisms dictated by interparticle interactions [91], nanocrystal geometry [92], external fields [87], and organic additives [93]. In general, some type of anisotropy in the interparticle interaction is needed to form mesocrystals. As demonstrated in Chapters 9 and 15, liquid cell electron microscopy excels at observing particle-mediated and non-classical growth mechanisms during nanoparticle formation.

Colfen and co-workers have proposed a general mechanism for mesocrystal formation by self-assembly of nanocrystals in solution [27]. In this model, nanocrystals are formed by nucleation and growth in the presence of a stabilizer that quenches their growth. Then crystallographic alignment of the nanocrystals proceeds by one or more of the following: (1) physical external fields or mutual alignment of crystal faces (i.e. oriented attachment), (2) formation of mineral bridges between neighboring nanocrystals, and (3) entropy-driven alignment [94].

Many aspects of mesocrystal formation are not well understood [88]. These include predicting mesocrystal formation conditions, understanding the role of anisotropies in controlling mesocrystal morphology and self-assembly, and developing a general theoretical mechanism for mesocrystal growth. Liquid cell electron microscopy coupled with advanced imaging techniques, such as electron holography (Section 23.4) and DTEM (Chapter 22), has immense potential to address these open questions. Here we discuss some of the issues and suggest how future liquid cell electron microscopy experiments could address them. We conclude with a short discussion of the potential obstacles for visualizing mesocrystal formation via liquid cell electron microscopy.

23.6.2 Towards Predicting Criteria for Particle Mediated Growth and Mesocrystal Formation

As with many nanocrystal synthesis routes, mesocrystal syntheses are mostly developed from empirical observations. Rational design of a mesocrystal synthesis route based on theoretical understanding is often inconceivable. Understanding the mechanisms is difficult due to a number of factors, including the wide range of spatial and temporal scales involved with mesocrystal formation [88], the often transient nature of mesocrystals, and the combined influence of kinetic factors and interparticle interactions [96]. The wide range of scales involved often necessitates multiple characterization techniques. Typical approaches use a combination of high resolution imaging techniques such as cryo-TEM and *in situ* ensemble spectroscopy or light scattering techniques [97]. Liquid cell electron microscopy provides both high resolution and *in situ* capabilities, so is therefore uniquely capable of capturing transient events during mesocrystal formation. For instance, we have recently directly visualized dynamic self-assembly pathways and kinetics for one-dimensional and branched Au nanoparticle chains (Figure 23.10a) [95]. In this case of far-from-equilibrium self-assembly, dynamic effects, specifically the nanoparticle mobility, determined the kinetic pathways and final morphology of nanoparticle chains (Figure 23.10b). Direct visualization of mesocrystal

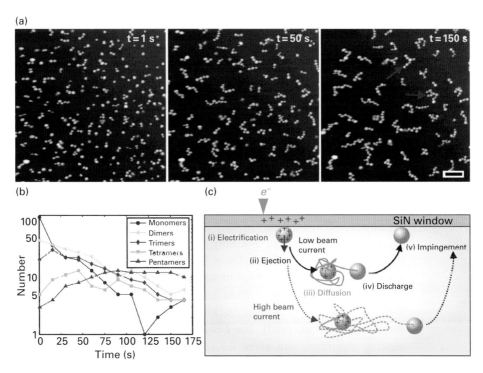

Figure 23.10. (a) Electron beam-induced nanoparticle self-assembly in the liquid cell. (b) Self-assembly kinetics extracted from *in situ* liquid cell electron microscopy movies. Arrows indicate branched chains. (c) Schematic representation of the hopping-mediated diffusion of a nanoparticle in the liquid cell. Solid lines indicate ejection, diffusion, and impingement of the nanoparticle at low beam currents, while the dashed lines indicate the same processes at high beam currents. Reproduced with permission from Ref. 95.

formation mechanisms will enable a better understanding of the physical principles involved and allow tailoring of mesocrystal synthesis routes based on physical understanding rather than empirical observations.

23.6.3 Direct Visualization of Anisotropic Interparticle Interactions

Mesocrystals can form with nearly perfect faceted surfaces having crystallographic order that is entirely different from their nanocrystal building blocks [94]. This observation indicates that classical crystallization mechanisms, e.g. nucleation followed by growth by monomer attachment, are not dominant during mesocrystal formation. It is thought that the faceted mesocrystal morphologies are due to anisotropies in the nanocrystal geometry [85, 92] and colloidal interactions [86, 98]; however, the exact roles of these anisotropies are not well known. Liquid cell electron holography (Section 23.4.2) could allow for direct visualization of anisotropic interactions such as electric or magnetic nanoparticle dipoles that are thought to lead to formation of various mesostructures [98, 99]. For instance, there is disagreement as to the relative contributions of electrostatic [100] and dipolar interactions [99] during the formation of one-dimensional nanocrystal chains that could be resolved by directly visualizing the electromagnetic fields surrounding the nanocrystals during self-assembly. We refer the reader to Chapters 9, 15, and 22 for examples of nanoparticle growth via non-classical anisotropic and aggregative mechanisms, such as oriented attachment, diffusion-limited aggregation, and dipolar-mediated attachment.

23.6.4 Obstacles for Visualizing Mesocrystal Formation in Real Time

23.6.4.1 Advanced Image Analysis Techniques for Particle Tracking

To improve understanding of mesocrystal growth mechanisms, quantitative liquid cell experiments must be performed and compared with theoretical models and *ex situ* experiments. This will likely involve acquiring *in situ* liquid cell STEM or TEM movies of nanocrystals during self-assembly and final mesocrystal formation. To date, quantitative data such as interparticle forces (Chapter 9), aggregation parameters (Chapter 22), orientational order parameters of nanoparticle superlattices [101], and Brownian motion of individual nanocrystals in liquid (Chapters 9 and 22) have been obtained from *in situ* liquid cell movies. Automated tracking of tens to hundreds of interacting objects is notoriously difficult [102], but current efforts in developing advanced multitarget tracking algorithms will allow for automated analysis of hundreds of nanoparticles. Chapter 22 outlines several examples where multitarget image analysis algorithms have already been implemented to extract various types of data from *in situ* movies of interacting nanoparticles. Future implementation of multitarget particle tracking algorithms is imperative to provide quantitative insights into the interparticle interactions and self-assembly kinetics that dictate mesocrystal formation.

23.6.4.2 Establishing Equivalent Aging Conditions in the Liquid Cell

Another obstacle for visualizing mesocrystal formation using liquid cell electron microscopy is creating conditions similar to those in *ex situ* aging experiments. In particular, Brownian motion of nanocrystals during aging allows them to rapidly sample many conformations with respect to neighboring nanocrystals, eventually leading to mutual crystallographic alignment and mesocrystal formation [85, 103]. However, the diffusion coefficients of nanoparticles in both graphene and Si_xN_y liquid cells are known to be significantly smaller than those predicted for bulk liquids (Chapters 9 and 13). Due to observations of sporadic jumps and beam current-dependent diffusion coefficients, nanoparticle diffusion during liquid cell electron microscopy is likely an electrostatically mediated surface diffusion process in most cases (Figure 23.10c) [95]. This reduced mobility is useful in permitting real-time imaging of the nanoparticles: if diffusion coefficients were two to three orders of magnitude higher, nanoparticle motion would be too rapid to allow imaging (Chapter 22). To establish an aging environment in liquid cells comparable to *ex situ* conditions, nanoparticle interactions with the fluid cell windows must be reduced. It has been suggested that graphene leads to weaker interaction than Si_xN_y [104], but systematic experiments have not been performed. Changing the chemistry of the window surface through functionalization may be one possible route to alter the nanoparticle–window interaction. Faster imaging strategies involving DTEM [15; Chapter 22] or direct electron detectors [105] will then be necessary to image the rapidly diffusing nanoparticles as they assemble into mesocrystals.

23.7 Conclusions

We have highlighted several biomaterial and biomolecular science research areas where liquid cell electron microscopy is expected to spur transformative research. Once the liquid environment maintained in the liquid cell is established as being representative of bench-top *ex situ* environments, the direct nanoscale observations afforded by liquid cell electron microscopy will have the potential to directly test our textbook understanding of biomolecular processes such as protein assembly and structure, and biomaterials processes such as biomineralization and biomimetics. These advances are contingent upon systematic studies of the obstacles and challenges currently associated with the liquid cell, including radiation damage, confinement effects, and temporal and spatial resolution limitations, all of which are intensive areas of research, as demonstrated in previous chapters. Development of new liquid cell hardware and electron microscopy tools will enable investigation, in the following decades, of many of the research avenues discussed here. And above all, the stigma of impracticality should not prevent scientists from trying new liquid cell electron microscopy experiments.

Acknowledgements

TP acknowledges support from the Department of Energy Office of Science Early Career Research Award, Biomolecular Materials Program. This work was supported by the U.S. Department of Energy, Office of Basic Energy Science, Division of Materials Sciences and Engineering. The research was performed at the Ames Laboratory, which is operated for the U.S. Department of Energy by Iowa State University under Contract No. DE-AC02-07CH11358.

References

1. D. A. C. Stapels, K. X. Ramyar, M. Bischoff *et al.*, *Staphylococcus aureus* secretes a unique class of neutrophil serine protease inhibitors. *Proc. Natl. Acad. Sci. USA*, **111** (2014), 13187–13192.
2. J. C. Kendrew, G. Bodo, H. M. Dintzis *et al.*, A three-dimensional model of the myoglobin molecule obtained by X-ray analysis. *Nature*, **181** (1958), 662–666.
3. J. Frank, Single-particle imaging of macromolecules by cryo-electron microscopy. *Annu. Rev. Biophys. Biomol. Struct.*, **31** (2002), 303–319.
4. A. Schmidt, M. Teeter, E. Weckert and V. S. Lamzin, Crystal structure of small protein crambin at 0.48 Å resolution. *Acta Crystallogr. Sect. F*, **67** (2011), 424–428.

5. X. C. Bai, I. S. Fernandez, G. McMullan and S. H. W. Scheres, Ribosome structures to near-atomic resolution from thirty thousand cryo-EM particles. *eLife*, **2** (2013), c00461.
6. A. Bartesaghi, D. Matthies, S. Banerjee, A. Merk and S. Subramaniam, Structure of beta-galactosidase at 3.2-angstrom resolution obtained by cryo-electron microscopy. *Proc. Natl. Acad. Sci. USA*, **111** (2014), 11709–11714.
7. R. M. Glaeser, Retrospective: Radiation damage and its associated "Information Limitations". *J. Struct. Biol.*, **163** (2008), 271–276.
8. R. F. Egerton, Control of radiation damage in the TEM. *Ultramicroscopy*, **127** (2013), 100–108.
9. B. L. Gilmore, S. P. Showalter, M. J. Dukes et al., Visualizing viral assemblies in a nanoscale biosphere. *Lab Chip*, **13** (2013), 216–219.
10. L. Reimer and H. Kohl, *Transmission Electron Microscopy: Physics of Image Formation* (New York: Springer, 2008).
11. J. M. Grogan, N. M. Schneider, F. M. Ross and H. H. Bau, Bubble and pattern formation in liquid induced by an electron beam. *Nano Lett.*, **14** (2014), 359–364.
12. N. M. Schneider, M. M. Norton, B. J. Mendel et al., Electron-water interactions and implications for liquid cell electron microscopy. *J. Phys. Chem. C*, **118** (2014), 22373–22382.
13. T. J. Woehl, K. L. Jungjohann, J. E. Evans et al., Experimental procedures to mitigate electron beam induced artifacts during in situ fluid imaging of nanomaterials. *Ultramicroscopy*, **127** (2013), 53–63.
14. U. M. Mirsaidov, H. Zheng, Y. Casana and P. Matsudaira, Imaging protein structure in water at 2.7 nm resolution by transmission electron microscopy. *Biophys. J.*, **102** (2012), L15–L17.
15. J. E. Evans and N. D. Browning, Enabling direct nanoscale observations of biological reactions with dynamic TEM. *Microscopy*, **62** (2013), 147–156.
16. K. L. Jungjohann, S. Bliznakov, P. W. Sutter, E. A. Stach and E. A. Sutter, In situ liquid cell electron microscopy of the solution growth of Au-Pd core-shell nanostructures. *Nano Lett.*, **13** (2013), 2964–2970.
17. H. A. Lowenstam and S. Weiner, *On Biomineralization* (New York: Oxford University Press, 1989).
18. K. Benzerara, F. Skouri-Panet, J. H. Li et al., Intracellular Ca-carbonate biomineralization is widespread in cyanobacteria. *Proc. Natl. Acad. Sci. USA*, **111** (2014), 10933–10938.
19. D. A. Bazylinski, Synthesis of the bacterial magnetosome: the making of a magnetic personality. *Int. Microbiol*, **2** (1999), 71–80.
20. M. Sumper and E. Brunner, Learning from diatoms: nature's tools for the production of nanostructured silica. *Adv. Funct. Mater.*, **16** (2006), 17–26.
21. T. J. Woehl, S. Kashyap, E. Firlar et al., Correlative electron and fluorescence microscopy of magnetotactic bacteria in liquid: toward in vivo imaging. *Sci. Rep.*, **4** (2014), 6854.
22. A. Arakaki, J. Webb and T. Matsunaga, A novel protein tightly bound to bacterial magnetic particles in *Magnetospirillum magneticum* strain AMB-1. *J. Biol. Chem.*, **278** (2003), 8745–8750.
23. N. Poulsen, M. Sumper and N. Kroger, Biosilica formation in diatoms: characterization of native silaffin-2 and its role in silica morphogenesis. *Proc. Natl. Acad. Sci. USA*, **100** (2003), 12075–12080.
24. T. Prozorov, D. A. Bazylinski, S. K. Mallapragada and R. Prozorov, Novel magnetic nanomaterials inspired by magnetotactic bacteria: topical review. *Mater. Sci. Eng. R.*, **74** (2013), 133–172.

25. C. Lang and D. Schueler, Biomineralization of magnetosomes in bacteria: nanoparticles with potential applications. In B. Rehm, ed., *Microbial Bionanotechnology* (Wymondham, UK: Horizon Bioscience, 2006) pp. 107–124.
26. T. Prozorov, P. Palo, L. Wang *et al.*, Cobalt ferrite nanocrystals: out-performing magnetotactic bacteria. *ACS Nano*, **1** (2007), 228–233.
27. H. Colfen and M. Antonietti, Mesocrystals: inorganic superstructures made by highly parallel crystallization and controlled alignment. *Angew. Chem. Int. Ed.*, **44** (2005), 5576–5591.
28. D. A. Bazylinski, A. J. Garrattreed and R. B. Frankel, Electron-microscopic studies of magnetosomes in magnetotactic bacteria. *Microsc. Res. Tech.*, **27** (1994), 389–401.
29. A. Komeili, Z. Li, D. K. Newmana and G. J. Jensen, Magnetosomes are cell membrane invaginations organized by the actin-like protein MamK. *Science*, **311** (2006), 242–245.
30. E. M. Pouget, P. H. H. Bomans, J. Goos *et al.*, The initial stages of template-controlled $CaCO_3$ formation revealed by cryo-TEM. *Science*, **323** (2009), 1455–1458.
31. D. A. Bazylinski and R. B. Frankel, Magnetosome formation in prokaryotes. *Nat. Rev. Micro.*, **2** (2004), 217–230.
32. D. Faivre and D. Schüler, Magnetotactic bacteria and magnetosomes. *Chem. Rev.*, **108** (2008), 4875–4898.
33. T. Prozorov, S. K. Mallapragada, B. Narasimhan *et al.*, Protein-mediated synthesis of uniform superparamagnetic magnetite nanocrystals. *Adv. Funct. Mater.*, **17** (2007), 951–957.
34. E. R. Epp, H. Weiss and A. Santomasso, The oxygen effect in bacterial cells irradiated with high-intensity pulsed electrons. *Rad. Res.*, **34** (1968), 320–325.
35. A. Komeili, H. Vali, T. J. Beveridge and D. K. Newman, Magnetosome vesicles are present before magnetite formation, and MamA is required for their activation. *Proc. Natl. Acad. Sci. USA*, **101** (2004), 3839–3844.
36. E. R. White, S. B. Singer, V. Augustyn *et al.*, In situ transmission electron microscopy of lead dendrites and lead ions in aqueous solution. *ACS Nano*, **6** (2012), 6308–6317.
37. S. Kashyap, T. J. Woehl, X. Liu, S. K. Mallapragada and T. Prozorov, Nucleation of iron oxide nanoparticles mediated by Mms6 protein in situ. *ACS Nano*, **8** (2014), 9097–9106.
38. ISO/ASTM51540-09, USA, 2009. Standard Practices for Use of Radiochromic Liquid Dosimetry System, ASTM International, West Conshohocken, PA, USA.
39. S. E. Fiester, S. L. Helfinstine, J. C. Redfearn, R. M. Uribe and C. J. Woolverton, Electron beam irradiation dose dependently damages the Bacillus spore coat and spore membrane. *Int. J. Microbiol.* (2012), 579593.
40. G. D. Ward, I. A. Watson, D. E. Stewart-Tull *et al.*, Bactericidal action of high-power Nd:YAG laser light on *Escherichia coli* in saline suspension. *J. Appl. Microbiol.*, **89** (2000), 517–525.
41. K. Nandakumar, H. Obika, A. Utsumi, T. Ooie and T. Yano, Molecular level damages of low power pulsed laser radiation in a marine bacterium *Pseudoalteromonas carrageenovora*. *Lett. Appl. Microbiol.*, **42** (2006), 521–526.
42. J. A. Tuszyn'ski, S. Portet, J. M. Dixon, C. Luxford and H. F. Cantiello, Ionic wave propagation along actin filaments. *Biophys. J.*, **86** (2004), 1890–1903.
43. H. F. Cantiello, C. Patenaude and K. Zaner, Osmotically induced electrical signals from actin filaments. *Biophys. J.*, **59** (1991), 1284–1289.
44. C. Merla, A. Paffi, F. Apollonio *et al.*, Microdosimetry for nanosecond pulsed electric field applications: a parametric study for a single cell. *IEEE Trans. Biomed. Eng.*, **58** (2011), 1294–1302.
45. J. M. Cowley, Twenty forms of electron holography. *Ultramicroscopy*, **41** (1992), 335–348.

46. P. Formanek, A. Lenk, H. Lichte *et al.*, Electron holography: applications to materials questions. *Annu. Rev. Mater. Res.*, **37** (2007), 539–588.
47. R. E. Dunin-Borkowski, M. R. McCartney, B. Kardynal *et al.*, Off-axis electron holography of exchange-biased CoFe/FeMn patterned nanostructures. *J Appl. Phys.*, **90** (2001), 2899–2902.
48. P. Simon, H. Lichte, P. Formanek *et al.*, Electron holography of biological samples. *Micron*, **39** (2008), 229–256.
49. R. E. Dunin-Borkowski, M. R. McCartney, M. Posfai *et al.*, Off-axis electron holography of magnetotactic bacteria: magnetic microstructure of strains MV-1 and MS-1. *Eur. J. Mineral.*, **13** (2001), 671–684.
50. T. Kasama, M. Posfai, R. K. K. Chong *et al.*, Magnetic properties, microstructure, composition, and morphology of greigite nanocrystals in magnetotactic bacteria from electron holography and tomography. *Am. Mineral.*, **91** (2006), 1216–1229.
51. E. T. Simpson, T. Kasama, M. Posfai *et al.*, Magnetic induction mapping of magnetite chains in magnetotactic bacteria at room temperature and close to the Verwey transition using electron holography. *J. Phys. Conf. Ser.*, **17** (2005), 108–121.
52. J. N. Longchamp, T. Latychevskaia, C. Escher and H. W. Fink, Non-destructive imaging of an individual protein. *Appl. Phys. Lett.*, **101** (2012), 093701.
53. T. Kawasaki, J. Endo, T. Matsuda, N. Osakabe and A. Tonomura, Applications of holographic interference electron microscopy to the observation of biological specimens. *J. Electron Microsc.*, **35** (1986), 211–214.
54. Y.-H. Pan, K. Sader, J. J. Powell *et al.*, 3D morphology of the human hepatic ferritin mineral core: new evidence for a subunit structure revealed by single particle analysis of HAADF-STEM images. *J. Struct. Biol.*, **166** (2009), 22–31.
55. H. Lichte, H. Banzhof and R. Huhle, Limitations in electron holography of magnetic microstructures. Proc. Int. Congr. Electr. Microsc., ICEM 14, Cancun, Mexico (1998), pp. 559–560.
56. M. Krack, H. Hohenberg, A. Kornowski *et al.*, Nanoparticle-loaded magnetophoretic vesicles. *J. Am. Chem. Soc.*, **130** (2008), 7315–7320.
57. H. Hopster and H. P. Oepen (eds.), *Magnetic Microscopy of Nanostructures* (Berlin: Springer, 2005).
58. A. S. Eggeman, A. K. Petford-Long, P. J. Dobson *et al.*, Synthesis and characterization of silica encapsulated cobalt nanoparticles and nanoparticle chains. *J. Magn. Magn. Mater.*, **301** (2006), 336–342.
59. M. Tanase and A. K. Petford-Long, In situ TEM observation of magnetic materials. *Microsc. Res. Tech.*, **72** (2009), 187–196.
60. G. H. Campbell, T. B. LaGrange, W. E. King *et al.*, The HCP to BCC phase transformation in Ti characterized by nanosecond electron microscopy. Solid-Solid Phase Transform. Inorg. Mater. 2005, Proc. Int. Conf., **2** (2005) 443–448.
61. Q. A. Pankhurst, J. Connolly, S. K. Jones and J. Dobson, Applications of magnetic nanoparticles in biomedicine. *J. Phys. D: Appl. Phys.*, **36** (2003), R167–R181.
62. G. Reiss and A. Huetten, Magnetic nanoparticles: applications beyond data storage. *Nat. Mater.*, **4** (2005), 725–726.
63. S. Förster, Amphiphilic block copolymers for templating applications. *Top. Curr. Chem.*, **226** (2003), 1–28.
64. T. Prozorov, Unpublished, 2013.
65. L. Zhang, S. I. Song, S. Zheng *et al.*, Nontoxic poly(ethylene oxide phosphonamidate) hydrogels as templates for biomimetic mineralization of calcium carbonate and hydroxyapatite architectures. *J. Mater. Sci.*, **48** (2013), 288–298.

66. D. Dobrunz, A. C. Toma, P. Tanner, T. Pfohl and C. G. Palivan, Polymer nanoreactors with dual functionality: simultaneous detoxification of peroxynitrite and oxygen transport. *Langmuir*, **28** (2012), 15889–15899.
67. P. Tanner, P. Baumann, R. Enea *et al.*, Polymeric vesicles: from drug carriers to nanoreactors and artificial organelles. *Acc. Chem. Res.*, **44** (2011), 1039–1049.
68. N. Goswami, R. Saha and S. K. Pal, Protein-assisted synthesis route of metal nanoparticles: exploration of key chemistry of the biomolecule. *J. Nanopart. Res.*, **13** (2011), 5485–5495.
69. D. M. Vriezema, M. C. Aragones, J. A. A. W. Elemans *et al.*, Self-assembled nanoreactors, *Chem. Rev.*, **105** (2005), 1445–1489.
70. S. Kashyap, T. Woehl, C. Valverde-Tercedor *et al.*, Visualization of iron-binding micelles in acidic recombinant biomineralization protein, MamC. *J. Nanomater.* (2014), 320124.
71. D. Karlin and R. Belshaw, Detecting remote sequence homology in disordered proteins: discovery of conserved motifs in the N-termini of Mononegavirales phosphoproteins. *PLoS One*, **7** (2012), e31719.
72. A. Heyman, I. Medalsy, O. Bet Or *et al.*, Protein scaffold engineering towards tunable surface attachment. *Angew. Chem. Int. Ed.*, **48** (2009), 9290–9294.
73. P. S. Ghosh and A. D. Hamilton, Noncovalent template-assisted mimicry of multiloop protein surfaces: assembling discontinuous and functional domains. *J. Am. Chem. Soc.*, **134** (2012), 13208–13211.
74. J. Diao, Crystal structure of a super leucine zipper, an extended two-stranded super long coiled coil. *Protein Sci.*, **19** (2010), 319–326.
75. M. T. Dedeo, K. E. Duderstadt, J. M. Berger and M. B. Francis, Nanoscale protein assemblies from a circular permutant of the tobacco mosaic virus. *Nano Lett.*, **10** (2010), 181–186.
76. S. E. Aniagyei, C. DuFort, C. C. Kao and B. Dragnea, Self-assembly approaches to nanomaterial encapsulation in viral protein cages. *J. Mater. Chem.*, **18** (2008), 3763–3774.
77. J. Sun, C. DuFort, M.-C. Daniel *et al.*, Core-controlled polymorphism in virus-like particles. *Proc. Natl. Acad. Sci. USA*, **104** (2007), 1354–1359.
78. L. L. Vatta, R. D. Sanderson and K. R. Koch, Magnetic nanoparticles: properties and potential applications. *Pure Appl. Chem.*, **78** (2006), 1793–1801.
79. H. Ai, C. Flask, B. Weinberg *et al.*, Magnetite-loaded polymeric micelles as ultrasensitive magnetic-resonance probes. *Adv. Mater.*, **17** (2005), 1949–1952.
80. C. C. Berry and A. S. G. Curtis, Functionalisation of magnetic nanoparticles for applications in biomedicine. *J. Phys. D: Appl. Phys.*, **36** (2003), R198–R206.
81. E. Chiancone, P. Ceci, A. Ilari, F. Ribacchi and S. Stefanini, Iron and proteins for iron storage and detoxification. *BioMetals*, **17** (2004), 197–202.
82. A. P. Busch, D. Rhinow, F. Yang *et al.*, Site-selective biomineralization of native biological membranes. *J. Mater. Chem. B*, **2** (2014), 6924–6930.
83. J. Baumgartner, G. Morin, N. Menguy *et al.*, Magnetotactic bacteria form magnetite from a phosphate-rich ferric hydroxide via nanometric ferric (oxyhydr)oxide intermediates. *Proc. Natl. Acad. Sci. USA*, **110** (2013), 14883–14888.
84. J. Baumgartner and D. Faivre, Magnetite biomineralization in bacteria. *Prog. Mol. Subcell. Biol.*, **52** (2011), 3–27.
85. R. L. Penn and J. F. Banfield, Imperfect oriented attachment: dislocation generation in defect-free nanocrystals. *Science*, **281** (1998), 969–971.
86. B. Gao, G. Arya and A. R. Tao, Self-orienting nanocubes for the assembly of plasmonic nanojunctions. *Nat. Nanotechnol.*, **7** (2012), 433–437.

87. Y. Nakagawa, H. Kageyama, Y. Oaki and H. Imai, Direction control of oriented self-assembly for 1D, 2D, and 3D microarrays of anisotropic rectangular nanoblocks. *J. Am. Chem. Soc.*, **136** (2014), 3716–3719.
88. R. Q. Song and H. Colfen, Mesocrystals-ordered nanoparticle superstructures. *Adv. Mater.*, **22** (2010), 1301–1330.
89. B. L. Sun, M. Wen, Q. S. Wu and J. Peng, Oriented growth and assembly of Ag@C@Co pentagonalprism nanocables and their highly active selected catalysis along the edges for dehydrogenation. *Adv. Funct. Mater.*, **22** (2012), 2860–2866.
90. J. Ihli, P. Bots, A. Kulak, L. G. Benning and F. C. Meldrum, Elucidating mechanisms of diffusion-based calcium carbonate synthesis leads to controlled mesocrystal formation. *Adv. Funct. Mater.*, **23** (2013), 1965–1973.
91. M. Niederberger and H. Colfen, Oriented attachment and mesocrystals: non-classical crystallization mechanisms based on nanoparticle assembly. *Phys. Chem. Chem. Phys.*, **8** (2006), 3271–3287.
92. C. Frandsen, B. A. Legg, L. R. Comolli *et al.*, Aggregation-induced growth and transformation of beta-FeOOH nanorods to micron-sized alpha-Fe_2O_3 spindles. *CrystEngComm*, **16** (2014), 1451–1458.
93. Y. Wang, A. E. DePrince, S. K. Gray, X. M. Lin and M. Pelton, Solvent-mediated end-to-end assembly of gold nanorods. *J. Phys. Chem. Lett.*, **1** (2010), 2692–2698.
94. H. Colfen and M. Antonietti, *Mesocrystals and Nonclassical Crystallization* (Chichester, UK: Wiley, 2008).
95. T. J. Woehl and T. Prozorov, The mechanisms for nanoparticle surface diffusion and chain self-assembly determined from real-time nanoscale kinetics in liquid. *J. Phys. Chem. C*, **119** (2015), 21261–21269.
96. N. D. Burrows, C. R. H. Hale and R. L. Penn, Effect of ionic strength on the kinetics of crystal growth by oriented aggregation. *Cryst. Growth Des.*, **12** (2012), 4787–4797.
97. R. L. Penn and J. A. Soltis, Characterizing crystal growth by oriented aggregation. *CrystEngComm*, **16** (2014), 1409–1418.
98. W. Ahmed, R. P. B. Laarman, C. Hellenthal *et al.*, Dipole directed ring assembly of Ni-coated Au-nanorods. *Chem. Commun.*, **46** (2010), 6711–6713.
99. J. Chai, X. Liao, L. R. Giam and C. A. Mirkin, Nanoreactors for studying single nanoparticle coarsening. *J. Am. Chem. Soc.*, **134** (2012), 158–161.
100. M. X. Yang, G. Chen, Y. F. Zhao *et al.*, Mechanistic investigation into the spontaneous linear assembly of gold nanospheres. *Phys. Chem. Chem. Phys.*, **12** (2010), 11850–11860.
101. J. Park, H. Zheng, W. C. Lee *et al.*, Direct observation of nanoparticle superlattice formation by using liquid cell transmission electron microscopy. *ACS Nano*, **6** (2012), 2078–2085.
102. C. Park, T. J. Woehl, J. E. Evans and N. D. Browning, Minimum cost multi-way data association for optimizing multitarget tracking of interacting objects, pattern analysis and machine intelligence. *IEEE Trans. Pattern Anal. Mach. Intell.*, **37** (2014), 611–624.
103. D. S. Li, M. H. Nielsen, J. R. I. Lee *et al.*, Direction-specific interactions control crystal growth by oriented attachment. *Science*, **336** (2012), 1014–1018.
104. J. M. Yuk, J. Park, P. Ercius *et al.*, High-resolution EM of colloidal nanocrystal growth using graphene liquid cells. *Science*, **336** (2012), 61–64.
105. H. G. Liao, D. Zherebetskyy, H. L. Xin *et al.*, Facet development during platinum nanocube growth. *Science*, **345** (2014), 916–919.

Index

aberration correction
   benefits for liquid cell microscopy, 165, 173, 434, 443, 452
   lifetime of corrected state, 441
acrosomal process, 366–68
   electron beam damage, 367
aerated water, 156
affinity capture technique, 357–59
   for reducing motion, 363
   procedure, 358
AirSEM, 121–23
analytical electron microscopy, 408–32
   challenges, 410, 412
   comparison of techniques, 411
   EELS, 413–20
   Fe mapping, 482
   for biological materials, 19
   principles, 408–9
   XEDS in closed cell SEM, 97
   XEDS in closed cell TEM, 420–30
atmospheric aerosols, 22
atmospheric scanning electron microscopy (ASEM)
   and inverted SEM, 106
   comparison with closed cell SEM, 106
   electrochemistry, 114
   for CLEM, 117–20
   principles, 106–8
   sample dish, 107
   temperature control, 115

bacteria
   atmospheric SEM, 119
   correlative light and electron microscopy, 481
   cryoelectron microscopy, 480
   early imaging, 7, 36
   EELS, 415
   magnetotactic, 480–82
   WetSTEM, 62
batteries, 237–52
   analytical electron microscopy, 250–52
   capacity, 244
   challenges, 238
   closed cell TEM, 238–52

coulombic efficiency, 244
C-rate, 244
Li-ion batteries, 63, 238
open cell TEM, 63–70, 238
battery materials and reactions
   air sensitivity, 242
   anode materials, 241
   cathode materials, 241
   cathodes, 74
   EELS, 250–52, 418
   electrolytes, 241, 246–47
   lithiation, 68, 247–48, 418, 467
   Li electrodeposition, 249–50, 466
   nanoparticle anodes, 68–70
   nanowire anodes, 63–68
   SEI, 68, 244
   Si, 68–70
   SnO_2, 66–67
biological macromolecules, 490–91
   conjugated DNA, 401–4
   electron beam damage, 400
   imaging strategies, 19
   lipoprotein discs, 469
biological materials and processes, 17–19, 356–69
   acrosomal process, 366–68
   analytical electron microscopy, 19
   axonal segmentation, 118
   bacterial cells in water, 119
   biological processes by snapshot imaging, 344
   biomimetics, 488–91
   biomineralization, 479–83
   breast cancer cells, 109–11, 343
   cell edges, 341
   cell trafficking, 118
   cellulose fibers, 123
   chemical stimulation, 24
   coliform bacteria, 7
   COS7 cells, 117, 339–41, 346
   electric field sensitivity, 483–84
   endoplasmic reticulum, 117
   epidermal growth factor receptor, 339–42
   ferritin, 485
   glioblastoma cells, 361

Index

biological materials and processes (cont.)
  holography, 484
  large numbers of cells, 342, 348
  leaf surface, 123
  liposomes, 360
  live cells, 17, 24, 346, 480–82
  lung cancer cells, 348
  magnetotactic bacteria, 480–82
  megakaryocytes, 111
  mesocrystals, 491–95
  micelles, 489
  muscle protein function, 371–89
  myosin, 378–88
  protein dimers, 490
  protein domains, 489–90
  protein structure in water, 476–79
  pseudopodia, 111
  ribosomal assemblies, 357–58
  RNA, 363
  root tips, 89
  rotavirus, 362–66
  STIM1 protein, 117
  temperature control, 24
  tissue blocks imaged in SEM, 109–10
  unlabeled biological materials, 357–68
  whole biological cells, 17, 334–51
  yeast, 345
biomimetics, 488–91
  block copolymers, 488–89
Brownian motion, 11, 16, 112, 177, 458, 479
bubbles
  detection by EELS, 416
  drift to center of window, 47
  formed during Joule heating, 128
  formed electrochemically, 227
  formed radiolytically, 16, 156
  in closed cell SEM, 99
  motion, 16
butterfly wing, 91–92

calcium carbonate
  attachment and crystallization, 328
  growth in organic matrix, 324–26, 328
  nucleation and growth, 322–24
  phases, 321
capsule SEM. See closed cell SEM
catalysis, 23
  photocatalysis, 87
  TiO$_2$, 87
cathodoluminescence, 121
  in AirSEM, 122
  in atmospheric SEM, 119–20
  in closed cell SEM, 86, 120–21
chromatic aberration correction, 446–52
  for energy filtered imaging, 174, 448
  for thick samples, 450
  in wide gap microscopes, 174, 452
  resolution improvement, 166, 174, 446

clays, 21
closed cell design for SEM
  2D window materials, 92–97
  ASEM sample dish, 107
  CLEM, 120–21
  commercial, 85–86
  electrochemistry, 92–93, 114
  fiber optics, 86, 92
  flow, 91
  for biological materials, 87–88
  for cathodoluminescence, 121
  for electrochemistry, 91
  for STEM, 121
  functionalize window, 87
  glass slide as window, 121
  graphene coated sample, 10, 94
  multiple orifices, 100
  principles, 86–87
  temperature control, 91, 115
  window transparency, 82, 87
closed cell design for TEM, 35–52
  affinity capture technique, 357–59
  analytical electron microscopy, 51
  ATP injection, 380–81
  carbon nanotubes, 46, 283
  cell failure, 46–47
  electrolyte bridge, 245
  electrolyte choice, 228
  flow and electrochemistry, 230, 242
  flow of nutrients, 336
  for anodic etching, 211–12
  for biomimetics, 488–91
  for electrochemistry, 49–50, 211–18, 238–42
  for extreme environments, 137
  for FEBIP, 294–97
  for holography, 485
  for liquid flow, 49
  for multiple stimuli, 51, 137
  for muscle protein function, 375–77
  for unlabeled biological materials, 357–68
  for whole biological cells, 338
  for XEDS, 421, 425
  graphene flow cell, 395–96
  graphene liquid cell, 45–46, 394
  heating, 50, 128
  hydration chamber, 371
  minimize drift, 129
  mixing, 323
  monolithic chips, 45
  non-microfabricated designs, 45–46
  paired chips, 39–44
  scrolls, 283
closed cell electrodes, 213–16
  beam-induced deposition, 241
  compatibility with lithium, 240
  FIB-cut sections, 215, 241
  in SEM, 92

Index

invisible areas, 229
materials, 215–16, 240–41
slurry printing, 215, 241, 248
small size, 230
thickness, 215
wires, 216
closed cell SEM, 78–100
AirSEM, 79, 121–23
applications, 87–90
ASEM, 79
CLEM, 120–21
comparison with closed cell TEM, 80
comparison with ESEM, 80
historical context, 8, 10
image formation, 80–83
inverted SEM, 106–20
resolution, 83–85
terminology, 79
upright ASEM, 123
closed cell TEM
comparison with cryoelectron microscopy, 361–68
historical context, 5
closed cells with paired chips, 39–44
alignment, 44
assembly, 216–17
glued chips, 41, 217
hermetic seal, 217
liquid introduction, 41, 217
o-ring clamping, 42–43
sealing, 41
spacer layer, 41
wafer bonded, 42
compressive sensing, 463–64
correlative light and electron microscopy, 10, 18
for cathodoluminescence, 119–20
for whole labeled cells, 338
in AirSEM, 121–23
in ASEM, 117–20
in closed cell S/TEM, 51, 336, 338–39, 342–44
in SEM, 116–23
magnetotactic bacteria, 480–82
corrosion, 13–14, 258–72
compositional analysis, 265
definition, 258
electrochemical techniques, 263
electrolyte, 268–69
electron beam effects, 263, 269
galvanic displacement, 259, 263
imaging conditions, 269
liquid cell design, 263, 270–71
localized, 259
of alloys, 265
of metal films, 266
of particles, 205, 266
of steels, 266–67

of thin sections, 266
pitting, 261, 264
sample preparation, 267
techniques, 258, 260
cryoelectron microscopy
artifacts, 478
comparison with closed cell S/TEM, 18, 348–50, 361–68
dose and radiation damage, 18, 341, 368, 477
for protein structure determination, 477–78
of acrosomal process, 367
of biomineralization, 480
of rotavirus, 364–66
reconstruction methods, 365, 478
Cu electrodeposition, 219–27

dendrites
electrochemical, 114, 226–27, 249–50, 466
nanoparticle, 201
detectors
and resolution, 164, 178, 460
backscatter, 82
direct electron detector, 368–69
dose fractionation, 478
gaseous secondary electron detector, 9
image plate, 379–80
phase plate, 368–69
time resolution, 369
diffraction
assessing beam damage, 367
battery materials, 239
calcium carbonate, 325
for phase identification, 323
liquid thickness, 324
diffusion
affinity capture technique, 357
and fast detectors, 369
anomalously slow, 16, 398–99
effect of solute, 112
effect of windows, 491–95
in graphene liquid cell, 398–99
modes of nanoparticle motion, 202
of nanoparticles, 201–2
DNA and RNA, 401–4
holography, 485
RNA transcription, 363
dose
assessing cell viability, 346, 481
calibrating, 184
comparison with cryoelectron microscopy, 341, 350, 478–79, 483
critical dose for protein function, 377–78
definition, 140
dose rate and electron flux, 140
due to single electron, 142
for damage of biomaterials, 341, 345
gray unit, 140

dose (cont.)
 in holography, 484
 lethal dose, 173, 344, 346
 minimizing, 181
 noise and contrast, 176–78
 tolerable dose, 18, 172, 479, 482–83
dynamic transmission electron microscope (DTEM), 460–63
 laser stimulation, 461
 pulsed source, 461
 spatial resolution, 461

EELS
 core loss in liquids, 413–14
 energy-filtered imaging, 410
 in graphene liquid cell, 400
 liquid thickness measurement, 51, 182, 416
 of bacteria, 415
 of battery materials, 238–52
 of biomolecules, 400
 principles, 409
 spectroscopic imaging, 409
 thickness effect, 413
 valence loss in liquids, 414–15
electrochemical measurements, 217–18, 242–43
 chronoamperometry, 243
 chronoamperometry (potentiostatic deposition), 218
 chronopotentiometry (galvanostatic deposition), 218
 cyclic voltammetry, 217, 243
 electrical impedance spectroscopy, 218, 243
 ultra-low currents, 238
electrochemistry, 12–14, 210–33, 243–44
 additives, 224
 Ag on TiO_2, 93
 analytical microscopy, 250–52
 and benchtop experiments, 13, 168, 229
 Au deposition, 114
 batteries. See battery materials and reactions
 beam effects, 230–32, 246–47
 corrosion, 258–72
 Cottrell equation, 221, 243
 counter electrode, 213
 Cu, 219–27
 dendrites, 226–27, 249–50, 466
 double layer, 23
 electrocatalysts, 253
 electrode design, 213–16
 $Fe(CN)_6$, 260
 future prospects, 22–23
 gas evolution, 225, 227
 growth models, 12
 imaging electrolyte, 215
 imaging lateral growth, 224
 in atmospheric SEM, 114
 in closed cell SEM, 92–93
 nucleation and growth models, 219
 number of electrodes, 216, 239
 ohmic drop, 216, 245
 plan view imaging, 219
 porous silicon formation, 211–12
 potentiostat, 217–18
 pseudo-reference electrode, 216
 pulse plating, 226
 reference electrode, 216
 working electrode, 213
electrolyte
 acidified copper sulphate, 219–27
 beam effects, 230–32, 246–47
 choice, 228
 detecting ion concentration, 227
 flow, 230
 ionic liquid, 64
 metal oxide, 65
 organic, 246, 249
 thickness, 229
electron beam effects, 11, 140–63, 194
 aerated water, 156
 alter surface chemistry, 194
 analytical electron microscopy, 412
 and flow, 478
 bubble formation, 99, 154–56
 changes in pH, 157–58
 charging, 94, 194
 chloride solutions, 160
 cryoelectron microscopy, 350, 477
 damage of biological materials, 341, 377–78, 482–83
 dewetting, 285
 diffraction spot intensities, 367
 droplet motion, 277
 effect of liquid flow, 99, 152
 effect of windows, 144
 etching, 159
 FEBIP, 291–310
 heating, 131, 146
 in electrochemistry, 230–32
 in SEM, 97–99
 ionic liquids, 61
 nanoparticle assembly, 112, 203
 nanoparticle growth, 14, 87, 90, 158, 193–94, 397
 nanoparticle growth in ionic liquids, 59–61
 nanoparticle motion, 204–5, 357
 organic electrolyte, 246–47
 plant cells, 89–90
electron holography, 484–86
electron microscopy of liquids
 closed cell strategies, 5, 8
 comparison of techniques, 19–20
 comparison with cryoelectron microscopy, 7
 future prospects, 21–25
 historical perspective, 4
 main issues, 10–11
 open cell strategies, 5, 8

energy-filtered imaging, 410, 417–20
　to improve resolution, 173, 419
energy transfer to liquids, 140–44
　backscattered electrons, 81
　continuous slow down approximation, 141
　heating, 144
　in SEM, 80–83, 98
　inelastic scattering, 408
　mean free path, 143, 413, 421
　multiple scattering, 144
　radiolysis products, 147
　range and straggle, 141
　secondary electrons, 81
　spurs, 143, 147–48
　stopping power, 141
　X-ray absorption path length, 421–22
ESEM, 8–10, 19, 78
　batteries, 73
　condensation of water, 15
　for FEBIP, 293
　historical context, 8
　terminology, 78
　thin water films, 62
　WetSTEM, 62
　whole biological cells, 18, 337
etching
　electrochemical stripping, 92, 222, 227
　focused electron beam-induced, 306–8
　porous Si, 211–12

ferritin, 400, 485
　holography, 486
flow. *See* liquid flow
fluid physics, 16, 22, 276–87
　anomalous interfacial layer, 277, 280
　condensation on butterfly wing, 91–92
　contact line dynamics, 278, 284–85
　dewetting, 277, 285
　double layer, 276
　evaporation, 113
　fluids in nanochannels, 282
　voids in thin films, 285
focused electron beam-induced processing (FEBIP), 291–310
　applications, 291–92
　FEBID, 292
　FEBIE, 292
　gas phase FEBIP, 291–93
　limitations, 292–93
　liquid phase FEBIP, 293–310
focused ion beam processing
　comparison with FEBIP, 293
　for cutting lamellae, 63, 241
　for patterning windows, 330, 394
　for welding materials to electrodes, 50, 64, 215
food science
　crystallization of ice cream, 136
　melting of chocolate, 134
fuel cells, 238, 252–53
　catalyst coarsening, 253
　closed cell TEM, 252–53
　operation, 238

geological materials, 21
　at extreme conditions, 21
graphene liquid cell, 393–405
　analytical electron microscopy, 412
　benefits, 45, 393, 405
　biomaterials, 400–4
　electron beam effects, 394
　fabrication, 45–46, 393–96
　first generation, 394
　for holography, 485
　for liquid phase FEBIP, 295
　liquid flow, 395–96
　nanoparticle growth, 197, 396–97
　nanoparticle motion, 398–99
　second generation, 394–95
　third generation, 395–96

holography, 486
　for biological materials, 484–86
　future prospects, 22–23
hydrated electrons, 11, 147, 158–59
　in FEBIP, 200, 309
　metal ion reduction, 158, 194

image analysis and simulation
　atomic resolution simulations, 179
　particle tracking, 253, 464–66, 494
　probe broadening simulations, 174
imaging parameters
　choice of kV, 180
　dark field TEM imaging, 8
　imaging mode, 8
　TEM or STEM, 181
ionic liquids
　definition, 57
　for FEBIP, 299
　nanoparticle growth, 59–61, 199
　thin films, 58
　use for passivation, 58, 62
　wetting of nanowires, 58
ionizing radiation, 140
iron oxyhydroxide, 197, 327–28

liquid flow
　and electron beam effects, 99, 152, 478
　closed cell design, 49
　closed cell design for SEM, 91
　for electrochemistry, 230, 242
　in graphene liquid cells, 395
　of nutrients, 336

liquid phase FEBID, 299–306
  alloys, 304–5
  droplet geometry in open cell, 297
  electrochemical reactions, 309–10
  hydrated electron kinetics, 304, 309
  liquid injectors, 298
  proximity effect, 302
  substrate effect, 309
  sulphides, 304–6
  transition metals, 299–303
liquid phase FEBIE, 306–8
  Cu, 308
  proximity effect, 309
  silicon nitride, 307–8
liquid phase FEBIP
  in closed cell, 294–97
  in ionic liquid, 299
  in open cell, 297–98
  in water, 297–99
  substrate effects, 295
liquid thickness
  and diffusion, 168, 229
  and radiolysis, 11, 144
  and resolution, 11, 35, 167–68, 175, 358
  effect on EELS signal, 413
  effect on XEDS, 421, 423–24
  measurement, 47, 182
  measurement by EELS, 182, 409, 415
  measurement by transmitted current, 184
  measurement by XEDS, 409, 427
  microwells, 19
  non-uniformity, 48
  spacer layer, 41
lithiation, 68
  EELS, 250–52
  of graphitic carbon, 245
  of nanowire in closed cell, 248
  of SnO_2, 66–67
  Si nanoparticle, 68–70
  Si nanowire, 71
  sodiation of SnO_2, 67
  volume change and fracture, 67–68, 247
Li electrodeposition, 249–50, 466
live biological cells
  assessing viability, 481–83
  imaging strategies, 17
  prospects, 24
  tolerable dose, 346, 482–83

magnetotactic bacteria
  assessing viability, 482
  correlative light and electron microscopy, 480–82
  electron beam effects, 482
  holography, 485
mesocrystals, 491–95
  growth physics, 491–92
microfabrication, 39–41

electrodes, 215
for closed cells, 37–45
historical context, 7
importance for liquid cell design, 36–37
microwells, 358–59
microscope design
  DTEM, 460–63
  for closed cell SEM, 8, 120–23
  for ESEM, 4
  for open cell TEM, 5–6
  for SEM, 78
  for WetSTEM, 337
mineralization and biomineralization, 16–17, 316–30, 479–83
  biomacromolecule/nanoparticle interactions, 490–91
  biomimetics, 488–91
  growth, 319
  importance, 316–17
  magnetotactic bacteria, 480–82
  nucleation, 317–19
  single cell organisms, 479
muscle protein function, 18
  cyclic movement, 383
  effect of actin, 387–88
  myosin head movement, 381–88
  response to ATP, 382–83
  sliding filament mechanism, 371–75
  spatially resolved displacement, 384–86

nanoparticle assembly, 202–3
  chains, 196, 492
  diffusion-limited aggregation, 203
  mesocrystals, 491–92
  superlattices, 203
nanoparticle growth physics, 14–15, 194–99
  Ag in ionic liquid, 59–61
  anisotropic interactions, 197, 493–94
  assembly into chains, 492
  assembly into mesocrystals, 491–92
  attachment and crystallization, 328
  calcium carbonate, 322–26
  coalescence, 15, 194–96, 326
  dendrites, 201
  effect of walls, 197
  facet-dependent growth rate, 197–98
  galvanic displacement, 200
  growth trajectories, 194–97
  iron oxyhydroxide, 327–28
  monomer attachment, 195, 319, 326
  oriented attachment, 197, 319, 326–27, 492
  reshaping, 196
  size oscillations, 199
  surfactants, 198
  Wulff construction, 197
nanoparticles
  Ag, 90, 195
  Ag on TiO_2, 87

Au, 197, 201
Au labels, 339, 341, 379
Au nanorods, 357
beam-induced assembly, 112, 202–3
beam-induced growth, 14, 192–94
beam-induced motion, 204–5, 357
Bi, 199
Brownian motion. *See* Brownian motion
calcium carbonate, 322, 328
coalescence, 398–99
core-shell, 200
diffusion, 201–2
diffusion in graphene liquid cell, 398–99
diffusion in nanoscale droplets, 279
growth at elevated temperature, 14, 193, 199
growth in graphene liquid cell, 396–97
growth in ionic liquids, 59–61
growth rate measurement, 193
heating effects, 133
heterogeneous, 200–1
hollow, 200
in plant cells, 88–89
interaction with proteins, 490
iron oxyhydroxide, 327–28
Kirkendall void formation, 133
labels for biomaterials, 117
labels for whole cells, 336
laser-induced formation, 462
lithiation of Si, 68–70
magnetite, 360, 480–82, 486, 490
magnetization, 487
need for controlled synthesis, 191
oxidative etching, 205, 263
Pd, 196, 200
Pt, 195, 197–98, 397
PtFe, 196, 199
PtPd, 197
quantum dot fluorescent labels, 342
shape evolution, 194, 197–98, 223
tracking algorithm, 464–66, 494
twinned, 201
uptake into biological cells, 347–48, 361
uptake into vesicles, 486
ZnO, 195
nanoscale droplets
  diffusion of nanoparticles, 279
  nucleation and growth, 280
  stick-slip motion, 277
nucleation
  Avrami kinetics, 222
  biomimetics, 488
  block copolymers, 488–89
  calcium carbonate, 322–26
  classical nucleation theory, 317, 324
  critical nucleus size, 222
  Cu, 219–24
  electrochemical, 219–24
  heterogeneous, 193, 324

  homogeneous, 193
  in organic matrix/template, 201, 324–26, 490
  on electrode, 215
  on existing nanoparticles, 200
  preferred sites, 221
  threshold dose, 194

open cell design
  challenges, 70–72
  for battery reactions in TEM, 63–70
  for liquid phase FEBIP, 297–98
open cell electron microscopy, 56–74
  ESEM, 57, 73, 78
  ETEM, 57
  principles, 56–57
  SEM, 56, 73
  TEM, 56–72
  Wet3TEM, 62
open cell SEM. *See* ESEM
open cell TEM, 57–72
  battery experiments, 58, 63–70
  comparison with closed cell, 70–72
  geometry for thin film experiments, 58
  historical context, 5
  in atmospheric science, 22
oriented attachment, 327–28
  iron oxyhydroxide, 197, 327
  jump to contact, 197, 328
  lattice resolution, 326

Pb electrodeposition, 225
phase plate, 24
phase transformations
  boiling, 133
  calcium carbonate, 322–24
  cooling of solder, 116
  freezing of saline solution, 135
  hydrothermal precipitation, 133
  lithiation of Si, 70–71
  lithiation of SnO_2, 67
  overview, 15
  sodiation, 68
phytotoxicity, 88–89
pressure
  effect on gas solubility, 155
  prospects for control, 21
protein structure
  comparison of techniques, 477–78

radiation chemistry, 140
radiolysis
  in biological imaging, 479
  in non-aqueous liquids, 160
  of water, 11
  scavengers, 147
radiolysis products
  diffusion, 149–50, 152
  G-value, 147

radiolysis products (cont.)
  homogeneous formation, 150
  hydrated electron, 147, 154, 158–59
  hydrogen, 154–56
  hydrogen ions, 157
  initial yield, 147
  oxygen, 155, 157
  reaction-diffusion equation, 148
  recombination, 149
  steady state concentration, 150, 152
resolution
  cryoelectron microscopy, 477
  overview, 10
  SEM, 11, 83–85, 121
  STEM, 174–77
  TEM, 165–74
  temporal, 456–62
rotavirus, 362–66
  attachment to membrane, 363
  imaging mRNA transcripts, 363
  internal structure, 365–66

sample design
  for battery reactions in open cell TEM, 58, 63
  for battery reactions in SEM, 73
  for live cell imaging, 344–45
  for WetSTEM, 62, 337
  for whole biological cells, 336–37, 340
  ionic liquid droplets in SEM, 57
  sample passivation with ionic liquids, 62
  thin foil, 267, 429
sample holder design
  control by potentiostat, 217–18, 239, 263
  examples, 43–44
  for anodic etching, 211–12
  for fluid mixing, 319–21
  for gas diffusion, 325
  for magnetization experiments, 487
  for open cell battery reactions, 63–70
  for XEDS, 422, 425
  glove box loading, 65, 229
  historical context, 6, 8
  laser heating, 130, 462
  liquid injection via glass capillary microelectrode, 380–81
  resistance furnace, 129
  temperature-controlled electrochemistry, 136
  thermoelectric device, 130
silicon nitride
  etching, 307–8
  for liquid phase FEBIP, 295–96
  materials properties, 39
  window fabrication, 39–41
  with patterned electrodes, 213–16
solder, 115
solid electrolyte interphase, 238, 244
  dendritic growth, 245

  formation, 245
solubility
  effect of pressure, 155
  of gases in water, 131, 154
spatial resolution in S/TEM
  aberration-corrected S/TEM, 434–53
  accelerating voltage, 487
  and sample thickness, 10, 166, 168, 450
  beam broadening, 177
  chromatic aberration limited, 166, 446
  contrast transfer function, 435–38
  determining factors, 165, 173
  diffraction limited, 165
  dose limited, 170–73, 176
  energy filtering, 173
  of labels in closed cell STEM, 336–37, 341
  probe broadening, 174
  spherical aberration limited, 166, 435, 444
  thin liquid in microwells, 358–59
  top–bottom effect, 168, 423
spatial resolution in SEM
  beam broadening, 85
  effect of accelerating voltage, 85
  effect of windows, 80–85
  in AirSEM, 122
  interaction volume, 80, 121, 298, 411
  minimum SEM beam current, 84
  noise limited, 83
  probing depth, 81
  through 2D window materials, 92–97
spherical aberration correction
  delocalization, 437
  effect on contrast transfer function, 435
  for nanoparticle growth, 397
  higher order aberrations, 440
  negative C_S imaging, 438
  resolution improvement, 166, 173, 437
  STEM, 444–46
  TEM, 435–44
steel, 421
  corrosion, 266, 430

temperature control, 127–38
  beam-induced heating, 131
  extreme conditions, 21
  for biological materials and processes, 489
  for nanoparticle growth, 199
  Joule heating, 128
  laser heating, 130
  Peltier heater, 130
  resistance furnace, 129
  thermal reservoir, 130
temperature measurement
  materials transformations, 132
  optical methods, 132
  resistance temperature detector, 129
  thermal diffusion models, 129

thermocouple, 130, 132
temporal resolution, 23, 177–79, 456–62
   benefits, 177, 456–57, 466
   dose fractionation, 178
   dose-limited, 178
   for biological processes, 368–69
   motion blur, 11, 177, 458, 470
   with pulsed electron source, 460
thin film growth physics
   dendrite formation, 113, 226–27, 249–50
   diffusion gradients, 227
   growth front stability, 226
   SEI formation, 244

WetSTEM, 9, 62, 337
   dose and damage, 342
   examples, 62
   of nanoparticle uptake, 348–49
   resolution, 9
   sample design, 337
   whole biological cells, 18, 341–42
wetting, 16, 19, 278
   ionic liquids, 58
   of windows, 168, 268
   void formation, 285
   water, 62
whole biological cells, 334–51
   bacteria, 415
   comparison of imaging techniques, 335–37, 342
   imaging strategies, 17, 334–37
   lethal dose, 18, 173
   live cells in closed cell STEM, 346
   preparation of cells on chips, 107, 336, 340–41
   viability testing, 346
window spacing and deflection, 35
   calculation, 47–48
   effect on resolution, 11, 326
   electrochemical experiments, 229
   graphene liquid cell, 394
   liquid phase FEBIP, 295
   measurement by EELS, 47, 182, 417
   microwells, 358–59
   partially filled liquid cell, 377
   radiolytic gas, 156
   thick liquid for biological imaging, 338
windows, 38–41
   carbon, 375–76
   design for holography, 485
   effect on SEM image, 80–85
   etched wells, 358–59
   for liquid phase FEBIP, 294–97
   graphene, 39, 93, 96
   graphene coated onto sample, 10, 94
   graphene oxide, 39, 95–96
   hexagonal boron nitride, 39
   materials properties, 38
   microfabrication, 39–41
   nitrocellulose, 5
   polyimide, 39, 83, 86, 294
   preparation of 2D materials, 95–96
   silicon nitride, 39, 107, 295, 337
   silicon nitride for SEM, 92–93
   silicon oxide, 39
   surface functionalization, 357–59
   thickness and resolution, 168
   to seal SEM column, 121
   transmittance in SEM, 82

XEDS
   composition of liquids, 428
   in closed cell SEM, 92, 94, 97
   liquid thickness, 423–24
   materials in water, 427
   principles, 409, 420
   spectroscopic imaging, 409
   through graphene, 97
   X-ray absorption path length, 422

Zn electrodeposition, 225